Advances in

POLYOLEFIN NANOCOMPOSITES

Advances in

POLYOLEFIN NANOCOMPOSITES

Edited by
Vikas Mittal

CRC Press
Taylor & Francis Group
Boca Raton London New York

CRC Press is an imprint of the
Taylor & Francis Group, an **informa** business

CRC Press
Taylor & Francis Group
6000 Broken Sound Parkway NW, Suite 300
Boca Raton, FL 33487-2742

First issued in paperback 2019

© 2011 by Taylor and Francis Group, LLC
CRC Press is an imprint of Taylor & Francis Group, an Informa business

No claim to original U.S. Government works

ISBN-13: 978-1-4398-1454-3 (hbk)
ISBN-13: 978-0-367-38320-6 (pbk)

Visit the Taylor & Francis Web site at
http://www.taylorandfrancis.com

and the CRC Press Web site at
http://www.crcpress.com

Contents

v

Preface

Polyolefins are very important materials for a large number of applications ranging from commodity use to high-end use. The generation of polyolefin nanocomposites is of great interest to improve the properties of existing polyolefins or conventional polyolefin composites, thereby expanding the spectrum of application of these materials. This also led to extensive research to generate clay-based nanocomposites ever since such nanocomposites for polyamide matrices were successfully demonstrated by researchers at Toyota in the early 1990s. However, owing to the apolar nature of polyolefin matrices, the dispersion of layered silicate clay platelets at the nanometer level in them is still a challenge. As the dispersion of platelets, or nanocomposite morphology, is directly responsible for the composite properties, optimum improvement in the composite properties at a low volume fraction of the filler is still to be achieved. Approaches like partial polarization of the matrix by the addition of low molecular compatibilizers or surfactants to improve interactions between the polymer and filler surface, or complete organophilization of the filler surface to facilitate its delamination in the polymer matrix by the action of shear, etc., have been reported to circumvent the problem of nonuniform filler delamination. Also, as the conventionally used ammonium ions–based filler surface modifications are susceptible to thermal degradation at the compounding temperatures used for polyolefin matrices, more thermally stable filler surface modifications have been developed that can resist the thermal degradation at high compounding temperatures. Improvements have also been made in the synthesis process of such nanocomposites. Although melt compounding is still the predominant method to generate polyolefin nanocomposites, the use of *in situ* polymerization in the presence of clay- or solution-blending methods is also gaining interest. Furthermore, theoretical insights in understanding the filler surface modification dynamics and the interface development in the nanocomposites have also been achieved. Not only montmorillonite-based fillers, but also other filler types like double layer hydroxides, sepiolite, nanotubes, etc., have been successfully incorporated in polyolefin matrices in recent years.

This book attempts to assimilate the various advances that have been made in the synthesis and properties of exfoliated polyolefin nanocomposites in recent years. Chapter 1 provides a brief overview of polyolefin nanocomposite technology and covers polymer matrices, filler particles, as well as methodologies for the generation of polyolefin-based nanocomposites. Chapter 2 describes nanocomposite materials using a blend of polyolefin matrices, thus providing additional functionalities for nanocomposite materials, which are otherwise not possible when only one type of polymer is used for the generation of composites. Chapter 3 focuses on the advances made in the synthesis of polyolefin nanocomposites using solution-blending methods. The generation of nanocomposites using reactive extrusion is covered in Chapter 4. Chapters 5 and 6 discuss the *in situ* polymerization of olefinic monomers in the presence of clay. They also elaborate on the effects of clay-supported catalysts as well as clay surface modifications on the composite synthesis. Chapter 7 compares the theoretical considerations with experimental results in light of clay surface modification and interface development in polyolefin nanocomposites. Chapters 8 and 9 focus on the other important filler categories of double layer hydroxides and nanotubes, respectively, for the generation of polyolefin nanocomposites. Chapter 10 describes the properties of polypropylene nanocomposites generated using the filler modified with thermally stable imidazolium-based modification. Chapter 11 covers the functional polyolefins used in the synthesis of polyolefin nanocomposites. Needle-like filler particles for the synthesis of polyolefin nanocomposites are described in Chapter 12. Chapter 13 focuses on the functional compatibilizers that are used to achieve compatibility between the organic and inorganic phases.

Finally, I would like to express my gratitude to the Taylor & Francis Group for their kind support during the project. I am indebted to my family, especially my mother, whose continuous support and motivation have made this work feasible. I dedicate this book to my dear wife, Preeti, for her valuable help in coediting the book as well as for her efforts in improving the quality of the book.

Vikas Mittal
Ludwigshafen, Germany

Editor

Vikas Mittal, PhD, studied chemical engineering at Punjab Technical University, Punjab, India. He received his master of technology degree in polymer science and engineering from the Indian Institute of Technology, Delhi, India. Subsequently, he joined the polymer chemistry group of Professor U. W. Suter in the Department of Materials at the Swiss Federal Institute of Technology, Zurich, Switzerland, where he worked on his doctoral degree with a focus on surface chemistry and polymer nanocomposites. He also worked jointly with Professor M. Morbidelli in the Department of Chemistry and Applied Biosciences on the synthesis of functional polymer latex particles with thermally reversible behavior.

After completing his doctoral research, he joined the Active and Intelligent Coatings Section of Sun Chemical Group Europe in London, where he worked on the development of water- and solvent-based coatings for food-packaging applications. He later worked at BASF Polymer Research in Ludwigshafen, Germany, as a polymer engineer and is still working there as a laboratory manager responsible for the physical analysis of organic and inorganic colloids.

His research interests include organic–inorganic nanocomposites, novel filler surface modifications, thermal stability enhancements, polymer latexes with functionalized surfaces, among others. He has authored more than 40 scientific publications, book chapters, and patents on these subjects.

Contributors

Spiros H. Anastasiadis
Institute of Electronic Structure and Laser
Foundation for Research and Technology,
   Hellas
Heraklion, Crete, Greece

Mathieu Bailly
Department of Chemical Engineering,
Queen's University
Kingston, Ontario, Canada

Fabio Bertini
Istituto per lo Studio delle Macromolecole
Consiglio Nazionale delle Ricerche
Milano, Italy

Emiliano Bilotti
School of Engineering and Materials Science
Queen Mary, University of London
London, England

Laura Boggioni
Istituto per lo Studio delle Macromolecole
Consiglio Nazionale delle Ricerche
Milano, Italy

Kiriaki Chrissopoulou
Institute of Electronic Structure and Laser
Foundation for Research and Technology,
   Hellas
Heraklion, Crete, Greece

Serena Coiai
Istituto di Chimica dei Composti
   OrganoMetallici
Consiglio Nazionale delle Ricerche
   (ICCOM-CNR) UOS Pisa
Pisa, Italy

Francis Reny Costa
Leibniz Institute of Polymer Research
   Dresden
Dresden, Germany

Jie Feng
Department of Polymer Engineering
University of Akron
Akron, Ohio

Sara Filippi
Dipartimento di Ingegneria Chimica
Università di Pisa
Pisa, Italy

Suat Hong Goh
Department of Chemistry
National University of Singapore
Singapore, Singapore

Gert Heinrich
Leibniz Institute of Polymer Research
   Dresden
Dresden, Germany

Hendrik Heinz
Department of Polymer Engineering
University of Akron
Akron, Ohio

Marianna Kontopoulou
Department of Chemical Engineering
Queen's University
Kingston, Ontario, Canada

Burak Kutlu
Leibniz Institute of Polymer Research
   Dresden
Dresden, Germany

Andreas Leuteritz
Leibniz Institute of Polymer Research
   Dresden
Dresden, Germany

Jia Ma
School of Engineering and Materials Science
Queen Mary, University of London
London, England

Pierluigi Magagnini
Dipartimento di Ingegneria Chimica
Università di Pisa
Pisa, Italy

Vikas Mittal
Department of Chemistry and Applied
  Biosciences
Institute of Chemical and Bioengineering
Swiss Federal Institute of Technology
Zurich, Switzerland

and

BASF SE
Ludwigshafen, Germany

Elisa Passaglia
Istituto di Chimica dei Composti
  OrganoMetallici
Consigloi Nazionale delle Ricerche
  (ICCOM-CNR) UOS Pisa
Pisa, Italy

Ton Peijs
School of Engineering and Materials Science
Queen Mary, University of London
London, England

Purv J. Purohit
BAM Federal Institute for Materials
  Research and Testing
Berlin, Germany

Zakir M.O. Rzayev
Faculty of Engineering
Department of Chemical Engineering
Hacettepe University
Ankara, Turkey

Ulrich Scheler
Leibniz Institute of Polymer Research
  Dresden
Dresden, Germany

Andreas Schoenhals
Federal Institute for Materials Research
  and Testing
Berlin, Germany

Susannah L. Scott
Department of Chemical Engineering
Department of Chemistry & Biochemistry
Mitsubishi Chemical Center for Advanced
  Materials
University of California
Santa Barbara, California

Incoronata Tritto
Istituto per lo Studio delle Macromolecole
Consiglio Nazionale delle Ricerche
Milano, Italy

Anastasia Vyalikh
Leibniz Institute of Polymer Research
  Dresden
Dresden, Germany

Udo Wagenknecht
Leibniz Institute of Polymer Research
  Dresden
Dresden, Germany

DeYi Wang
Center for Degradable and Flame-
  Retardant Polymeric Materials
State Key Laboratory of Polymer Materials
  Engineering
College of Chemistry
Sichuan University
Chengdu, China

and

Leibniz Institute of Polymer Research
  Dresden
Dresden, Germany

1

Polyolefin Nanocomposites Technology*

Vikas Mittal

CONTENTS

1.1 Introduction

Polyolefins like polyethylene (PE) and polypropylene (PP) are two of the most commonly used polymers with a wide range of applications. Inorganic fillers are conventionally added to these polymers in order to reduce costs as well as to improve the mechanical or thermal properties of the polymers. The use of higher extents of fillers improves the composite properties but make the composites more opaque as well as bulky, which hinders the use of such materials for applications like food packaging, etc. With the advent of polymer nanocomposites by Toyota researchers [1,2], in which the layered silicate filler platelets were delaminated to the nanoscale in the polyamide matrix, the research on polyolefin nanocomposites has also grown exponentially. Polymer nanocomposites are organic–inorganic hybrid materials that have at least one dimension of the filler phase in nanometer scale, and owing to large interfacial contacts between the polymer and the filler, a completely different morphology is generated at the interface and, as a result, only a small amount of filler is required to enhance the properties significantly. Thus, unlike conventional composites, the nanocomposite materials have almost the same density as the parent polymers, and the transparency of the polymer matrices is also retained. The filler is required to be surface modified by organic molecules to achieve its nanoscale dispersion in the polymer matrix as it improves the compatibility between the filler and the polymer matrices. The nanoscale dispersion of the filler has been reported to be much more uniform in the case of polar polymers, thus resulting in tremendous enhancement of composite properties [3–8]. However, the filler dispersion in the case of polyolefins is not very straightforward owing to the absence of any interactions between the organic and inorganic phases. As a result, the complete delamination of the layered silicate platelets in the polyolefin matrices has by far not been achieved (except when special means are employed) and thus the properties

---

* This review work was carried out at the Institute of Chemical and Bioengineering, Department of Chemistry and Applied Biosciences, ETH Zurich, Zurich, Switzerland.

of the resulting composites are also not optimally enhanced owing to their dependence on the morphology of the composites. The following paragraphs briefly explain the properties and applications of polyolefins followed by the layered silicate fillers commonly used for the synthesis of nanocomposites. The synthesis of polyolefin nanocomposites and the current status of the synthesis methods with a focus to achieve homogenous distribution and dispersion of the filler in the polymer are subsequently dealt with in later sections.

1.2 Polyolefins

High-molecular-weight polymers from propylene and other olefins were produced by G. Natta in 1954 following the work of K. Ziegler using the "Ziegler-type" catalysts.

By variations on the form of the catalysts used, Natta was able to produce a number of different types of high-molecular-weight PPs, which differed extensively in their properties. One form, now known as isotactic PP, had a higher softening point, rigidity, and hardness, while another form, the atactic polymer, was amorphous and had little strength [9–12]. Montecatini marketed the isotactic PP grade in 1957, which followed rapid commercial exploitation of the polymer. PP has been used in tremendous amounts in a number of applications since then, namely, in fibers, films, and injection moldings. PP is a linear hydrocarbon polymer containing little or no unsaturation. The presence of a methyl group attached to alternate carbon atoms on the chain backbone can alter the properties of PP as compared to PE in a number of ways. A slight stiffening of the chain occurs in the structure owing to the pendant alternate methyl group and it also interferes with the molecular symmetry. The first effect leads to an increase in the crystalline melting point whereas the interference with molecular symmetry tends to depress it. In the case of the most regular PPs, the net effect is a melting point around 50°C higher than that of the most regular PE grades. Some aspects of chemical behavior are also affected by the methyl group. For example, the tertiary carbon atom provides a site for oxidation causing PP to be less stable than PE to the influence of oxygen. In addition, thermal and high-energy treatment leads to chain scission rather than cross-linking. The most significant influence of the methyl group is its generation of different tacticity in the products, ranging from completely isotactic and syndiotactic structures to atactic molecules. The morphological structure of PP is rather complex and at least four different types of spherulite structures have been observed. The properties of the polymer depend on the size and type of crystal structure formed, which in turn is dependent on the relative rates of nucleation to crystal growth. The ratio of these two rates can be controlled by varying the rate of cooling and by the incorporation of nucleating agents. In general, the smaller the crystal structures, the greater the transparency and flex resistance and lesser rigidity and heat resistance.

PP has a number of advantages. It has a lower density (0.90 g cm^{-3}) and low cost. It has a higher softening point and hence a higher maximum service temperature and dimensional stability. Articles can withstand boiling water and can be subjected to many steam sterilizing operations. The moldings could be sterilized in hospitals for over 1000 h at 135°C in both wet and dry conditions without severe damage. PP is also free from environmental stress cracking problems. The only exception is concentrated sulfuric and chromic acids and with aqua regia. It has a higher brittle point. It has high mechanical properties and is environmentally friendly. It has a strong barrier for water vapor. However, it is more susceptible to oxidation and high oxygen permeation.

PE is another polyolefin, which constitutes roughly one-third of the total plastic production of the world [13,14]. The world production of low-density polyethylene (LDPE) and linear low-density polyethylene (LLDPE) was 30 million tons in 2001 and is expected to increase till 2010 up to 45 million tons. For high-density polyethylene (HDPE), the world production was 23 million tons in 2001 and will reach 38 million tons in 2010 [14]. PE is conventionally synthesized by following either low-pressure or high-pressure polymerization, the products of which differ markedly in properties. Polymerization at high pressure leads to branched chains and the polymer has a low density 0.915–0.935 g cm^{-3} with crystallinity between 40% and 50%. On the other hand, when ethylene is polymerized at low pressure, the chain branching is eliminated. The generated material has a crystallinity of 60%–80% and a density of 0.95–0.965. PE, like PP, has also many good properties, which make it a material of choice for a number of applications and owing to this reason, the composites of PE are continuously studied to expand the number of applications. HDPE exhibits good thermal stability, water vapor barrier, and a low glass-transition temperature, but high crystallinity, which make it suitable for packaging frozen foods. In the packaging laminate, often a number of layers are adhered together and each layer has a special function like water vapor barrier, oxygen barrier, mechanical strength, printability, etc. Polyolefins like PP and PE are employed in such applications for water vapor barrier. However, if additional functionality of mechanical strength and the oxygen barrier can also be generated in these materials by the incorporation of inorganic fillers, then the laminate thickness can be reduced, thus saving the material costs [15,16]. Similarly, other application areas can also benefit from these properties of the composites.

1.3 Layered Silicates

Fillers of a wide range of chemical nature as well as physical dimensions are incorporated into polymers to achieve polymer composites. When one dimension of the filler is in nanoscale where the other two dimensions are in microscale, the filler particles have the shape of platelets. Such fillers are represented by alumino-silicates or 2:1 layered silicates. The particles in such layered silicates consist of stacks of 1 nm thick aluminosilicate layers (or platelets) with a regular gap in between (interlayer). Each layer consists of a central Al-octahedral sheet fused to two tetrahedral silicon sheets. Montmorillonite (from the family of smectites) is one of the most commonly used silicate in the family of layered silicates for the synthesis of nanocomposites [17,18]. The general formula of montmorillonites is $M_x(Al_{4-x}Mg_x)Si_8O_{20}(OH)_4$. Isomorphic substitutions of aluminum by magnesium in the octahedral sheet generate negative charges, which are compensated for by alkaline-earth- or hydrated alkali-metal cations, as shown in Figure 1.1 [19,20]. Based on the extent of the substitutions in the silicate crystals, a term called layer charge density is defined. Montmorillonites have a mean layer charge density of 0.25–0.5 equiv. mol^{-1}.

The electrostatic and van der Waals forces holding the layers together are relatively weak in smectites and these stacks swell easily in water. This results in the delamination of 1 nm thick layers with high aspect ratio in water. This delamination offers an opportunity to organically modify the surface of platelets by exchanging the surface alkali or alkaline earth metal cations with organic long alkyl chain ammonium ions [21,22]. This exchange is required as the high energy hydrophilic surfaces of unmodified platelets are incompatible with many polymers, whose low energetic surfaces are hydrophobic. Alkyl ammonium

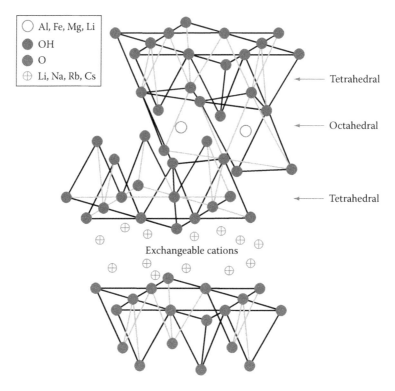

FIGURE 1.1
Structure of the 2:1 layered silicates. (Reproduced from Pavlidoua, S. and Papaspyrides, C.D., *Prog. Polym. Sci.*, 33, 1119, 2008; Beyer, G., *Plast. Addit. Compound.*, 4, 22, 2002.)

ions like octadecyltrimethylammonium, dioctadecyldimethylammonium, benzylhexa-decyltrimethylammonium, etc., have been conventionally used for the organic modification of silicates. The organic modification of filler surface renders the platelets organophilic and hydrophobic thus making them more compatible with the polymer matrices. The modification also increases the basal plane or interlayer spacing (*d*-spacing), which helps in intercalating the polymer chains in between the interlayers during composite formation.

As mentioned above, montmorillonites have the layer charge density in the range 0.25–0.5 equiv. mol^{-1} depending on the extent of substitutions as well as source. There are other silicate materials, which have much higher layer charge densities thus indicating that a large amount of organic material can be exchanged on the surface in these materials. In the case of montmorillonite, a larger area per cation is available on the surface, which leads to a lower interlayer spacing in the modified montmorillonite after surface ion exchange with alkyl ammonium ions. Vermiculites have medium charge densities of 0.5–0.8 equiv. mol^{-1} and the chemical constitution of its unit cell is $(Mg,Al,Fe)_3(Al,Si)_4O_{10}$ $(OH)_2Mg_x(H_2O)_n$ [18,22–24]. Owing to the higher charge density, the platelets of vermiculite do not fully swell in water; however, the cation exchange still leads to much higher basal plane spacing in the modified montmorillonite. On the other hand, the minerals like mica though have much higher charge densities of 1 equiv. mol^{-1}, which can lead to much higher interlayer between the platelets in the stacks after surface modification; however, owing to very strong electrostatic forces present in the interlayers due to the increased number of ions, these minerals do not swell in water and thus do not allow the cation exchange.

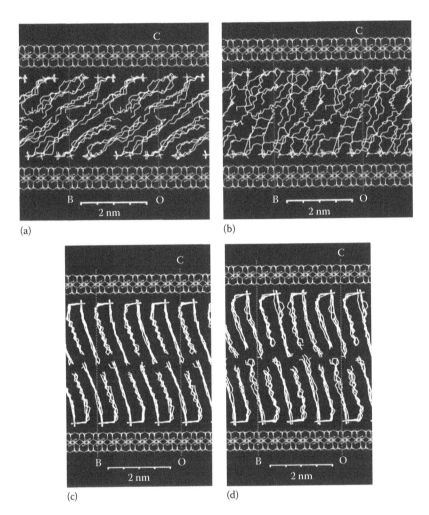

FIGURE 1.2
Snapshots of octadecyltrimethylammonium (C18) modified mica and dioctadecyldimethylammonium (2C18) modified mica in MD after 400 ps. (a) C18-mica, 20°C; (b) C18-mica, 100°C; (c) 2C18-mica, 20°C; and (d) 2C18-mica, 100°C. A major conformational change (corresponding to two phase transitions) in C18-layered mica can be seen, whereas in 2C18-mica only some more gauche conformations are present at 100°C and no order–disorder transition occurs due to close packing. (Reproduced from Heinz, H. et al., *J. Am. Chem. Soc.*, 125, 9500, 2003.)

The structure of the interlayer in organically modified silicates is believed to have the cationic head group of the alkylammonium molecule residing at the layer surface to counter the negative charge on the surface, thus leaving the organic tail radiating away from the surface [25]. Molecular dynamics simulations have also been used to gain further insights into the structures of the layered silicate as well as positioning and behavior of ammonium ions after cation exchange and as a function of temperature. Figure 1.2 depicts these more realistic simulated models of silicate layers before and after alkylammonium surface exchange [26]. The exchange was carried out using octadecyltrimethylammonium and dioctadecyldimethylammonium ions and the effect of temperature on the dynamics of the modified silicates was studied.

If all the dimensions of the filler are in nanometer scale, then the fillers have the form of spherical nanoparticles. If two dimensions of the filler are in nanoscale, whereas the third has microscale dimensions, the fillers are more in the shape of nanotubes, whiskers, etc. Polyolefin nanocomposites with both of these filler categories have also been reported, though majority of the research effort is still focused on the layered silicates as fillers. In all these different categories of the fillers, the basic requirements of the filler modifications and interface compatibilization, however, are similar and the advances in one filler category can be modified for the use in other filler category with less effort.

1.4 Polyolefin—Layered Silicate Nanocomposites

As the properties of the nanocomposites are affected by the interface between the organic and inorganic components, uniform dispersion and distribution of the filler at the nanoscale is of utmost importance. In the modified form, the filler platelets are present in the form of loosely held stacks owing to the residual polarity of the platelet surfaces after ion exchange. The polyolefin chains during the synthesis of the nanocomposites are expected to intercalate in the interlayers and, thus, delaminate the filler platelets from these stacks. The melt intercalation method is the most commonly used method for the synthesis of polyolefin nanocomposites. In this method, the polymer is melted at high temperature in a compounder or extruder and the filler is then added to the polymer melt under the action of shear. The method benefits from the fact that it does not require any solvent as used in the other modes of nanocomposite synthesis. The method is also well suited for the common polymer-processing equipments. The thermodynamics driving the intercalation of molten polymer chains inside a modified layered silicate filler have been studied by Vaia and Giannelis in their lattice model theory [27]. The authors suggested that the polymer intercalation in the inorganic filler is a result of interplay of entropic and enthalpic factors. The overall entropy of the polymer chains is negatively impacted when the polymer chains intercalate the silicate layers. On the other hand, the decrease in the entropy of the chains may be compensated by the increase in conformational freedom of the tethered alkyl surfactant chains as the inorganic layers separate, due to the less confined environment. It was observed that as small increases in the gallery spacing may not strongly influence the total entropy change, intercalation is rather driven by the changes in total enthalpy. In the model, the enthalpy of mixing was classified by two components: apolar interactions, which are generally unfavorable and arise from interaction between polymer and surfactant aliphatic (apolar) chains, and polar interactions, which originate from the polar silicates interacting with the polymer chains (if the polymer chains are polar or are deliberately functionalized with polar groups) [28]. The enthalpy of mixing can thus be rendered favorable by maximizing the magnitude and number of favorable polymer–surface interactions while minimizing the magnitude and number of unfavorable apolar interactions between the polymer and the aliphatic chains introduced along the modified layer surfaces. However, these considerations are valid for the case of polar polymer as in the case of apolar polymers like polyolefins, no positive interaction between the polymer chains and the filler surface can be expected. From this model, it implies thus that the intercalation of the polyolefin chains can be improved if the matrix polyolefin is partially polarized by the addition of surfactants or compatibilizers, then the positive interactions between the polymer matrix and the clay surface can be generated, which would lead to the subsequent intercalation of polyolefin chains also inside the filler interlayers.

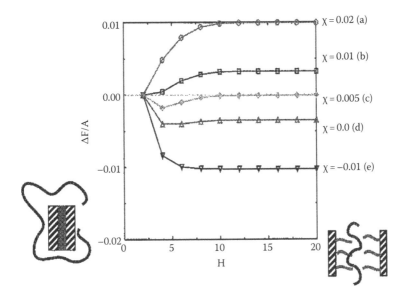

FIGURE 1.3
Free energy per unit area as a function of surface separation for five different values of χ. (Reproduced from Balazs, A.C. et al., *Macromolecules*, 31, 8370, 1998.)

Theoretical studies implying the use of self-consistent field (SCF) models have also predicted that the self-consistent potential or tendency of exfoliation is a function of the grafting density of the tethered surfactants on the filler surface and the Flory Huggins interaction parameter or χ values [29–31]. These models, like the lattice model theory, also suggested that favorable enthalpic interactions between the organically modified montmorillonite and the polymer chains can overwhelm the entropic losses and lead to effective intermixing of polymer and clay. It was suggested that in general for polymers, the mixtures of long-chain homopolymers with the organically modified clays to be thermodynamically stable, χ must be less than zero. Figure 1.3 explains this phenomenon [31]. For χ > 0 (cases a and b), free energy change would be greater than zero and consequently, the corresponding mixture would be immiscible. For χ ~ 0 (cases c and d), the plots show distinct local minima for free energy change, which indicates that the mixture forms an intercalated structure. Lowest energy state (case e) is the case where the polymer chains enter the interlayer thus distancing the filler platelets further apart leading to the exfoliation of the filler in the polymer. For polyolefins (χ = 0), such structures were predicted to exhibit only an intercalated morphology without exfoliation. The studies also suggested that the thermodynamic state of the system could be improved on increasing the length of the tethered surfactants to the surface. This was due to the fact that more interlayer distance generated among the filler platelets reduces the effective electrostatic interactions between the clay sheets, which then can be expected to interact with the polymer chains. It was also predicted by using molecular dynamic simulations of the modified filler surfaces that for a given density of alkyl chains on the surface, long chains were predicted to form a more homogenous phase than the short ones [26]. Thus, even in the absence of any attractive interaction between the long polymer chains and the surfactant molecules (χ = 0), i.e., at theta conditions, the increase in the basal plane spacing by incorporating longer surfactant chains can help in achieving more delamination of the filler in polyolefins. Grafting density of the tethered surfactant molecules was also predicted to significantly influence the final morphology of the composite. It was opined that too loose and too packed clay platelets were found unfavorable to result in effective mixing in polymers [32]. It was also reported in the theoretical studies that the exchange of surface modifications with chemical architecture similar to that of polyolefins also does not ensure the complete exfoliation of filler in the matrix as the polymer chains and the surfactant chains still lack any attractive interactions [33], and owing to the lack of interactions between the polar filler surface and the polymer chains, the system does not have a driving force, which would push the polymer chains in the filler interlayers. In such cases, it is only the delamination of the filler by the action of shear, which leads to the dispersion of the filler. In such a case, the filler platelets are only kinetically trapped inside the polymer matrix even though their dispersion was not thermodynamically stable [15,16]. Therefore, to achieve filler exfoliation by following filler modification approach, proper optimization of the organic monolayer structure in combination with the mechanical shear is of utmost requirement.

Based on the above-mentioned theoretical consideration, two general routes for the synthesis of polyolefin nanocomposites have been reported in the literature. The first based on the matrix functionalization method aims to improve the positive interactions between the surface of the filler and the polar part of the polymer matrix. This is achieved by the addition of an amphiphilic compatibilizer (mostly PP- or PE-grafted maleic anhydride (PP-*g*-MA, PE-*g*-MA)) or by the chemical functionalization of the polymer matrix to attain polar groups on the polymer chains. Block copolymers have also been used as compatibilizers. The other method is based on the optimization of the filler surface modification, which would lead to the delamination of the filler in the polymer matrix by the action of the shear forces, thus this method aims to completely organophilize the filler surface and does not use the amphiphilic compatibilizers used in the other approach. Various ammonium ions modifications with long chains, higher chain densities, etc., have been reported for this purpose, which lead to higher basal plane spacing in the filler than the conventional ammonium ion modifications. Polymerization of oligomers from the surface of the fillers has also been developed. In some instances, ammonium-terminated polyolefins have also been reported to be ionically exchanged on the surface of the filler. Other functional copolymers have also been similarly exchanged in the surface of the filler to enhance the interactions between the filler surface and polymer chains. These methods of nanocomposite synthesis are based on melt intercalation mode, but other methods of nanocomposite synthesis have also been reported. In situ polymerization of the polyolefins either in the presence of filler platelets or grafted directly from the surface of filler platelets have also been reported. Solution mixing of the organic–inorganic phases with the aid of a solvent has also been used as an alternative approach for nanocomposite synthesis, though it is less common owing to the large amount of solvent required for the process. It is also worth mentioning that owing to high compounding temperatures used during melt compounding, occasionally there is also degradation of polymer as well as surface modification. Therefore, care is required while selecting the compounding temperature and time. To avoid the significant thermal degradation of surface modifications, thermally stable surface modifications have also been developed. The following paragraphs briefly explain these methods of polyolefin nanocomposite synthesis in the light of few representative literature studies.

As mentioned above, the copolymers or nonionic surfactants (or compatibilizers) like PP-*g*-MA or PE-*g*-MA are used to achieve better compatibility between the polar clay interlayers and apolar polyolefins matrices [34–44]. Figure 1.4 demonstrates the schematic of compounding of PP with the organically modified montmorillonite in the presence of PP-*g*-MA. It has been observed in almost all the reported studies that the filler

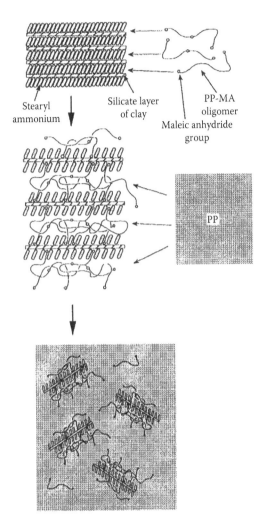

FIGURE 1.4
Schematic of polymer intercalation in the silicates in the presence of PP-*g*-MA. (Reproduced from Kawasumi, M. et al., *Macromolecules*, 30, 6333, 1997.)

exfoliation and mechanical properties were enhanced on increasing the extent of copolymers. The properties were also reported to be independent of the molecular weight of the compatibilizer. The extent of the grafting of the chains with maleic anhydride was a deciding factor to ensure compatibility between the organic and inorganic phases. Figure 1.5 shows the x-ray diffractograms of the PE nanocomposites with increasing extents of block copolymer of PE and PE glycol (PE-*b*-PEG) [45]. The increase in the extent of the intercalation is clearly visible as the diffraction peaks correspondingly shift to the lower diffraction angles. PE-*b*-PEG used in this study contained 33 methylene groups and 2.6 ethylene oxide units per molecule on an average. The filler in this study was treated with dioctadecyldimethylammonium and its fraction was always fixed at 3 vol%. Figure 1.6 also demonstrates the transmission electron microscopy images of the PP nanocomposites synthesized with 3 vol% of the filler modified with dioctadecyldimethylammonium and 2 wt% of PP-*g*-MA compatibilizer [46]. The copolymer in this case had a number average molecular weight of 3900 g mol^{-1} and 4 wt% maleic anhydride. The images indicate good dispersion of filler platelets at the nanometer level in the PP matrix; however, some thin tactoids were also present indicating that the composite morphology was a mix of

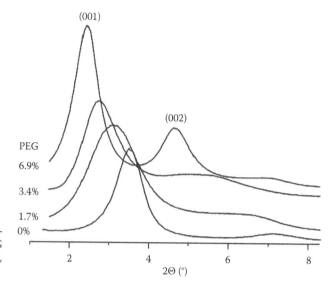

FIGURE 1.5
X-ray diffractograms of the PE nanocomposites with varying amounts of PE-*b*-PEG copolymer. (Reproduced from Osman, M.A. et al., *Polymer*, 46, 8202, 2005.)

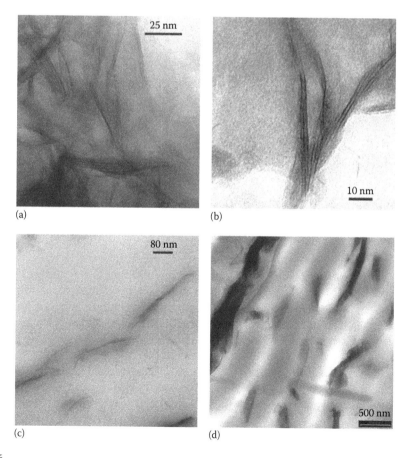

FIGURE 1.6
(a)–(d) TEM images of PP nanocomposites containing 2 wt% of PP-*g*-MA compatibilizer and 3 vol% of dioctadecyldimethylammonium-modified silicates. (Reproduced from Mittal, V., *J. Appl. Polym. Sci.*, 107, 1350, 2008.)

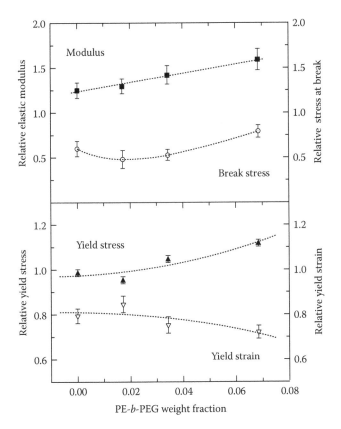

FIGURE 1.7

Mechanical performance of PP nanocomposites with varying amounts of PE-*b*-PEG compatibilizer. (Reproduced from Osman, M.A. et al., *Polymer*, 46, 8202, 2005.)

exfoliated and un-intercalated platelets. The platelets were also observed to be significantly misaligned, bent, and folded.

Mechanical properties of the PE composites (the x-ray diffraction patterns of which were demonstrated in Figure 1.5) were also reported by Osman et al. [45]. The tensile modulus was observed to increase as a function of the compatibilizer in the system, which correlated well also with the x-ray diffraction findings of increase in polymer intercalation as a function of compatibilizer fraction. As shown in Figure 1.7, the improvement in the tensile modulus of the nanocomposites at 7 wt% compatibilizer content was much significant as compared to the composites without any compatibilizer. The break stress as well as yield stress of the nanocomposites also increased as function of compatibilizer fraction. There was a marginal decrease in the elongation at break at higher fractions of compatibilizer. However, it is also important to consider that the improvement in the mechanical properties of the nanocomposites does not mean that automatically other properties would also improve. An example is the barrier properties of the composites, which are affected more significantly by the interfacial interactions or compatibility between the filler surface and the polymer matrix. Figure 1.8 shows that barrier properties of the PP nanocomposites with increasing amount of compatibilizer content [46]. The oxygen permeation in the compatibilized composites was either the same as the composite without compatibilizer or was slightly worse. It was reported that the incompatibility of the polymer matrix with the apolar surfactant chains ionically bound to the surface of the filler may lead to the generation of micro voids at the interface, which actually lead to the increase in the permeation through the composites. However, the filler is also exfoliated owing to the

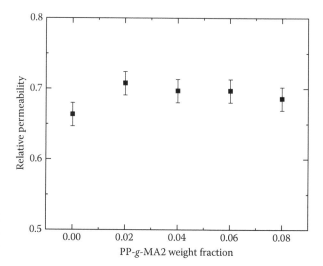

FIGURE 1.8
Barrier properties of PP nanocomposites with increasing amounts of compatibilizer fraction in the composite. (Reproduced from Mittal, V., *J. Appl. Polym. Sci.*, 107, 1350, 2008.)

intercalation of compatibilizer chains thus improving the aspect ratio of the platelets in the composites and thus barrier properties. Thus, a combined effect of two opposing factors leads to almost no change in the barrier properties of these composites. Decrease in the mechanical properties of the composites after a certain amount of compatibilizer has also been reported [46]. In this case, it was suggested that the increased exfoliation due to the addition of compatibilizer is initially helpful in enhancing the composite properties, but at higher fractions of compatibilizer, matrix plasticization is predominant and hence dissolves the effect of exfoliation of filler.

Hasegawa et al. also reported x-ray silent nanocomposites using PP-*g*-MA as polymer matrix (MA content 0.2 wt% and Mw of 210,000) with silicate organically modified with octadecyl amine [47]. These hybrids were prepared as masterbatches, which were diluted with PP to give the final composites in the hope that this will lead to an exfoliated structure. However, the PP chains did not diffuse in the interlayers and Young's modulus of the PP nanocomposites increased by 20% only and the elongation at break and tensile strength decreased [37]. Hydroxy modified PP has also been used to generate the hybrid with the organically modified clay and it acted similarly as the PP-*g*-MA compatibilizer [34]. The authors suggested that this hybrid can then be compounded with high-molecular-weight PP to achieve exfoliated nanocomposites.

Successes in the partial exfoliation of the filler in the polymer matrices have also been achieved by following the filler functionalization approach in which no compatibilizers were added for the composite synthesis. As mentioned earlier, this technique relies on the organophilization of the filler surface so that the platelets are only loosely held owing to the significant elimination of electrostatic forces between them. These loosely held platelets can then be exfoliated in the polymer melt under the action of shear. Special modifications and modification methods are required for this approach as the platelets modified with conventional ammonium ions still retain partial polar; thus, the electrostatic forces holding them together are still present. As the electrostatic forces between the platelets are a function of the interlayer separation, increasing the basal plane spacing in the modified fillers beyond that achieved with the conventional ammonium ions has been suggested for this purpose. As the apolar polymer chains in this case are not expected to have any thermodynamic interaction either with the clay surface or with the surface modification, kinetic mode of filler delamination and polymer entrapment in the delaminated platelets

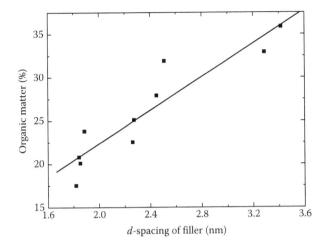

FIGURE 1.9
Correlation between the organic matter ionically bound to the layered silicate surface and the resulting basal plane spacing of the modified filler. (Reproduced from Mittal, V., *J. Thermoplast. Compos. Mater.*, 20, 575, 2007.)

is used as an alternative [15,16,48,49] in this approach. The composite is reported to be stable during the various processing steps of molding and extrusion, even though the filler is only kinetically trapped.

The basal plane spacing of the modified fillers is the function of organic matter ionically bound to the surface of the platelets, as shown in Figure 1.9 [16]. The organic matter ionically exchanged to the surface of the platelets can be increased by either using high cation exchange capacity fillers, or by increasing the length of the alkyl chains in the surface modification or by increasing the density of long alkyl chains in the surface modification molecules so that they are placed more vertically on the platelet surface. Figure 1.10 shows the schematic of a filler platelet surface modified with tetraoctadecylammonium ions [50]. Owing to higher cross-sectional area of the surface modification molecules, the basal

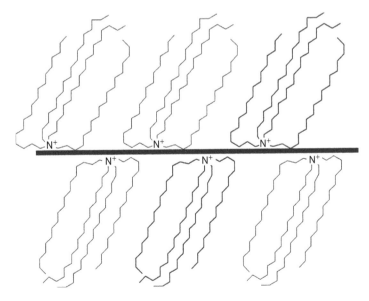

FIGURE 1.10
Representation of filler platelets surface modified with tetraoctadecylammonium. (Reproduced from Mittal, V., Need of new surface modifications, in Mittal, V. (ed), *Polymer Nanocomposites: Advances in Filler Surface Modification Techniques*, Nova Science Publishers, New York, 1–14, 2009.)

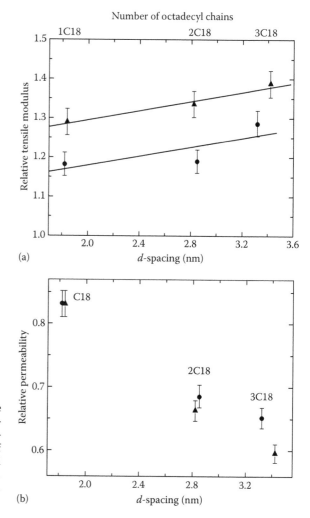

FIGURE 1.11

(a) Relative tensile modulus and (b) relative oxygen permeability through the PP nanocomposites containing 3 vol% of the filler with modifications consisting of increasing number of octadecyl chains in the structure. (Reproduced from Osman, M.A. et al., *Macromol. Chem. Physic.*, 208, 68, 2007; Mittal, V., *J. Thermoplast. Compos. Mater.*, 20, 575, 2007.)

plane spacing is much higher than corresponding low chain density surface modifications. Figure 1.11 shows the impact of increasing chain density in the surface modification on the relative tensile modulus as well as relative oxygen permeability of the PP nanocomposites [15,16]. The filler fraction was always fixed to be 3 vol% in order to relate the impact on the composite properties solely to the number of octadecyl chains in the modification. As is clearly evident, increasing the basal plane spacing by increasing the chain density of the surface modification also leads to the enhancement of the tensile modulus of the composites. The oxygen permeation through the composites was accordingly reduced. These effects were reported to be a result of increasing exfoliation of the filler platelets as with increasing basal plane spacing in the filler interlayers, they become more susceptible to be delaminated by shear in the compounder. Thus, a partial exfoliation of the filler was achieved even without the addition of conventionally added low-molecular-weight compatibilizers.

It was also reported that the cleanliness of the filler surface is also of utmost importance in ensuring the optimum enhancement of composite properties [16]. The modified fillers procured commercially may have the presence of excess surface ammonium ions present

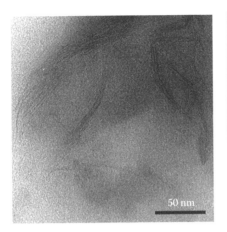 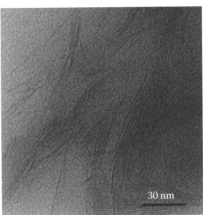

FIGURE 1.12
Transmission electron micrographs of PP nanocomposites. The dark lines represent the cross sections of aluminosilicates. (Reproduced from Mittal, V., *J. Thermoplast. Compos. Mater.*, 20, 575, 2007.)

unbound to the surface as pseudo bilayers [51]. These excess molecules can have a detrimental effect on the composite properties owing to the low-temperature degradation of
these molecules [52–54]. This early degradation may affect the molecular weight of the
polymer by inducing unwanted side reactions and thus disturb the interface between the
organic and inorganic components of the composite. By using multiple washing protocols
of the modified filler and with the aid of high-resolution thermogravimetric analysis, it
was reported that the partial filler exfoliation as well as significant enhancement in the
composite properties can be achieved even if the filler is modified with dioctadecyldimethylammonium ions and no compatibilizer is used [15,16,48,49]. Figure 1.12 shows the
TEM micrographs of PP nanocomposites, which contained 3 vol% of the filler modified
with dioctadecyldimethylammonium ions [16]. Single platelets and thin stacks (with two
to three platelets) are observed to be present along with thicker stacks indicating a mixed
morphology of the composites. The mechanical as well as oxygen barrier properties of the
composites as a function of filler fraction were also reported as demonstrated in Figure
1.13 [15,16]. The tensile modulus of the composites increased by 45% at 4 vol% of the filler,
whereas the oxygen permeation through the composites was observed to decrease by 40%
at the same amount of filler in the composites. The basal plane spacing of the filler did not
increase during the composite generation, but the delamination of the filler stacks took
place due to shear as confirmed by the TEM investigations as well as enhancement in the
composite properties.

As the exchange of preformed long alkyl ammonium ions on the surface of the filler
platelets suffers from solubility as well as steric hindrance problems, the achievement
of basal plane spacing beyond a certain extent cannot be increased. To attain further
high interlayer distances, therefore, different approaches have been reported in the
literature. One such approach is the synthesis of oligomers grafted from the surface
of the filler platelets. These modified fillers can then be compounded with the polyolefins. One has to be careful that the oligomers should not have too high molecular
weight as it may cause compatibility problems with the matrix polymer during compounding. Figure 1.15 reports one such example, where brushes of lauryl methacrylate were grafted from the surface of the platelets by first immobilizing an initiator
on the surface. The initiator was subsequently used to graft the polymer brushes.

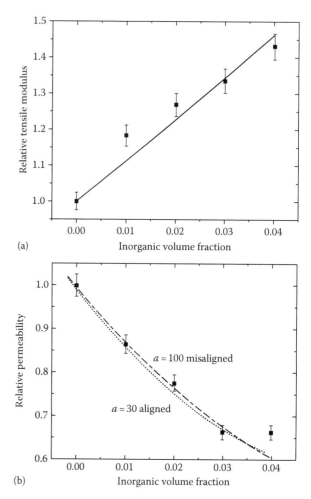

FIGURE 1.13

(a) Relative tensile modulus and (b) relative oxygen permeability through the 3 vol% propylene nanocomposites as a function of filler volume fraction in the composites. The theoretical considerations of the aspect ratio calculation from the experimental data by considering aligned and misaligned platelets has also been presented in Figure 1.13b. (Reproduced from Osman, M.A. et al., *Macromol. Chem. Physic.*, 208, 68, 2007; Mittal, V., *J. Thermoplast. Compos. Mater.*, 20, 575, 2007.)

The living conditions of polymerization by using nitroxide were used in order to control the molecular weight of the grafts as well as to eliminate the unwanted termination reactions. In fact, when the polymerization was carried out in the absence of nitroxide, very insignificant grafting took place, as shown in Figure 1.14 [55]. Curve I corresponds to the filler modified with the initiator whereas curves II and III, respectively, represent the fillers after the grafting reactions by using nonliving and living polymerization conditions. By using the living polymerization conditions, the basal plane spacing of the filler could be increased from 1.90 nm for the unreacted filler to roughly 4 nm for the grafted filler platelets. The amount of the grafting can also be further increased by the addition of subsequent batch of monomer. Graft copolymers can also be generated on the surface of the filler platelets by using various living polymerization methods like nitroxide-mediated polymerization, atom transfer radical polymerization, radical addition fragmentation chain transfer polymerization, etc. Such fillers represent high-potential materials for complete exfoliation in polyolefins when compounded with them at high temperatures under shear.

Similarly, other surface reactions on the filler surface can also be carried out, which lead to higher basal plane spacing. In one such study, esterification reactions on the surface of

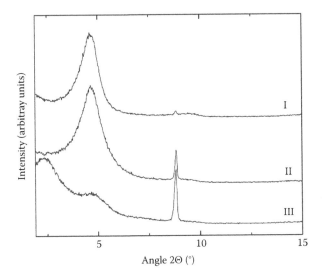

FIGURE 1.14
X-ray diffractograms of filler before grafting (I) and filler after grafting using conventional free radical polymerization (II) and nitroxide-mediated living polymerization (III). (Reproduced from Mittal, V., *J. Colloid Interface Sci.*, 314, 141, 2007.)

the filler platelets were carried out [56]. The filler platelets were first modified with surface modifications carrying hydroxyl groups. These reactive groups were subsequently reacted with long alkyl chain carboxylic acid. The reactions led to an increase of the basal plane spacing of 1.79 nm for the unreacted filler to 5.27 nm for the filler after esterification reaction as shown in Figure 1.15. Other modification routes like physical adsorption of the organic molecules on the surface of pre-modified alumino-silicates have also been suggested [57]. In this study, the silicate platelets were modified first with dioctadecyldimethylammonium ions followed by the physical adsorption of the long alkyl chain polar molecules on the surface. The ion exchange of ammonium ions leads to the gaps on the surface of platelets owing to the lower cross-sectional area of the ammonium ions as compared to area available per charge on the montmorillonite surface. Therefore, the physical adsorption tends to cover these gaps in order to eliminate the residual electrostatic forces of attraction between the platelets. These adsorbent molecules can adsorb by forming H bonds with the OH groups present either in the inside structure of clay crystals or on the edges of the platelets. Also, the adsorption has been reported to take place on the pre-adsorbed water molecules in the clay

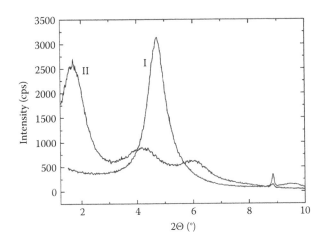

FIGURE 1.15
X-ray diffractograms of the clay modified with benzyl(2-hydroxyethyl)methyloctadecyl-ammonium (I) and filler after esterification with dotriacontanoic (lacceroic) acid (II). (Reproduced from Mittal, V., *J. Colloid Interface Sci.*, 315, 135, 2007.)

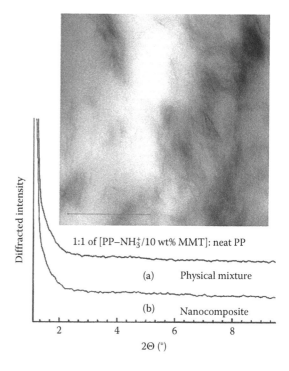

FIGURE 1.16
X-ray diffraction patterns of (a) the physical mixture of ammonium terminated PP/MMT and i-PP, and (b) the same mixture after static melt intercalation. Bright-field TEM image of the nanocomposite (corresponding to (b)). (Reproduced from Wang, Z. M. et al., *Macromolecules*, 36, 8919, 2003.)

interlayers [58–61]. It was reported that when high-molecular-weight polymer like poly(vinyl pyrrolidone) was adsorbed on the surface, not only the organophilization of the surface was achieved, but the thermal stability of the modified montmorillonites was also significantly enhanced.

Ammonium-terminated PP has also been exchanged on the filler surface in order to obtain exfoliated nanocomposites with PP [62]. The binary mixture of ammonium-terminated PP with montmorillonite generated by annealing was observed to have completely featureless patterns in the x-ray diffraction. This hybrid was further mixed/blended (at a 50/50 weight ratio) with a neat i-PP (Mn = 110,000 and Mw = 250,000 g mol^{-1}). As shown in Figure 1.16, the initial exfoliated structure was maintained even after mixing with i-PP. The TEM image shown in the inset also confirms this observation. The authors reported that the i-PP polymer chains largely served as diluents in the ternary system.

Wilkie et al. reported the exchange of functional copolymer on the surface of the filler platelets [63]. In this study, ammonium salt of copolymer of styrene and vinylbenzene chloride was synthesized. This copolymer after ion exchange with the silicate surface was compounded with PE as well as PP. Figure 1.17 shows the TEM micrographs of the PP nanocomposites. The authors observed that nanodispersion of the filler was achieved by analyzing the low magnification images. In the case of PP, the nanocomposites was observed to have more intercalated morphology, but for PE, more extent of single layers were observed in the micrographs leading to the classification of the morphology as a mixture of intercalated and exfoliated. The molecular weight of the copolymer ion exchanged on the filler surface was in the range 5000–6000 g mol^{-1}; therefore, a large amount to this hybrid was required to be added to polyolefin matrix in order to achieve the required amount of inorganic filler content in the composites.

In situ formation of the polyolefin nanocomposites has also been developed in great details. Polymerization filling technique for the synthesis of polyolefin nanocomposites

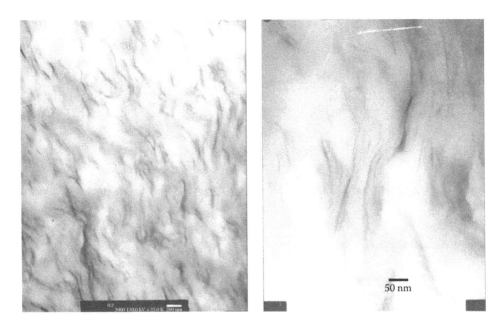

FIGURE 1.17
TEM micrographs of the PP nanocomposites containing filler modified with ammonium salt of copolymer of styrene and vinylbenzene chloride. (Reproduced from Su, S. et al., *Polym. Degrad. Stabil.*, 83, 321, 2004.)

was reported [64], in which a Ziegler–Natta or any other coordination catalyst is anchored to the surface of the layered silicates. This then can be directly used or the polymerization of olefins like ethylene, propylene, etc., from the surface of the silicate. Here, the immobilization of the catalyst is carried out by the electrostatic interactions of the catalytic materials with the MAO initially anchored to the filler surface. In one such instance, when the polymerization of ethylene was carried out in the absence of any chain transfer agent, ultrahigh-molecular-weight PE-silicate nanocomposite was achieved, which was not possible to process further. However, by adding hydrogen to the system, the molecular weight of the polymer matrix can be reduced and the processability also improves. Similarly, the use of palladium-based complex and synthetic fluorohectorite as polymerization catalyst and inorganic component for the generation of PE nanocomposites has been reported [65]. Titanium-based Ziegler–Natta catalysts were similarly immobilized on the inner surfaces of montmorillonites and organic salts with hydroxyl groups for the modification of montmorillonite were used [66]. The hydroxyl groups acted as reactive sites for anchoring catalyst between the clay layers. Heinemann et al. [67] also reported the polymerization of ethylene in the presence of modified layered silicates. Surface modifications like dimethyldistearylammonium and dimethylbenzylstearylammonium were used for the study. The composites were also prepared with the non-modified fillers. It was reported that nanocomposites were only observed when the modified clays were used, whereas in the case of non-modified fillers, only microcomposites were formed. Another study on the in situ polymerization of ethylene with polymer grafting to the surface approach was reported [68]. In this study, triisobutylaluminum was reacted with hydroxyl groups within the clay galleries followed by the washing of the excess of this compound. Subsequently, the clay was reacted with a vinyl alcohol (ω-undecylenylalcohol). The polar hydroxyl groups reacted with the alkyaluminum compounds on the surface of the clay galleries. The vinyl groups thus chemically linked to the silicate surface were copolymerized with

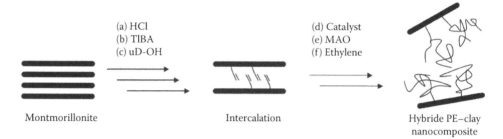

FIGURE 1.18
Schematic of in situ polymerization of ethylene in the presence of montmorillonite-containing vinyl groups on the surface. (Reproduced from Shin, S.-Y.A. et al., *Polymer*, 44, 5317, 2003.)

ethylene inside the clay galleries using a coordination catalyst, as shown in Figure 1.18 [68]. The authors reported that the exfoliation of the filler platelets in the polymer could be achieved and the PE chains were chemically bound to the surface of the filler during copolymerization reaction.

As mentioned earlier, the use of other fillers like spherical nanoparticles as well as nanotubes to achieve polyolefin nanocomposites has also been vigorously explored.

When electrical and electronic properties are also of interest apart from mechanical and thermal properties, carbon nanotubes can be of better advantages. Nanotubes are inert in nature and, therefore, also require surface modification in order to achieve compatibility with the polymer matrices. Thus, the nanoscale dispersion of the nanotubes is as important and challenging as the layered silicates as the properties are dependant on the generated morphology in the composites. In a representative study, Teng et al. [69] studied the incorporation of nanotubes in a variety of PP matrices differing in their melt flow indices. Figure 1.19 shows the TEM micrograph of the nanocomposites using the polymer with the lowest melt flow index. The nanotubes were observed to have the diameters of in the range of 15–50 nm. The dispersion of filler was observed to be in random arrays and showed an interconnected structure, which also exhibited partial aggregation in the PP matrix.

FIGURE 1.19
TEM micrograph of the PP nanotube nanocomposite containing 5 wt% of the nanotubes. The black spots indicate the nanotubes perpendicular to the sample. (Reproduced from Teng, C.-C. et al., *Composites: Part A*, 39, 1869, 2008.)

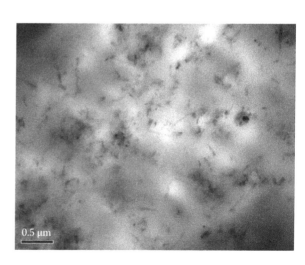

References

1. Yano, K., Usuki, A., Okada, A., Kurauchi, T., and Kamigaito, O. 1993. Synthesis and properties of polyimide-clay hybrid. *Journal of Polymer Science, Part A: Polymer Chemistry* 31:2493–2498.
2. Kojima, Y., Fukumori, K., Usuki, A., Okada, A., and Kurauchi, T. 1993. Gas permeabilities in rubber-clay hybrid. *Journal of Materials Science Letters* 12:889–890.
3. Lan, T., Kaviratna, P. D., and Pinnavaia, T. J. 1994. On the nature of polyimidie-clay hybrid composites. *Chemistry of Materials* 6:573–575.
4. Lan, T. and Pinnavaia, T. J. 1994. Clay-reinforced epoxy nanocomposites. *Chemistry of Materials* 6:2216–2219.
5. Burnside, S. D. and Giannelis, E. P. 1995. Synthesis and properties of new poly(dimethyl siloxane) nanocomposites. *Chemistry of Materials* 7:1597–1600.
6. Wang, Z. and Pinnavaia, T. J. 1998. Nanolayer reinforcement of elastomeric polyurethane. *Chemistry of Materials* 10:3769–3771.
7. Messersmith, P. B. and Giannelis, E. P. 1995. Synthesis and barrier properties of poly(ε-caprolactone)-layered silicate nanocomposites. *Journal of Polymer Science, Part A: Polymer Chemistry* 33:1047–1057.
8. Yano, K., Usuki, A., and Okada, A. 1997. Synthesis and properties of polyimide-clay hybrid films. *Journal of Polymer Science, Part A: Polymer Chemistry* 35:2289–2294.
9. Katz, H. S. and Milewski, J. V. 1987. *Handbook of Fillers for Plastics*. New York: Van Nostrand.
10. Rothon, R. 1995. *Particulate-Filled Polymer Composites*. Harlow, U.K.: Longman.
11. Brydson, J. A. 1975. *Plastic Materials*. London, U.K.: Newnes-Butterworths.
12. Karger-Kocsis, J. 1995. *Polypropylene Structure, Blends and Composites*. London, U.K.: Chapman & Hall.
13. Elias, H.-G. 2003. *Anwendungen von Polymeren*, vol. 4 of *Makromolekuele*. Weinheim, Germany: Wiley-VCH.
14. Robertson, G. L. 1993. *Food Packaging-Principles and Practice*, Vol. 6 of *Packaging and Converting Technology*. New York: Marcel Dekker, Inc.
15. Osman, M. A., Mittal, V., and Suter, U. W. 2007. Poly(propylene)-layered silicate nanocomposites: Gas permeation properties and clay exfoliation. *Macromolecular Chemistry and Physics* 208:68–75.
16. Mittal, V. 2007. Polypropylene-layered silicate nanocomposites: Filler matrix interactions and mechanical properties. *Journal of Thermoplastic Composite Materials* 20:575–599.
17. Bailey, S. W. 1984. *Reviews in Mineralogy*. Blacksburg, VA: Virginia Polytechnic Institute and State University.
18. Brindley, G. W. and Brown, G. 1980. *Crystal Structures of Clay Minerals and Their X-Ray Identification*. London, U.K.: Mineralogical Society.
19. Pavlidoua, S. and Papaspyrides, C. D. 2008. A review on polymer-layered silicate nanocomposites. *Progress in Polymer Science* 33:1119–1198.
20. Beyer, G. 2002. Nanocomposites: A new class of flame retardants for polymers. *Plastic Additives and Compounding* 4:22–27.
21. Theng, B. K. G. 1974. *The Chemistry of Clay-Organic Reactions*. New York: Wiley.
22. Jasmund, K. and Lagaly, G. 1993. *Tonminerale und Tone Struktur*. Darmstadt, Germany: Steinkopff.
23. Tjong, S. C. and Meng, Y. Z. 2003. Preparation and characterization of melt-compounded polyethylene/vermiculite nanocomposites. *Journal of Polymer Science, Part B: Polymer Physics* 41:1476–1484.
24. Xu, J., Li, R. K. Y., Xu, Y., Li, L., and Meng, Y. Z. 2005. Preparation of poly(propylene carbonate)/organo-vermiculite nanocomposites via direct melt intercalation. *European Polymer Journal* 41:881–888.

25. Vaia, R. A., Teukolsky, R. K., and Giannelis, E. P. 1994. Interlayer structure and molecular envi-
    ronment of alkylammonium layered silicates. *Chemistry of Materials* 6:1017–1022.
26. Heinz, H., Castelijns, H. J., and Suter, U. W. 2003. Structure and phase transitions of alkyl chains
    on mica. *Journal of American Chemical Society* 125:9500–9510.
27. Vaia, R. A. and Giannelis, E. P. 1997. Lattice of polymer melt intercalation in organically-modified
    layered silicates. *Macromolecules* 30:7990–7999.
28. Alexandre, M. and Dubois, P. 2000. Polymer layered silicate nanocomposites: Preparation, prop-
    erties and uses of a new class of materials. *Materials Science and Engineering R: Reports* 28:1–63.
29. Balazs, A. C., Singh, C., Zhulina, E., and Lyatskaya, Y. 1999. Modeling the phase behavior of
    polymer/clay nanocomposites. *Accounts of Chemical Research* 32:651–657.
30. Ginzburg, V. V., Singh, C., and Balazs, A. C. 2000. Theoretical phase diagrams of polymer/clay
    composites: The role of grafted organic modifiers. *Macromolecules* 33:1089–1099.
31. Balazs, A. C., Singh, C., and Zhulina E. 1998. Modeling the interactions between polymers and
    clay surfaces through self-consistent-field theory. *Macromolecules* 31:8370–8381.
32. Hasegawa, R., Aoki, Y., and Doi, M. 1996. Optimum graft density for dispersing particles in
    polymer melts. *Macromolecules* 29:6656–6662.
33. Edgecombe, S. R., Gardiner, J. M., and Matsen, M. W. 2002. Suppressing autophobic dewetting
    by using a bimodal brush. *Macromolecules* 35:6475–6477.
34. Kato, M., Usuki, A., and Okada, A. 1997. Synthesis of polypropylene oligomer-clay intercala-
    tion compounds. *Journal of Applied Polymer Science* 66:1781–1785.
35. Kawasumi, M., Hasegawa, N., Kato, M., Usuki, A., and Okada, A. 1997. Preparation and
    mechanical properties of polypropylene-clay hybrids. *Macromolecules* 30:6333–6338.
36. Usuki, A., Kato, M., Okada, A., and Kurauchi, T. 1997. Synthesis of polypropylene-clay hybrid.
    Journal of Applied Polymer Science 63:137–138.
37. Hasegawa, N., Kawasumi, M., Kato, M., Usuki, A., and Okada, A. 1998. Preparation and
    mechanical properties of polypropylene-clay hybrids using a maleic anhydride-modified poly-
    propylene oligomer. *Journal of Applied Polymer Science* 67:87–92.
38. Reichert, P., Nitz, H., Klinke, S., Brandsch, R., Thomann, R., and Muelhaupt, R. 2000.
    Poly(propylene)/organoclay nanocomposite formation: Influence of compatibilizer function-
    ality and organoclay modification. *Macromolecular Materials and Engineering* 275:8–17.
39. Manias, E., Touny, A., Wu, L., Strawhecker, K., Lu, B., and Chung, T. C. 2001. Polypropylene/
    montmorillonite nanocomposites. Review of synthetic routes and materials properties.
    Chemistry of Materials 13:3516–3523.
40. Zhang, Q., Fu, Q., Jiang, L., and Lei, Y. 2000. Preparation and properties of polypropylene/
    montmorillonite layered nanocomposites. *Polymer International* 49:1561–1564.
41. Oya, A., Kurokawa, Y., and Yasuda, H. 2000. Factors controlling mechanical properties of clay
    mineral/polypropylene nanocomposites. *Journal of Materials Science* 35:1045–1050.
42. Xu, W., Liang, G., Wang, W., Tang, S., He, P., and Pan, W. P. 2003. PP-PP-*g*-MAH-Org-MMT
    nanocomposites. I. Intercalation behavior and microstructure. *Journal of Applied Polymer Science*
    88:3225–3231.
43. Ellis, T. S. and D'Angelo, J. S. 2003. Thermal and mechanical properties of a polypropylene
    nanocomposite. *Journal of Applied Polymer Science* 90:1639–1647.
44. Zhang, J., Jiang, D. D., and Wilkie, C. A. 2006. Polyethylene and polypropylene nanocompos-
    ites based on a three component oligomerically-modified clay. *Polymer Degradation and Stability*
    91:641–648.
45. Osman, M. A., Rupp, J. E. P., and Suter, U. W. 2005. Effect of non-ionic surfactants on the exfolia-
    tion and properties of polyethylene-layered silicate nanocomposites. *Polymer* 46:8202–8209.
46. Mittal, V. 2008. Mechanical and gas permeation properties of compatibilized polypropylene-
    layered silicate nanocomposites. *Journal of Applied Polymer Science* 107:1350–1361.
47. Hasegawa, N., Okamoto, H., Kawasumi, M., Kato, M., Tsukigase, A., and Usuki, A. 2000.
    Polyolefin-clay hybrids based on modified polyolefins and organophilic clay. *Macromolecular
    Materials and Engineering* 280–281:76–79.

48. Osman, M. A., Rupp, J. E. P., and Suter, U. W. 2005. Gas permeation properties of polyethylene-layered silicate nanocomposites. *Journal of Material Chemistry* 15:1298–1304.

49. Osman, M. A., Rupp, J. E. P., and Suter, U. W. 2005. Tensile properties of polyethylene-layered silicate nanocomposites. *Polymer* 46:1653–1660.

50. Mittal, V. 2009. Need of new surface modifications. In *Polymer Nanocomposites: Advances in Filler Surface Modification Techniques*, ed. V. Mittal, pp. 1–14. New York: Nova Science Publishers.

51. Mittal, V. 2008. Effect of the presence of excess ammonium ions on the clay surface on permeation properties of epoxy nanocomposites. *Journal of Materials Science* 43:4972–4978.

52. Xie, W., Gao, Z., Pan, W. P., Hunter, D., Singh, A., and Vaia, R. 2001. Thermal degradation chemistry of alkyl quaternary ammonium montmorillonite. *Chemistry of Materials* 13:2979–2990.

53. Morgan, A. B. and Harris, J. D. 2003. Effects of organoclay Soxhlet extraction on mechanical properties, flammability properties and organoclay dispersion of polypropylene nanocomposites. *Polymer* 44:2313–2320.

54. Shah, R. and Paul, D. R. 2006. Organoclay degradation in melt processed polyethylene nanocomposites. *Polymer* 47:4075–4084.

55. Mittal, V. 2007. Polymer chains grafted "to" and "from" layered silicate clay platelets. *Journal of Colloid and Interface Science* 314:141–151.

56. Mittal, V. 2007. Esterification reactions on the surface of layered silicate clay platelets. *Journal of Colloid and Interface Science* 315:135–141.

57. Mittal, V. and Herle, V. 2008. Physical adsorption of organic molecules on the surface of layered silicate clay platelets: A thermogravimetric study. *Journal of Colloid and Interface Science* 327:295–301.

58. Zhou, Q., Frost, R. L., He, H., Xi, Y., and Liu, H. 2007. Adsorbed *para*-nitrophenol on HDTMAB organoclay-A TEM and infrared spectroscopic study. *Journal of Colloid and Interface Science* 307:357–363.

59. Zhou, Q., He, H., Frost, R. L., and Xi, Y. 2007. Adsorption of *p*-nitrophenol on mono-, bi-, and trialkyl surfactant-intercalated organoclays: A comparative study. *Journal of Physical Chemistry C* 111:7487–7493.

60. Zhou, Q., Frost, R. L., He, H., and Xi, Y. 2007. Changes in the surfaces of adsorbed *para*-nitrophenol on HDTMA organoclay—The XRD and TG study. *Journal of Colloid and Interface Science* 307:50–55.

61. Zhou, Q., Frost, R. L., He, H., Xi, Y., and Zbik, M. 2007. TEM, XRD, and thermal stability of adsorbed paranitrophenol on DDOAB organoclay. *Journal of Colloid and Interface Science* 311:24–37.

62. Wang, Z. M., Nakajima, H., Manias, E., and Chung, T. C. 2003. Exfoliated PP/Clay nanocomposites using ammonium-terminated PP as the organic modification for montmorillonite. *Macromolecules* 36:8919–8922.

63. Su, S., Jiang, D. D., and Wilkie, C. A. 2004. Poly(methyl methacrylate), polypropylene and polyethylene nanocomposite formation by melt blending using novel polymerically-modified clays. *Polymer Degradation and Stability* 83:321–331.

64. Dubois, P., Alexandre, M., Hindryckx, F., and Jerome, R. 1998. Homogeneous polyolefin-based composites. *Journal of Macromolecular Science: Reviews in Macromolecular Chemistry and Physics C* 38:511–565.

65. Bergman, J. S., Chen, H., Giannelis, E. P., Thomas, M. G., and Coates, G. W. 1999. Synthesis and characterization of polyolefin–silicate nanocomposites: A catalyst intercalation and in situ polymerization approach. *Journal of Chemical Society Chemical Communications* 21:2179–2180.

66. Jin, Y.-H., Park, H.-J., Im, S.-S., Kwak, S.-Y., and Kwak, S. 2002. Polyethylene/clay nanocomposite by in situ exfoliation of montmorillonite during Ziegler–Natta polymerization of ethylene. *Macromolecular Rapid Communications* 23:135–140.

67. Heinemann, J., Reichert, P., Thomann, R., and Muelhaupt, R. 1999. Polyolefin nanocomposites formed by melt compounding and transition metal catalyzed ethene homo- and copolymerization in the presence of layered silicates. *Macromolecular Rapid Communications* 20:423–430.
68. Shin, S.-Y. A., Simon, L. C., Soares, J. B. P., and Scholz, G. 2003. Polyethylene-clay hybrid nanocomposites: In situ polymerization using bifunctional organic modifiers. *Polymer* 44:5317–5321.
69. Teng, C.-C., Maa, C.-C. M., Huang, Y.-W., Yuen, S.-M., Weng, C.-C., Chen, C.-H., and Su, S.-F. 2008. Effect of MWCNT content on rheological and dynamic mechanical properties of multi-walled carbon nanotube/polypropylene composites. *Composites: Part A* 39:1869–1875.

2

Nanocomposite Blends Containing Polyolefins

Mathieu Bailly and Marianna Kontopoulou

CONTENTS

2.1 Introduction

Polyolefins offer a wide range of properties, from plastic to elastomeric, depending on their structure. Blending technology is used to achieve an even wider range of properties by combining two polyolefins, or a polyolefin with another thermoplastic, resulting in improvements in mechanical strength, toughness, processability, thermal stability, aging resistance, etc. The properties of a blend depend to a large extent on the degree of phase separation of its components, namely, if they are miscible, partially miscible, or immiscible. Generally speaking, miscible blends, being single-phase systems, have intermediate properties with respect to the two components. Polyolefin-based blends are for the most part thermodynamically immiscible, consisting of two or more phases and displaying a combination of the properties of the components, often with synergistic effects [1]. The microstructure of immiscible blends depends on many factors: temperature, composition, interfacial tension, rheology of the individual components, and compounding method. Microstructure, in turn, controls the physical properties [2,3]. Extensive efforts have been made to improve the compatibility between the two polymeric components in order to prevent particle coalescence and maximize the degree of reinforcement induced by the addition of the dispersed phase.

It is common practice to add fillers, such as talc or glass fibers, to a thermoplastic matrix to achieve cost reduction and mechanical reinforcement, as well as to enhance various properties such as electrical conductivity, thermal properties, and dimensional stability. Large amounts of conventional micron-size fillers are typically required in these formulations, which results in deterioration of processability and surface appearance.

More recently nanoscale fillers such as clay platelets, silica, nano-calcium carbonate, titanium dioxide, and carbon nanotube nanoparticles have been used extensively to achieve reinforcement, improve barrier properties, flame retardancy and thermal stability, as well as synthesize electrically conductive composites. In contrast to micron-size fillers, the desired effects can be usually achieved through addition of very small amounts (a few weight percent) of nanofillers [4]. For example, it has been reported that the addition of 5 wt% of nanoclays to a thermoplastic matrix provides the same degree of reinforcement as 20 wt% of talc [5]. The dispersion and/or exfoliation of nanofillers have been identified as a critical factor in order to reach optimum performance. Techniques such as filler modification and matrix functionalization have been employed to facilitate the breakup of filler agglomerates and to improve their interactions with the polymeric matrix.

Unfortunately, the improvements in mechanical strength and modulus achieved upon introduction of the nanofillers usually come at the expense of the toughness. On the contrary, adding a dispersed rubbery phase to a thermoplastic matrix will enhance the toughness but at the expense of the mechanical strength and stiffness. Hence the concept of embedding nanofillers together with a rubbery phase into a thermoplastic matrix has been developed to take advantage of the desirable properties of both components by forming ternary nanocomposites or polymer blend nanocomposites.

If ternary nanocomposites can merge the advantages of both polymer nanocomposites and polymer blends, they also hold the complexity of both fields, as many factors have the potential to impact their morphology and final properties. This is one of the reasons why polymer blends containing nanofillers have been the subject of intense investigation during the last decade. In addition to the considerations pertaining to the microstructure of polymer blends and the nanofiller dispersion mentioned above, ternary nanocomposites present a unique set of challenges. Some of the issues encountered in the literature are related to the selective localization of the nanofillers and their effects on the microstructure and physical properties of immiscible and partially miscible polymer blends. The localization of the nanofillers in the matrix, the interphase between the two polymers, or in the secondary polymeric phase dictates their microstructure, which in turn controls the final properties; therefore, numerous studies have been dedicated toward the understanding of filler partitioning and the thermodynamic considerations involved [6]. Furthermore, the introduction of fillers in a polymer blend can be seen as an innovative and efficient way to enhance the compatibility between the two polymers, instead of using expensive compatibilization agents [7–14].

Ternary nanocomposites contain three components; therefore, countless combinations of materials and compositions are encountered in the literature and in practice. Polyolefins are almost always encountered as a component in nanocomposite blends, being the matrix, the dispersed phase, or both. Generally, semicrystalline thermoplastics such as polypropylene (PP) and polyethylene (PE) are very popular as the matrix component, while the dispersed phase consists of either another thermoplastic [15–17] or an elastomer [5,18–23]. Polyolefin-based thermoplastic elastomers, such as ethylene/propylene rubber, and other ethylene-based copolymers frequently comprise the elastomeric dispersed phase. Depending on the targeted property, different types of nanofillers are added to the blend formulations.

In this chapter, the latest advances and findings in the field of ternary nanocomposites will be reviewed, including thermodynamic and kinetic considerations for filler partitioning, the different types of morphologies that can be achieved, and the resulting microstructure and mechanical properties. Lastly, some systems of industrial interest will be presented. The topic of polymer blend nanocomposites is vast, given the wide array of nanofillers and polymer matrices that have been studied in the literature. For the sake of brevity, the discussion in this chapter will be limited to nanocomposite blends containing siliceous nanofillers, such as montmorillonite clay and nanosilica. While the focus is on nanocomposite blends containing a polyolefin component, for the sake of completeness, some other blend systems are mentioned as well.

2.2 Filler Localization

Filler particles distribute unevenly between the two phases in the overwhelming majority of thermodynamically immiscible blends [6]. Depending on thermodynamic and kinetic considerations that will be discussed in this section, the fillers may be located inside the matrix, within the dispersed phase, or at the interface. The resulting microstructure can be characterized as (a) *encapsulated* if the fillers are embedded into the dispersed phase (Figure 2.1a), (b) *segregated* if the fillers and the minor phase are dispersed within the matrix (Figure 2.1b), and (c) *core–shell* when the fillers remain at the interface and therefore form a "shell" surrounding the dispersed phase (Figure 2.2).

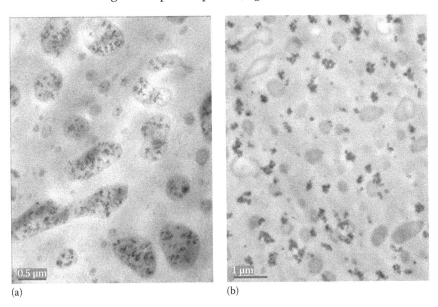

(a)   (b)

FIGURE 2.1
TEM images showing encapsulated and segregated morphologies, respectively: (a) PP/maleated EPR 80/20 blend containing 5 wt% silica in the dispersed EPR phase and (b) PP/ethylene-octene copolymer 80/20 containing 5 wt% silica in the matrix. The elastomeric phase appears darker. Blend system studied in Ref. [24]. (From Liu, Y. and Kontopoulou, M., *J. Vinyl Addit. Technol.*, 13, 147, 2007.)

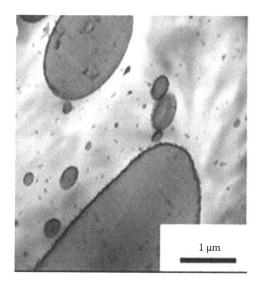

FIGURE 2.2
TEM image showing the morphology of a PP/PS 70/30
blend containing 3 wt% of hydrophobic silica, located
at the PP/PS interface. (Reprinted from Elias, L. et al.,
Polymer, 48, 6029, 2007.)

The basis for understanding filler segregation in two-phase systems was established by
the science of low-viscosity fluid emulsions. Adsorption of solid particles at the interface
between two fluids has been shown to be a promising alternative to conventional surfactants for stabilizing the emulsions [25]. The mechanisms through which solid particles contribute to stabilization are still under investigation, but there seems to be a consensus in
the literature on the fact that the particles at the interface prevent coalescence by acting as
physical barriers. As a result, the global area of the dispersed phase is reduced compared to
a pristine emulsion, thus lowering the macroscopic or "effective" interfacial tension.

Although the science of low-viscosity emulsions has set the fundamentals of a thermodynamic approach, significant differences with their high-viscosity polymer melts' counterparts exist. Unlike emulsions, where the preferred localization of the solid particles is at
the interface, in polymer blends a particular organization of the filler might be preferred,
depending on the targeted properties. For example, as described in Section 2.4, to achieve
improvements in mechanical strength, modulus, and toughness, the well-dispersed fillers
must be localized within the matrix [27].

It is generally acknowledged that two distinct factors contribute to the final localization
of the fillers: thermodynamic effects, based on the affinity that the fillers might have with
the polymers and kinetic effects that include the mode of preparation of the composites
and the ability of the fillers to migrate to a different location, i.e., from one phase to the
other or from one phase to the interface. These factors are critical in the prediction and
control of filler localization and are described in detail below.

2.2.1 Thermodynamic Effects

When fillers are introduced into a polymer blend, they generally tend to migrate to the
phase with which they have the most affinity, in a thermodynamic sense. The affinity
between polymers and fillers has been described qualitatively via the study of the surface
chemistry of both phases under the terminology of polymer/polymer and polymer/filler
interactions.

Attraction between the polymeric phases and the fillers can occur through either physical
or chemical interactions. Physical interactions have been extensively reported in rubber/
silica and rubber/carbon black composites [28,29] and usually involve physical adsorption

of the polymer chains onto the surface of the particles. Hydrogen bonding and acid–base interactions are other examples of physical interactions, which can result in strong associations between the polymer and the fillers. Chemical interactions are established if the fillers and the matrix are associated via covalent bonding. This can be achieved by using coupling agents or by functionalizing the matrix in order to establish a reaction with the functional groups present at the particle surface [27].

The theoretical framework applied to quantitatively characterize the polymer/filler interactions uses the concept of surface tension. The surface free energy of a condensed phase, being either liquid (surface tension γ_l) or solid (surface free energy or activity γ_s), represents the strength of the intermolecular interactions in the bulk. It can be expressed as a sum of two components: γ^d (the dispersive component) describing London-type interactions between fluctuating dipoles and γ^{sp} (the specific component) that includes all the other interactions, such as polar interactions, γ^p, hydrogen bonds, γ^h, and acid–base interactions γ^{ab} [30]. The surface free energy of a solid or surface tension of a liquid is thus

$$\gamma = \gamma^d + \gamma^{sp} \tag{2.1}$$

$$\text{with } \gamma^{sp} = \gamma^p + \gamma^h + \gamma^{ab} \tag{2.2}$$

Two methods can be used for the assessment of γ and its components: contact angle measurements and inverse gas chromatography (IGC) [31]. Chibowski and Perea-Carpio [32] reviewed the problems encountered when attempting to determine the surface free energy of powered solids, like silica particles, using the contact angle technique. Wu reviewed the different techniques that can be employed to measure the surface tension of polymer melts [30]. These techniques are based on the pendant and sessile drop techniques that require density data or contact angle measurements.

Furthermore, the interfacial tension or interfacial energy, γ_{12} quantifies the interactions that occur at the interface between two phases 1 and 2. It is associated to the work of cohesion W_c, which represents the energy required to create two new surfaces out of a single material (Figure 2.3a), and the work of adhesion (W_a), which represents the energy required to separate two materials that were previously adhered (Figure 2.3b).

$$W_c^i = 2\gamma_i \quad \text{with } i = 1,2 \tag{2.3}$$

and

$$W_a = \gamma_1 + \gamma_2 - \gamma_{12} \tag{2.4}$$

Logically, the greater the attraction between the two surfaces, the greater the work of adhesion will be. From Equation 2.4, it is also deduced that the smaller the interfacial tension,

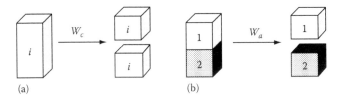

(a)   (b)

FIGURE 2.3
(a) Work of cohesion and (b) work of adhesion.

the greater the work of adhesion. It is, therefore, of crucial interest to have access to surface and interfacial tension data in order to predict the adhesion between two phases.

Good introduced an interaction parameter Φ [33], which is the ratio of the free energy of adhesion and the geometric mean of the free energies of cohesion:

$$\frac{W_a}{\sqrt{W_c^1 W_c^2}} = \phi \tag{2.5}$$

Equation 2.5 combined with Equations 2.3 and 2.4 gives

$$\gamma_{12} = \gamma_1 + \gamma_2 - 2\phi\sqrt{\gamma_1\gamma_2} \tag{2.6}$$

Experimental data of the surface and interfacial tensions showed that the interaction parameter was equal to 1 for systems in which dispersion forces or polar interactions were predominant [33], so that

$$\gamma_{12} = \gamma_1 + \gamma_2 - 2\sqrt{\gamma_1\gamma_2} \tag{2.7}$$

Equation 2.7 is also referred as the Girifalco–Good equation.

Alternative equations have been proposed by Owens and Wendt [34]:

$$\gamma_{12} = \gamma_1 + \gamma_2 - 2\sqrt{\gamma_1^d \, \gamma_2^d} - 2\sqrt{\gamma_1^p \, \gamma_2^p} \tag{2.8}$$

and by Wu [30]:

$$\gamma_{12} = \gamma_1 + \gamma_2 - \frac{4\gamma_1\gamma_2}{\gamma_1 + \gamma_2} \tag{2.9}$$

Given that surface tension data are not always available at the temperature of interest, the expression derived by Guggenheim [35] can be used to take into account the variation of temperature [26]. It is also possible to experimentally measure interfacial tensions via the same techniques applicable for surface tension measurements [30].

Once values of interfacial tensions are determined, they can be used to predict the partitioning tendency of the fillers. Sumita et al. [36] were the first to propose a criterion based on the calculation of the interfacial tensions between the different components. This criterion has been applied in many recent publications [26,27,37] and is based on the tendency of the system to minimize its total free energy. As a result, fillers will tend to go to the phase with which they have the strongest affinity, corresponding to the lowest interfacial tension.

The wetting coefficient is defined as

$$\omega = \frac{\gamma_{\text{filler}-2} - \gamma_{\text{filler}-1}}{\gamma_{12}} \tag{2.10}$$

where subscripts 1 and 2 stand for phases 1 and 2 of the polymer blend. Based on this criterion, the filler particles distribute within the phase 1 if $\omega > 1$, within phase 2 if $\omega < 1$, and at the interface if $-1 < \omega < 1$.

As expected, the greater affinity of the fillers toward a certain phase is strongly correlated with the relative value of the interfacial tension, which in turn depends on the nature of the polymer/filler interactions. Table 2.1 summarizes the values obtained for the wetting

TABLE 2.1

Wetting Coefficient for Different Blends/Filler Combinations the Filler Localization Was Characterized by Electron Microscopy

Polymer 1	Polymer 2	Filler	ω_{12} Owens–Wendt	ω_{12} Girifalco–Good	ω_{12} Wu	Filler Localization
PMMA	PP	Carbon black	0.75–0.3	−19.63	−10.24	Interface [36]
PMMA	PE	Carbon black	−0.1–0.28	−20.98	−19.63	Interface [36]
PE	PP	Carbon black	3.5–3.75	−8.82	−8	PE phase [36]
PS	PP	Hydrophilic silica	4.87	−6.74	−6.18	PS phase [26]
PS	PP	Hydrophobic silica	−1.13	0.62	0.62	Interface [26]
EVA	PP	Hydrophilic silica	8	12.63	11.4	EVA phase [37]
EVA	PP	Hydrophobic silica	0.72	−0.33	−0.33	Interface [37]
BR	EPDM	Carbon black	5.15			BR phase [42]
EVA	PLA	Carbon black	2.69	94.2	80	EVA phase [43]
PBA	PP	CaCO$_3$-g-PBA			3.30	PBA phase [39]
PP	EC[a]	CaCO$_3$			−3.54	PP phase [39]
PP	EC[a]	CaCO$_3$-stearic			0.76	Interface [39]
PP	EC[a]	Hydrophilic silica	−8.3			EC phase [27]
PP	EC[a]	Hydrophobic silica	1.1			PP phase [27]

Source: Adapted from Fenouillot, F., Cassagnau, P., and Majesté, J., *Polymer* 50, 1333, 2009.

[a] Ethylene-octene copolymer.

coefficient in various ternary nanocomposites and the final localization of the fillers, as determined by microscopy. The values of ω can be dramatically different due to the type of equation chosen to estimate the interfacial tensions, but generally leads to similar predictions.

As Table 2.1 shows, the concept of the wetting coefficient has been successfully applied in filled polymer blends containing various fillers, such as carbon black [36], silica [26,27,37], or nano-CaCO$_3$ particles [38,39]. Limitations of this criterion include strong discrepancies in interfacial tensions, due to the lack of data in the literature regarding polymer/filler interfaces and issues of extrapolation to the appropriate temperature. Also, this criterion assumes that thermodynamic equilibrium has been reached, which is not always the case experimentally due to the limited processing time.

Alternatively, some papers have based their analysis on the values of the work of adhesion, W_a and the interfacial tensions rather than ω [40,41] to predict the filler localization. Ma et al. [39] compared three different methods to predict the morphology of composites containing nano-CaCO$_3$, based on interfacial tension data, estimation of the work of adhesion, and estimation of the wetting coefficient. They reported that the wetting coefficient is the most accurate tool to predict the phase structure, by comparing with the actual localization of the nano-CaCO$_3$ particles observed using SEM.

Based on the considerations presented above, preferential localization can be theoretically obtained by adjusting the affinity of the fillers toward the targeted phase, i.e., by enhancing the polymer/filler interactions. Various strategies have been developed in order to do so, usually consisting of modifying the surface chemistry of either the targeted polymeric phase or the fillers. Examples of these strategies include functionalizing one of the two polymers [27,40,19,44], adding a compatibilizer [24,45], and modifying the surface chemistry of the particles [26,37,40,41,46]. Research by Kontopoulou et al. [19,24,27,44,45] has shown that nanoclays and nanosilica tend to migrate spontaneously to the functionalized polyolefin phase, irrespective of the blend composition. Figure 2.4a shows a 70/30 maleated EPR/PP blend filled with 5 wt% organically modified clay, where all the clay resides in

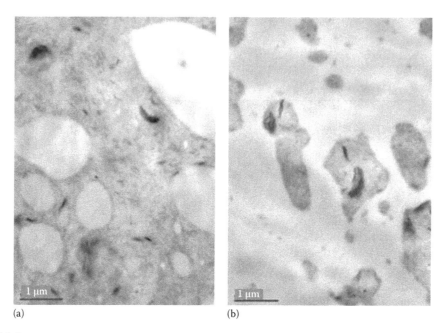

(a)   (b)

FIGURE 2.4
TEM images of (a) 70/30 maleated EPR/PP and (b) 30/70 maleated EPR/PP containing 5 wt% clay. The elastomer phase is stained and appears darker.

the maleated EPR matrix [19]. In the inverse composition shown in Figure 2.4b, all the clay is localized in the maleated EPR dispersed phase, resulting in a very high concentration of the clay in the dispersed domains. Moreover, adding a maleated PP as compatibilizer in PP/ethylene–octene copolymer blends containing nanosilica resulted in selective localization of nanosilica to the PP [45], as shown in Figure 2.5.

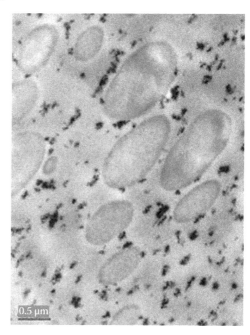

FIGURE 2.5
TEM image of a 80/20 PP/ethylene-octene elastomer blend containing 5 wt% silica. A maleated PP compatibilizer has been added to the composite.

2.2.2 Kinetic Effects

Based on the thermodynamic considerations mentioned above, when the components of a ternary nanocomposite are introduced simultaneously in the compounding device, the fillers should localize within the phase with which they have the largest affinity. However, in some cases, experimental observations showed disagreements between predictions and experiment, shedding light on the importance of multiple other factors that do not solely depend on the thermodynamic affinity between the polymers and the fillers. These factors are commonly classified as "kinetic effects" and are related to (1) the physical properties of the polymers, (2) the time during which the components are mixed, and (3) the compounding sequence.

First of all, physical properties such as the melting temperature and the viscosity of the polymers strongly dictate to a large extent the medium in which the fillers will be localized. Generally speaking, when there are significant differences in the viscosities or melting temperatures, the fillers tend to localize within the less viscous polymer, or within the polymer with the lowest melting point [47,48].

Additionally, the sequence followed when the components introduced into the compounding device can potentially determine the filler localization, regardless of any thermodynamic considerations. The blending sequence can be varied by changing the order of introduction of the components: (1) all together at the same time, (2) the two polymers first followed by the fillers, (3) the matrix and the fillers first followed by the dispersed phase, and (4) the dispersed phase and the fillers first and then the matrix. One of the reasons to choose procedures (3) and (4) is to force the fillers to localize within the phase that is not thermodynamically favored. Dasari et al. demonstrated the importance of the compounding procedure by testing the four blending sequences on Nylon-66/SEBS-*g*-MA/organoclay composites at an 80/15/5 wt% composition [49]. When procedures (1) and (2) were used, the organoclay platelets were equally distributed between the two phases, whereas in (3) and (4) they were located in the phase with which they were blended first. Obviously, the different morphologies that they obtained had an impact on the properties of the composites, as discussed in the next section. On the contrary, Elias et al. [26] prepared PP/PS/silica at a 70/30/3 wt% composition using the blending sequences (1) and (3) and observed the same morphology in both cases, with the silica particles located within the dispersed PS phase, after 5 min of mixing. This observation agreed well with the theoretical predictions based on the wetting coefficient ($\omega = 4.87$). Recently, Bailly and Kontopoulou [27] compounded PP/ethylene-octene copolymer/surface-modified silica (80/20/5 wt%) using sequences (1) and (3) during 8 min, and also confirmed the accuracy of the predictions made with the wetting coefficient, as the silica particles were localized within the matrix regardless of the blending sequence. Vermant et al. [50] also came to the same conclusion in PDMS/PIB/hydrophobic silica (70/30/1 wt%) composites.

The latter studies seem to indicate that a thermodynamic equilibrium is reached only after a few minutes of mixing, since varying the blending sequence does not affect the final localization of the fillers, which was predicted successfully by the wetting coefficient. This is especially interesting in the case of the work of Elias et al. [26] since the particles migrated from the PP phase, in which they were first introduced, to the dispersed PS phase in only 5 min. The authors suggested that if the mixing had been interrupted shortly after the introduction of the silica, the particles would have still been in the PP phase, while later they would have been found at the interface. The mechanism of particle transfer from one phase to another was studied in detail in another paper by Elias et al. [37] and summarized by Fenouillot et al. [6] in a review article.

On the other hand, Kontopoulou et al. [44] observed that exfoliated nanoclay fillers migrated spontaneously to the maleated EPR phase within the first minute of melt compounding, no matter whether is constituted the dispersed phase (shown in Figures 2.6a through c), or the continuous phase (Figure 2.6d through f). To eliminate the issue of disparity in melting points between the elastomer and the PP, the two polymers were precompounded for 4 min prior to introducing the filler, to ensure that both components were melted prior to addition of the filler.

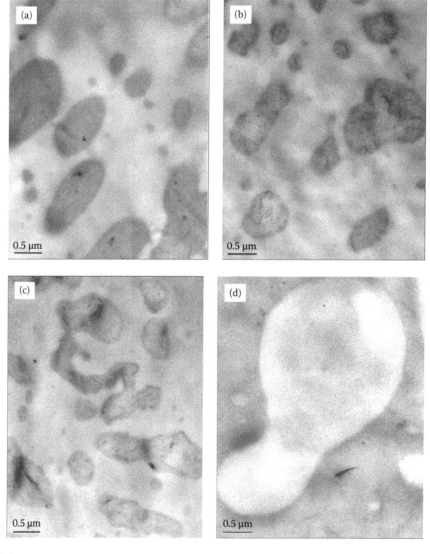

FIGURE 2.6
Evolution of morphology at various compounding times: (a)–(c) 70/30 PP/EPR-*g*-MA; (a) unfilled, (b) 1 min, (c) 2 min; (d)–(f) 30/70 PP/EPR-*g*-MA; (d) unfilled

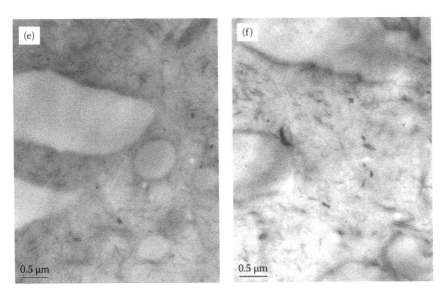

FIGURE 2.6 (continued)
(e) 1 min, and (f) 2 min. (Reprinted from Kontopoulou, M et al., *Polymer*, 48, 4520, 2007.)

Longer compounding times only impacted the degree of exfoliation of the organoclay platelets and the size of the dispersed phase. The latter finding suggests that the presence of fillers strongly affects the morphology development in immiscible blends. This is the topic of the analysis presented in Section 2.3.

2.3 Effects of Nanofillers on Blend Morphology

The images presented in Figure 2.6 imply that the presence of exfoliated clay platelets, or well-dispersed nanoparticles results in changes of the blend morphology. The dimensions of nanoparticles are comparable to, or smaller than, the domain sizes of many multiphase blends. This may enable them to interfere with the process of particle breakup and coalescence during compounding, thus resulting in changes of the blend morphology. In contrast to the traditional considerations of blend composition, viscosity ratio, mixing procedure, i.e., shear rate, mixing time, and mixing temperature that are well understood, the mechanisms through which nanoparticles affect the morphology of immiscible and partially miscible blends are still debated and the topic of intense investigation.

2.3.1 Effect of Nanoparticles on Droplet-Matrix Morphology

Substantial reductions in the size of the domains of the dispersed phase, as well as narrowing of the particle size distribution, have been reported extensively, in various immiscible blends containing nanoclay or nanosilica fillers [9,11–14,26,27,45,51]. As an example in the functionalized PP/ethylene-octene copolymer (80/20 wt%) system prepared by Bailly and Kontopoulou, the average particle size of the dispersed phase was reduced by about

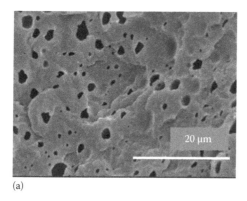
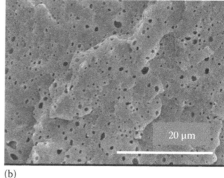

(a)   (b)

FIGURE 2.7
Effect of nanosilica addition on the morphology of a 80/20 PP/ethylene-octene copolymer blend: (a) unfilled blend; (b) composite containing 7 wt% surface-modified silica. The PP matrix has been grafted with a silane. The holes correspond to the etched dispersed phase. (Reprinted from Bailly, M. and Kontopoulou, M., *Polymer*, 50, 2472, 2009.)

50% upon the addition of 7 wt% of hydrophobic silica nanoparticles, as it can be seen on the SEM images of Figure 2.7.

Theories that describe the reduction of the size of the dispersed phase in the presence of nanoparticles vary, depending on whether the filler is located in the continuous phase, in the dispersed phase, or at the interphase between the two blend components. Compatibilizing effects due to polymer adsorption on the filler surface, as well as reduction in the interfacial tension between the two phases in the presence of the filler, are the generally accepted mechanisms when the fillers are located at the interface [11,13,26]. Ray et al. [11] showed that upon addition of only 0.5 wt% of organically modified clay, the interfacial tension decreased from 5.1 to 3.4 mN/m for a PS/PP blend and from 4.8 to 1.1 mN/m for PS/PP-g-MA, suggesting a possible interfacial activity of the clay that is localized at the interface in similar fashion to classical compatibilizers.

However, these mechanisms are obviously not dominant when the fillers reside in the matrix, as shown through estimations of the interfacial tension using the Palierne model [44]. First of all, it should be kept in mind that when the filler partitions in one phase, the reduction in dispersed particle size may be attributed to a compositional effect. In the presence of the nanofiller, the ratio of compositions is altered. For example, 5 wt% of a filler localized in the matrix will correspond to a higher ratio of matrix over the dispersed phase, which in turn may affect the morphology. This effect should be more pronounced for relatively high filler concentrations that are not commonly encountered.

The changes in morphology when the filler partitions preferentially in one of the phases have been discussed extensively in the context of bulk continuum properties, such as altered viscosity ratios in the presence of nanoparticles. In the presence of exfoliated clay, the viscosity of the blend component increases substantially, thus altering the viscosity ratio and affecting the balance between droplet breakup and coalescence [9]. Furthermore, nanofillers present in the continuous phase may improve stress transfer to the dispersed phase and/or retard droplet coalescence during compounding [23,51]. In many cases, however, a quick inspection of the rheological data can reveal that the change in viscosity at shear rates relevant to compounding is not always substantial, and thus may not be sufficient to explain the large reductions in particles size.

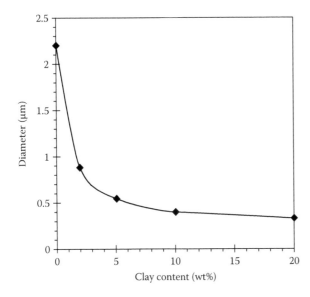

FIGURE 2.8
Effect of organoclay content on dispersed phase particle size in a maleated EPR/PP 70/30 blend.

In fact, it is highly likely that blends filled with nanofillers do not follow the well-known bulk dynamics of immiscible polymer blends during shear flow. Given that the interparticle or inter-aggregate distances are in many cases of similar order of magnitude, or even smaller, than the size of the droplets of the dispersed phase, most probably the droplets of the dispersed phase are forced to follow a highly tortuous path, corresponding to conditions of highly confined flow. It has been shown that confinement may lead to a shift in the critical capillary number to higher numbers, meaning that droplet breakup is facilitated in shear flow [52].

In addition to particle breakup, the coalescence process may be affected as well. It has been speculated that exfoliated clay platelets or well-dispersed nanoparticles may hinder particle coalescence by acting as physical barriers [19,22]. Furthermore, it has been suggested that an immobilized layer, consisting of the inorganic nanoparticles and bound polymer, forms around the droplets of the dispersed phase [50]. The reduced mobility of the confined polymer chains that are bound to the fillers likely causes a decrease in the drainage rate of the thin film separating two droplets [44]. If this is the case, this phenomenon should be dependent on filler concentration; this is shown in Figure 2.8, which shows the effect of nanoclay fillers on the dispersed particle size of a 70/30 maleated EPR/PP blend [19].

On the other hand, higher filler concentrations would be needed to obtain the same effect in the case of low-aspect fillers, such as nanosilica. This is consistent with the observations of Bailly and Kontopoulou, who showed that a reduction in size was only obtained above 7 wt% nanosilica content (see also Figure 2.7).

Figure 2.9 shows recent results obtained in our laboratory, demonstrating the coarsening of an unfilled 60/40 PP/ethylene-octene copolymer upon annealing at various times, and comparing it to that of a blend containing 5 phr silica. It is clear that the morphology development is very different in the absence of shear in these two blends, indicating that indeed the nanofiller affects the droplet coalescence rate.

2.3.2 Effect of Nanoparticles on Co-Continuous Morphologies

Even though fillers are usually introduced into polymer blends of droplet-matrix morphology, a few reports dealing with filled co-continuous blends can be found in the

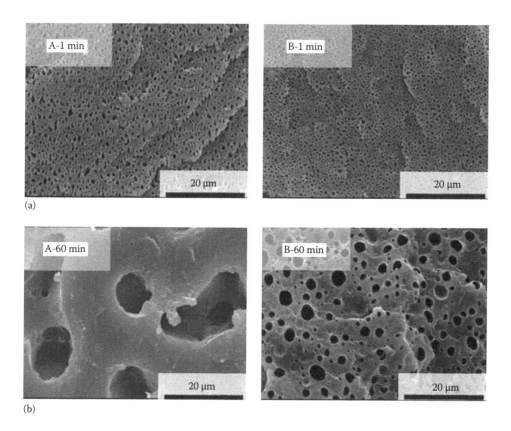

FIGURE 2.9
Effect of annealing on the morphology of (PP/maleated PP)/ethylene-octene copolymer 60/40 blends:
(a) unfilled blend and (b) blend containing 5 phr silica. The label on the micrographs represents the annealing time. The ethylene-octene copolymer has been extracted.

literature [16,53–58]. Most of these reports mention that nanofillers favor the formation
of co-continuous structures in various blend combinations.

Li and Shimizu [16] studied polyphenylene oxide (PPO)/polyamide (PA6) 50/50 blends
reinforced with organoclay and found that the unfilled blend presented a discontinuous
morphology in which PPO was dispersed into a PA6 matrix. The introduction of 2 wt%
of well-exfoliated organoclay within the PA6 phase reduced the size of the PPO domains.
But when an amount higher than 5 wt% was used, the matrix-domain structure was transformed into a co-continuous morphology. The authors attributed this transformation to
the effects of the clay platelets on the rheological properties of the blend components: With
PPO being much more viscous than PA6 at the beginning, the viscosity ratio was 200.
Upon addition of 5 wt% clay, which resulted in an increase in the viscosity of the PA6
phase, the viscosity ratio was reduced to 7.5.

Lee et al. [58] investigated the effects of the addition of silica nanoparticles in PP/
ethylene-octene copolymer 50/50 blends containing a maleated PP to ensure a fine dispersion of the nanofiller in the PP phase. Upon addition of very small amounts of the filler
(1 wt%), they observed a transformation to a much finer morphology as seen on the TEM
images of Figure 2.10. When up to 5 wt% of nanoparticles was introduced, the co-continuous
blend morphology was transformed to a morphology consisting of a continuous PP
phase, containing very high amounts of the dispersed elastomer. In this work, changes in

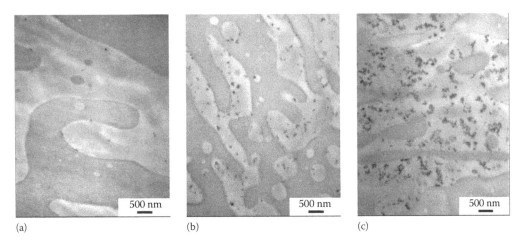

FIGURE 2.10
TEM images of (PP/maleated PP 90/10)/ethylene-octene copolymer 50/50 composites, (a) 0 wt% SiO_2; (b) 1 wt% SiO_2; and (c) 5 wt% SiO_2. Scale bar: 500 nm. Dark domains represent the stained ethylene-octene copolymer phase. (Reprinted from Lee, S.H. et al., *Polymer*, 51, 1147, 2010.)

viscosity ratio were not the reason for this observation, as revealed by rheological characterization. It was speculated, therefore, that changes in the dynamics of droplet breakup in the presence of the nanofiller were responsible for the altered morphologies. The structural changes affected significantly the mechanical properties that are the topic of the following section.

2.4 Mechanical Properties

The relationship between microstructure and physical properties of ternary nanocomposites has been the subject of intense investigation due to their very complex nature. In this section, the mechanical properties of polymer blend nanocomposites are reviewed in the context of their microstructure. Following this, the various strategies that have been employed to obtain improved filler dispersion and interfacial interactions are mentioned.

Generally, the overall mechanical performance of composites depends on (1) the nature of the components, (2) the quality of the interface between the components, (3) their architecture, and (4) their preparation procedure. The effects of preparation method on the final microstructure have already been covered in Section 2.2, so only the effects of the different microstructures on the mechanical properties and the importance of the interfacial properties are discussed.

2.4.1 Effects of Filler Localization on Mechanical Properties

As mentioned in Section 2.2, fillers may localize in different phases, depending on thermodynamic and kinetic considerations, leading to three main types of morphologies: encapsulated, core–shell, or segregated. The mechanical properties of the composites in turn will vary according to the morphology. The following discussion summarizes some of the key findings, for systems that are commonly encountered in the literature. These include almost always blends consisting of a rigid phase (typically polyamide or polyolefin) and an elastomeric phase that is most commonly polyolefin based, reinforced with nanoclays or nanosilica.

2.4.1.1 Polyamide (PA)-Based Nanocomposite Blends

Dasari et al. published a series of articles [49,59,60] investigating the differences in reinforcement and toughening depending on the filler localization. In their first paper [49], they prepared and characterized 85/15 PA66/SEBS-*g*-MA blends containing 5 wt% organoclay by using the different blending sequences described in Section 2.2. They obtained significant enhancements of the impact strength when the clay was localized in the matrix (+88%), but at the expense of the flexural modulus (−11%) and flexural strength (−22%). On the contrary, the localization of the organoclay inside the dispersed phase had a detrimental effect on both notched impact strength and flexural properties. To explain the reduction in toughness, the authors proposed that the organoclay stiffens the SEBS-*g*-MA phase and therefore reduces its ability to cavitate.

Mert and Yilmazer [61] prepared PA66/EBA-*g*-MA/organoclay (93/5/2 wt%) composites using various blending sequences. They observed significant improvements in impact strength compared to the neat PA66 when the three components were introduced simultaneously, resulting in a segregated morphology. They attributed their results to the good dispersion level of the clay platelets and the elastomeric domain. Nevertheless, a 20% reduction in Young's modulus was also noted. The same group obtained mixed results when investigating the effects of clay treatment on the same composites: for the best type of clay identified, they managed to improve slightly Young's modulus (+20%) and the elongation at break (+20%) but the impact strength remained almost constant and the tensile strength decreased slightly (−5%) [62].

Gonzalez et al. [63] prepared PA6/SEBS-*g*-MA/organoclay composites of various compositions and morphologies, based on the compounding method. Compounding the PA6 with SEBS-*g*-MA first, prior to addition of the filler, resulted in a compatibilizing effect attributed to the reaction between the amine end groups of PA6 and the succinic anhydride groups of SEBS-*g*-MA. The resulting segregated morphology where the organoclay resided in the PA6 phase had the best properties. As expected, addition of the elastomer caused a decrease in Young's modulus and yield stress, as seen in Figure 2.11. However, these properties improved significantly when clay was added at all SEBS-*g*-MA concentrations. Addition of clay reduced the notched impact strength substantially pointing to the need of introducing higher amounts of the elastomer in order to maintain the same ductility (Figure 2.11).

Wang et al. [64] observed a sharp brittle–ductile transition around 10–20 wt% of rubber regardless of the amount of clay in PA6/EPDM-*g*-MA/organoclay systems. They related this finding to the critical interparticle distance. The fact that the addition of organoclay reduced the impact strength was attributed to a "blocking effect" by the clay platelets. Finally, Kelnar et al. [65] managed to increase the tensile strength at yield (+15%), Young's modulus (+40%), and impact strength (+270%) on 95/5 PA6/EPR-*g*-MA blends containing 5 wt% organoclay. They achieved this result through an in-situ compatibilization approach, with the clay being located within the PA6 matrix.

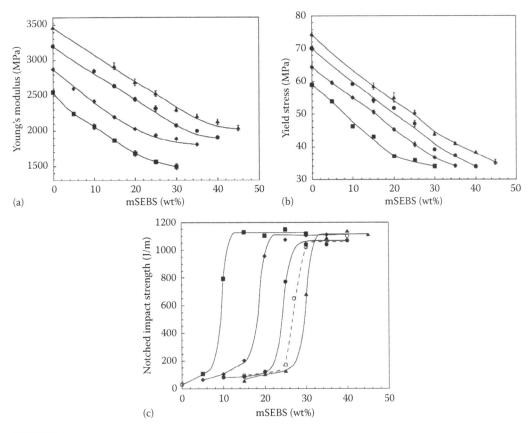

FIGURE 2.11
Mechanical properties of PA6/SEBS-*g*-MA/organoclay as a function of organoclay and elastomer contents (a) Young's modulus, (b) yield stress, and (c) impact strength. Organoclay contents are (■) 0 wt%, (♦) 1.5 wt%, (●) 3 wt%, (▲) 4.5 wt%. (Reprinted from Gonzalez, I. et al., *Eur. Polym. J.*, 44, 287, 2008.)

2.4.1.2 Polyolefin-Based Nanocomposite Blends

Lee et al. [5] discussed the mechanisms of reinforcement and toughening in a segregated morphology of PP/Ethylene copolymer/clay systems (70/30 containing up to 7 wt% clay). They noticed similar trends as the ones shown in Figure 2.10. The most widely used arguments put forward to explain the improvements in mechanical properties were the reduction of the size of the domains, which caused a toughening effect, whereas the strengthening effect was related to the state of dispersion of the clay platelets. Lee and Goettler [18] prepared and tested PP/EPDM/organoclay composites (70/30 with 6 wt% clay). They obtained the best reinforcement when PP and clay were blended first, whereas poor tensile properties were obtained for the encapsulated morphology case. Su and Huang [66] prepared and characterized PP/SEBS/organoclay nanocomposites. They introduced up to 25 wt% of SEBS and up to 7 wt% of organoclay to reach the same conclusions mentioned above.

While most of the literature cited above deals with composites containing nanoclays, nanosilica has also been used as a reinforcing agent, especially in polyolefin-based systems. Yang et al. [40] studied PP/EPDM/silica composites at various compositions, up to 40 wt%

of EPDM and up to 5 wt% of silica. They obtained core–shell morphologies when using hydrophobic particles regardless of the blending sequence, and a segregated morphology or a mix of segregated and core–shell morphologies with hydrophilic particles. The latter led to the best toughening due to the formation of a "filler network structure" consisting of silica particles simultaneously dispersed in the PP matrix and agglomerated around the EPDM domains. The toughening mechanism identified by the authors was the overlap of the stress volume that could be achieved because of the special morphology obtained. Hui et al. [67] used various compounding sequences on LDPE/EVA-based elastomer/silica (40/60 containing 3 wt% silica) to obtain segregated and encapsulated morphologies, with LDPE as the continuous matrix. Surprisingly, they obtained the best reinforcement when the elastomer and fillers were blended first, leading to an encapsulated morphology. The arguments put forward to explain the enhanced properties by the authors were that the silica particles dispersed more finely in the polar EVA phase compared to the LDPE phase, thus reinforcing the weaker phase of the blend. Wang and Liu [68] showed that the optimal composition in PP/SBR/silica nanocomposites was at 5.4 wt% of silica and 7.6 wt% of SBR. It is worthwhile to point out that the amount of elastomer used by these authors is relatively low compared to the average found in the literature, which is around 5 wt% of nanofiller and 20 wt% of elastomer. This certainly highlights the importance and difficulty of identifying the optimum compositions in these systems. Finally, Liu and Kontopoulou [24] were able to change the partitioning of nanosilica by adding PP-*g*-MA to the matrix or by using a functionalized elastomer EPR-*g*-MA in PP/EPR/silica systems (80/20 with 5 wt% silica). Their results clearly indicated that when an encapsulated morphology is obtained, Young's modulus and tensile stress are slightly decreased, as opposed to a segregated morphology.

2.4.1.3 Implications in Rubber-Toughened Systems

Based on the results reviewed above, it is obvious that desirable properties can be obtained when strategically partitioning the filler in a preferred phase. This is particularly true in rubber-toughened systems, which typically consist of a rigid matrix and a finely dispersed elastomer. As reviewed by Zou et al. [69] for silica-reinforced composites and by Ray and Okamoto [70] for layered silicate-reinforced composites, addition of nanofillers in polymeric matrices typically offers great enhancement in strength and modulus, but at the expense of ductility and toughness, whereas impact-modified polymers gain in toughness but loose in strength [71]. This has led to the concept of "preferential" reinforcement of the rigid matrix phase, which utilizes nanofillers to selectively reinforce the matrix, while leaving the elastomeric phase and its toughening capability intact. The success of this approach largely depends on the localization of the filler within the polymer blend.

This concept has been demonstrated by the work of Liu and Kontopoulou [45] and Bailly and Kontopoulou [27] on silica-reinforced thermoplastic olefins (TPOs). Their data have shown that when silica resides in the PP matrix, the improvements in Young's and flexural modulus are accompanied by unchanged or even slightly improved impact strength (Figure 2.12). The effects on impact properties were particularly beneficial when surface-modified silica (mSiO$_2$) was used.

Furthermore, the morphological transformation of PP/ethylene-octene copolymer blends from co-continuous to droplet-matrix reported by Lee et al. [58] at compositions close to the phase inversion (Figure 2.10) was accompanied with the appearance of a yield stress, and a higher modulus than the unfilled blend, while the ductility was maintained.

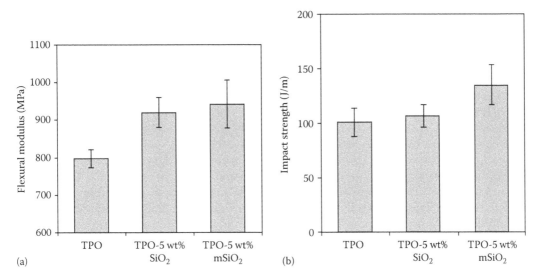

FIGURE 2.12
Influence of nanosilica (SiO$_2$) and elastomer concentrations on the mechanical properties of TPOs, comprising of a PP/PP-*g*-MA matrix and a dispersed ethylene-octene copolymer elastomer, at a 80/20 ratio by weight. Silica content is 5 wt%: (a) flexural modulus and (b) impact strength. (Data from Liu, Y. and Kontopoulou, M., *Polymer*, 47, 7731, 2006.)

2.4.2 Strategies for Improvement of Mechanical Properties

In addition to the considerations based on morphology that have been analyzed above, further strategies for improving mechanical properties are similar to those employed in binary systems, and include filler treatment, addition of compatibilizers, and functionalization of the polymers. These are briefly reviewed below.

2.4.2.1 Filler Treatment

Addition of nanofillers above a critical loading has been found to deteriorate the mechanical properties due to the formation of agglomerates [72]. Khare and Burris [73] recently addressed the issue of quantification of nanofiller dispersion via TEM images by proposing a method based on the estimation of the free-space length. To avoid filler aggregation and the ensuing deterioration of the mechanical properties, surface treatments are commonly employed to reduce the filler/filler interactions. In the case of silica, these treatments basically involve replacing some of the hydroxyl groups present at the surface by alkyl chains that can naturally interact with nonpolar polymers such as TPOs. Many different groups can be grafted on the surface of silica particles, as reviewed by Rong et al. [74]. In nanocomposites containing nanoclays, surface treatment consists of replacing the cations located within the interlayer spacing by organic cations such as alkylammonium in order to render the platelets more hydrophobic.

In the case of segregated and encapsulated morphologies, it is important to note that since the fillers reside preferentially in one of the two phases, the actual filler loading contained in that particular phase is increased. For example, in a 70/30 polymer blend assuming that 5 wt% of a filler is introduced and resides in the dispersed phase, its actual concentration in the dispersed phase would be 15 wt%. The increase in filler loading inevitably results in filler aggregation, which can be very detrimental. The issue of filler treatment becomes therefore even more important in these systems.

A few studies investigated the effects of surface treatments on the mechanical properties of ternary nanocomposites. Isik et al. [62] compared the reinforcement ability of three types of organoclays in 95/5 PA66/EBA-g-MA composites containing 2 wt% organoclay. The three types of organoclay used were differentiated according to their degree of hydrophobicity. They found that the most hydrophilic clay provided the best reinforcement in PA66-based binary nanocomposites, whereas the most hydrophobic one was better for PA66/EBA-g-MA-based ternary nanocomposites. These findings were attributed to the different nature of the PA66 matrix, which is very polar and therefore compatible with hydrophilic fillers, compared to the less polar EBA-g-MA.

In silica-reinforced systems, Yang et al. [40] observed a larger increase in impact strength for nontreated compared to surface-treated silica particles in PP/EPDM systems. While they acknowledged that this result was highly unusual, they attributed it to the special morphology of their composites. Bailly and Kontopoulou [27] studied PP/ethylene-octene copolymer/silica composites and used three different types of silica particles, including nontreated and silane-treated particles, the latter differentiated by the degree of modification, i.e., the fraction of silanol groups that have been substituted by alkyl chains and the length of the alkyl chain. The best dispersion and the best mechanical properties were obtained for the particles that were the most heavily modified. This was attributed to the ability of the alkyl groups grafted at the surface of the particles to reduce the filler/filler interactions by breaking up the hydrogen bonds between the silanol groups, which make the silica particles naturally prone to agglomeration. The same conclusions were reached by Liu and Kontopoulou [45] on similar composites.

2.4.2.2 Matrix and Elastomer Functionalization

Functionalization of the polymer has been widely employed in binary nanocomposites to improve the polymer/filler interactions and thus maximize the load transfer. Functionalization also serves to enhance the compatibility between the two components of polymer blend. Many grades of functionalized polymers are now available, including maleated grades and silane-grafted polymers. Examples of functionalized matrices studied in ternary nanocomposite studies include PP-g-MA [41], PP-g-VTEOS [27] and examples of functionalized elastomers include SEBS-g-MA [8,49,65], EPR-g-MA [19,44,65], POE-g-MA [75], and EPDM-g-MA [64]. It is also important to note that only nonpolar matrices and elastomers require functionalization, as opposed to polar polymers like PA6, which present natural interactions with polar fillers such as silica particles and clay platelets.

Gonzalez et al. [8] reported the effects of different maleic anhydride (MA) contents in PA6/SEBS-g-MA/organoclay composites, containing 0.5–1.5 wt% grafts. They found that the size of the SEBS-g-MA domains clearly decreased with increasing MA content as SEM observations showed. This reduction was accompanied with improvements of the ductility but minor changes were seen in strength and stiffness as well as impact strength. Kelnar et al. [65] compared the efficiency of EPR and EPR-MA as an impact modifier in a PA6 matrix reinforced with organoclay and observed that the impact strength was greatly enhanced (+25%) when using the maleated EPR. Yang et al. [41] used two different types of matrices, either PP or PP-g-MA (0.9 wt%) in PP/EPDM/silica composites. They measured the contact angles of the different components and estimated the work of adhesion for each pair. There was no significant improvement when using PP-g-MA compared to PP in terms of adhesion, regardless of the type of silica particles used.

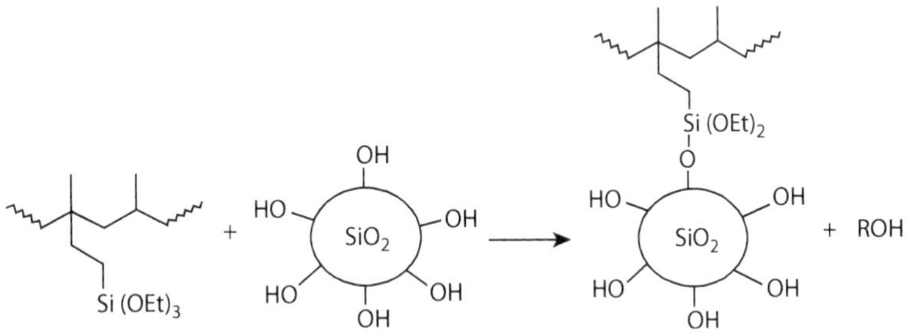

FIGURE 2.13
Hydrolysis reaction between the silanol groups located at the surface of the silica with the PP-*g*-VTEOS matrix. (Reprinted from Bailly, M. and Kontopoulou, M., *Polymer*, 50, 472, 2009.)

Recently, Bailly and Kontopoulou [27] assessed the effects of matrix functionalization in PP/ethylene-octene copolymer/silica composites by using two matrices: a silane-grafted PP and a degraded derivative with similar rheological properties. Although the dispersion of modified silica particles as well as that of the dispersed polyolefin elastomer phase was not substantially different in the two systems, the mechanical properties were significantly enhanced when the functionalized PP was used. This was attributed to the chemical reaction that occurs between the silane group grafted on PP and the hydroxyl groups of the silica particles depicted in Figure 2.13. As a result, polymer/filler interactions were greatly enhanced, leading to the improved strength and toughness.

2.4.2.3 Addition of Compatibilizers and Coupling Agents

Just like functionalization, addition of compatibilizers or coupling agents is a strategy employed for nonpolar matrices and elastomers—such as those involved in TPOs and thermoplastic vulcanizates (TPVs)—in which the fillers or the elastomeric phase present a poor adhesion with the matrix. In nanocomposite technology, the compatibilizer is usually made of the same polymer as the matrix, but with a functional group grafted on its backbone—maleic anhydride in the vast majority of systems—that is able to interact with the fillers [76,77]. As a result, dispersion of the nanofillers is facilitated and accompanied by improvements in strength and stiffness. In polymer blends, a popular approach is to form a block or graft copolymer at the domain interface during processing by in situ reaction of functional groups (reactive compatibilization) [78] or to just add a functionalized derivative of the matrix [79]. As a result of the presence of a compatibilizer, smaller domains have been found, leading to improved toughness and ductility.

In ternary nanocomposites, compatibilizers have been mostly used to improve the adhesion at the polymer/filler interface rather than to modify the polymer/elastomer interface. Mishra et al. [80] compounded PP/EPDM/organoclay (75/25/5 wt%) and added PP-*g*-MA (1 wt% MA) as a compatibilizer with a clay/PP-*g*-MA ratio of 1/3. They characterized the interlayer spacing of the clay platelets by XRD and observed that it increased from 3.4 to 4.3 nm for systems without and with compatibilizer. This was attributed to a better diffusion of the PP-*g*-MA chains inside the interlayer spacing thanks to their functional groups. Numerous other authors prepared and characterized ternary composites with a compatibilizer. Examples include Lim et al. [81] and Lee et al. [5] on PP/PP-*g*-MA/POE/ organoclay systems, Mehta et al. [23] on PP/PP-*g*-MA/EPR/organoclay systems, and Liu and Kontopoulou [24,45] on PP/PP-*g*-MA/ethylene-octene copolymer/silica composites. It should be noted that the compatibilizer itself may affect the properties of the matrix and there should be an optimum content, beyond which the mechanical properties of the matrix may be compromised. Liu and Kontopoulou [45] selected an amount of PP-*g*-MA for compounding with the PP matrix following an investigation of the mechanical properties of PP/PP-*g*-MA composites. They were able to reach a trade-off between a sufficient concentration of PP-*g*-MA to guarantee a good dispersion of the silica particles, while not compromising the mechanical properties of the PP matrix. Indeed, they reported that with increasing PP-*g*-MA concentration, the tensile properties of the binary composites start to drop. Eventually, a 90/10 ratio was selected and used in ternary nanocomposites.

Coupling agents have also been used in ternary nanocomposites, for example, in a paper by Hui et al. [67] who used a silane-coupling agent on LDPE/EVA elastomer (40/60)/3 wt% silica composites. They observed dramatic improvements in the mechanical properties (increase of 41% for the tensile strength) that were attributed to the reduced agglomeration tendency of the particles, as proved by AFM.

2.5 Summary

Polyolefins have found widespread applications in polymer blend nanocomposite formulations. In this chapter, polymer blend nanocomposites, or ternary nanocomposites, are described as a class of composite materials composed of a two-phase polymer blend, reinforced with rigid inorganic nanofillers. The parameters governing the localization of the fillers were discussed, including thermodynamic effects depending on the extent of affinity of the fillers toward the polymers and kinetic effects that include the role of the viscosities of the polymers and the mixing time. Depending on the localization of the fillers and their state of dispersion as well on the state of dispersion of the minor polymeric phase, several morphologies could be found in the literature. The mechanical properties of polyamide-based and TPO-based ternary nanocomposites reinforced with either silica nanoparticles or clay platelets were reviewed. It was concluded that the mechanical properties of ternary nanocomposites are dictated by their morphology, as well as by the strength of the polymer/interactions. In a few systems, a synergistic effect could be observed, characterized by superior mechanical properties of the ternary nanocomposite compared to the neat matrix.

References

1. Utracki, L. A. 1989. *Polymer Alloys and Blends: Thermodynamics and Rheology*. Munich, Germany: Hanser.
2. Wu, S. H. 1988. A generalized criterion for rubber toughening—The critical matrix ligament thickness. *Journal of Applied Polymer Science* 35:549–561.
3. Wu, S. H. 1985. Phase-structure and adhesion in polymer blends—A criterion for rubber toughening. *Polymer* 26:1855–1863.
4. Wu, C. L., Zhang, M. Q., Rong, M. Z., and Friedrich, K. 2002. Tensile performance improvement of low nanoparticles filled-polypropylene composites. *Composites Science and Technology* 62:1327–1340.

5. Lee, H., Fasulo, P. D., Rodgers, W. R., and Paul, D. R. 2005. TPO based nanocomposites. Part 1. Morphology and mechanical properties. *Polymer* 46:11673–11689.

6. Fenouillot, F., Cassagnau, P., and Majesté, J. 2009. Uneven distribution of nanoparticles in immiscible fluids: Morphology development in polymer blends. *Polymer* 50:1333–1350.

7. Chow, W. S., Mohd Ishak, Z. A., Karger-Kocsis, J., Apostolov, A. A., and Ishiaku, U. S. 2003. Compatibilizing effect of maleated polypropylene on the mechanical properties and morphology of injection molded polyamide 6/polypropylene/organoclay nanocomposites. *Polymer* 44:7427–7440.

8. Gonzalez, I., Eguiazabal, J. I., and Nazabal, J. 2005. Compatibilization level effects on the structure and mechanical properties of rubber-modified polyamide-6/clay nanocomposites. *Journal of Polymer Science Part B-Polymer Physics* 43:3611–3620.

9. Hong, J. S., Namkung, H., Ahn, K. H., Lee, S. J., and Kim, C. 2006. The role of organically modified layered silicate in the breakup and coalescence of droplets in PBT/PE blends. *Polymer* 47:3967–3975.

10. Ray, S. S. and Bousmina, M. 2005. Effect of organic modification on the compatibilization efficiency of clay in an immiscible polymer blend. *Macromolecular Rapid Communications* 26:1639–1646.

11. Ray, S. S., Pouliot, S., Bousmina, M., and Utracki, L. A. 2004. Role of organically modified layered silicate as an active interfacial modifier in immiscible polystyrene/polypropylene blends. *Polymer* 45:8403–8413.

12. Ray, S. S. and Bousmina, M. 2005. Compatibilization efficiency of organoclay in an immiscible polycarbonate/poly(methyl methacrylate) blend. *Macromolecular Rapid Communications* 26:450–455.

13. Voulgaris, D. and Petridis, D. 2002. Emulsifying effect of dimethyldioctadecylammonium-hectorite in polystyrene/poly(ethyl methacrylate) blends. *Polymer* 43:2213–2218.

14. Wang, Y., Zhang, Q., and Fu, Q. 2003. Compatibilization of immiscible poly(propylene)/polystyrene blends using clay. *Macromolecular Rapid Communications* 24:231–235.

15. Gelfer, M. Y., Song, H. H., Liu, L., Hsiao, B. S., Chu, B., Rafailovich, M., Si, M., and Zaitsev, V. 2003. Effects of organoclays on morphology and thermal and rheological properties of polystyrene and poly(methyl methacrylate) blends. *Journal of Polymer Science Part B: Polymer Physics* 41:44–54.

16. Li, Y. and Shimizu, H. 2004. Novel morphologies of poly(phenylene oxide) (PPO)/polyamide 6 (PA6) blend nanocomposites. *Polymer* 45:7381–7388.

17. Mehrabzadeh, M. and Kamal, M. R. 2004. Melt processing of PA-66/clay, HDPE/clay and HDPE/PA-66/clay nanocomposites. *Polymer Engineering and Science* 44:1152–1161.

18. Lee, K. Y. and Goettler, L. A. 2004. Structure-property relationships in polymer blend nanocomposites. *Polymer Engineering and Science* 44:1103–1111.

19. Austin, J. R. and Kontopoulou, M. 2006. Effect of organoclay content on the rheology, morphology, and physical properties of polyolefin elastomers and their blends with polypropylene. *Polymer Engineering and Science* 46:1491–1501.

20. Ma, X., Liang, G., Liu, H., Fei, J., and Huang, Y. 2005. Novel intercalated nanocomposites of polypropylene/organic–rectorite/polyethylene–octene elastomer: Rheology, crystallization kinetics, and thermal properties. *Journal of Applied Polymer Science* 97:1915–1921.

21. Xiaoyan, M., Guozheng, L., Haijun, L., Hailin, L., and Yun, H. 2005. Novel intercalated nanocomposites of polypropylene, organic rectorite, and poly(ethylene octene) elastomer: Morphology and mechanical properties. *Journal of Applied Polymer Science* 97:1907–1914.

22. Khatua, B. B., Lee, D. J., Kim, H. Y., and Kim, J. K. 2004. Effect of organoclay platelets on morphology of nylon-6 and poly(ethylene-ran-propylene) rubber blends. *Macromolecules* 37:2454–2459.

23. Mehta, S., Mirabella, F. M., Rufener, K., and Bafna, A. 2004. Thermoplastic olefin/clay nanocomposites: Morphology and mechanical properties. *Journal of Applied Polymer Science* 92:928–936.

24. Liu, Y. and Kontopoulou, M. 2007. Effect of filler partitioning on the mechanical properties of TPO/nanosilica composites. *Journal of Vinyl and Additive Technology* 13:147–150.

25. Binks, B. P. 2002. Particles as surfactants—Similarities and differences. *Current Opinion in Colloids and Interface Science* 7:21–41.
26. Elias, L., Fenouillot, F., Majeste, J. C., and Cassagnau, P. 2007. Morphology and rheology of immiscible polymer blends filled with silica nanoparticles. *Polymer* 48:6029–6040.
27. Bailly, M. and Kontopoulou, M. 2009. Preparation and characterization of thermoplastic olefin/ nanosilica composites using a silane-grafted polypropylene matrix. *Polymer* 50:2472–2480.
28. Leblanc, J. L. 2002. Rubber-filler interactions and rheological properties in filled compounds. *Progress in Polymer Science* 27:627–687.
29. Wang, M. J. 1998. Effect of polymer-filler and filler-filler interactions on dynamic properties of filled vulcanizates. *Rubber Chemistry and Technology* 71:520–589.
30. Wu, S. H. 1982. *Polymer Interface and Adhesion*. New York: Marcel Dekker.
31. Zhang, W. and Leonov, A. I. 2001. IGC study of filler–filler and filler–rubber interactions in silica-filled compounds. *Journal of Applied Polymer Science* 81:2517–2530.
32. Chibowski, E. and Perea-Carpio, R. 2002. Problems of contact angle and solid surface free energy determination. *Advances in Colloid and Interface Science* 98:245–264.
33. Girifalco, L. A. and Good, R. J. 1957. A theory for the estimation of surface and interfacial energies. I. Derivation and application to interfacial tension. *The Journal of Physical Chemistry* 61:904–909.
34. Owens, D. K. and Wendt, R. C. 1969. Estimation of the surface free energy of polymers. *Journal of Applied Polymer Science* 13:1741–1747.
35. Guggenheim, E. A. 1945. The principle of corresponding states. *Journal of Chemical Physics* 13:253–261.
36. Sumita, M., Sakata, K., Asai, S., Miyasaka, K., and Nakagawa, H. 1991. Dispersion of fillers and the electrical-conductivity of polymer blends filled with carbon-black. *Polymer Bulletin* 25:265–271.
37. Elias, L., Fenouillot, F., Majesté, J., Martin, G., and Cassagnau, P. 2008. Migration of nanosilica particles in polymer blends. *Journal of Polymer Science Part B: Polymer Physics* 46:1976–1983.
38. Ma, C. G., Mai, Y. L., Rong, M. Z., Ruan, W. H., and Zhang, M. Q. 2007. Phase structure and mechanical properties of ternary polypropylene/elastomer/nano-CaCO3 composites. *Composites Science and Technology* 67:2997–3005.
39. Ma, C. G., Zhang, M. Q., and Rong, M. Z. 2007. Morphology prediction of ternary polypropylene composites containing elastomer and calcium carbonate nanoparticles filler. *Journal of Applied Polymer Science* 103:1578–1584.
40. Yang, H., Zhang, Q., Guo, M., Wang, C., Du, R., and Fu, Q. 2006. Study on the phase structures and toughening mechanism in PP/EPDM/SiO2 ternary composites. *Polymer* 47:2106–2115.
41. Yang, H., Zhang, X., Qu, C., Li, B., Zhang, L., Zhang, Q., and Fu, Q. 2007. Largely improved toughness of PP/EPDM blends by adding nano-SiO2 particles. *Polymer* 48:860–869.
42. Ibarra-Gomez, R., Marquez, A., Valle, L. F. R. D., and Rodriguez-Fernandez, O. S. 2003. Influence of the blend viscosity and interface energies on the preferential location of CB and conductivity of BR/EPDM blends. *Rubber Chemistry and Technology* 76:969–978.
43. Katada, A., Buys, Y. F., Tominaga, Y., Asai, S., and Sumita, M. 2005. Relationship between electrical resistivity and particle dispersion state for carbon black filled poly (ethylene-co-vinyl acetate)/poly(L-lactic acid) blend. *Colloid and Polymer Science* 284:134–141.
44. Kontopoulou, M., Liu, Y., Austin, J. R., and Parent, J. S. 2007. The dynamics of montmorillonite clay dispersion and morphology development in immiscible ethylene–propylene rubber/polypropylene blends. *Polymer* 48:4520–4528.
45. Liu, Y. and Kontopoulou, M. 2006. The structure and physical properties of polypropylene and thermoplastic olefin nanocomposites containing nanosilica. *Polymer* 47:7731–7739.
46. Yang, H., Li, B., Wang, K., Sun, T. E., Wang, X., Zhang, Q., Fu, Q., Dong, X., and Han, C. C. 2008. Rheology and phase structure of PP/EPDM/SiO2 ternary composites. *European Polymer Journal* 44:113–123.
47. Clarke, J., Clarke, B., Freakley, P. K., and Sutherland, I. 2001. Compatibilising effect of carbon black on morphology of NR-NBR blends. *Plastics Rubber and Composites* 30:39–44.

48. Zhou, P., Yu, W., Zhou, C., Liu, F., Hou, L., and Wang, J. 2007. Morphology and electrical properties of carbon black filled LLDPE/EMA composites. *Journal of Applied Polymer Science* 103:487–492.

49. Dasari, A., Yu, Z., and Mai, Y. 2005. Effect of blending sequence on microstructure of ternary nanocomposites. *Polymer* 46:5986–5991.

50. Vermant, J., Cioccolo, G., Golapan Nair, K., and Moldenaers, P. 2004. Coalescence suppression in model immiscible polymer blends by nano-sized colloidal particles. *Rheologica Acta* 43:529–538.

51. Zhang, Q., Yang, H., and Fu, Q. 2004. Kinetics-controlled compatibilization of immiscible polypropylene/polystyrene blends using nano-SiO2 particles. *Polymer* 45:1913–1922.

52. Vananroye, A., Puyvelde, P. V., and Moldenaers, P. 2006. Effect of confinement on droplet breakup in sheared emulsions. *Langmuir* 22:3972–3974.

53. Li, Y. and Shimizu, H. April. 2009. Structural control of co-continuous poly(l-lactide)/poly(butylene succinate)/clay nanocomposites. *Journal of Nanoscience and Nanotechnology* 9:2772–2776.

54. Filippone, G., Dintcheva, N. T., Acierno, D., and La Mantia, F. P. 2008. The role of organoclay in promoting co-continuous morphology in high-density poly(ethylene)/poly(amide) 6 blends. *Polymer* 49:1312–1322.

55. Wu, D., Wu, L., Zhang, M., Zhou, W., and Zhang, Y. 2008. Morphology evolution of nanocomposites based on poly(phenylene sulfide)/poly(butylene terephthalate) blend. *Journal of Polymer Science Part B: Polymer Physics* 46:1265–1279.

56. Martin, G., Barres, C., Sonntag, P., Garois, N., and Cassagnau, P. 2009. Co-continuous morphology and stress relaxation behaviour of unfilled and silica filled PP/EPDM blends. *Materials Chemistry and Physics* 113:889–898.

57. Zou, H., Ning, N., Su, R., Zhang, Q., and Fu, Q. 2007. Manipulating the phase morphology in PPS/PA66 blends using clay. *Journal of Applied Polymer Science* 106:2238–2250.

58. Lee, S. H., Kontopoulou, M., and Park, C. B. 2010. Effect of nanosilica on the co-continuous morphology of polypropylene/polyolefin elastomer blends. *Polymer* 51:1147–1155.

59. Dasari, A., Yu, Z., Yang, M., Zhang, Q., Xie, X., and Mai, Y. 2006. Micro- and nano-scale deformation behavior of nylon 66-based binary and ternary nanocomposites. *Composites Science and Technology* 66:3097–3114.

60. Dasari, A., Yu, Z., Mai, Y., and Yang, M. 2008. The location and extent of exfoliation of clay on the fracture mechanisms in nylon 66-based ternary nanocomposites. *Journal of Nanoscience and Nanotechnology* 8:1901–1912.

61. Mert, M. and Yilmazer, U. 2009. Comparison of polyamide 66–organoclay binary and ternary nanocomposites. *Advances in Polymer Technology* 28:155–164.

62. Isik, I., Yilmazer, U., and Bayram, G. 2008. Impact modified polyamide-6/organoclay nanocomposites: Processing and characterization. *Polymer Composites* 29:133–141.

63. Gonzalez, I., Eguiazabal, J. I., and Nazabal, J. 2008. Effects of the processing sequence and critical interparticle distance in PA6-clay/mSEBS nanocomposites. *European Polymer Journal* 44:287–299.

64. Wang, K., Wang, C., Li, J., Su, J., Zhang, Q., Du, R., and Fu, Q. 2007. Effects of clay on phase morphology and mechanical properties in polyamide 6/EPDM-g-MA/organoclay ternary nanocomposites. *Polymer* 48:2144–2154.

65. Kelnar, I., Kotek, J., Kapralkova, L., and Munteanu, B. S. 2005. Polyamide nanocomposites with improved toughness. *Journal of Applied Polymer Science* 96:288–293.

66. Su, F. and Huang, H. 2009. Mechanical and rheological properties of PP/SEBS/OMMT ternary composites. *Journal of Applied Polymer Science* 112:3016–3023.

67. Hui, S., Chaki, T. K., and Chattopadhyay, S. 2008. Effect of silica-based nanofillers on the properties of a low-density polyethylene/ethylene vinyl acetate copolymer based thermoplastic elastomer. *Journal of Applied Polymer Science* 110:825–836.

68. Wang, W. and Liu, T. 2008. Mechanical properties and morphologies of polypropylene composites synergistically filled by styrene-butadiene rubber and silica nanoparticles. *Journal of Applied Polymer Science* 109:1654–1660.

69. Zou, H., Wu, S., and Shen, J. 2008. Polymer/silica nanocomposites: Preparation, characterization, properties, and applications. *Chemical reviews* 108:3893–3957.
70. Ray, S. S. and Okamoto, M. 2003. Polymer/layered silicate nanocomposites: A review from preparation to processing. *Progress in Polymer Science* 28:1539–1641.
71. Liang, J. Z. and Li, R. K. Y. 2000. Rubber toughening in polypropylene: A review. *Journal of Applied Polymer Science* 77:409–417.
72. Bikiaris, D. N., Papageorgiou, G. Z., Pavlidou, E., Vouroutzis, N., Palatzoglou, P., and Karayannidis, G. P. 2006. Preparation by melt mixing and characterization of isotactic polypropylene/SiO$_2$ nanocomposites containing untreated and surface-treated nanoparticles. *Journal of Applied Polymer Science* 100:2684–2696.
73. Khare, H. S. and Burris, D. L. 2010. A quantitative method for measuring nanocomposite dispersion. *Polymer* 51:719–729.
74. Rong, M. Z., Zhang, M. Q., and Ruan, W. H. 2006. Surface modification of nanoscale fillers for improving properties of polymer nanocomposites: A review. *Materials Science and Technology* 22:787–796.
75. Lim, S., Dasari, A., Wang, G., Yu, Z., Mai, Y., Yuan, Q., Liu, S., and Yong, M. S. 2010. Impact fracture behaviour of nylon 6-based ternary nanocomposites. *Composites Part B* 41:67–75.
76. Hasegawa, N., Kawasumi, M., Kato, M., Usuki, A., and Okada, A. 1998. Preparation and mechanical properties of polypropylene-clay hybrids using a maleic anhydride-modified polypropylene oligomer. *Journal of Applied Polymer Science* 67:87–92.
77. Pavlidou, E., Bikiaris, D., Vassiliou, A., Chiotelli, M., and Karayannidis, G. 2005. Mechanical properties and morphological examination of isotactic polypropylene/SiO2 nanocomposites containing PP-*g*-MA as compatibilizer. *Second Conference on Microelectronics, Microsystems and Nanotechnology* 10:190–193.
78. Macosko, C. W., Jeon, H. K., and Hoye, T. R. 2005. Reactions at polymer-polymer interfaces for blend compatibilization. *Progress in Polymer Science* 30:939–947.
79. Liu, N. C., Xie, H. Q., and Baker, W. E. 1993. Comparison of the effectiveness of different basic functional groups for the reactive compatibilization of polymer blends. *Polymer* 34:4680–4687.
80. Mishra, J. K., Hwang, K., and Ha, C. 2005. Preparation, mechanical and rheological properties of a thermoplastic polyolefin (TPO)/organoclay nanocomposite with reference to the effect of maleic anhydride modified polypropylene as a compatibilizer. *Polymer* 46:1995–2002.
81. Lim, J. W., Hassan, A., Rahmat, A. R., and Wahit, M. U. 2006. Rubber-toughened polypropylene nanocomposite: Effect of polyethylene octene copolymer on mechanical properties and phase morphology. *Journal of Applied Polymer Science* 99:3441–3450.

3

Polyolefin Nanocomposites by Solution-Blending Method

Sara Filippi and Pierluigi Magagnini

CONTENTS

3.1 Introduction

Despite the enormous potential applications of polymer/layered silicate nanocomposites (PLSN) [1], their commercialization, which was estimated to exceed 500 kt/year by 2009 [2], grew slowly in the last years mainly due to difficulties in dispersing the nanosized silicate layers within the polymer bulk at the industrial scale. Commercial organoclays consist of powders whose particles have dimensions in the 1–30 µm range, and each particle contains an average of over 1 million platelets. Thus, it is a real challenge to practically reach the high level of clay exfoliation that would grant the performance improvements theoretically anticipated for these materials. Recent studies carried out by small-angle x-ray scattering, light scattering, and electron imaging have shown that large-scale morphological disorder is present in nanocomposites, regardless of the level of dispersion, leading to substantial lowering of mechanical properties, compared to predictions based on idealized nanocomposite morphology [3].

On a thermodynamic ground, nanoscale mixing of layered silicates with polymers is only possible if the chemical structure of both components is such to grant favorable energetic interactions [4–8]. However, the morphology of experimental nanocomposites will be dependent on kinetics as well, and, consequently, on the pathway of their preparation.

In industry, melt compounding and in situ polymerization are generally preferred as they are compatible with existing processing technologies and equipment, and because they do not require the use of environment unfriendly organic solvents [2,9,10]. A discussion of the difficulties of adapting the commercial melt-processing techniques to the preparation of polymer nanocomposites with improved dispersion of silicate platelets, and of the proposed solutions to such problems will be found in other chapters of this book.

The use of solution blending, the third important preparation procedure, is generally limited to research laboratories. Despite the more elaborated protocols, involving the preparation of the polymer and clay solutions, their blending, and the final solvent removal step, this method offers some advantages, e.g., that of requiring milder temperature conditions with limited danger of degradation of the organic clay modifier. For obvious reasons, solution blending has been especially used to prepare nanocomposites based on water-soluble polymers, such as polyvinyl alcohol, polyethylene oxide, polyacrylic acid, polyvinyl pyrrolidone, etc. [11]. On the contrary, this procedure was rarely employed for polyolefin-based composites, not only because these polymers are commercially processed in the molten state at moderate temperatures, but also because of their considerable degree of crystallinity and low solubility. Nevertheless, a study of the structure and morphology of polyolefin nanocomposites prepared by solution blending, as compared to, e.g., melt-compounded counterparts, may be useful for a deeper understanding of the different factors influencing the nanostructure formation in PLSN.

3.2 The Solution-Blending Method

Solution blending is a solvent-assisted process by which a preformed polymer is filled with nanosized clay particles (either individual silicate platelets or thin intercalated layers stacks). The basic principle of the procedure is quite simple: a solvent capable of dissolving the polymer and swelling the clay is selected; the polymer solution and the clay suspension are mixed with mechanical and/or ultrasonic stirring; the solvent is removed by evaporation (e.g., film casting or freeze drying) or by precipitation in a non-solvent; the polymer/clay mixture is dried and, often, subjected to thermal treatment. As such, this preparation pathway can be applied to all soluble thermoplastic polymers. However, it was also adapted to the preparation of nanocomposites based on insoluble polymers. For example, polyimide-based nanocomposites were obtained by preparing an organoclay-filled polyamic acid film by solution blending and film casting, and treating the product at 300°C to trigger the imidization reaction [12].

Probably because of the apparent conceptual simplicity of the preparation of polymer–clay nanocomposites by the solution blending procedure, many of the available literature reports concern predominantly empirical investigations focused on the composite properties, and the experimental techniques are often described paying scant attention to important details of the synthetic protocol. However, the mechanisms involved in the solution blending process are rather intricate and the successful achievement of PLSN with optimum morphology depends on a great number of factors. For example, as will be shown in more detail below, the important effects of the thermal after-treatments on the structure and morphology of solution-blended composites has been often underestimated, or even overlooked. On the other hand, as pointed out by Li and Ishida [13], different

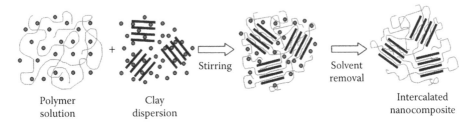

FIGURE 3.1
Schematic representation of the "polymer–solvent exchange" solution intercalation process.

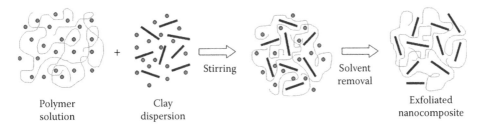

FIGURE 3.2
Scheme of the preparation of an exfoliated nanocomposite by the "exfoliation–adsorption" mechanism.

mechanisms might be considered to describe the solvent-assisted PLSN preparation. These authors distinguish a "polymer–solvent exchange" mechanism, which, in agreement with Pavlidou and Papaspyrides [10], can be schematically illustrated, as shown in Figure 3.1, and an "exfoliation–adsorption" mechanism, formerly suggested by Alexandre and Dubois [9], a version of which is exemplified by the cartoon in Figure 3.2. Clearly, both schemes represent particular or even extreme situations. For example, a fully exfoliated nanocomposite has been considered as the final product in Figure 3.2, but an intercalated nanocomposite or even a microcomposite comprising unintercalated tactoids can in fact be obtained if the silicate layers can reassemble during the solvent removal step. Nevertheless, the two schematic cartoons lend themselves to a discussion of the many factors playing a role in solvent-assisted nanostructure formation.

3.2.1 Polymer Solution

In both schemes, a polymer solution with random coiled macromolecules is represented: it is implied that the solvent is capable of dissolving the polymer, but no indication is given on whether the solvent is a "good" or "theta" or "poor" solvent for the polymer, in Flory's sense [14]. In other words, the intensity of the polymer–solvent interactions, which play an important role in the subsequent steps of the solution blending process, is not specified. And in most of the papers discussing the structure and the properties of PLSN prepared from solution, the rationale behind the choice of the adopted solvent (or solvents) is not illustrated.

3.2.2 Clay Dispersion

Here, on the contrary, the two schemes depict profoundly different situations with respect to solvent–clay interactions. In Figure 3.1, a clay dispersion comprising solvent-intercalated tactoids is considered, whereas complete exfoliation of the organoclay is assumed in

Figure 3.2. It is well known that the extent of interaction between solvent molecules and clay platelets depends on many factors, including the chemical structures (Hansen's solubility parameters) of both components, the concentrations, the temperature, the presence of excess modifier in the organoclay, etc. For example, the commercial montmorillonite (MMT) organically modified with an excess of an $M_2(HT)_2$ surfactant (dimethyl dihydrogenated tallow ammonium chloride) (Cloisite® 15A) was shown to be fully exfoliated in chloroform, both before and after the extraction of the excess surfactant with boiling ethyl alcohol, whereas in benzene, toluene, and *p*-xylene it formed swollen tactoids (with—three to six layers), which became considerably thinner for the ethanol-extracted organoclay [15,16]. It should be noticed, however, that even with these (less interacting) aromatic solvents, the Cloisite 15A particles possess a disk-like geometry with a very high aspect ratio, so that a very large amount of the exchanged surface area is available [17].

The results of a study of suspensions of Cloisite 6A (modified with a larger excess of the same $M_2(HT)_2$ surfactant) in *p*-xylene, tetrahydrofuran or chloroform, carried out by small angle neutron scattering (SANS), ultra-small angle neutron scattering (USANS), small angle x-ray scattering (SAXS), and rheological measurements suggest that surfactant is liberated from the surface when the organoclay is suspended in the solvent, and its content evolves to an equilibrium amount granting charge balance for each layer [18]. It was also found that the surfactant-coated silicate sheets partly overlap one another to form plate-like structures with an average number of sheets per stack ranging from $\langle N \rangle = 1.25$ to $\langle N \rangle = 2$, and lateral dimensions of 3–6 µm (cf. Figure 3.3). Thus, the schemes of clay dispersion shown in Figures 3.1 and 3.2 are both crude simplifications. In particular, the scheme of Figure 3.1 does not account for the finding that, in the plate-like structures suggested in Ref. [18], the frequency of single sheets is nearly 80%.

In conclusion, a detailed knowledge of the interactions of any given organoclay with different solvents, as well as of their dependence on concentration, is needed in order to make reliable hypotheses on the morphology of the organoclay dispersions [19]. However, a detailed comparison of the results found by the use of different solvents for the preparation of given polymer/clay composites was rarely made [20,21]. Quite obviously, the mechanism implicitly meant in the "solution intercalation" definition, i.e., replacement of solvent molecules by polymer chains within the clay galleries, such as that depicted in Figure 3.1 and postulated in many papers and reviews (cf. e.g., Ref. [10]), is formally impossible if the silicate is dispersed in the solvent as individual exfoliated layers, or even as plate-like structures such as that shown in Figure 3.3.

3.2.3 Mixing of Polymer Solution and Clay Dispersion

To predict the morphology of the three-component mixture produced by blending the polymer solution with the clay dispersion is the most difficult task. This is particularly so if different solvents are used for polymer solution and organoclay dispersion. The situation

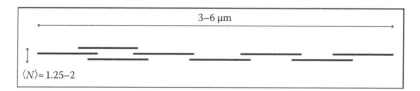

FIGURE 3.3
Structure of solvent-dispersed organoclay particles suggested by scattering parameters [18].

depicted in Figure 3.2 may be valid in the particular event that polymer–solvent and clay–solvent interactions have comparable intensities. In this case, no appreciable morphological change is expected if polymer is added to the clay dispersion or, conversely, organoclay is added to the polymer solution. Clearly, this is a purely idealized situation. In fact, if the polymer–solvent affinity exceeds that between solvent and clay, there will be a tendency of the latter to reassemble, forming solvent-intercalated tactoids. If, on the contrary, the solvent is "poor" (for the polymer) and the organoclay–solvent interactions are strong, there will be a tendency of the polymer to coagulate. Finally, if the polymer–clay interactions are stronger than both the polymer–solvent and clay–solvent ones, formation of suspended intercalated tactoids, as schematically shown in Figure 3.1, will be favored. Actually, this is just the hypothesis on which the polymer–solvent exchange mechanism referred to in many papers and reviews and illustrated in Figure 3.1 rests. The entropy increase of the solvent molecules leaving the clay galleries is thought to balance the entropy loss associated with the confinement of the intercalating polymer chains. However, an enthalpic contribution due, e.g., to polar polymer–clay interactions is required for the expected replacement mechanism to be operative.

3.2.4 Solvent Removal

This may be an extremely critical step. In both the above schemes, solvent removal is envisioned as a process that causes no changes of the relative localization of polymer macromolecules and clay particles. This is likely to be so for the mechanism illustrated in Figure 3.1. In fact, the solvent molecules must simply be removed from the polymer solution and the polymer-intercalated clay tactoids may well remain unaffected by the treatment. In this case, therefore, the final structure of the silicate stacks will probably be independent of the way solvent removal is carried out. For the exfoliation–adsorption mechanism illustrated in Figure 3.2, on the contrary, the kinetics of solvent removal may be expected to play a role because the effects of the different intensities of the polymer–solvent and clay–solvent interactions will probably be magnified as the solvent is removed and the concentration increases. The polymer–clay dispersion will remain unaltered, as Figure 3.2 suggests, only under particular conditions. For example (1) when solvent removal is carried out very rapidly (composite recovered by precipitation under vigorous stirring), (2) when the mixed solution is frozen and the solvent is removed by sublimation (freeze drying), and (3) when the solution is very viscous and the mobility of the macromolecules is severely hindered. However, even if this particular situation prevails, the question of whether the exfoliated nanocomposite thus formed is thermodynamically stable or, conversely, will undergo important morphology changes when subsequently processed should be addressed.

The solvent-removal technique will also influence the mutual orientation of the anisometric clay particles (either single layers or thin tactoids). Precipitation in a non-solvent, or freeze drying, will expectedly yield composites with randomly oriented clay platelets. On the other hand, the composites prepared by film casting, or multilayer film casting [22,23], will be characterized by the preferred orientation of such platelets parallel to the film surface. This particular morphology can be of interest, e.g., in view of the production of films with enhanced barrier properties.

3.2.5 Thermal After-Treatments

Thermal after-treatments that are often required to complete solvent removal [13], as well as those undergone during melt processing, if any, do also affect the final morphology of the polymer/clay composites prepared by solution blending. In fact, it has been repeatedly shown that the intercalation of polymer chains within the clay galleries can take place quite rapidly (with respect to the timescale of the thermal treatments), at temperatures higher than, or close to, the polymer melting point, even in the absence of applied stresses [24–30]. Therefore, in order to clearly distinguish the effects of solution blending and static annealing on the final morphology of the nanocomposites, it should be useful to carry out a complete solvent removal at low temperature and an accurate morphological analysis of the product both before and after the subsequent thermal treatment. This detailed investigation was rarely made and, therefore, some of the literature data cannot be unambiguously interpreted.

3.3 Polyolefin Nanocomposites Prepared by Solution Blending

In the following paragraphs, a short review of the literature reports on polyolefin nanocomposites prepared by solution blending will be made. Attention will be mainly paid to composites based on polyethylene (PE) and polypropylene (PP). However other polymer matrixes, such as olefin copolymers, including functionalized PE and PP, will also be considered shortly.

3.3.1 Polyethylene Nanocomposites

Polyethylene, the most important commodity plastic, with a yearly production of ca. 70 Mt, accounts for about 40% of the total volume of world production of polymeric materials and is sold in the market in three main grades, high density (HDPE), low density (LDPE), and linear low density (LLDPE), in comparable amounts. The degree of crystallinity of commercial HDPE ranges between 50% and 80%, with a density of 940–965 kg m^{-3}; that of LLDPE varies between 30% and 45%, with a density of 915–925 kg m^{-3}; and that of LDPE may be between 20% and 40%, with a density of 910–935 kg m^{-3}. Ethylene polymers with even lower degrees of crystallinity are also produced by copolymerization of ethylene with higher proportions of substituted olefin comonomers. The layered silicate nanocomposites based on the PEs have been prepared mainly by melt compounding and in situ polymerization. Due to their extreme hydrophobic character, all ethylene grades have no affinity for layered silicates. After modification of native clays with alkyl-substituted ammonium ions, the entropic penalty due to confinement of the PE chains may be counterbalanced by the increased conformational freedom of the tethered alkyl groups of the modifier as the layers separate; however, there is no favorable excess enthalpy to promote intercalation [4–6]. Therefore, as discussed below and, in more detail, in other chapters of this book, either microcomposites or slightly intercalated nanocomposites were obtained when PEs were compounded in the melt with commercial organically modified clays [31,32]. On the contrary, intercalated or even exfoliated nanocomposites have been prepared using PE copolymers with lowered hydrophobicity, such as ethylene-vinyl acetate (EVA), ethylene-acrylic or methacrylic acid (EAA, EMAA), ethylene-glycidyl methacrylate (EGMA), ethylene-maleic anhydride (EMA), etc., because in this case, the polymer–silicate interactions may be even more favorable than the surfactant–silicate interactions [32,33]. The extent to which dispersion of the clay platelets within the functionalized ethylene polymers and copolymers occurs depends on many factors including the type, the amount, and the distribution of the polar groups along the chains, the architecture of the macromolecules (either linear or branched), the clay loading, the preparation pathway, etc. The degree of crystallinity of the matrix polymer is also important because, in the solidified specimens, the clay particles tend to concentrate in the amorphous phase. In some cases, polymer crystallization was even found to act as the driving force for macromolecules to diffuse out of the silicate galleries, thus leading to reversible de-intercalation [34].

Solution blending was used in a limited number of cases for ethylene polymers and copolymers.

Jeon et al. [35] prepared a HDPE composite filled with 20 wt% of a MMT, with a cation exchange capacity (CEC) of 1.19 mequiv. g^{-1}, organically modified with a $H_3(C_{12})_1$ ammonium surfactant (dodecyl ammonium chloride). The two components were dispersed in an 80/20 wt/wt mixture of xylene and benzonitrile at 110°C for 30 min and the composite was recovered as a powder by precipitation in an excess of vigorously stirred tetrahydrofuran (THF). It was then repeatedly washed with fresh THF, dried under vacuum at 70°C for 7 h, and compression molded at 130°C. After staining with a RuCl-sodium hypochlorite solution, the composite was sectioned at room temperature with an ultramicrotome. The XRD pattern of the composite showed a reflection corresponding to a d_{001} spacing of 1.77 nm. This was compared with the $d_{001} = 1.65$ nm measured for the organoclay and the minimal increase of spacing led the authors to conclude that little mixing had occurred between the HDPE chains and the alkyl groups of the clay modifier. Actually, since the thickness of an extended PE chain was calculated to be about 0.35 nm [36], an expansion of only 0.12 nm is probably too small to be ascribed to intercalation of HDPE chains [32]. The XRD pattern of the composite also showed, at $2\theta \approx 21.6°$ and 24.0°, strong 110 and 200 reflections suggesting a high degree of crystallinity. The ratio of the observed intensities of the two reflections ($I_{110}/I_{200} > 2$) was typical for unoriented PE. This was expected on the basis of the preparation of the specimen, involving precipitation in a non-solvent and compression molding [30]. Interestingly, the TEM micrograph of the stained composite (Figure 3.4) showed that the HDPE lamellae lay parallel to the nearest clay tactoids. In summary, the solution blending procedure employed in this work did probably fail to produce an intercalated HDPE nanocomposite; however, the clay agglomerates were efficiently broken into individual tactoids with very high aspect ratio, and these made the polymer lamellae grow parallel to their surface. The latter result was considered as an indication that improved mechanical properties of HDPE might be obtained by the addition of organoclays.

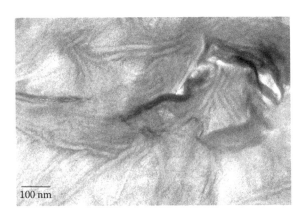

100 nm

FIGURE 3.4
TEM micrograph of an ultrathin section of HDPE filled with 20 wt% MMT modified with dodecyl ammonium ions. Stacked particles are undulated and mutually parallel to the HDPE lamellar crystals. (Reprinted from Jeon, H.G. et al., *Polym. Bull.*, 41, 107, 1998. With permission from Springer.)

In a series of papers by Qu et al., the preparation of nanocomposites comprising maleic anhydride grafted PE (EMA) [37] or LLDPE [38,39] as matrixes and organically modified layered double hydroxides (OLDH) as nanofillers, carried out by solution blending, has been described. For example, MgAl-LDH (hydrotalcite) was calcined at 500°C for 6h and then suspended in an aqueous solution of sodium dodecyl sulfate and refluxed for 6h to yield the OLDH. EMA (with 0.8 wt% MA) and 2 wt% or 5 wt% OLDH were then refluxed in xylene for 24h and the solution was poured with stirring in excess ethanol. The product was dried under vacuum at 100°C for 24h. XRD and TEM analyses confirmed that this preparation protocol yielded an exfoliated nanocomposite [37]. Similar procedures were used for the preparation of LLDPE nanocomposites filled with organically modified (by sodium dodecyl sulfate) MgAl-LDH or ZnAl-LDH [38–40]. The XRD patterns and TEM micrographs demonstrated that the LLDPE nanocomposites with 2.5–10 wt% modified MgAl-LDH are intercalated (or exfoliated/intercalated for the lower filler loadings), whereas those containing modified ZnAl-LDH are exfoliated in the whole composition range. The latter finding was explained considering that the ZnAl-LDH layers are easily broken during the refluxing step.

The same authors also applied the solution blending procedure to the preparation of LLDPE composites with a MMT organically modified with $M_3(C_{16})_1$ (trimethyl hexadecyl ammonium ions). Intercalated nanocomposites were obtained, characterized by interlayer distances d_{001} of 3.54–3.85 nm, for organoclay loadings of 10–5 wt%, respectively (Figure 3.5).

As shown in Figure 3.5, the organoclay had a strong basal reflection at $2\theta \approx 3.5°$, corresponding to a spacing $d_{001} \approx 2.53$ nm, plus two weaker reflections at $2\theta \approx 5.3°$ and $7.3°$, probably due to higher orders. It should be observed that, for the montmorillonites modified with $M_3(C_{16})_1$, lower figures of the interlayer spacing were given in the literature. Thus, in Ref. [41], a value of d_{001} equal to 1.8 nm ($2\theta \approx 4.91°$) was reported. Although some deviations among the interlayer distance values may be ascribed to differences in the CEC of the native MMTs or to the use of an excess of surfactant, the value of the spacing reported in this work [40] for the organoclay appears excessive. In fact, for the organoclays modified by $M_3(C_{18})_1$ (a surfactant with a slightly longer alkyl group) with concentrations ranging from one to four times the CEC of the native MMT, the d_{001} spacing was found to increase from 1.97 to 2.03 nm [42]. Independent of this, however, the gallery expansion ($\Delta d_{001} \approx 1$ nm) reported by Qu et al. [40] for the nanocomposites is indeed considerable and would be

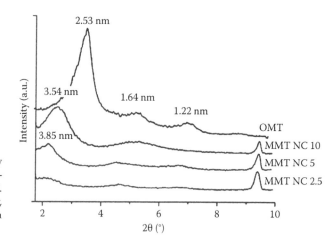

FIGURE 3.5
XRD patterns for the $M_3(C_{16})_1$ organoclay (OMT) and the LLDPE/OMT nanocomposites with 2.5, 5, and 10 wt% OMT. (Reprinted from Qiu, L. et al., *Polymer*, 47, 922, 2006. With permission from Elsevier.)

even stronger if the 2.53 nm figure given for the OMT spacing should be changed into the more realistic value of 2 nm or less. Thus, extensive intercalation of LLDPE chains within the interlayer galleries of this organoclay did apparently occur during the solution blending procedure used by these authors.

Direct comparison of the results found by Qu et al. for the LLDPE/clay nanocomposites [40] with those found for the HDPE composite by Jeon et al. [35] cannot be made because the differences between the two systems and procedures are too many. However, besides the solvent, which had much higher polarity in Ref. [35], one might consider the different degree of crystallinity of the two polymer matrixes to be another important factor.

HDPE and LLDPE have been used as matrixes of polymer/clay nanocomposites prepared by melt compounding in a number of studies. HDPE was always found to be reluctant to intercalate the galleries of commercial organoclays, unless a suitable compatibilizer was added, or the HDPE itself was functionalized, e.g., by grafting, with appropriate polar groups [32]. Thus, the results found by Jeon et al. [35] are in line with expectation and, on the other hand, they confirm that solution blending may be very effective in breaking-up the clay agglomerates into thin tactoids and dispersing them homogeneously within the polymer bulk.

A comparison of the results by Qu et al. [40] for the LLDPE/organoclay nanocomposites with those described in the literature for similar systems prepared in the melt is more intriguing. For example, Hotta and Paul [31] melt-compounded LLDPE, added with different amounts of LLDPE-g-MA as compatibilizer, with organoclays modified with $M_2(HT)_2$ or $M_3(HT)_1$, using a corotating twin-screw extruder at a set temperature of 200°C. Nanocomposites were injection molded at a barrel temperature of 190°C and a mold temperature of 30°C. (Note that the $M_3(HT)_1$ organoclay is very similar to the one, modified with $M_3(C_{16})_1$, employed by Qu et al. [40]. In fact, hydrogenated tallow (HT) is a blend of saturated alkyl groups with approximate composition: 65% C_{18}, 30% C_{16}, and 5% C_{14}.) As it is shown in Figures 3.6 and 3.7, for none of the investigated compositions, with either

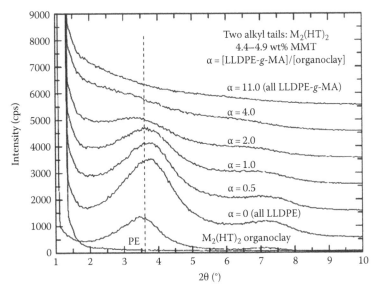

FIGURE 3.6
XRD patterns for the $M_2(HT)_2$ organoclay and its nanocomposites with LLDPE for different contents of LLDPE-g-MA. MMT content = 4.4–4.9 wt%. (Reprinted from Hotta, S. and Paul, D.R., *Polymer*, 45, 7639, 2004. With permission from Elsevier.)

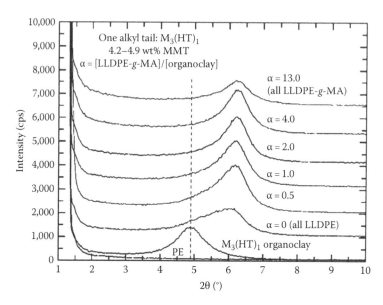

FIGURE 3.7
XRD patterns for the $M_3(HT)_1$ organoclay and its nanocomposites with LLDPE for different contents of LLDPE-*g*-MA. MMT content = 4.2–4.9 wt%. (Reprinted from Hotta, S. and Paul, D.R., *Polymer*, 45, 7639, 2004. With permission from Elsevier.)

organoclay, a shift to smaller angles of the XRD basal reflection suggesting intercalation could be observed. With the $M_2(HT)_2$ organoclay, only the intensity of the 001 reflection at 2θ ≈ 3.6° decreased as the amount of LLDPE-*g*-MA increased, until the reflection disappeared completely for a LLDPE-*g*-MA concentration of about 50% (Figure 3.6). The morphology of the LLDPE/LLDPE-*g*-MA/$M_2(HT)_2$ composites is therefore characterized by an increasing level of exfoliation and by the presence of a gradually decreasing population of unintercalated clay stacks. On the other hand, as shown in Figure 3.7, the reflection of the $M_3(HT)_1$ clay at 2θ ≈ 4.9° (d_{001} ≈ 1.8 nm), after compounding with the LLDPE or the LLDPE/LLDPE-*g*-MA blends, shifted to wider angles (2θ ≈ 6.4°) and did not disappear even when pure LLDPE-*g*-MA was used as the matrix. Such shift of the basal reflection of the $M_3(HT)_1$ clay, which was also observed for melt-compounded composites based on different polymer matrixes, such as LDPE [43], EAA [44], EMAA [45], and EMAA ionomers [41], has been attributed to the collapse of the interlayer spacing of the silicate stacks due to thermal degradation of the organic modifier. From their study of melt-compounded LLDPE/organoclay composites, Hotta and Paul concluded that nanocomposites based on the organoclays having two alkyl tails, such as those modified with $M_2(HT)_2$, are superior, in terms of clay dispersion and mechanical properties, to those based on the organoclays with one alkyl tail, such as that modified with $M_3(HT)_1$, not only because of the higher thermal stability of the former clay, but also because, contrary to the latter, it undergoes a high level of exfoliation when the LLDPE matrix is compatibilized with appropriate amounts of LLDPE-*g*-MA.

The fact that no interlayer spacing collapse of the $M_3(C_{16})_1$ organoclay was observed by Qu et al. in their work [40] should probably not be a surprise because the highest temperature adopted in their protocol was that (140°C) of the xylene refluxing step, and even the $M_3(HT)_1$ organoclay was shown to undergo no significant degradation when melt-compounded with LDPE at this temperature [43]. On the other hand,

Frost et al. [42] have measured, for the organoclay modified with a 100% excess of $M_3(C_{18})_1$, i.e., with a milliequivalent exchange ratio (MER) equal to twice the CEC of the native MMT, an initial decomposition at 192.5°C. More intriguing is the finding that intercalation of LLDPE chains within the galleries of the $M_3(C_{16})_1$ clay, confirmed by the appearance of an XRD reflection at lower 2θ ($d_{001} \approx 3.5$–3.8 nm), takes place with a solution blending procedure, whereas no intercalation at all was observed during melt compounding for both the $M_3(HT)_1$ and $M_2(HT)_2$ organoclays (the latter was not degraded). We cannot offer any convincing interpretation for these apparently conflicting observations. Clearly, the expected lack of thermodynamic driving force for PE intercalation [4–6] finds an experimental confirmation in the results by Hotta and Paul [31]; on the other side, the very interesting claims by Qu et al. [40] with respect to the LLDPE composites prepared from solution have not been exploited in subsequent studies by others.

3.3.2 Polypropylene Nanocomposites

PP is the second most important commodity plastic, with a production exceeding 45 Mt/year. It is not surprising, therefore, that a huge number of investigations have been carried out on the synthesis and characterization of PP-based nanocomposites with mono-, bi-, and tri-dimensional nanofillers. One of the first reports on the preparation of PP/organoclay nanocomposites came from the Toyota laboratories in the 1990s [46]. It showed that exfoliated nanocomposites could be obtained by melt compounding a dry blend of organoclay, PP, and a PP-MA oligomeric compatibilizer (which should be miscible with PP). Actually, thermodynamic arguments like those discussed above with respect to intercalation of the PE chains within the galleries of organoclays modified with surfactants carrying one or more long alkyl chains are also valid for PP [47,48]. Although exfoliated nanocomposites based on neat PP and Cloisite 15A were apparently obtained with a solution technique followed by a 2 h melt kneading step in a Brabender Plasticorder at 170°C and 70 rpm [49], it is generally accepted that introduction of polar or polarizable groups in the PP chain is needed to enhance the interactions between polymer and silicate layers surface, thus providing an enthalpy contribution to promote intercalation. On the other hand, early attempts in which solution blending was used for the preparation of master batches to be subsequently diluted with neat PP by melt compounding had already been made by Oya et al. over 10 years ago [50–52]. These authors had adopted a somewhat complex procedure consisting of two or three steps: in the first step (optional), the clay mineral (organically modified hectorite, MMT, or mica) was intercalated, in toluene, with a radical initiator and a polar monomer (diacetone acrylamide), which was then polymerized by increasing the temperature to 75°C for 1 h; in the second step, PP-MA (with 10 wt% MA) was added to the clay solution, blending was continued for 1 h at 100°C, and the hybrid was recovered by precipitation in methanol and used as masterbatch; in the last step, neat PP was melt compounded with the masterbatch in a twin screw extruder and processed to yield the specimen for subsequent characterizations. Although considerable fragmentation of the mineral particles was demonstrated by TEM analyses for all three clays, yielding satisfactory dispersion of thin tactoids within the polymer matrix and slightly improved mechanical properties, an appreciable expansion of interlayer spacing was only observed for the hectorite clay when the three-step procedure was adopted. In all other examples, including that involving a two-step procedure applied to hectorite, the d_{001} spacing measured by XRD for the composites was equal to, or lower than, the spacing of the original organically modified clay.

The preferred protocol so far employed to produce PP/organoclay nanocomposites is certainly that of using PP copolymers, with small concentrations of polar comonomers, either as polymer matrixes or as compatibilizers [9–11]; however, other synthetic strategies were also tested. One such strategy consists of increasing the number of alkyl chains of the modifier, so as to enhance the organoclay interlayer spacing thereby facilitating shear-driven exfoliation [48]. However, the possibility that the nanocomposite morphology produced under high shear conditions may be thermodynamically unstable and that the organoclay layers may reassemble in case of reprocessing should be considered. The second strategy consists of lowering the enthalpic interactions of the silicate surface with the modifying ammonium ions, so as to favor those with the PP chains. This route was tested by Manias et al. [47] by using semi-fluorinated surfactants. They found that this new semi-fluorinated MMT was intercalated and partly exfoliated when melt compounded, either statically (i.e., with no applied stress) or under shear conditions, with neat PP. The morphology of the statically annealed product was similar to that of analogous nanocomposites prepared from traditional organoclays and PP samples functionalized with polar groups derived from *p*-methylstyrene, maleic anhydride, or hydroxyl-containing styrene.

More interesting, for the scope of the present review, is that the same authors [47] prepared similar hybrids by the solution-blending method in order to compare their structure and morphology with those of the melt-compounded materials. Specifically, PP (or functionalized PP) and the organoclay were blended in a common solvent (1,3,5-trichlorobenzene), which was then evaporated leaving the clay layers trapped in an almost exfoliated manner in the polymer matrix. This was demonstrated by the complete absence of low angle reflections in the XRD patterns. Apparently, the solution blending preparation adopted in this work obeyed quite closely the mechanism schematically illustrated by the cartoon in Figure 3.2. Subsequently, the hybrids were either annealed statically at 180°C, or melt kneaded in a twin-head mixer at the same temperature, and the XRD patterns were recorded at time intervals until no further variation occurred (≤30 min). The hybrid prepared from neat PP and the organoclay with semi-fluorinated surfactant showed a very rapid change of the XRD pattern, upon melt compounding in a twin-head mixer. A basal reflection of increasing intensity appeared at the same angular position recorded for the analogous sample prepared by static annealing. After 8 min of kneading, no further change was observed, thus showing that the intercalated structure corresponds to thermodynamic equilibrium for this system. As suggested in Section 3.2.4, the question of whether the clay dispersion produced by the adopted solution blending procedure was retained after melt processing was addressed in this work, and the highly exfoliated morphology observed just after solvent removal turned out to be thermodynamically unstable. Interestingly, even the composite prepared from solution using neat PP and an MMT organically modified with $M_2(C_{18})_2$ (dimethyl dioctadecyl ammonium ions) was found to be highly exfoliated after solvent removal. However, when this nanocomposite was statically annealed at 180°, the dispersed layers collapsed toward immiscible/intercalated structures within 10–15 min. Another interesting behavior was observed for the nanocomposite of PP-MA (~0.5 wt% MA) and the $M_2(C_{18})_2$ organoclay. Whereas the composite prepared by annealing unassisted by shear displayed (by XRD and TEM) a morphology with exfoliated layers and intercalated tactoids, that produced by solution blending was fully exfoliated, and this structure was shown to be thermodynamically stable by melt reprocessing. Clearly, during the preparation of the same composite by static annealing, thermodynamic equilibrium had only been approached.

Based on theoretical prediction by Balazs et al. [6], suggesting that an increase in the chain length of the organic groups tethered on clays would facilitate dispersion of the clay sheets in a polymer, Chung et al. synthesized ammonium ion terminated PP samples (PP-t-NH$_3^+$Cl$^-$) and used them to prepare fully exfoliated nanocomposites by static melting with pristine MMT-Na$^+$ or with the M$_2$(C$_{18}$)$_2$ organoclay [53,54]. Subsequent dilution of the above nanocomposites with neat PP, carried out by melt blending, did not alter their exfoliated structure, as shown by XRD and TEM, thus confirming that PP-t-NH$_3^+$ ions were ionically tethered on the silicate surface, leading to thermodynamically stable exfoliated nanocomposites. For comparison, several functionalized PP polymers, containing functional groups randomly distributed along the chain, were shown to produce, by static melting with M$_2$(C$_{18}$)$_2$ organoclay, intercalated nanocomposites. An illustration of the two structure types is shown, respectively, by Figure 3.8a and b.

The same basic principle of tethering PP chains to the clay surface was applied by Zhang et al. [55] using an original adaptation (simultaneous grafting intercalation) of the solution blending procedure. These authors modified an MMT-Na$^+$ using a mixture of surfactants with ~50 mol% M$_2$(C$_{18}$)$_2$, ~20 mol% M$_2$(C$_{16}$)$_2$, and ~30 mol% [(2-methacryloyloxy)ethyl] trimethyl ammonium chloride. The obtained organoclay was blended with PP in a boiling xylene solution, in the presence of a small amount of dicumyl peroxide (DCP), for 2 h under flowing nitrogen; the suspension was then added with an antioxidant and evaporated at room temperature for several hours, until a viscous gel resulted. This was finally dried at 80°C under vacuum. The hybrid was then processed at 210°C with a Mini-Max Molder. XRD, SEM, and TEM analyses indicated that the product had a mixed exfoliated/intercalated morphology. Solution intercalation is thought to have occurred concurrently with grafting of the unsaturated groups of the clay modifier onto the PP chains. Clearly, grafting could have taken place in more than one position on each PP chain. Thus, an intermediate kind of structure, such as that schematically shown in Figure 3.8c, was apparently achieved in this work. It should be noticed that ionic tethering of the PP chains on the clay prevails in structures (a) and (c), whereas weaker, yet multiple, polar interactions bind the functionalized PP chains within the clay galleries in structure (b). In Zhang's work [55], in agreement with expectation, no hint of intercalation was noticed in a control preparation carried out with neat PP and an organoclay modified with a blend of M$_2$(C$_{18}$)$_2$ and M$_2$(C$_{16}$)$_2$ in a 70/30 mole ratio.

A reactive solution blending procedure was also adopted by Sirivat et al. [56] to prepare exfoliated/intercalated PP/MMT nanocomposites. In this work, sodium-MMT was first functionalized by treatment with appropriate amounts of amino silanes (either

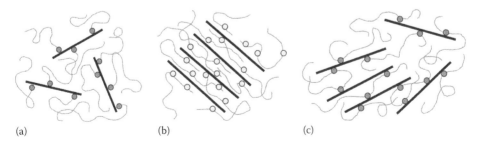

(a)   (b)   (c)

FIGURE 3.8
Schematic representation of the morphology of PP/organoclay nanocomposites prepared by (a) ammonium ion–terminated PP and MMT-Na$^+$, (b) static annealing of functionalized PP (e.g., with MA) and M$_2$(C$_{18}$)$_2$ organoclay, (c) solution blending with simultaneous grafting of the unsaturated surfactant onto the PP chains.

3-aminopropyl diisopropylethoxy silane, or 3-aminopropyl dimethylethoxy silane). The functionalization was made in water or dimethylacetamide (DMAC) for half an hour at 90°C–100°C, using neutral conditions to prevent ionization of the amino groups that would lead to the displacement of the MMT sodium ions. After filtration, washing, and drying, the amino functional clay was dispersed in DMAC and the suspension was slowly added into a hot xylene solution of PP-*g*-MA, and the mixture was stirred at 125°C for 15 min. The solution blending treatment led to amidization between the amino groups of the modified MMT and the anhydride rings of PP-*g*-MA, with covalent tethering of polymer chains to the inorganic surface. The product, recovered by filtration of the mixture (cooled to room temperature) and methanol washings, was then blended with neat PP by melt mixing to yield exfoliated/intercalated nanocomposites. According to the XRD and TEM data reported in the paper [56], a scheme similar to that of Figure 3.8c can be assumed for these nanocomposites.

Another example of solvent-assisted nanocomposite preparation involving modification of the PP chains to enhance their interaction with the clay surface is offered by the paper of Wu et al. [57]. These authors grafted hexamethylenediamine (HMDA) onto PP-*g*-MA (with 0.485 wt% MA) according to the reaction below, carried out in xylene solution at 120°C for 2 h, and used the product (PP-*g*-HMA) to prepare the nanocomposites by solution blending. PP-*g*-HMA was repeatedly washed with deionized water and dried under vacuum at 100°C for 12 h.

PP-*g*-MA   **HMDA**   **PP-*g*-HMA**

The PP-*g*-HMA nanocomposites, with 1, 3, and 5 wt% of an MMT modified by cation exchange with cetyl pyridinium chloride, were prepared in xylene solution at 120°C for 6 h. The powdery products were sandwiched between cover glasses and melted at 200°C to form thin films, which were then cooled with a rate of 20°C min^{-1}. The XRD and TEM analyses showed that the nanocomposites possess a mixed exfoliated/intercalated morphology, with a level of exfoliation that increases as the clay loading is lowered. Since PP-*g*-HMA contains more than one functional group per chain, the structure of the nanocomposites is probably similar to that of Figure 3.8c.

According to Avella et al. [58], exfoliated PP nanocomposites filled with 1 and 3 wt% of an MMT organically modified with $H_3(C_{16-18})_1N^+$ (reported as stearyl ammonium ions) were produced by a solution blending procedure (*o*-dichlorobenzene, 180°C, 1 h; precipitation in cold ethanol; vacuum drying; compression molding at 210°C). However, this claim was only based on the lack of discernible low angle reflections in the XRD patterns. An XRD peak shifted to wider angles, with respect to the basal organoclay reflection, was instead seen for a composite loaded with 5 wt% filler. Therefore, the results of this work should be confirmed by accurate TEM analyses before being accepted.

A recent paper by Chiu and Chu [59] describes the preparation, carried out by solution blending, of PP nanocomposites with three different commercial organoclays, Cloisite 15A (15A), Cloisite 20A (20A), and Cloisite 30B (30B). These organoclays are produced by Southern Clay Products Inc. from Cloisite Na^+ by cation exchange with $M_2(HT)_2$, for 15A and 20A, and with $M_1T_1(HE)_2$ (methyl tallow bis-2-hydroxyethyl), for 30B. PP and 5 wt% organoclays were blended in 1,2,4-trichlorobenzene (TCB), at a concentration of about 3 wt%, under vigorous stirring at 130°C for 60 min. The hybrids were recovered by evaporation of TCB on stainless dishes kept at 140°C, followed by treatment under vacuum at the same temperature for at least 12 h. The XRD patterns of the nanocomposites showed no clearly discernible low angle reflections, except for very broad and weak diffractions centered at $2\theta \approx 2.4°$ for PP/15A, $2\theta \approx 3.0°$ for PP/20A, and $2\theta \approx 6.4°$ for PP/30B. The first two diffraction peaks can be ascribed to (disorderly) intercalated silicate stacks, whereas the third is probably due to the collapse of the interlayer spacing of 30B caused by the degradation of the organic modifier [60]. The TEM micrographs taken on the PP/15A sample showed the presence of uniformly distributed thin multilayered structures and single layers. The dispersion of the 20A and 30B fillers was found to be slightly coarser. No significant effect of the clays on the thermal properties of the composites was demonstrated by DSC, except for a slightly increased crystallization temperature, whereas the increase of thermal stability of the nanocomposites was shown by TGA to parallel the sequence of the thermal stabilities of the fillers (20A > 15A > 30B).

The main conclusion drawn by these authors [59] is that nanocomposites with mixed exfoliated and intercalated morphology can be obtained from neat PP, with no compatibilizer, by the solution-blending method. However, this generalized claim is probably overdrawn. In fact, no check of the morphological stability of the nanocomposites was apparently made in their work. As recalled before, Manias et al. [47] had already applied the solution blending procedure, using very similar solvent (1,3,5-trichlorobenzene) and conditions (solvent removal by evaporation), to the preparation of a PP composite with an $M_2(C_{18})_2$ organoclay, and the product was find to possess an XRD silent pattern. However, upon melting (with or without shear), the clay layers were shown to reassemble into immiscible/intercalated stacks.

In summary, despite the few, apparently conflicting results in the available literature, solution blending of neat PP with common organoclays does not seem to promise significant improvements, with respect to the more attractive melt-compounding operation. All available reports confirm that extensive clay fragmentation and uniform dispersion of the resulting tactoids throughout the polymer matrix can be achieved by the use of appropriate solvents, so that solution blending has been employed in some studies as the first step of nanocomposite synthesis. However, this practice does not seem to be compatible with the requirements of a commercial production. In some instances, reactive solution blending techniques aimed at improving the interactions between PP and clay have been used. To this end, either the PP or the clay surface has been modified by the introduction of appropriate functionalities. This practice too can hardly be of interest on a commercial scale. Interestingly, some of the reported solution blending techniques, in particular those carried out by the use of chlorinated aromatic solvents [47,58,59], have been shown to lead to nanocomposites with a high level of exfoliation, in agreement with the scheme of Figure 3.2. However, in one case [47] it was experimentally proven that the morphology of as received composites does not correspond to thermodynamic equilibrium.

3.3.3 Nanocomposites from Olefin Copolymers

Most commercial grades of HDPE, LLDPE, and PP are in fact copolymers of ethylene or propylene with small amounts of other α-olefins, which are randomly added into their backbones so as to optimize some properties, e.g., impact resistance. Larger comonomer contents drastically lower the degree of crystallinity of polyolefins and yield products with elastomeric properties. Ethylene–propylene copolymers (EPM) and ethylene–propylene–diene crosslinkable terpolymers (EPDM) are used as polyolefin elastomers (POE), either neat or blended with polyamides or PP to form impact-resistant resins. On the other hand, the copolymerization of ethylene with, e.g., norbornene yields the so-called cyclic olefin copolymers (COC), characterized by low degree of crystallinity, high glass transition temperature, low moisture uptake, and excellent optical properties. In addition to copolymerization with other olefins, ethylene and propylene can also be copolymerized with a wide range of polar monomers yielding a number of specialty resins, including ionomers, for numerous applications. For example, EVA, ethylene-butyl acrylate (EBA), etc., are employed as hot melts in the packaging industry as quick-sealing adhesives, for journal and book binding, and in the woodworking industry. A very rich scientific literature is available on the nanocomposites based on olefin copolymers, particularly those filled with layered silicates by the melt-compounding method. On the contrary, limited attention was paid to nanocomposites produced by solution blending. A review of the scientific literature on nanocomposites of olefin copolymers prepared from solution is given below.

Cyclic polyolefin copolymer (COC)/layered silicate nanocomposites were prepared by the solution mixing process by Wu and Wu [61] in view of their application as flexible display substrates. Due to their hydrocarbon character, and the presence of rigid norbornene units in their backbone, COCs were expected to be reluctant to intercalation within the interlayer organoclay galleries. Therefore, in this work, MMT-Na$^+$ was organically modified with M$_3$(C$_{16}$)$_1$ (trimethylhexadecyl ammonium ions) by the usual cation exchange protocol, and the product was further modified by treatment with a 4/1 mixture of styrene and methyl methacrylate, followed by polymerization initiated by potassium persulfate. The solvent-assisted preparation of the nanocomposites was made by blending the modified clay with COC in xylene solution at room temperature, under ultrasonic stirring for 24 h. Films of neat COC and of the composites were then prepared onto a glass substrate by spin-casting, with a thickness of 100 μm. XRD analysis showed that the original spacing of the native silicate (d_{001} = 1.25 nm) was expanded to 3.68 nm after clay modification, whereas no low angle reflection was seen for the nanocomposites with clay loadings of 0.5–3 wt%, suggesting complete destruction of the silicate stacking order. Disorderly intercalated morphology of these nanocomposites, with average interlayer distance of about 5 nm, was in fact demonstrated by TEM. The nanocomposite films displayed mechanical and barrier properties significantly improved, with respect to those prepared from neat COC. However, no information was provided with respect to the thermodynamic stability of the structure and morphology of these nanocomposites.

Zhang et al. [62] filled isobutylene-isoprene rubber (IIR) with an organoclay in order to enhance its barrier properties in view of extreme applications in aerospace and high vacuum systems. Since an IIR latex was not commercially available, the nanocomposite preparation could not be made by the convenient co-coagulation process; thus, solution blending and melt compounding were adopted. Melt compounding was carried out in a two-roll mill under unspecified conditions. Solution blending was made in toluene with vigorous stirring for 24 h and the composites were recovered by solvent evaporation. The composites were then added with the cure ingredients and vulcanized at 160°C for the optimum cure time determined by torque rheometry. As shown in Figure 3.9, the interlayer spacing (d_{001} ≈ 2.2 nm) of the organoclay (modified with an unspecified surfactant) was expanded to d_{001} ≈ 4.2 nm for both solution-blended and melt-compounded nanocomposites. An intercalated morphology, with limited clay exfoliation, was confirmed by TEM for both types of composites. The structural and morphological similarity of the nanocomposites prepared by the two procedures is certainly not surprising as both materials underwent a prolonged

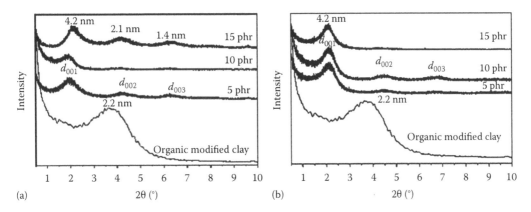

FIGURE 3.9
XRD patterns of the organoclay and the IIR nanocomposites, with different silicate loadings, prepared by (a) solution blending and (b) melt compounding. (Reprinted from Liang, Y. et al., *Polym. Test.*, 24, 12, 2005. With permission from Elsevier.)

high-temperature vulcanization step. This thermal after-treatment, which is applied to most rubber-based nanocomposites, does not allow a reliable comparison of the effect of the preparation procedures on nanostructure formation. It is interesting, however, that the lateral dimensions of the silicate platelets and tactoids were shown by TEM to be considerably smaller for the melt-compounded composites, probably due to the much higher shear stress acting on the system during melt kneading. As a consequence of this, the mechanical and barrier properties of the solution-blended composites were found to be superior to those of the melt-compounded counterparts.

EPDM/organoclay intercalated nanocomposites were prepared by Ma et al. [63] by blending the rubber with a chloroform dispersion of an $M_3(C_{16})_1$ organoclay, further modified by in situ polymerization of dimethyldichlorosilane, and recovering the composite by precipitation into excess methanol. An expansion of the interlayer spacing from $d_{001} \approx 1.97$ nm to $d_{001} \approx 3.56$ nm was obtained when the organoclay grafted with polydimethylsiloxane was solution blended with EPDM. On the contrary, fully exfoliated nanocomposites were obtained, by the same preparation procedure, with more polar matrixes, such as styrene-butadiene rubber (SBR), polystyrene (PS), or poly(vinyl chloride) (PVC).

EPDM-based nanocomposites were also prepared from solution by Srivastava et al. [64] using MgAl-LDH organically modified with sodium dodecyl sulfate as nanofiller (OLDH). EPDM and different amounts of OLDH were solution blended in boiling toluene under stirring for 12 h. Then DCP was added as a cross-linking initiator and toluene was removed by evaporation under reduced pressure. The composites were finally compression molded at 150°C for 45 min. The XRD analysis (Figure 3.10) showed that the basal reflection of OLDH was completely absent for all the nanocomposites, regardless of the filler loading (2–8 wt%). Good dispersion of LDH particles 2–4 nm thick and 50–100 nm wide, suggesting a high level of exfoliation, was also shown by TEM and atomic force microscopy analyses. The same authors [65] also prepared partially exfoliated EPDM/EVA45/OLDH nanocomposites by a similar solution blending procedure. Additionally, completely exfoliated nanocomposites filled with varying amounts of OLDH were synthesized from 1:1 wt/wt blends of LDPE and EVA28 by the same research group [66].

The solution-blending method was also employed by Srivastava et al. to prepare EPDM/ EVA/organoclay ternary nanocomposites [67,68]. EVA, with 45 wt% vinyl acetate (VA), and EPDM were blended in a 50/50 wt/wt ratio with 2–8 wt% $H_3(C_{16})_1$ organoclay in hot toluene

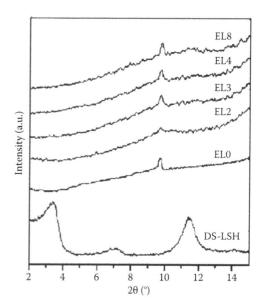

FIGURE 3.10
XRD spectra of OLDH (DS-LDH) and the EPDM/LDH composites (EL0–8) with 0–8 wt% filler. (Reprinted from Acharya, H. et al., *Nanoscale Res. Lett.*, 2, 1, 2007. With permission from Springer Science + Business Media.)

under stirring for 2 h. Then, DCP initiator was added to the solution and toluene evaporation was carried out under vacuum. The composite was finally compression molded at 145°C for 45 min to induce cross-linking. XRD showed that the $d_{001} \approx 2.04$ nm of the organoclay had been expanded to 3.59 nm for the composites, independent of the clay loading, and TEM images confirmed that these materials possess a morphology with intercalated clay stacks and some individual silicate platelets. SEM analysis of the fractured surfaces of composite samples etched with ethyl acetate, a good solvent for EVA, showed that an increase of the clay loading gradually changed the sea-island morphology of the EPDM/EVA blend into that of a co-continuous mixture of immiscible phases. The concomitant reduction of the phase size was interpreted as being due to a reduction of the interfacial tension. However, no clear evidence of the localization of the clay particles was drawn from the microscopic analyses. Also, no direct information on whether the organoclay intercalation was due to EPDM or EVA macromolecules, or both, was apparently obtained.

EVA represents the largest-volume segment of the ethylene copolymers market. The resin has good clarity and gloss, barrier properties, low-temperature toughness, and good resistance to UV radiation. Depending on the VA content, EVA grades can be tailored for uses as plastics, thermoplastic elastomers, and rubbers. The applications of EVA range from flexible packaging and wire and cable insulation, to footwear foams and melt adhesives. For most of these applications, additives are often employed to improve both mechanical properties and fire retardancy of the EVA materials. Thus, considerable attention was paid to the possibility of improving the properties of EVA through its silicate nanocomposites. Most of the available literature reports are dealt with EVA/organoclay nanocomposites prepared by melt compounding. In general, an increase of the VA content of the copolymer was shown to enhance the level of clay exfoliation. As far as the effect of the number of long alkyl chains of the organic modifier of clays, there seems to be general agreement that the organoclays with two-tailed surfactant, e.g., $M_2(HT)_2$ or $M_2(C_{18})_2$, lead to intercalated nanocomposites with d_{001} spacing of about 3.6–4.2 nm. For example, the Cloisites 20A, 15A, and 6A, modified with an increasing excess of the same $M_2(HT)_2$ surfactant, have interlayer spacing of 2.42, 3.15, and 3.6 nm, respectively, and were always found to afford intercalated, or partially exfoliated products with $d_{001} \approx 3.8$–$4.2\,nm$ [69–74]. On the contrary, for one-tailed organoclays, conflicting reports may be found in the literature. In fact, some authors [60,70,72,73,75] reported almost complete exfoliation for Cloisite 30B, modified with the $M_1T_1(HE)_2$ surfactant, and concluded that this organoclay is more compatible than, e.g., 20A, with EVA, especially for the higher VA contents, whereas the results by others [74], apparently supported by molecular dynamics simulation [76], showed that organoclays modified by two long alkyl tails have a better interaction with EVA. The origin of the discrepancy is probably ascribable to the lower thermal stability of the one-tailed organo-clays. In particular, 30B and other organoclays modified with one alkyl chain, such as, e.g., $H_3(C_{18})_1$ or $M_3(C_{18})_1$, were found quite often to undergo reaggregation of the silicate layers, with reappearance of a 1.2–1.4\,nm XRD peak, especially when melt compounding was carried out under severe temperature conditions [60,70,72,74,77,78]. Surfactant decomposition via Hoffmann elimination can also form protons on the clay surface, which may catalyze deacetylation of EVA. Thus, solution blending has been considered as an alternative pathway to produce exfoliated EVA nanocomposites, especially when thermally labile organo-clays are used.

In a series of papers by Srivastava et al. [79–84], the preparation by solution blending of nanocomposites based on EVA copolymers with 12, 28, and 45 wt% VA (EVA12, EVA28, and EVA45) and an organoclay modified with dodecyl ammonium chloride ($H_3(C_{12})_1$) has been described. The organophilic clay was prepared from an MMT-Na$^+$ with a 0.764\,meq g^{-1} CEC by treatment with dodecyl amine and hydrochloric acid, washed twice with hot water, dried under vacuum, ground, and sieved with a 270-mesh sieve. The organoclay was then dispersed under stirring in DMAC at 90°C for 3\,h. EVA12 and EVA28 were stirred in toluene at 70°C until a jelly-like solution was obtained. EVA45 was dissolved in DMAC at 85°C. The copolymer solution was then added with an appropriate amount of the organoclay dispersion and stirring was continued for 1.5–2\,h at 70°C–80°C. Dicumyl peroxide (DCP) was then added into the mixture and the evaporation of the solvent (or solvents) was made at ~70°C under reduced pressure. The obtained solid was finally dried under vacuum at 100°C–110°C for 2–3\,h and compression molded at 150°C for 45\,min. The organoclay loadings were in the range 1–8\,wt%. The absence of XRD reflections in the $2\theta = 3°$–$10°$ range for the composites with 1–6\,wt% organoclay was interpreted as suggestive of delamination of the organoclay and formation of exfoliated nanocomposites. When the clay loading was increased to 8\,wt%, however, the XRD patterns of the EVA12 and EVA28 composites displayed a reflection, which was ascribed to the aggregation of aluminosilicate layers. For example, in Figure 3.11, the XRD patterns of the native MMT-Na$^+$, the $H_3(C_{12})_1$ organoclay, and the relevant EVA12 nanocomposites are shown [82]. The $d_{001} = 1.19\,nm$ spacing of the native clay is expanded to 1.58\,nm after modification with the organic surfactant.

Neat EVA12 and the nanocomposites with 2–6\,wt% organoclay have silent XRD spectra, whereas a fairly strong reflection appears for the composite with 8\,wt% filler. Although it is known that delaminated silicate layers of exfoliated nanocomposites tend to aggregate into intercalated stacks when the filler loading increases, the observed abrupt change in the XRD patterns caused by an increase of the organoclay content from 6 to 8\,wt% is rather surprising. Another intriguing aspect of the XRD pattern of the composite with 8\,wt% organoclay, which was not highlighted and discussed by the authors, is that the observed reflection has the same angular position of that of the organoclay itself. This suggests that only delaminated silicate layers and unintercalated stacks coexist in this material. TEM analyses of the nanocomposites demonstrated that the silicate particles were homogeneously dispersed within the EVA matrix. For the EVA12, EVA28, and EVA45 nanocomposites, the presence of silicate layers with average thickness of 12–15, 3–5, and

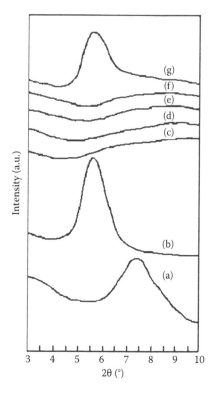

FIGURE 3.11
XRD patterns of (a) MMT-Na$^+$, (b) H$_3$(C$_{12}$)$_1$ organoclay, (c) neat EVA12, (d), (e), (f), and (g) EVA12/organoclay nanocomposites with 2, 4, 6, and 8 wt% filler, respectively. (Reprinted from Pramanik, M., *Macromol. Res.*, 11, 260, 2003. With permission from the Polymer Society of Korea.)

2–4 nm, respectively, was reported from the TEM micrographs [80–82]. It should be emphasized that the EVA/organoclay nanocomposites prepared in this series of works [79–84] underwent long-lasting thermal treatments at 100°C–150°C, after being recovered from the DMAC or toluene/DMAC solution, in order to grant the DCP-initiated cross-linking. Moreover, according to a recent paper by Mishra and Luyt [85], DCP seems to initiate EVA grafting on the clay surface, which results in an exfoliated morphology. Therefore, it is not clear whether the structure and morphology of the EVA nanocomposites prepared by Srivastava et al. [79–84] are the result of the solvent-assisted preparation or of the subsequent thermal treatments. It is noteworthy, however, that no discernible hint of organoclay degradation was observed for the materials prepared in these studies.

Srivastava et al. did also prepare EVA18, EVA28, EVA45, and EVA 60 nanocomposites filled with OLDH (dodecyl sulfate modified LDH) from toluene solution [86–89]. Blending was carried out at 100°C for 6 h, the solvent was removed under reduced pressure and the products were compression molded at 150°C for 45 min. The nanocomposites with up to 3 wt% OLDH were shown to be fully exfoliated by XRD and TEM; at higher filler loadings, partial agglomeration of the OLDH particles was observed. For these systems too, the dispersion of mineral particles was found to be improved by an increase of the VA concentration.

A surfactant-free method for the preparation of exfoliated EVA/silicate nanocomposites has been developed recently by Sogah et al. [90,91]. The process comprises two steps: the first step involves the reactive solution blending of MMT-Na$^+$ (with a CEC ≈ 0.90 mequiv. g^{-1}) with preformed random copolymers of VA and 2-(acryloyloxy)ethyltrimethyl ammonium chloride (AETMC); the second step, also made by solution blending, consists of a dilution with EVA of the masterbatch prepared in the first step. As AETMC is more reactive than

VA, the three copolymers containing 0.37, 1.00, and 2.14 mol% AETMC, indicated as PVAcA, PVAcB, and PVAcC, respectively, were prepared by slowly adding a methanol solution of AETMC into a methanol solution of VA at 60°C under nitrogen flow. The polymerization, initiated by AIBN, was continued for 1.5–3 h. Then, the viscous solution was diluted with acetone and poured into excess hexane under stirring. The tacky precipitate was purified by dissolution in methanol and precipitation in cold water. The structure of the three PVAc copolymers is indicated in the scheme below.

$$\left[\mathrm{CH_2{-}CH}\right]_x \left[\mathrm{CH_2{-}CH}\right]_y$$

PVAcA y = 0.37 mol%
PVAcB y = 1.00 mol%
PVAcC y = 2.14 mol%

Exfoliated or exfoliated/intercalated PVAc/MMT nanocomposites with up to 20 wt% silicate were prepared by slowly adding a water dispersion of MMT-Na$^+$ into a methanol solution of the PVAc and stirring the mixture at 50°C for 4 h. Evaporation of the methanol with a rotatory evaporator yielded an aqueous suspension, which was filtered. The white solid was dried under vacuum at 50°C. The XRD patterns of the nanocomposites with 5, 10, and 20 wt% silicate are shown in Figure 3.12, together with those of the MMT-Na$^+$ and of the composites containing neat polyvinyl acetate (PVAc) as the matrix. The latter polymer yields an intercalated composite with a d_{001} reflection corresponding to 2.1 nm, whose intensity increases with an increase of the MMT loading. The XRD patterns of the PVAcC composites show no basal reflections, even for a silicate content of 20 wt%, suggesting complete exfoliation. On the contrary, for the composites of the copolymers with lower AETMC content (PVAcA and PVAcB), an XRD peak corresponding to a spacing of 2–2.4 nm, suggesting partial intercalation, appears when the silicate concentration increases. Thus, the extent of exfoliation of the masterbatches, as suggested by XRD, depends on the silicate loading, as well as on the milli-equivalents of cationic sites in the copolymers. Interestingly, a quantitative examination of the data showed that low angle reflections were absent in the XRD spectra when the number of cationic sites in the copolymers was more than 50% of the number of exchangeable cations in the clay.

Scanning transmission electron microscopy (STEM) of the composites fully confirmed the XRD indications. For example, the STEM micrograph in Figure 3.13 shows that mostly single silicate sheets are visible in the PVAcA-10 composite (containing only 43% ammonium cations relative to the number of exchangeable cations in the MMT) despite the presence of a broad shoulder in the XRD (Figure 3.12B), which is probably due to a small amount of stacked silicate layers.

As mentioned before, the second step of the preparation was carried out by blending a THF solution of the PVAcB-20 masterbatch with a solution of the appropriate amount of EVA (with a VA content of 39 wt%) in the same solvent. Blending was continued for 4 h at 50°C with stirring. After cooling to room temperature, the mixture was evaporated under vacuum to dryness. The XRD and TEM analyses of the nanocomposites diluted to a silicate content of 1–5 wt% demonstrated the high level of clay exfoliation. In particular, even the broad shoulder present in the XRD of the PVAcB-20 masterbatch (Figure 3.12C) disappeared after dilution with EVA, which suggests that the nanocomposites have highly exfoliated morphology.

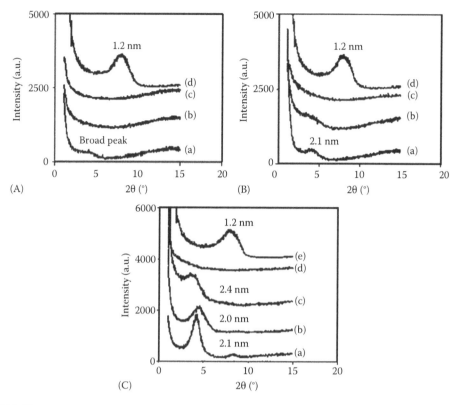

FIGURE 3.12
XRD patterns of (A): (a) PVAc-05, (b) PVAcA-05, (c) PVAcB-05, (d) MMT; (B): (a) PVAc-10, (b) PVAcA-10, (c) PVAcB-10, (d) MMT; (C): (a) PVAc-20, (b) PVAcA-20, (c) PVAcB-20, (d) PVAcC-20, (e) MMT. (Reprinted from Shi, Y. et al., *Chem. Mater.*, 19, 1552, 2007. With permission from American Chemical Society.)

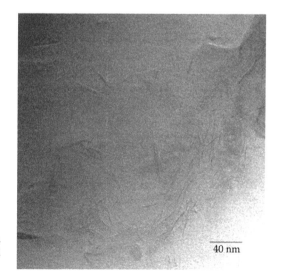

FIGURE 3.13
STEM image of the PVAcA-10 nanocomposite. (Reprinted from Shi, Y. et al., *Chem. Mater.*, 19, 1552, 2007. With permission from American Chemical Society.)

An example of the use of solution blending to prepare EVA/organoclay nanocomposite films with enhanced barrier properties is provided by a recent paper by Ramazani et al. [92]. EVA18 and EVA28 were added into a $CHCl_3$ dispersion of Cloisite 15A and the mixture was stirred at 40°C for 3 h. The solution was then poured on a glass mold with a 0.5 mm thick border and left to evaporate at room temperature. The obtained dry films were used for XRD, TEM, and oxygen permeability characterizations. The EVA18/15A composites with 3, 5, and 7 wt% filler were shown to exhibit XRD peaks corresponding to 6.8, 5.86, and 4.62 nm. The EVA28/15A films showed no reflection in the 1°–10° 2θ range, for an organoclay concentration of 3 wt%, and XRD peaks corresponding to 6.3 and 5.4 nm, for the 5 and 7 wt% clay loadings, respectively. Thus, a considerable expansion of the interlayer spacing of the 15A organoclay (3.15 nm) was found for all composites prepared, and its extent was found to increase with an increase of the VA content of the EVA copolymer and with a reduction of the clay loading. As it was mentioned before, most of the melt-compounding preparations of EVA/15A nanocomposites described in the scientific literature afforded intercalated morphologies characterized by interlayer spacings in the 3.6–4.2 nm range, practically independent of the clay loading. It cannot be excluded, therefore, that the composite films prepared in this work [92] were not in a thermodynamically stable equilibrium condition. Nevertheless, it is interesting that considerably enhanced oxygen barrier properties were found for these nanocomposite films, probably because of the preferred orientation of the silicate layers parallel to the film surface.

Ramazani et al. also modified their solution-blending procedure to adapt it to the preparation of 15A nanocomposite films with a polymer matrix consisting of an HDPE/EVA18 blend [93]. Both a one-step procedure (blending in xylene at 100°C for 3 h) and a two-step procedure (preparation of an EVA/15A composite in $CHCl_3$, as described before, and blending the dry binary composite with HDPE in xylene at 100°C) were employed to prepare the composite films. As in the previous work, the hot solution was evaporated at room temperature on a glass mold to produce films about 60 μm thick. The XRD analysis showed that the films of a 10/85/5 HDPE/EVA18/15A composite, prepared by the one-step and two-step procedures, possess an intercalated morphology with d_{001} spacing of 3.9 and 4.2 nm, respectively, in good agreement with the literature data for melt-compounded EVA/15A composites. It was also shown that the $d_{001} \approx 4.2$ nm of the composites prepared by the two-steps method is independent of the clay loading, in the investigated 3–7 wt% range. Thus, the conclusions drawn in this work on the structure of the nanocomposites are at variance with those of the previous study [92]. Since HDPE was added to EVA18 in a small amount (10 wt%), it seems reasonable to guess that the difference of the diffractometric data obtained for the two types of composites, i.e., EVA/15A [92] and HDPE/EVA/15A [93], is simply due to the fact that the preparation procedure used in the second work involved a higher temperature (100°C) treatment in xylene, which favored a better approach to a thermodynamically stable condition.

An interesting application of the solution-blending method to the preparation of nanocomposite drug delivery systems was illustrated in a paper by Saltzman et al. [94]. EVA (with unspecified VA content) was blended with three different organosilicates in CH_2Cl_2 until a homogeneous dispersion was obtained. The employed silicate loadings were 0, 5, 10, and 20 wt%. The organosilicates were Cloisite 20A (MMT modified with $M_2(HT)_2$, $d_{001} \approx 2.6$ nm), Somasif MAE® ($d_{001} \approx 2.9$ nm), and Somasif MAE300 ($d_{001} \approx 3$ nm) (the latter are synthetic mica, with different lateral dimensions, also modified with $M_2(HT)_2$). Appropriate amounts of a dioxane solution of a drug (dexamethasone) were then added, and the mixture was pipetted into a chilled teflon mold of diameter 2 cm and depth 1 cm. The mold was placed in a freezer at −80°C for 20 min and then, for 48 h, at −20°C. The resulting pellets were then freeze-dried for 72 h. XRD of the composites revealed an increase of the d_{001} spacing for all three organosilicates, implying polymer intercalation. The angular position of the basal reflection was $2\theta \approx 2.2°$ ($d_{001} \approx 4$ nm) for all composites, regardless of silicate type and loading. It was found, however, that the mechanical properties and the rate of drug release from the matrix copolymer depend on the aspect ratio of the dispersed particles, i.e., on both the lateral dimensions of the silicate layers and the extent of exfoliation. Thus, an appropriate choice of the organosilicate and of its content can allow to achieve any desired combination of drug release rate and mechanical properties.

Nanocomposites based on blends of HDPE and PE-*g*-MA, with an MMT modified by $M_3(C_{16})_1$, were prepared in two steps by Liang et al. [95]. The first step consisted of the preparation of PE-*g*-MA/$M_3(C_{16})_1$ masterbatches (with 15–50 wt% organoclay) by melt compounding in a roller mixer at ~150°C for 15 min. A masterbatch with 20 wt% organoclay was also prepared by solution blending in xylene, at 130°C for 30 min, with an unspecified solvent removal method. In the second step, the masterbatches were diluted with HDPE by melt kneading at ~155°C for 15 min. The products were then compression molded for 30 min at 160°C into 4 mm thick plates. From an XRD study of two 82/15/3 HDPE/PE-*g*-MA/organoclay composites prepared by either method, these authors concluded that the pathway consisting of a first step carried in solution was more effective and suggested that the reason for this could be either kinetic (enhanced mobility of the polymer chains in solution) or thermodynamic (entropic gain associated with desorption of a multitude of solvent molecules). However, since no direct comparison of the morphology of the two masterbatches of equal composition prepared with different procedures was made, this conclusion should be supported by further evidence before being accepted.

The structure and the morphology of nanocomposites based on random ethylene-acrylic acid copolymers (EAA), prepared from solution and in the melt (with and without shear), were comparatively investigated by Filippi et al. [29]. EAA6 and EAA11, with 6 and 11 wt% AA, respectively, were filled with 15A or 20A (5 and 10 wt%). The preparation by melt compounding was carried out in a Brabender mixer, at 120°C and 60 rpm for 10 min, and the products were compression molded at 100°C–120°C. The preparation from solution was made in xylene or toluene (occasionally, other solvents such as 1,2,4-trichlorobenzene, chloroform, and xylene/benzonitrile mixtures, were used) at temperatures close to the boiling point, under stirring for 2 h, followed by precipitation in appropriate non-solvents (acetone, ethanol, pentane, etc.), or by room-temperature solvent evaporation, and final drying under vacuum at room temperature. The preparation by static melting was carried out by thermal treatment of tablets of the products prepared from solution, or of dry blends of polymer and clay powders, in a mold at temperatures close to, or slightly higher than the EAA melting point (~100°C). The melt-compounded composites were also redissolved in hot xylene and precipitated in ethanol to assess their deintercalation capability. The results of the XRD analyses are summarily exemplified in Figure 3.14 by the patterns recorded for some of the EAA11/15A composites investigated. Very similar results were found for composites containing EAA6 and/or 20A and when solution blending was made with different solvents or non-solvents.

As shown by traces (b) and (c), the d_{001} spacing of 15A is reduced from ~3.10 to 2.42 nm if the clay is dispersed in xylene, precipitated into ethanol, filtered, washed with pure ethanol, and dried at room temperature. An XRD pattern very similar to trace (c) was also obtained for 20A after the same treatment. This shows that the excess surfactant present in these commercial clays is completely extracted during a solvent treatment similar to that used for the preparation of the solution-blended composites. Pattern (d) demonstrates that solution blending of EAA with 15A fails to yield intercalation when no thermal treatment is made on the product recovered by precipitation in a non-solvent (similar results were also

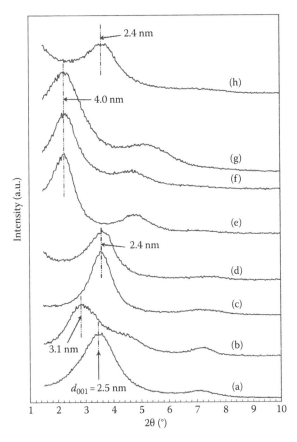

FIGURE 3.14
XRD patterns of (a) 20A; (b) 15A; (c) 15A after dispersion in xylene and precipitation in ethanol; (d) EAA11/15A, with 10 wt% clay, prepared by solution blending in xylene, precipitation in ethanol, and room-temperature drying; (e) same as (d) after static melting at 120°C; (f) EAA11/15A, with 10 wt% clay, prepared by static melting of a dry blend of polymer and clay powders; (g) EAA11/15A, with 10 wt% clay, prepared by melt compounding; (h) same as (g) after dissolution in hot xylene, precipitation in ethanol, and room-temperature drying.

found when solvent removal was made by room-temperature evaporation and vacuum drying). In fact, the angular position of the basal reflection demonstrates that the silicate layers, which were almost fully exfoliated in the starting xylene dispersion [15], reassembled into unintercalated clay stacks, with no excess surfactant within the galleries, during the solvent-removal step. However, simple static melting of this product led to fast intercalation, with a d_{001} spacing expansion to about 4 nm, as demonstrated by pattern (e). Fast and complete intercalation was also obtained by the static melting of a dry blend of polymer and clay powders, as shown by trace (f). Notice that the average size of the organoclay particles in the above dry blend is presumably much larger than that of the tactoids or small agglomerates present in the solution-blended material. The XRD pattern of the melt-compounded sample (trace (g)) is very similar, except for a slightly broader basal reflection, to those found for the statically melted composites described above. This demonstrates that an intercalated structure with an interlayer spacing $d_{001} \approx 4$ nm is produced upon melt blending, both with and without shear, and suggests that this structure corresponds to thermodynamic equilibrium for the EAA/15A and EAA/20A systems. It is also interesting that, as demonstrated by pattern (h), complete deintercalation was observed when a melt-compounded sample with an XRD pattern similar to trace (g) was dissolved in xylene and precipitated into ethanol. This finding indicates that intercalation of EAA chains within the organoclay galleries is granted by reversible polar interactions, rather than by chemical reactions leading to stable chemical bonds as suggested by others [96].

TEM analyses showed that melt-compounded $EAA/M_2(HT)_2$ composites have homogeneous dispersion of thin intercalated tactoids with several individual silicate platelets

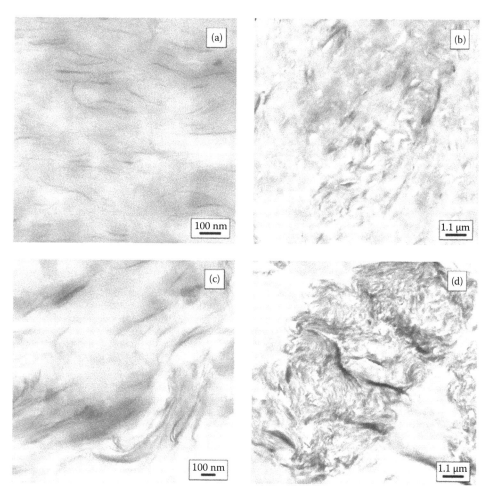

FIGURE 3.15

TEM images of EAA11/15A composites prepared by (a) melt compounding, (b) and (c) static melting of the product obtained from solution, and (d) static melting of a dry blend of polymer and clay powders. (Reprinted from Filippi, S. et al., *Eur. Polym. J.*, 43, 1645, 2007. With permission from Elsevier.)

(Figure 3.15a). The dispersion of clay particles is not as good for the composite prepared from solution (Figure 3.15b), although a number of exfoliated layers are visible in the higher magnification image (Figure 3.15c). The micrograph of the sample prepared by static melting of a dry blend of polymer and clay powders confirms that this material is in fact a microcomposite comprising large agglomerates of silicate stacks, though intercalated by polymer chains as demonstrated by XRD (Figure 3.15d).

A sketch illustrating the structure changes occurring in EAA/organoclay mixtures during melt compounding (upper line) and solution blending (lower line) is shown in Figure 3.16. Contrary to that shown in Figure 3.2, the present scheme emphasizes the formation of unintercalated silicate stacks during the solvent removal step. The sketch also pictures the de-intercalation undergone by the melt-compounded nanocomposite when it is dissolved and then recovered by precipitation in a non-solvent. The fast intercalation caused by melting the microcomposite comprising unintercalated clay tactoids under static conditions is also shown in the scheme. Ready intercalation occurring in unsheared polymer/clay mixtures

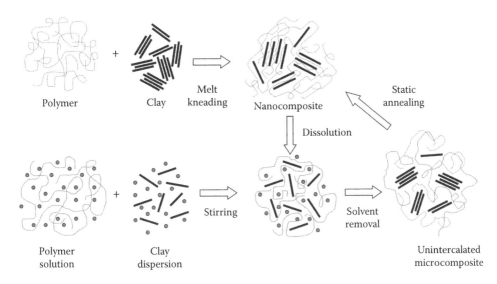

FIGURE 3.16
Scheme of the structural changes occurring in EAA/15A mixtures during melt compounding, solution blending, solvent de-intercalation, and static annealing.

demonstrates that the compatibility of the EAA chains with the $M_2(HT)_2$ clays is considerable. However, solvent–clay interactions must be even more effective if, as observed, no replacement of solvent molecules by polymer chains occurs in solution, as it is commonly understood when the process is indicated as "solution intercalation", but, on the contrary, the polymer chains are displaced from the galleries by the solvent molecules.

A study of nanocomposites based on HDPE-g-MA (HDMA) and 20A, also prepared with three different techniques, i.e., solution blending, melt compounding, and static annealing, showed that the conclusions drawn in the previous work [29] are valid for this system too [30]. The results of the XRD characterization of the HDMA/20A composites are exemplified in the cumulative plot of Figure 3.17. Patterns (b) and (c) were recorded for composites, containing 10 wt% organoclay, prepared from a xylene solution and recovered, respectively, by precipitation in pentane and by room-temperature evaporation under reduced pressure, and extensively dried under vacuum at room temperature. Both patterns contain a broad reflection corresponding to the original d_{001} spacing of 20A. This demonstrates that the silicate layers reassembled into unintercalated stacks during solvent removal. The different intensity of the basal reflections of the two samples may be due to their different physical state (the specimen isolated by precipitation was in powder form, whereas that recovered by evaporation was a thin film) and/or to a difference in the extent of clay platelets reassembling. Patterns (b) and (c) suggest that morphology of the composites comprises individual silicate layers and unintercalated clay tactoids.

The effect of static annealing on the structure of the samples prepared from solution and of powdered polymer/clay dry blends is illustrated by traces (d)–(f). Tablets of the solution-blended composite were placed in a mold preheated to 150°C, or 190°C, and annealed 1 min at this temperature before being quenched in ice water. Patterns (d) and (e) demonstrate that the basal reflection of the original specimen (pattern (b)) was strongly reduced in intensity after a short annealing at 150°C and disappeared completely after a 1 min treatment at 190°C. On the contrary, the latter treatment, when applied to a polymer/clay dry blend, led to a shift to lower angles of the basal reflection, with an increase of the d_{001} spacing to about 2.8 nm. Thus, a short annealing at 190°C caused complete delamination of the

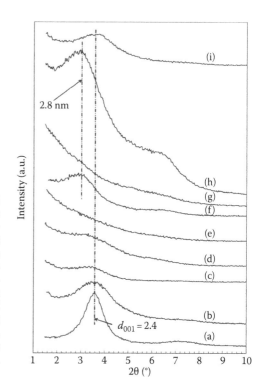

FIGURE 3.17
XRD patterns of (a) 20A after dispersion in xylene and precipitation in ethanol; (b) HDMA/20A, with 10 wt% clay, prepared by solution blending in xylene, precipitation in pentane, and room-temperature drying; (c) HDMA/20A, with 10 wt% clay, prepared by solution blending in xylene and room-temperature solvent evaporation; (d) and (e) same as (b) after 1 min static annealing at 150 and, respectively, 190°C; (f) HDMA/20A, with 10 wt% clay, prepared by 1 min static annealing at 190°C of a dry blend of polymer and clay powders; (g) and (h) HDMA/20A, with 15 and, respectively, 25 wt% clay, prepared by melt compounding; (i) same as (g) after dissolution in hot xylene, precipitation in ethanol, and room-temperature drying.

thin 20A tactoids contained in the solution-blended composites, whereas it only led to fast intercalation of the much larger clay agglomerates present in the dry blend. Intercalated silicate stacks with the same d_{001} spacing of 2.8 nm were also present in melt-compounded composites with high filler loadings, as demonstrated for a sample with 25 wt% 20A by pattern (h), whereas almost complete exfoliation was suggested by XRD for the melt-compounded composites with up to 15 wt% organoclay, as shown by pattern (g). Finally, pattern (i) shows that reassembling of the clay layers into unintercalated tactoids takes place also when a fully exfoliated melt-compounded sample (XRD pattern (g) is dissolved in xylene, recovered by precipitation in a non-solvent, and dried at room temperature. This result suggests that the anhydride functional groups of HDMA form polar bonds with the silicate surface rather than reacting irreversibly to give chemical linkages.

In agreement with the results found for the EAA/clay composites described before [29], TEM images of the different HDMA/20A composites (not shown here) demonstrated that the dispersion of clay particles was excellent for the melt-compounded materials, slightly worse for the solution-blended ones, and definitely poor for the microcomposites obtained by static annealing of a blend of powders. However, another interesting morphological difference was observed with reference to the orientation of the clay platelets and the HDMA crystals in the compression-molded samples obtained from melt-compounded and solution-blended composites. In fact, as demonstrated by the XRD scans in Figure 3.18, the ratio I_{110}/I_{200} of the intensities of reflections 110 and 200 (at $2\theta = 21.6°$ and $24.0°$, respectively) is ca. 2.5 for the solution-blended composite, suggesting lack of orientation of the PE crystals, whereas it decreases dramatically to about 0.1 for the melt-compounded nanocomposite.

Famulari et al. [97] showed that such strong change of the I_{110}/I_{200} intensity ratio is due to an orientation of the HDMA crystallites, with their crystallographic *a*-axis orthogonal

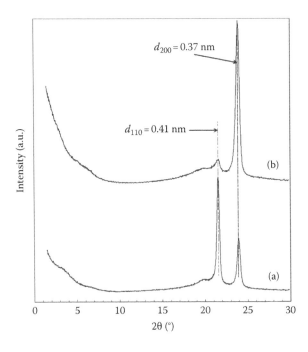

FIGURE 3.18
XRD patterns of compression molded (at 190°C) HDMA nanocomposites with 15 wt% 20A prepared by (a) solution blending and (b) melt compounding.

to the surface of the compression-molded disk, while the *b*- and *c*-axes are coplanar. Since no orientation was observed for similar specimens prepared from neat HDMA, it was concluded that the orientation of the polymer crystals was induced by the preferred orientation of the clay platelets caused by flow in the melt-mixing and compression-molding steps. The orientation of both the clay layers and the HDMA crystallites was clearly demonstrated by transmission WAXS [97], as well as by SAXS, as shown in Figure 3.19c. Figure 3.19 also shows the two-dimensional SAXS transmission patterns of HDMA/20A samples prepared by solution blending, both as such (image a) and after four compression-molding cycles at 190°C (image b). No orientation of the anisometric clay particles is discernible

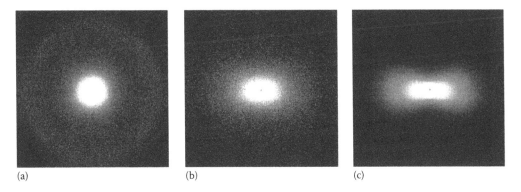

(a) (b) (c)

FIGURE 3.19
Two-dimensional SAXS transmission patters of 90/10 HDMA/20A composite tablets mounted vertical and edge on with respect to the incident beam. The samples were prepared by (a) solution blending in hot xylene, precipitation in excess acetone, followed by drying and compression molding at room-temperature; (b) same as (a) after four compression-molding cycles at 190°C; and (c) melt compounding and compression molding at 190°C. (Reprinted from Filippi, S. et al., *Eur. Polym. J.*, 44, 987, 2008. With permission from Elsevier.)

in the solution-blended composite recovered by precipitation in a non-solvent, whereas an increasing orientation develops upon repeated compression-molding cycles. However, the flow-induced parallel alignment of the clay platelets taking place during the kneading step, with the formation of nematic-like domains, leads, during the subsequent compression molding, to an even more pronounced orientation parallel to the specimen surface, as demonstrated by pattern (c).

The different morphology of solution-blended and melt-compounded composites, with respect to the spatial orientation of the silicate platelets, is visualized even more clearly by the SEM images shown in Figure 3.20. The micrographs were taken on cryofractured fragments of compression-molded disks of HDMA/20A composites prepared by the two methods, after burning away the organic material by heating the samples in air to 900°C with a rate of 10°C min^{-1} in a TGA apparatus. Interestingly, the inorganic residue of the fragment of melt-compounded material retained the shape it had before burning away the polymer, whereas that of the solution-blended composite was extremely friable and crumbled easily. Micrographs (a) and (b) provide a clear demonstration of the extensive local silicate layers parallelism, which, on the contrary, is completely absent in the solution-blended material, as shown by images (c) and (d).

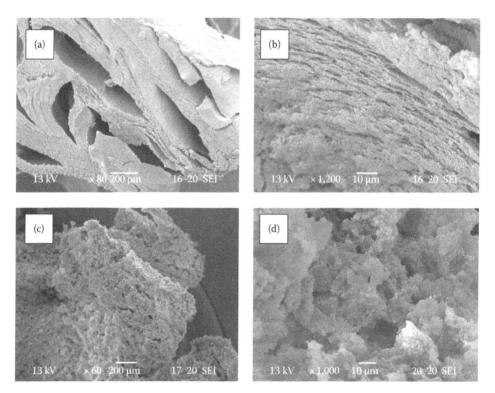

FIGURE 3.20
SEM micrographs of the inorganic residues of fragments of compression-molded HDMA/20A composites prepared by (a) and (b) melt compounding; (c) and (d) blending in hot xylene, precipitation in acetone, room-temperature drying, and static fusion at 190°C. (Reprinted from Filippi, S. et al., *Eur. Polym. J.*, 44, 987, 2008. With permission from Elsevier.)

3.4 Conclusions

Solution blending cannot compete with melt compounding for the preparation of polyolefin/organoclay nanocomposites at the commercial scale, except in the case of possible niche operations, exemplified by the production of cast films for drug delivery systems illustrated in [94]. Some reasons are that (1) environment-unfriendly organic solvents should be employed; (2) large amounts of solvents and high temperatures might be needed because of the high degree of crystallinity and low solubility of most olefin polymers; and (3) polyolefins can conveniently be processed in the molten state at moderate temperatures, with limited danger for the thermal stability of the organoclays. Nevertheless, a number of features of the solution blending process might be advantageous, especially for laboratory preparations. In fact, the operative protocols, though involving a larger number of steps, are quite simple, do not require expensive equipment, can be applied under mild temperature conditions and lend themselves to the preparation of a large number of small size specimens. Moreover, solution blending grants very effective fragmentation of the organoclay agglomerates and excellent dispersion of the resulting thin particles, either platelets or tactoids, within the polymer bulk. With melt compounding, the same effects are only obtained by subjecting the molten polymer/clay mixture to considerable stresses, which may have negative effects on their stability, and may lower the aspect ratio of the silicate platelets. For these reasons, in a number of studies, solution blending was employed as a first step of the nanocomposite synthesis, to produce highly filled masterbatches that can be subsequently diluted with neat polymer by melt compounding. Solution blending was also successfully employed for the production of highly exfoliated nanocomposites by coupling the physical mixing process with chemical reactions leading to tethering of polymer chains to the silicate surface.

In principle, the method by which a nanocomposite is prepared from a polymer (or a polymer blend) and a commercial organoclay, e.g., solution blending or melt compounding, has just an effect on the mechanism and the kinetics of the mixing process, whereas the thermodynamic aspects are only dependent on the nature of the components being mixed, i.e., the chemical structure of both the polymer chains and the organic modifier tethered on the silicate surface. Thus, comparing the structure and morphology of nanocomposites prepared by either method may provide important information on the dynamics of nanostructure formation in these systems. To this end, however, scrupulous attention should be paid to the many factors playing a role in the process so as to avoid ambiguous results and erroneous conclusions. For example, as it was shown in this review, in addition to the chemical structure of the polymer (and the compatibilizer, if any) and to the nature of the employed organoclay, other important factors influence the results of a solution blending preparation, such as the nature of the solvent (or solvents) used to dissolve the polymer and disperse the organoclay, the way solvent removal is made to recover the product, and the thermal treatments carried out on the latter before the structural and morphological characterization. Moreover, since the prepared nanocomposites must often be reprocessed to produce the final articles, the question of whether their structure and morphology are thermodynamically stable or, conversely, tend to undergo important changes upon melting, should be carefully considered and firmly settled.

References

1. Sanchez, C., Julián, B., Belleville, P., and Bopal, M. 2005. Applications of hybrid organic-inorganic nanocomposites. *Journal of Materials Chemistry* 15:3559–3592.
2. Utracki, L. A. 2008. Polymeric nanocomposites: Compounding and performance. *Nanoscience and Nanotechnology Letters* 8:1582–1596.
3. Schaefer, D. W. and Justice, R. S. 2007. How nano are nanocomposites? *Macromolecules* 40:8501–8517.
4. Vaia, R. A. and Giannelis, E. 1997. Lattice model of polymer melt intercalation in organically-modified layered silicates. *Macromolecules* 30:7990–7999.
5. Vaia, R. A. and Giannelis, E. 1997. Polymer melt intercalation in organically-modified layered silicates: Model predictions and experiment. *Macromolecules* 30:8000–8009.
6. Balazs, A. C., Singh, C., and Zhulina, E. 1998. Modeling the interactions between polymers and clay surfaces through self-consistent field theory. *Macromolecules* 31:8370–8381.
7. Zhulina, E., Singh, C., and Balazs, A. C. 1999. Attraction between surfaces in a polymer melt containing telechelic chains: Guidelines for controlling the surface separation in intercalated polymer-clay composites. *Langmuir* 15:3935–3943.
8. Ginzburg, V. V., Singh, C., and Balazs, A. C. 2000. Theoretical phase diagrams of polymer/clay composites: The role of grafted organic modifiers. *Macromolecules* 33:1089–1099.
9. Alexandre, M. and Dubois, P. 2000. Polymer-layered silicate nanocomposites: Preparation, properties and uses of a new class of materials. *Materials Science and Engineering* 28:1–63.
10. Pavlidou, S. and Papaspyrides, C. D. 2008. A review on polymer-layered silicate nanocomposites. *Progress in Polymer Science* 33:1119–1198.
11. Koo, J. H. 2006. *Polymer Nanocomposites: Processing, Characterization, and Applications*. New York: McGraw Hill Nanoscience and Technology Series.
12. Yano, K., Usuki, A., Okada, A., Kurauchi, T., and Kamigaito, O. 1993. Synthesis and properties of polyimide-clay hybrid. *Journal of Polymer Science, Part A: Polymer Chemistry* 31:2493–2498.
13. Li, Y. and Ishida, H. 2003. Solution intercalation of polystyrene and the comparison with poly(ethyl methacrylate). *Polymer* 44:6571–6577.
14. Flory, P. J. 1974. Spatial configuration of macromolecular chains, Nobel lecture, December 11, http://nobelprize.org/nobel_prizes/chemistry/laureates/1974/flory-lecture.pdf
15. Ho, D. L., Briber, R. M., and Glinka, C. J. 2001. Characterization of organically modified clays using scattering and microscopic techniques. *Chemistry of Materials* 13:1923–1931.
16. Ho, D. L. and Glinka, C. J. 2003. Effects of solvent solubility parameters on organoclay dispersions. *Chemistry of Materials* 15:1309–1312.
17. Hanley, H. J. M., Muzny, C. D., Ho, D. L., and Glinka, C. J. 2003. A small-angle neutron scattering study of a commercial organoclay dispersion. *Langmuir* 19:5575–5580.
18. King, H. E., Milner Jr., S. T., Lin, M. Y., Singh, J. P., and Mason, T. G. 2007. Structure and rheology of organoclay suspensions. *Physics Reviews E* 75:021403, 1–20.
19. Burgentzlé, D., Duchet, J., Gérard, J. F., Jupin, A., and Fillon, B. 2004. Solvent-based nanocomposite coatings: I. Dispersion of organophilic montmorillonite in organic solvents. *Colloid and Interface Science* 278:26–39.
20. Dan, C. H., Kim, Y. D., Lee, M., Min, B. H., and Kim, J. H. 2008. Effect of solvent on the properties of thermoplastic polyurethane/clay nanocomposites prepared by solution mixing. *Journal of Applied Polymer Science* 108:2128–2138.
21. Bonzanini Romero, R., Paula Leite, C.A., and do Carmo Gonçalves, M. 2009. The effect of the solvent on the morphology of cellulose acetate/montmorillonite nanocomposites. *Polymer* 50:161–170.
22. Malwitz, M. M., Lin-Gibson, S., Hobbie, E. K., Butler, P. D., and Schmidt, G. 2003. Orientation of platelets in multilayered nanocomposite polymer films. *Journal of Polymer Science, Part B: Polymer Physics* 41:3237–3248.
23. Stefanescu, E. A. et al. 2006. Supramolecular structures in nanocomposite multilayered films. *Physical Chemistry Chemical Physics* 8:1739–1746.

24. Vaia, R. A., Jandt, K. D., Kramer, E. J., and Giannelis, E. P. 1995. Kinetics of polymer melt intercalation. *Macromolecules* 28:8080–8085.
25. Vaia, R. A., Jandt, K. D., Kramer, E. J., and Giannelis, E. P. 1996. Microstructural evolution of melt intercalated polymer-organically modified layered silicates nanocomposites. *Chemistry of Materials* 8:2628–2635.
26. Yoon, J. T., Jo, W. H., Lee, M. S., and Ko, M. B. 2001. Effects of comonomers and shear on the melt intercalation of styrenics/clay nanocomposites. *Polymer* 42:329–336.
27. Lertwimolnun, W. and Vergnes, B. 2005. Influence of compatibilizer and processing conditions on the dispersion of nanoclay in a polypropylene matrix. *Polymer* 46:3462–3471.
28. Homminga, D., Goderis, B., Hoffman, S., Reynaers, H., and Groeninckx, G. 2005. Influence of shear flow on the preparation of polymer layered silicate nanocomposites. *Polymer* 46:9941–9954.
29. Filippi, S., Mameli, E., Marazzato, C., and Magagnini, P. 2007. Comparison of solution-blending and melt intercalation for the preparation of poly(ethylene-*co*-acrylic acid)/organoclay nanocomposites. *European Polymer Journal* 43:1645–1659.
30. Filippi, S., Marazzato, C., Magagnini, P., Famulari, A., Arosio, P., and Meille, S. V. 2008. Structure and morphology of HDPE-g-MA/organoclay nanocomposites: Effects of the preparation procedures. *European Polymer Journal* 44:987–1002.
31. Hotta, S. and Paul, D. R. 2004. Nanocomposites formed from linear low density polyethylene and organoclays. *Polymer* 45:7639–7654.
32. Mainil, M., Alexandre, M., Monteverde, F., and Dubois, P. 2006. Polyethylene organo-clay nanocomposites: The role of the interface chemistry on the extent of clay intercalation/exfoliation. *Journal of Nanoscience and Nanotechnology* 6:337–344.
33. Wang, K. H., Choi, M. H., Koo, C. M., Choi, Y. S., and Chung, I. J. 2001. Synthesis and characterization of maleated polyethylene/clay nanocomposites. *Polymer* 42:9819–9826.
34. Sun, L., Ertel, E. A., Zhu, L., Hsiao, B. S., Avila-Orta, C. A., and Sics, I. 2005. Reversible de-intercalation and intercalation induced by polymer crystallization and melting in a poly(ethylene oxide)/organoclay nanocomposite. *Langmuir* 21:5672–5676.
35. Jeon, H. G., Jung, H.-T., Lee, S. W., and Hudson, S. D. 1998. Morphology of polymer/silicate nanocomposites. High density polyethylene and a nitrile copolymer. *Polymer Bulletin* 41:107–113.
36. Flory, P. J. 1988. *Statistical Mechanics of Chain Molecules.* New York: Oxford University Press.
37. Chen, W. and Qu, B. 2003. Structural characteristics and thermal properties of PE-*g*-MA/MgAl-LDH exfoliated nanocomposites synthesized by solution intercalation. *Chemistry of Materials* 15:3208–3213.
38. Chen, W., Feng, L., and Qu, B. 2004. Preparation of nanocomposites by exfoliation of ZnAl layered double hydroxides in nonpolar LLDPE solution. *Chemistry of Materials* 16:368–370.
39. Chen, W. and Qu, B. 2004. LLDPE/ZnAl-LDH exfoliated nanocomposites: Effects of nanolayers on thermal and mechanical properties. *Journal of Materials Chemistry* 14:1705–1710.
40. Qiu, L., Chen, W., and Qu, B. 2006. Morphology and thermal stabilization mechanism of LLDPE/MMT and LLDPE/LDH nanocomposites. *Polymer* 47:922–930.
41. Shah, R. K., Hunter, D. L., and Paul, D. R. 2005. Nanocomposites from poly(ethylene-*co*-methacrylic acid) ionomers: Effect of surfactant structure on morphology and properties. *Polymer* 46:2646–2662.
42. Xi, Y., Ding, Z., He, H., and Frost, R. L. 2004. Structure of organoclays—An x-ray diffraction and thermogravimetric analysis study. *Journal of Colloid Interface Science* 277:116–120.
43. Shah, R. K. and Paul, D. R. 2006. Organoclay degradation in melt processed polyethylene nanocomposites. *Polymer* 47:4075–4084.
44. Filippi, S., Marazzato, C., Magagnini, P., Minkova, L., Tzankova Dintcheva, N., and La Mantia, F. P. 2006. Organoclay nanocomposites from ethylene-acrylic acid copolymers. *Macromolecular Materials and Engineering* 291:1208–1225.
45. Shah, R. K., Kim, D. H., and Paul, D. R. 2007. Morphology and properties of nanocomposites formed from ethylene/methacrylic acid copolymers and organoclays. *Polymer* 48:1047–1057.

46. Kawasumi, M., Hasegawa, N., Kato, M., Usuki, A., and Okada, A. 1997. Preparation and mechanical properties of polypropylene-clay hybrids. *Macromolecules* 30:6333–6338.

47. Manias, E., Touny, A., Wu, L., Strawhecker, K., Lu, B., and Chung, T. C. 2001. Polypropylene/montmorillonite nanocomposites. Review of the synthetic routes and materials properties. *Chemistry of Materials* 13:3516–3523.

48. Mittal, V. 2007. Polypropylene-layered silicate nanocomposites: Filler matrix interactions and mechanical properties. *Journal of Thermoplastic Composite Materials* 20:575–599.

49. Mani, G., Fan, Q., Ugbolue, S. C., and Yang, Y. 2005. Morphological studies of polypropylene-nanoclay composites. *Journal of Applied Polymer Science* 97:218–226.

50. Kurokawa, Y., Yasuda, H., and Oya, A. 1996. Preparation of a nanocomposite of polypropylene and smectite. *Journal of Materials Science Letters* 15:1481–1483.

51. Kurokawa, Y., Yasuda, H., Kashiwagi, M., and Oya, A. 1997. Structure and properties of a montmorillonite/polypropylene nanocomposite. *Journal of Materials Science Letters* 16:1670–1672.

52. Oya, A., Kurokawa, Y., and Yasuda, H. 2000. Factors controlling mechanical properties of clay mineral/polypropylene nanocomposites. *Journal of Material Science* 35:1045–1050.

53. Chung, T. C. 2002. Synthesis of functional polyolefin copolymers with graft and block structures. *Progress in Polymer Science* 27:39–85.

54. Wang, Z. M., Nakajima, H., Manias, E., and Chung, T. C. 2003. Exfoliated PP/clay nanocomposites using ammonium terminated PP as the organic modification for montmorillonite. *Macromolecules* 36:8919–8922.

55. Xie, S., Zhang, S., and Wang, F. 2004. Synthesis and characterization of poly(propylene)/montmorillonite nanocomposites by simultaneous grafting-intercalation. *Journal of Applied Polymer Science* 94:1018–1023.

56. Keener, B. D., Hudson, S. D., Li, Y., Moran, I. W., Phongphour, C., and Sirivat, A. 2005. Preparation, morphology, mechanical properties and fracture resistance of nanocomposites comprising montmorillonite and polypropylene. *Material Research Innovations* 9:559–576.

57. Wu, J.-Y., Wu, T.-M., Chen, W.-Y., Tsai, S.-J., Kuo, W.-F., and Chang, G.-Y. 2005. Preparation and characterization of PP/clay nanocomposites based on modified polypropylene and clay. *Journal of Polymer Science, Part B: Polymer Physics* 43:3242–3254.

58. Avella, M., Cosco, S., and Errico, M. E. 2005. Preparation of isotactic polypropylene/organoclay nanocomposites by solution mixing methodology: Structure and properties relationships. *Macromolecular Symposia* 228:147–154.

59. Chiu, F.-C. and Chu, P.-H. 2006. Characterization of solution-mixed polypropylene/clay nanocomposites without compatibilizers. *Journal of Polymer Research* 13:73–78.

60. Benali, S. et al. 2008. Study of interlayer spacing collapse during polymer/clay nanocomposite melt intercalation. *Journal of Nanoscience and Nanotechnology Letters* 8:1707–1713.

61. Wu, T.-M. and Wu, C.-W. 2005. Surface characterization and properties of plasma-modified cyclic olefin copolymer/layered silicate nanocomposites. *Journal of Polymer Science, Part B: Polymer Physics* 43:2745–2753.

62. Liang, Y., Wang, Y., Wu, Y., Zhang, H., and Zhang, L. 2005. Preparation and properties of isobutylene-isoprene rubber (IIR)/clay nanocomposites. *Polymer Testing* 24:12–17.

63. Ma, J., Xu, J., Ren, J.-H., Yu, Z.-Z., and Mai, Y.-W. 2003. A new approach to polymer/montmorillonite nanocomposites. *Polymer* 44:4619–4624.

64. Acharya, H., Srivastava, S. K., and Bhowmick, A. K. 2007. A solution blending route to ethylene propylene diene terpolymer/layered double hydroxide nanocomposites. *Nanoscale Research Letters* 2:1–5.

65. Kuila, T., Srivastava, S. K., and Bhowmick, A. K. 2009. Ethylene vinyl acetate/ethylene propylene diene terpolymer-blend-layered double hydroxide nanocomposites. *Polymer Engineering and Science* 49:585–591.

66. Kuila, T., Srivastava, S. K., Bhowmick, A. K., and Saxena, A. K. 2008. Thermoplastic polyolefin based polymer-blend-layered double hydroxide nanocomposites. *Composites Science and Technology* 68:3234–3239.

67. Acharya, H., Srivastava, S. K., and Bhowmick, A. K. 2006. Ethylene propylene diene terpolymer/ethylene vinyl acetate/layered silicate ternary nanocomposite by solution method. *Polymer Engineering and Science* 46:837–843.
68. Acharya, H., Kuila, T., Srivastava, S. K., and Bhowmick, A. K. 2008. Effect of layered silicate on EPDM/EVA blend nanocomposite: Dynamic mechanical, thermal, and swelling properties. *Polymer Composites* 29:443–450.
69. Alexandre, M. et al. 2001. Preparation and properties of layered silicate nanocomposites based on ethylene vinyl acetate copolymers. *Macromolecular Rapid Communications* 22:643–646.
70. Riva, A., Zanetti, M., Braglia, M., Camino, G., and Falqui, L. 2002. Thermal degradation and rheological behaviour of EVA/montmorillonite nanocomposites. *Polymer Degradation Stability* 77:299–304.
71. Gelfer, M. et al. 2002. Manipulating the microstructure and rheology in polymer-organoclay composites. *Polymer Engineering and Science* 42:1841–1851.
72. Li, X. and Ha, C. S. 2003. Nanostructure of EVA/organoclay nanocomposites: Effects of kinds of organoclays and grafting of maleic anhydride onto EVA. *Journal of Applied Polymer Science* 87:1901–1909.
73. Chaudhary, D. S., Prasad, R., Gupta, R. K., and Bhattacharya, S. N. 2005. Clay intercalation and influence of crystallinity of EVA-based clay nanocomposites. *Thermochimica Acta* 433:187–195.
74. Cui, L., Ma, X., and Paul, D. R. 2007. Morphology and properties of nanocomposites formed from ethylene-vinyl acetate copolymers and organoclays. *Polymer* 48:6325–6339.
75. Costache, M. C., Jiang, D. D., and Wilkie, C. A. 2005. Thermal degradation of ethylene-vinyl acetate copolymer nanocomposites. *Polymer* 46:6947–6958.
76. Zhang, Q., Ma, X., Wang, Y., and Kou, K. 2009. Morphology and interfacial action of nanocomposites formed from ethylene vinyl acetate copolymers and organoclays. *Journal of Physical Chemistry B* 113:11898–11905.
77. Zanetti, M., Camino, G., Thomann, R., and Mülhaupt, R. 2001. Synthesis and thermal behaviour of layered silicate-EVA nanocomposites. *Polymer* 42:4501–4507.
78. Duquesne, S., Jama, C., Le Bras, M., Delobel, R., Recourt, P., and Gloaguen, J. M. 2003. Elaboration of EVA-nanoclay systems: Characterization, thermal behaviour and fire performance. *Composites Science and Technology* 63:1141–1148.
79. Pramanik, M., Srivastava, S. K., Samantaray, B. K., and Bowmick, A. K. 2001. Preparation and properties of ethylene vinyl acetate-clay hybrids. *Journal of Materials Science and Letters* 20:1377–1380.
80. Pramanik, M., Srivastava, S. K., Samantaray, B. K., and Bowmick, A. K. 2002. Synthesis and characterization of organosoluble, thermoplastic elastomer/clay nanocomposites. *Journal of Polymer Science, Part B: Polymer Physics* 40:2065–2072.
81. Pramanik, M., Srivastava, S. K., Samantaray, B. K., and Bowmick, A. K. 2003. Rubber-clay nanocomposite by solution blending. *Journal of Applied Polymer Science* 87:2216–2220.
82. Pramanik, M., Srivastava, S. K., Samantaray, B. K., and Bowmick, A. K. 2003. EVA/clay nanocomposite by solution blending: Effect of aluminosilicate layers on mechanical and thermal properties. *Macromolecular Research* 11:260–266.
83. Pramanik, M., Acharya, H., and Srivastava, S. K. 2004. Exertion of inhibiting effect by aluminosilicate layers on swelling of solution blended EVA/clay nanocomposite. *Macromolecular Materials and Engineering* 289:562–567.
84. Srivastava, S. K., Pramanik, M., and Acharya, H. 2006. Ethylene/vinyl acetate copolymer/clay nanocomposites. *Journal of Polymer Science, Part B: Polymer Physics* 44:471–480.
85. Mishra, S. B. and Luyt, A. S. 2008. Effect of organic peroxides on the morphological, thermal and tensile properties of EVA-organoclay nanocomposites. *eXPRESS Polymer Letters* 2:256–264.
86. Kuila, T., Acharya, H., Srivastava, S. K., and Bhowmick, A. K. 2007. Synthesis and characterization of ethylene vinyl acetate/Mg-Al layered double hydroxide nanocomposites. *Journal of Applied Polymer Science* 104:1845–1851.

87. Kuila, T., Acharya, H., Srivastava, S. K., and Bhowmick, A. K. 2008. Effect of vinyl acetate content on the mechanical and thermal properties of ethylene vinyl acetate/Mg-Al layered double hydroxide nanocomposites. *Journal of Applied Polymer Science* 108:1329–1335.

88. Kuila, T., Srivastava, S. K., and Bhowmick, A. K. 2009. Rubber/LDH nanocomposites by solution blending. *Journal of Applied Polymer Science* 111:635–641.

89. Kuila, T., Acharya, H., Srivastava, S. K., and Bhowmick, A. K. 2009. Ethylene vinyl acetate/Mg-Al LDH nanocomposites by solution blending. *Polymer Composites* 30:497–502.

90. Shi, Y., Peterson, S., and Sogah, D. Y. 2007. Surfactant-free method for the synthesis of poly(vinyl acetate) masterbatch nanocomposites as a route to ethylene vinyl acetate/silicate nanocomposites. *Chemistry of Materials* 19:1552–1564.

91. Shi, Y., Kashiwagi, T., Walters, R. N., Gilman, J. W., Lyon, R. E., and Sogah, D. Y. 2009. Ethylene vinyl acetate/layered silicate nanocomposites prepared by a surfactant-free method: Enhanced flame retardant and mechanical properties. *Polymer* 50:3478–3487.

92. Shafiee, M. and Ramazani, S. A. A. 2008. Investigation of barrier properties of poly(ethylene vinyl acetate)/organoclay nanocomposite films prepared by phase inversion method. *Macromolecular Symposia* 274:1–5.

93. Dadfar, S. M. R., Ramazani, S. A. A., and Dadfar S. M. A. 2009. Investigation of oxygen barrier properties of organoclay/HDPE/EVA nanocomposite films prepared using a two-step solution method. *Polymer Composites* 30:812–819.

94. Cypes, S. H., Saltzman, W. M., and Giannelis, E. P. 2003. Organosilicate-polymer drug delivery systems: controlled release and enhanced mechanical properties. *Journal of Controlled Release* 90:163–169.

95. Liang, G., Xu, J., Bao, S., and Xu, W. 2004. Polyethylene/maleic anhydride grafted polyethylene/organic-montmorillonite nanocomposites. I. Preparation, microstructure, and mechanical properties. *Journal of Applied Polymer Science* 91:3974–3980.

96. Xu, Y., Fang, Z., and Tong, L. 2005. On promoting intercalation and exfoliation of bentonite in high-density polyethylene by grafting acrylic acid. *Journal of Applied Polymer Science* 96:2429–2434.

97. Famulari A. et al. 2007. Clay-induced preferred orientation in polyethylene/compatibilized clay nanocomposites. *Journal of Macromolecular Science, Part B Physics* 46:355–371.

4

Polyolefin Nanocomposites by Reactive Extrusion

Zakir M.O. Rzayev

CONTENTS

4.1 Introduction

In the last decade, an increasing number of research groups in academia and industry focused their research efforts on thermoplastic and thermoset polymer–layered silicate, polymer/silica hybrid, and polymer/nanoparticle (metal oxides, nanotubes, etc.) nanocomposites. This considerable scientific and engineering interest has been stimulated by the possibility of the significant improvements in physical, chemical, mechanical, thermal, and other important properties of nanocomposite materials. The results of these studies, especially on polymer/silicate nanocomposites were summarized and discussed in several reviews [1–6] and books [7–9]. Polymer/silicate systems are a class of organic–inorganic hydrides, composed of organic polymer matrix, in which layered silicate particles of nanoscale dimension are embedded owing to which these nanocomposites show enchased mechanical properties [10–14], gas barrier properties [15–18], and improved chemical resistance [19,20], thermal stability [2,6,13,21], low flammability [2,22–24], and resistance to corrosion [20,25].

Many industrial applications have emerged since the discovery of organo-silicates. Particularly, the ability of organo-clays to swell and delaminate in organic solvents has led to their widespread use as rheological control agents; as additives in paints, greases, inks, and oil-well drilling fluids; as reinforcement/filler in plastics and rubber; and as additives to enhance flammability of plastics [23,24]. Now organo-clays, especially organo-montmorillonite (MMT), are the most widely utilized silicates in polymer nanotechnology. In clay structures of MMT type, the single silicate layers have a thickness of about 1 nm and typically lateral dimensions of several hundred nanometers up to micrometers. These layers are arranged in stacks with regular interstices called interlayer galleries. MMT is a layered silicate, crystal lattice, which consists of two silica tetrahedral and one aluminum octahedral sheet with a plate-like structure of 1 nm thickness and 100 nm length [26]. The isomorphous substitution of central metal ion Al^{3+} by lower-valent metal ions, i.e., Mg^{2+}, Mn^{2+}, results in a charge imbalance on the surface of each platelet. The negative charge imbalance is neutralized by the absorption of hydratable cations, i.e., Na^+, Ca^{2+}, being responsible for the hydrophilic nature of MMT [8]. Physical and chemical structures, and molecular dynamics of surface-modified clay, especially alkane monolayers self-assembled on mica platelets, were investigated in detail by Osman et al. [27,28] using XRD, DSC, FTIR, and NMR spectroscopy methods. According to the authors, the alkyl chains preferentially assume an all-*trans* conformation at low temperatures, leading to a highly ordered two-dimensional lattice. With increasing temperature, the all-*trans* conformation is gradually transformed to a mixture of *trans*- and *gauche*-conformers.

MMT has hydrophilic surface with exchangeable metal ions and does not disperse very well in organic polymers, especially in hydrophobic polymers such as polyolefins (POs), polystyrene, halogen-containing polymers, etc. Because MMT undergoes surface modification with a wide range of functional organic compounds, including predominantly long chain alkylamines, its reactive compatibility (preferably via H-bonding) with conventional polymers and their maleic anhydride (MA) (or its structural analogs) functionalized derivatives can be improved through ion exchange reactions. These surface-modified MMTs are generically referred to as organo-MMT, nanoclay, or intercalant. To synthesize hydrophobic and nonpolar PO nanocomposites, many researchers typically use MA and/or its isostructural analogs or other functional monomers for the preparation of chemically modified POs through grafting or graft (co)polymerization reactions, which can easily intercalate the organo-modified MMT [29–31]. In general, PO nanocomposites can be prepared through melt (predominantly by reactive extrusion) intercalation process using thermoplastic polymers as matrix polymers and their MA (or other functional monomers)-grafted derivatives as reactive compatibilizers, and various types of organo-silicates as intercalants.

In the last decade, considerable progress was observed in the field of PO/compatibilizer (predominantly on the base of PO-*g*-MA)/organo-surface-modified clay nanocomposites. Polyethylene (PE), polypropylene (PP), and ethylene–propylene (EP) rubber are one of the most widely used POs as matrix polymers in the preparation of nanocomposites [3,4,6,30–52]. The PO silicate/silica (other clay minerals, metal oxides, carbon nanotubes, or other nanoparticles) nanocomposite and nanohybrid materials, prepared using intercalation/exfoliation of functionalized polymers in situ processing and reactive extrusion systems, have attracted the interest of many academic and industrial researchers because they frequently exhibit unexpected hybrid properties synergistically derived from the two components [9,12,38–43]. One of most promising composite systems are nanocomposites based on organic polymers (thermoplastics and thermosets), especially POs, functionalized POs, and inorganic clay minerals consisting of silicate layers [15,16,29,31,33,43–52].

One the other hand, MA and its isostructural analogs (fumaric acid and its amides ad esters, N-substituted maleimides, citraconic and itaconic acids/anhydrides and their esters, amides and imides, and other unsaturated dicarboxylic acids and their functional derivatives) as polyfunctional electron-acceptor monomers, which have difficulty copolymerizing while easily copolymerizing with donor-acceptor type of vinyl, allyl, and acrylic monomers, are being widely used in the synthesis of reactive macromolecules with linear, hyperbranched, cross-linked, and self-assembled structures, and high-performance engineering and bioengineering materials by their radical copolymerization [53–61], complex-radical alternating copolymerization [62–73], radical controlled/living copolymerization [74–87], cyclocopolymerization [88–93], and terpolymerization [94–103] reactions. These functional monomers are also used for the graft modification of various thermoplastic polymers (POs, copolymers of α-olefins, polystyrene, etc.), natural polymers and synthetic rubber, etc. In the last decade, grafting of MA onto various thermoplastic polymers and preparation of high-performance engineering materials and nanocomposites by using reactive extruder systems and in situ compatibilization of polymer blends have been significantly developed, some results of which have important commercial applications. Thus, the chemistry of the matrix PO polymers, functionalized POs—compatibilizers and organic surface-modified clays—as well as in situ processing conditions (temperature, shear rate, screw configuration, screw speed, residence time, etc.) are important factors in the formulation of PO nanocomposites.

In this review, we will predominantly describe the various aspects of PO/organo-silicate (or silica) nanocomposites and the most significant results reported recently in this field. General principle of preparing the PO nanocomposites through reactive extrusion in situ processing can be schematically represented as follows (Figure 4.1).

As seen from this scheme, functionalized POs play an important role as reactive compatibilizers in the preparation of PO nanocomposites and various PO/thermoplastics (synthetic and natural polymers) reactive blends and their hybrid nanocomposites in melt by reactive extrusion in situ processing. Therefore, before discussing the results on PO nanocomposites, it is necessary to describe here some types and methods (predominantly extrusion methods) for the synthesis of reactive polymer compatibilizers by the functionalization of POs with MA and its isostructural analogs.

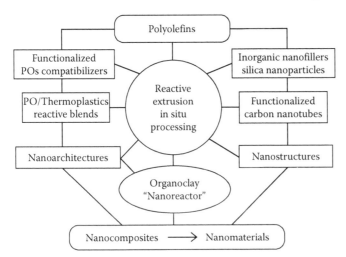

FIGURE 4.1
General schematic of the synthetic routes of PO nanocomposites through reactive extrusion in situ processing.

4.2 Functionalization of Polyolefins

History of graft modification of POs, especially PP with MA dates back to the 1969s [104] when a method was developed for reacting MA on a particular isotactic PP below its melting point. Since the 1960s, dibutyl maleate and acrylic acid have been grafted onto POs in screw extruders [105]. Ide and Hasegawa [106] have reported grafting of MA onto isotactic PP in the molten state using benzoyl peroxide as an initiator and a Brabender plastograph. This graft copolymer was used in blends of polyamide-6 and PP as a reactive compatibilizer. Swiger and Mango [107] studied the grafting of MA onto low density polyethylene (LDPE) backbone in the presence of dicumyl peroxide (DCP) in the melt in a bath reactor. Cha and White [109] reported MA modification of POs in an internal mixer (batch reactor) and a twin extruder. They have presented a detailed kinetic model for grafting processes in a reactor and twin-screw extruder systems. For the evaluation of grafting mechanism, functionalization of PP with MA have been carried out both in solution [106,109–113] and in the molten state [109,110,114–118] using various extruder systems. Taking into consideration low costs and operating facility, such grafting reactions were preferably carried out in the melt via reactive processing.

4.2.1 Grafting in Melt by Reactive Extrusion

The modification of POs in twin-screw extruder systems is achieved to produce new materials is an inexpensive and rapid way to obtain new commercially valuable polymers. This reactive extrusion technology is an increasingly important method of producing sizable quantities of modified polymers for various industrial applications [32,108,112]. The most investigations were related with MA grafting of PP and PE in the melt using various types of extruders and mixers [108,109,111–115,117–126]. Now twin-extruders act as continuous flow reactors for polymers playing an important role in functionalization of POs [127–134] to produce high-performance engineering PO thermoplastics, nanocomposites, and nanomaterials. A reactive extrusion process for the functionalization (free radical grafting) of PP with MA in the presence of supercritical carbon dioxide was studied by Dorscht and Tzoganakis [127]. They observed that the use of carbon dioxide led to improved grafting when high levels of MA were used. The influence of residence time on degree of MA grafting onto PP in an internal mixer and a twin-screw extruder was studied by Cha and White [108] through measuring reaction yields with respect to reaction time in the internal mixer as well as along the screw axis in the extruder using various peroxide-type initiators. Bettini and Agnelli [117,118] investigated the effects of the monomer and initiator concentrations, rotor speed and reaction time on the grafting reaction of MA onto PP. According to the authors, the increase in rotor speed leads to better mixing of MA in the reaction mass, an increase in the production of macroradicals and sublimation of MA monomer, due to the increase in reaction temperature.

Periodical and patent publications on the synthesis of PO graft copolymers including POs with grafted MA, maleates, fumarates, maleimides, etc., by reactive extrusion or other forms of melt phase processing were summarized and discussed by Moad [132]. Publications in this area through 1990 were also summarized and described in the book edited by Xanthos [128].

The modification of PE through free-radical grafting of itaconic acid onto LDPE by reactive extrusion have also been reported. Thus, many aspects of the grafting mechanism of itaconic acid (IAc) onto PO macromolecules have been investigated by

Jurkowski et al. [135–140]. Authors determined and optimized the most important formulation and technological factors so as to control the course of the grafting process, the grafting efficiency, the course of the by-process including cross-linking of PE macromolecules and the poly(PE-*g*-IA) physical structure. IAc was grafted onto PP/LDPE blends by Pesetskii et al. [121] in melt with 2,5-dimethyl-2,5-di(*tert*-butyl peroxy)-hexane as a initiator using a reactive extrusion. According to the authors, grafting efficiency increased by introducing LDPE into PP.

The synthesis of the graft copolymers of PP (powder and granular) with acrylic acid (AA), maleic and citraconic (CA) anhydrides, and poly[PP-*g*-(MA-alt-AA)] using grafting in solution and reactive extrusion techniques and their main characteristics, as well as results of the structure–composition–property relationship have been reported by Rzayev et al. [122,123,141–145]. The structure, macrotacticity, crystallinity, crystallization, and thermal behavior of synthesized i-PP grafts depend on the monomer unit concentration in polymers. Although reactivity of CA radicals is lower than MA, grafting reaction with CA proceeds more selectively than those for MA. This can be explained by relatively low electron-acceptor properties of citraconic double bond due to a CH_3 group in the α-position, as well as by steric effect of this group [145]. This inhibition effect of chain scission of the α-methyl groups may be due to the formation of quaternary carbon atom in CA-grafted linkage as shown in the following scheme (Figure 4.2) [144,145].

We have reported [122,123,141–144] the synthesis of poly(PP-*g*-MA)s with different compositions as the precursor for the preparation of nanocomposites by radical grafting reaction of powder and granular PP with MA in melt by reactive extrusion using DCP as initiator. It was demonstrated that the structure, macrotacticity, crystallinity, crystallization rate, and thermal behavior of PP changed with grafting and depended on the grafting degree. The MA grafting efficiency of powder PP was higher than that obtained for the granular form of PP. We have also been investigating the grafting of MA onto powder PP in the thermal oxidative conditions using reactive extrusion [122,123]. In these conditions, the degradation of PP is not a controllable process and essentially depends on many factors, such as temperature, screw speed, and other extrusion parameters. The introduction of MA into this system allows the process to reasonably control the chain degradation reactions [123].

The above-mentioned investigations of free-radical grafting of MA and its isostructural analogs onto different POs (LDPE, HDPE, PP, EPR, EPDM, etc.) indicated that side reactions (cross-linking, degradation, oligomerization of monomer, β-scission of C–C bonds in

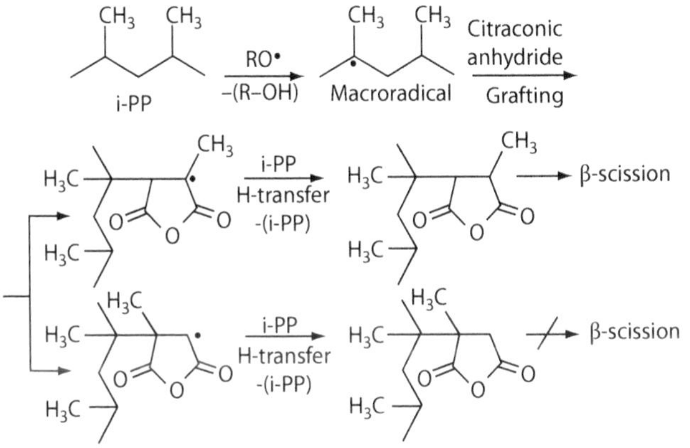

FIGURE 4.2
Schematic representation of an inhibition effect of chain scission in grafting of citraconic anhydride onto isotactic polypropylene. (Reproduced from Devrim, Y., *Polym. Bull.*, 59, 447, 2007.)

the main chain, etc.) occurred simultaneously with the grafting reactions, which caused some changes in the molecular structure of POs, depending on the type of polymers used. The use of various types of reactive extrusion systems in the grafting process provides important advantages in the production of high-performance commercial materials, especially PO nanocomposites and nanomaterials.

4.2.2 Graft Copolymerization

To increase the grafting levels onto POs, mixtures of different monomers, including donor–acceptor monomer mixture of styrene and MA, were utilized [146,147]. It was reported that there are some monomers, predominantly styrene, capable of acting as a donating electron to MA, leading to an enhancement of MA grafting efficiency onto POs. Spontaneous bulk polymerization of MA/styrene mixture in the presence of POs, such as PE, PP, poly(ethylene-*co*-propylene), *cis*-1,4-polybutadiene, may be carried out in various types of mixing apparatus and extruders [148–153]. Dong and Liu [146] studied the styrene-assisted free-radical graft polymerization of MA onto PP in supercritical CO_2. They showed that the addition of styrene drastically increased the MA functionality degree, which reached a maximum when the molar ratio of MA and styrene was 1:1. According to the authors, styrene, an electron-donor monomer, could interact with MA through charge-transfer complexes to form the alternating copolymer, which could then react with PP macroradicals to produce branches by termination between radicals. Li et al. [147] also found that the addition of styrene as a second monomer in the melt-grafting system assisted in increasing the grafting degree of MA on PP. According to the authors, supercritical graft copolymerization is an advantageous process compared with known conventional methods (melt-grafting technology, solution graft copolymerization, etc.). Abd El-Rehim et al. [153] reported the radiation-induced graft copolymerization of MA and styrene with PE, and some reactions of graft copolymers with metal salts and amine-containing compounds. The effect of various donor–acceptor-type comonomers on grafting conversion of MA onto LLDPE [82,149,154] and PP [146,155–157] was also studied. It was observed that grafting yields to PP decreases in the following raw: styrene ≫ α-methylstyrene > MMA > vinyl acetate > (no comonomer) > *N*-vinyl-2-pyrrolidone. It was proposed that higher yields can be explained by the formation of a charge-transfer complex between comonomers and MA [63,132,156].

4.3 Polyolefin Reactive Blends and In Situ Processing

In polymer blends, immiscibility of the components results in incompatibility of the phases. Compatibility and adhesion between different polymeric phases can be improved by the addition of suitable block or graft copolymers that act as interfacial agents. An alternative is to generate these copolymers in situ during blend preparation through polymer–polymer graft reactions by using functionalized polymers [158–161]. Recently, extruders have increasingly been utilized as chemical reactor [162–164]. A new trend, called "reactive extrusion," has been developed in the technology of polymer production and processing. No matter in what reactor the chemical process occurs, it is subject to the basic thermodynamic laws [165].

4.3.1 Polyolefin/Polyamide Reactive Blends

POs functionalized with MA exhibit enhanced adhesion to polar materials like polyamide (PA), metals, and glass, and are used as compatibilizing agents. For instance, the improvement of impact properties upon blending of PA-6 with various maleated POs is associated with the formation of a polyamide (PA)-PO graft copolymer in situ during melt blending [166–171]. Pesetskii et al. [139] reported preparation and properties of reactive blends of poly(PE-*g*-itaconic acid) with polyamide 6 (PA6). They show that IA grafted onto PE improves adhesion of interphases in the blends, increases the impact strength of the materials, and improves their processability. According to the authors, variations in the ratios of the polymers in the PP/LDPE-*g*-IA systems led to both nonadditive and complex changes in the viscoelastic properties as well as mechanical characteristics for the composites. Abacha and Fellahi [168,169] reported the synthesis of PP-*g*-MA and evaluation of its effect on the properties of glass fiber reinforced nylon 6/PP blends. They found that the incorporation of the compatibilizer (MA grafted PP) enhances the tensile strength and the modulus, as well as the Izod impact properties of the prepared blends. The reactions of MA units in graft copolymers with the amine end groups of PA6 and PA66 were studied in detail by Marechal et al. [166] and well documented by many researchers. Blends of industrially relevant polymers such as PA6, poly(E-*co*-P) rubber, poly(E-*co*-P)-*g*-MA (80/20/0–80/0/20; w/w/w) [170–173] were selected for the studies of the changes in the rheological behavior, morphology, and chemical conversion of along a compounding co-rotating twin-screw extruder [173,174]. Authors demonstrated that the dynamic viscosity and storage modulus decrease substantially upon melting of the components, since this process is induced by a significant increase in interfacial area through strong H-bonding as a result of in situ grafting reaction of amine (from PA6) and anhydride (from maleated rubber) groups during extrusion processing.

Wu et al. [175] investigated the effects of PP-*g*-MA (0.6%) compatibilizer content on the crystallization of PA12/PP blends and their morphology. They found that an in situ reaction occurred between the MA units of compatibilizer and the amide end groups of PA12. According to the authors, the in situ interfacial reaction in the modified blend component resulted in compatibilization connected with higher finely dispersed blend morphology and the appearance of fractionated crystallization. Tjong and Meng [173,176,177] used reactive PP-*g*-MA as a compatibilizer for PA6/liquid crystalline copolyester (LCP) and PP/LCP blends and prepared PA-6/LCP/PP-*g*-MA blends by extrusion blending of pellets followed by injection molding at 295°C. They prepared MA compatibilized blends of PP and LCP either by the direct injection molding (one-step process), or by twin-screw extrusion blending, after which specimens were injection molded (two-step process). Filippi et al. [178–180] reported the reactive compatibilization of PA blends with commercial HDPE-*g*-MA containing different amount MA unit, and some thermoplastic elastomers grafted with MA [poly(SEBS-*g*-MA)s]. They found that all these compatibilizer precursors (CPs) react during blending, with the functional groups of PA to produce CP-*g*-PA copolymers, though different kinetics and at different yields. The results of authors confirm that the anhydride functional groups possess considerably higher efficiency, for the reactive compatibilization of LDPE/PA blends, than that of the ethylene-acrylic acid and ethylene-glycidyl methacrylate copolymers. Chen and Harrison [181] used six different CPs, including two grades of poly(SEBS-*g*-MA), to compatibilize 80/20 blends of PE with an amorphous PA with the aim of producing PE films reinforced with PA fibers for balloon applications. Armat and Moet [182] showed that poly(SEBS-*g*-MA) is good CP for the reactive compatibilization of 25/75 LDPE/PA blends.

4.3.2 Polyolefin/Rubber Reactive Blends

Supertough blends of PA (nylon-6) with maleated PO elastomers, such a MA-grafted ethylene-propylene rubber [poly(E-*co*-P-*g*-MA)] and styrene/hydrogenated butadiene/styrene triblock copolymer as the high performance engineering materials, have become commercially important materials of considerable scientific interest [183–187]. Borggreve et al. [187] used a MA-grafted ethylene-propylene-diene rubber (EPDM) to reach the same goal. An essential feature of these materials in the graft copolymer generated from the reaction of the grafted MA unit with polyamide amine end groups during the melt-blending process. Many researchers [188–195] investigated the poly(SEBS-*g*-MA) reactive compatibilization of i-PP/PA blends by mechanical, morphological, thermal, and rheological analyses. Huang et al. [195] showed that the twin-screw extruder produced smaller particles with a more narrow distribution of sizes than the single-screw extruder.

The compatibilization of PO elastomers/PA blends involving the grafting of MA onto the elastomer phase was also reported by several researchers [196–198]. It was shown that MA-grafted elastomers improved the impact strength and reduced the dispersed phase size in those blends. Maleation of metallocene poly[ethylene-*co*-1-octene (25 wt.%)] (PEO elastomer) carried out with DCP as an initiator at 200°C and 120 rpm in melt by co-rotating twin-screw extruder by Yu et al. [199]. They reported the effect of the addition of grafted PEO on the impact strength, yield strength, and modulus of nylon 6, and demonstrated that the incorporation of poly(PEO-*g*-MA) ensured good compatibility between nylon 11 and PEO particles. The effects of the MA graft ratio and functionality of a series PEO on the mechanical properties and morphology of binary nylon-11/PEO-*g*-MA and ternary nylon-11/PEO/PEO-*g*-MA reactive blends were investigated in detail by Li et al. [200]. According to the authors, the optimal MA content for the preparation of blend with maximum value of Izod impact strength (~800 J/m) is 0.56%.

Melt grafting of poly(ethylene-*co*-propylene-*co*-norbornene) (EPDM) with MA and the use of EPDM-*g*-MA as compatibilizer in polymer blends was reported to be successful in improving the properties of the blends [201]. Enhancement of the impact strength of PP/EPDM blend upon grafting the EPDM with MA {poly[EPDM-*g*-MA(1.5 wt.%)]} was reported by Zhao and Dai [202]. Gonza'les-Montiel et al. [193] observed that the effect of MA grafting level in polyamide-6/poly(E-*co*-P) rubber (EPR) blend to result in improved mechanical properties and morphology of the blend. Similar increase of interfacial adhesion in PP/EPDM/poly[EPDM-*g*-MA(0.3 wt.%)] blends was observed by Purnima et al. [203]. van Duin et al. [204] demonstrated that melt blending of PA6 and PA6,6 with poly(EPDM-*g*-MA) and poly(St-alt-MA) results in the coupling of PA with MA-containing polymer via an imide linkage and is accompanied by PA degradation. They showed that the degree of crystallinity of the PA phase is decreased only when the size of the PA phase between the MA-containing polymer domains approaches the PA crystalline lamellar thickness.

4.3.3 Polyolefin/Natural Polymers Reactive Blends

POs are widely used as a matrix in wood fiber-plastic composites [205–207]. MA graft copolymers, such as maleated PPs, are applied as coupling agents, for the preparation of composites useful in the automobile and packaging industries [208,209], and for outdoor application because when maleated PP is used, the fiber surface becomes hydrophobic and the water uptake decreases significantly [209]. In situ chemical modification–interfacial interaction was observed by several researchers [82,210,211] in the maleated

PO/cellulose (or starch) blend composite systems. The properties of such composites depend on their interfacial adhesion as well as properties of the individual components. Tasanatanachai et al. [209] prepared the high oxygen permeability film, so-called breathable film, by reactive blending and processing of LLDPE, natural rubber, and epoxidized polyisoprene with a small content of MA as a reactive coagent. Authors proposed that the prepared films may be utilized for the agricultural applications to enhance growth and production of fruit trees, flowers, vegetables, and even grains, especially during the growing stage of plants.

4.4 Polyolefin/Organo-Silicate Nanocomposites

PO–clay nanocomposites have not been reported up to 1996 because the silicate layers of the clay having polar hydroxy groups are incompatible with hydrophobic POs. In the last decade, considerable progress has been achieved in the field of PO/compatibilizer (predominantly on the base of MA-grafted POs)/organo-surface-modified clay nanocomposites. PE, PP, and EP rubber are one of the most widely used POs matrix polymers in the preparation of nanocomposites. Initial works on the PO/ layered silicate nanocomposite preparation have been carried out in recent years [2–6,15,18,31,33,43,45,49–51,212–244]. One of the most promising composite systems is nanocomposites based on POs, including MA-grafted POs, and inorganic clay minerals consisting of silicate layers. According to Beecroft and Ober [212], the formation of nanocomposites is enabled by the use of organo-modified clay material in which the counter ions of the silicate layers located in the interstitial regions are exchanged with bulky alkylammonium ions; such organo-modified clays are available commercially in several variations, tailor-made for different applications, and compatible with different polymeric materials. Authors proposed that due to the small dimensions of the filler particles and the resulting high surface-to-volume ratio, nanocomposites typically have, even at low filler concentrations, a high fraction of interfacial regions with a major influence on the physical properties and, therefore, on the material performance. Böhning [49] investigated polymer/clay nanocomposite materials based on PP-*g*-MA and two different organophilic modified clays by dielectric relaxation spectroscopy. They found that in contrast to ungrafted PP, MA-grafted PP shows a dielectrically active relaxation process. Authors assumed that the MA groups, which act as compatibilizers enabling the clay dispersion in the nonpolar matrix, are preferentially located in the interfacial regions between polymer and clay sheets, which indicate a significantly enhanced molecular mobility in those regions.

Many attempts to synthesize PE [23,28,33,37,213–217,232–234,237–244] or PP [5,11,30–32, 47–52,123,218–231,235–272] nanocomposites have been described by many researchers. It was earlier observed that, because of hydrophobic nature of PP, the modification of the silicate was not sufficient to produce intercalated or exfoliated structures. Thus, in the majority of the studies, a compatibilizer containing polar groups, such as PP-*g*-MA [47,49,123–127,247,252,265–273] is generally introduced as the third component to compensate the polarity difference between PP and the filler. Both the strong interaction of the compatibilizer with the nanolayers and its miscibility with the PP polymer are very important factors for achieving exfoliation and homogeneous dispersion of the nanolayers in the hybrids.

4.4.1 Polyethylene Nanocomposites

PE is one of the most widely used PO for the preparation of nanocomposites. Alexandre et al. [236] reported the preparation of PE/layered silicate nanocomposites by the polymerization-filling technique. Shin et al. [33] used bifunctional organic modifiers to prepare PE/clay hybrid nanocomposites by in situ polymerization. Zhang and Wilkie [23] obtained low-density PE/clay nanocomposites with good flammability properties via melt mixing in a Brabrender mixer. Wang et al. [237–239] prepared linear low-density PE/clay nanocomposites by grafting the polar monomer MA to the backbone. Gopakumar et al. [240] also obtained PE/clay nanocomposites by direct-melt intercalation and studied its isothermal crystallization kinetics. Preparation and characterization of poly(PE-*g*-MA-*co*-styrene)/ MMT nanocomposite were reported by Zhou et al. [241,242]. Because of the difference in character between PE and inorganic MMT, authors used two steps for the preparation of nanocomposites: (1) choosing organic MMT and (2) modifying PE via grafting with MA and styrene together. They showed that the crystallization rate of PE improved when PE was grafted with MA and styrene; the thermal stability improved, and the tensile strength of the nanocomposites reached the maximum value (31.7 MPa) when the loading of organic MMT was 3 wt.%.

Recently, Sibolda et al. [243] used the tailored compounds, based on alkyl-substituted derivatives of succinic anhydride, acid and dipotassium salt to evaluate their role for intercalation of MMT clay and the formation of nanocomposites based on PE in the presence of PE-*g*-MA as compatibilizer. They found that these tailored compounds can be intercalated into the layers of MMT and expansion of the lattice was confirmed by wide angle x-ray diffraction and FTIR spectra. Masenelli-Varlot et al. [244] proposed a procedure to quantitatively characterize the microstructure of PE-75/PE-*g*-maleic acid (Mac)-20/organo-MMT-5 nanocomposites and to study their structure–mechanical properties relationship using TEM, wide angle x-ray scattering (WAXS) and dynamic mechanical analysis methods. PE-*g*-Mac graft copolymer (Mn = 10,400 g/mol) was used to ensure better interaction between the clay and PE. Organo-MMT (Nanofill 15, Süd Chemie Co.), surface modified with dimethylditallow alkyl ammonium (cation exchange rate 30%) was used as a nanofiller. They prepared biaxially oriented nanofilms by melt intercalation using a twin-screw extruder system and then by blow extrusion into a 100 µm thick film. According to the authors, a microstructure of nanocomposites in terms of (1) the dispersed state in clay, (2) relative orientation of clay and PE crystalline lamellae, and (3) PE/clay interfacial adhesion strength, which would lead to the highest reinforcement below and above glass-transition temperature (Tg). They explained this fact by a network formation of the organo-MMT platelets and the PE crystalline lamellae.

4.4.2 Polypropylene Nanocomposites

Though PP finds extensive use in various applications, its use is limited because of poor gas permeability characteristics and low thermal and dimensional stability [247]. First Usuki et al. [11–13,48,247,248] reported a novel approach to prepare a PP–clay nanocomposite by using a functional oligomer, MA-modified oligopropylene (acid number 52 mg KOH/g, softening point 145°C and Mw = 30,000 g/mol) by GPC as a compatibilizer. In the last case, the authors used organophilic clay, i.e., organo-MMT, silicate layers of which were intercalated with stearyl ammonium. They demonstrated that MA-modified oligopropylene improved the dispersibility of the clays in the PP-clay composites; the dynamic storage modulus of these composites is higher than PP up to 130°C, and 1.8 times higher than that of PP at 80°C. According to the authors, the following critical factors are important for clay dispersion: strong interaction of organo-MMT with PP-*g*-MA and compatibility of the MA-grafted PP macromolecules with the matrix PP. They [11,12,247–249] also found that functionalized POs intercalate into the galleries of organophilic clay, and they successfully prepared PP-clay nanocomposites based on MA-modified PP and organophilic clay. In the PP-clay nanocomposites, the silicate layers of the organophilic clay exfoliate and homogeneously disperse at the nanometer level in the PP matrix. This approach, introducing MA units to nonpolar polymer chains, is applicable to other POs, such as PE, E-P, and E-P-diene rubbers.

Many researchers prepared three types of PP-*g*-MA-layered silicate nanocomposites with different dispersion states of layered silicate (deintercalated, intercalated, and exfoliated states) from two kinds of PPs with different molecular weights, organic modified layered silicate to study the effect of the final morphology of the nanocomposites on the rheological and mechanical properties [222–230,248]. Recently, Xu et al. [51] presented a systematic effort to understand and control the structure in PP-layered silicate nanocomposites by utilizing a PP-grafted MA compatibilizer. They showed how one can control the structure, from a phase-separated PP microcomposite to a completely exfoliated nanocomposite, by varying the ratio of PP-*g*-MA compatibilizer to organoclay.

Okamoto et al. [11] described the formation of a house of cards structure in PP/clay nanocomposite melt under elongational flow by TEM analysis. According to the authors, both strong strain-induced hardening and rheopexy features are originated from the perpendicular alignment of the silicate layers to the stretching direction. A similar effect was observed by other researchers: Xu et al. [250] for oriented PP/MMT clay nanocomposites using solid-state NMR and TEM to quantitatively examine the evolution of clay morphology upon equibiaxial stretching. La Mantia et al. [251] observed a pronounced increase of the mechanical properties of PE/clay nanocomposite drawn fibers due to a flow-induced intercalated/exfoliated morphology transition.

Mittal et al. [252] reported organo-clay exfoliation and gas permeation properties of PP/PP-*g*-MA/clay nanocomposites prepared in the presence of organo-MMTs with different surface areas exchanged by alkylammonium ions, carrying alkyl chains of different length. They observed an increase in the basal-plane spacing (*d*-spacing) of the organo-MMTs and their exfoliation, and gas-barrier properties of composites with increasing length and number of the alkyl chains. The oxygen permeation coefficient of the nanocomposites was found to be a nonlinear function of the volume fraction of the inorganic part of the organo-MMT.

The results of nonisothermal crystallization kinetics of PP and PP-*g*-MA nanocomposites, studied by DSC at various cooling rates, showed that maleated PP could accelerate the overall nonisothermal crystallization process of PP [253]. Ton-That et al. [254,255] indicated that the crystallization of the PP matrix played a significant role in the intercalation of organo-MMT (Cloisite 15A clay preintercalated with dimethyl dihydrogenated tallow ammonium chloride) in the matrix. According to the authors, the exfoliation of PP nanocomposites is complex, and in order to achieve exfoliation by melt compounding, one needs to overcome different challenges in terms of chemistry, thermodynamics, crystallization, and processing. Authors concluded that for obtaining nanocomposites with better dispersion/performance, it is necessary to increase the gallery distance of the clay (probably not smaller than 20 Å) prior to melt compounding. This favors the penetration of compatibilizer and PP macromolecules into the interlayer of the clay during melt mixing. Ratnayake and Haworth [256] studied the influence of low-molecular-weight polar additives, such as amide-type slip agent and PP-*g*-MA (1 wt.%) reactive compatibilizer, on PP-clay (surface modified with dimethyl dihydrogenated tallow ammonium) nanocomposites, in terms of intercalation and degree of exfoliation achievable by melt-state mixing processes. According to the authors, the interaction between polar amine group of this additive and the polar sites on the clay-filler surface appears to be the driving force for the intercalation. Peltola et al. [257] examined the effect of screw speed of the co-rotating twin-screw extruder on the clay exfoliation and PP/oligo-PP-g-MA (softening point 154°C–158°C, acid number 37–45 mg KOH/g, and density 0.89–0.93 g/cm³)/organo-MMT (Nanofiller 1.44P) nanocomposite properties. The main result of this study was that nanocomposites showed both intercalated and exfoliated structures depending on the screw speeds of extruder; the dispersion of silicate layers was greatly influenced by the screw speed. However, as noted by the authors, even when the silicate layers were highly exfoliated, there was no remarkable effect on mechanical properties of the nanocomposite.

According to Utracki and Simha [4,32], owing to the wide variety of commercially available PP-g-MA grades (molecular weight, polydispersity, number of MA groups per chain, contamination by coreactants, homopolymerized MA, etc.), its selection for polymer nanocomposite compatibilization is far from simple. According to the authors, the main concern is the balance between the degree of maleation and molecular weight. The use of primary ammonium ions may be important as three hydrogens of a RNH_3^+ cation may react with the MA group (covalent bonding). They also studied the pressure–volume–temperature dependencies of PP melt and PP/organo-MMT nanocomposites to determine compressibility and the thermal expansion coefficient [32]. Nanocomposites containing 0, 2, and 4 wt.% of organo-MMT (Cloisite-15A, or C15) and 0, 2, and 4 wt.% of a compatibilizer (PP-g-MA or PP-g-glycidyl methacrylate) with anhydride or epoxy functionalities, which can easily interact with clay via its –OH or amine groups, were investigated at 177°C–257°C and a pressure of 0.1–190 MPa. The incorporation of organo-MMT (2 wt.%) into PP resulted in the reduction of specific volume by $\Delta V \approx 1\%$, but that of free volume (hole) fraction by $\Delta h \approx 5\%$.

Kurokawa et al. [258–260] developed a novel but somewhat complex procedure for the preparation of PP/clay nanocomposites and studied some factors controlling mechanical properties of PP/clay mineral nanocomposites. This method consisted of the following three steps: (1) a small amount of polymerizing polar monomer, diacetone acrylamide, was intercalated between clay mineral [hydrophobic hectorite (HC) and hydrophobic MMT clay] layers, surface of which was ion exchanged with quaternary ammonium cations, and then polymerized to expand the interlayer distance; (2) polar maleic acid-grafted PP (m-PP), in addition was intercalated into the interlayer space to make a composite (master batch, MB); (3) the prepared MB was finally mixed with a conventional PP by melt twin-screw extrusion at 180°C and at a mixing rate of 160 rpm to prepare nanocomposite. Authors observed that the properties of the nanocomposite strongly dependent on the stiffness of clay mineral layer. Similar improvement of mechanical properties of the PP/clay/m-PP nanocomposites was observed by other researchers [50,261].

Tang et al. [50] reported synthesis and thermal parameters of PP/MMT nanocomposites. Zhang et al. [262] prepared PP/MMT nanocomposites by melt intercalation using organo-MMT and conventional twin-screw extrusion. The silicate layers of MMT clay were dispersed at the nanometer level in the PP matrix, as revealed by x-ray and TEM results. They found that the impact strength is greatly improved at lower content of MMT. Parija et al. [263] utilized PP, PP-g-MA (1%), and organophilic MMT (Nanomer 1.30P) as base polymer, compatibilizer, and reinforcing agent, respectively, for nanocomposite preparation. According to the authors, nanomer and compatibilizer loadings play an important role in producing PP/layered silicate nanocomposites with superior mechanical properties and lower density by separating nanolayers through mechanical shear.

Morgan and Gilman [264] analyzed several polymer-layered clay (melamine ammonium or dimethyl dehydrogenated tallow ammonium-modified MMT) nanocomposites, including thermoset (cynate esters) and thermoplastics (polystyrene, polyamide, and maleated PP), by TEM and XRD in an effort to characterize the nanoscale dispersion of the layered silicate. Authors showed that the overall nanoscale dispersion of the clay in the polymer is best described by TEM analysis. Wenyi et al. [265] prepared Organo-MMT/PP-*g*-MA/ PP nanocomposites by melt blending with twin-screw extruder. Authors showed that the mechanical properties, melting point, crystallization temperature, thermal stability, and the total velocity of the crystallization of nanocomposites increased with the addition of organo-MMT, which was dispersed and intercalated by macromolecular chains in the matrix on the nanometer scale.

Tidjani et al. [266,267] prepared PP-*g*-MA-based nanocomposites and described the influence of the presence of oxygen during melt blending/preparation, causing a partial degradation of the organic cations used to modify the clay fillers, on the resulting polymer/layered silicate nanocomposites as well as on their thermal stability. Ramos et al. [268] prepared nanocomposites based on isotactic PP and modified bentonite clay by melt intercalation. They carried out thermooxidation of compression-molded films of the nanocomposites at 110°C for up to 165 h. According to the authors, the PP compounds based on the modified clay had a higher thermal stability in the solid state than those with the pristine clay due to a higher dispersion of clay particles reducing oxygen diffusion through the samples. However, authors observed that the thermal degradation of PP with the modified clay was more significant than the pure polymer. According to the authors, this can be resulted from the acidic nature of the clay, interaction between the clay and PP stabilizers, and decomposition of the organic salt during processing.

Tjong et al. [269–272] prepared MA-compatibilized PP hybrid nanocomposites reinforced with various fiber reinforcements, such as whiskers, liquid-crystalline polymers, and glass fibers, by in situ compounding in a twin-screw extruder. They also developed a novel approach for the preparation of polymer nanocomposites utilizing a reactive modifying reagent such as MA monomer that acts both as a modifying additive for the polymer matrix and as a swelling agent for the vermiculite silicate nanofiller. Authors observed that the tensile moduli and strengths of the nanocomposites tend to increase dramatically with the addition of small amounts of vermiculite clay. Toshniwal et al. [273] reported a method for the preparation of dyeable PP fibers via nanotechnology. They prepared PP/nanoclay (Cloisite-15A) nanocomposite fibers with three clay loadings (2, 4, and 6 wt.%) by solution mixing followed by melt spinning using a single- and/or twin-screw extruder. According to the authors, the modification of PP with organoclay increased the modulus and the strength of the PP nanofibers.

4.4.3 One-Step Method of Preparation

Rzayev et al. [123] developed a one-step method for preparing nanocomposites based on powdered isotactic polypropylene (i-PP) as a matrix polymer (melt-flow index (MFI) = 7.2 g/10 min), oligo(i-PP-*g*-MA) (MFI = 35–70 g/10 min) as a reactive compatibilizer and dodecylamine-surface-modified MMT clay. This method involves grafting MA monomer onto i-PP chains in the melt state under controlled thermal degradation conditions and intercalative compounding of the obtained oligo(i-PP-*g*-MA) with i-PP and organo-MMT by reactive twin-screw extrusion. The mechanism of formation and properties of PP/oligo(PP-*g*-MA)s/organo-MMT nanocomposites were investigated by FTIR, XRD, and thermal analysis (DSC, TGA, and DTA). The results indicated that the formation of nanostructured morphologies proceeded through the formation of strong H-bonding and amidization/imidization reactions between the anhydride units of exfoliated grafted i-PP chains and alkylamines within the organo-MMT interlayers. Synthesis and characterization of i-PP/oligo(i-PP-g-MA)s/organo-MMT silicate nanocomposites included the following important stages: (1) preparation of oligo(i-PP-g-MA)s with different composition by grafting MA onto powder i-PP besides using twin-screw reactive extrusion in the controlled degradation conditions, and their characterization; (2) synthesis of nanocomposites in melt by reactive extrusion under the given processing conditions; (3) study of nanocomposite composition–property (thermal and crystallization behavior) relationship; and (4) analysis of various models and mechanisms of formation of the intercalated and exfoliated nanoarchitectures through H-bonding and amidization/imidization in situ reactions [123,274–277]. Both the grafting and in situ modification reactions were carried out in melt by using a specially designed lab-size twin-screw extruder with the following temperature profile of the extruder zones: 165°C, 170°C, 175°C, and 180°C. The screw speed was set at around 14–30 rpm; grafting of MA onto i-PP was carried out in a wide range of i-PP/MA/DCP ratios in air under controlled degradation conditions using non-compounded powder i-PP while the process of the i-PP80/oligo(i-PP-g-MA)s15/organo-MMT5 nanocomposite preparation was realized by using compounded i-PP and MA-grafted i-PPs with different grafted unit contents (0.15, 0.86, and 1.22 mol%).

The formation of intercalated/exfoliated nanocomposite architectures was confirmed by the comparative FTIR, XRD, and DSC analyses of the i-PP-80/oligo(i-PP-g-MA)s-15/organo-MMT-5 nanocomposites and their individual components. The comparative analysis of nanocomposites and their individual components indicated that MFI values of the nanocomposites and graft oligomers significantly depended on the content of MA units. MFI decreased with the increasing amount of MA grafting or with the increasing fraction of MA monomer in i-PP/MA mixture undergoing grafting under thermal oxidative degradation conditions. DSC curves of oligo(i-PP-g-MA)s and their nanocomposites, including melting and recrystallization processes, showed that the T_m and T_c values, especially enthalpy of these transitions (ΔH), depended on the grafting degree of copolymers but these parameters insignificantly changed with the increasing MA content in the nanocomposites. However, the observed difference between these values ($\Delta = T_m - T_c$) for virgin i-PP (54.7°C) and nanocomposites (47.5°C) is significant. Thus, the recrystallization process proceeded relatively rapidly in the nanocomposites as compared to virgin i-PP and its grafted oligomers. This fact can be explained by the easy reorientation of nanostructural fragments in the studied i-PP-80/Oligo(i-PP-g-MA)s-15/Organo-MMT-5 nanocomposites. On the other hand, crystallinity (χ_c) of nanocomposites, calculated from DSC (enthalpy values of melting, ΔH) and XRD patterns, is higher (around 42%–45%) than those for i-PP (30.6%) and its grafted oligomers (around 33%–41%). This is a reasonable argument confirming the formation of highly oriented nanostructural architecture in the synthesized nanocomposites [123].

X-ray powder diffraction (XRD) was employed to check the effect of grafting on the crystallization and formation of nanostructural linkages. The following changes in XRD patterns were observed: (1) increase in the *d*-spacing of a strong silicate peak as compared to the original Organo-MMT clay from 25.08 to 34.76 2θ values, (2) shift of this peak to lower position of 2θ (from 3.47° to 1.56°), and (3) appearance of characteristic XRD peaks for the i-PP chains in the nanocomposite with relatively narrow widths and high intensity. Moreover, the relatively lower molecular weight (MFI = 61.0–35.2 g/10 min, Mw = 3010–3560 g/mol) of grafted copolymers compared to powder i-PP homopolymer (MFI = 7.2 g/10 min, Mw = 1.85 × 10⁵ g/mol, Petrochemical Inc., Izmir, Turkey) allows them to easily intercalate and

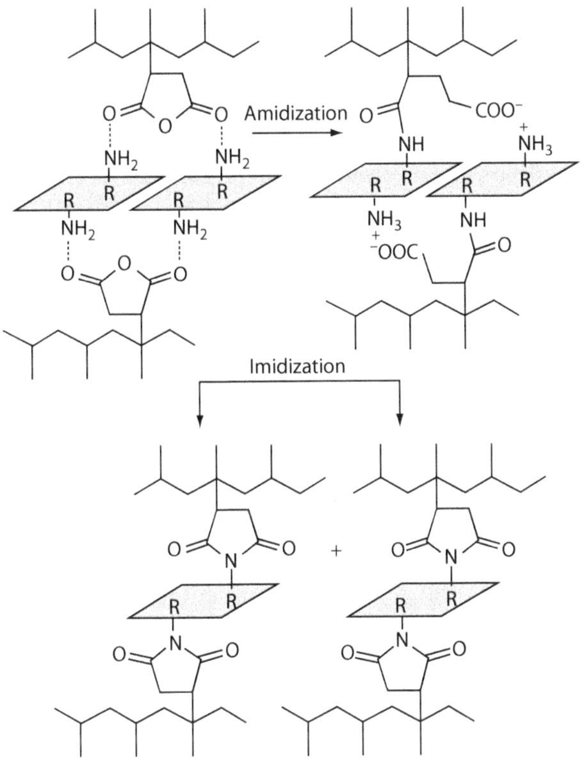

FIGURE 4.3
Schematic representation of the intercalation/exfoliation processing via strong H-bonding and amidization/imidization in situ reactions in PP/oligo(PP-*g*-MA)/organo-MMT nanosystem. (Reproduced from Rzayev, Z.M.O. et al., *Adv. Polym. Technol.*, 26, 41, 2007.)

exfoliate between silicate interlayer galleries, where their anhydride-containing macromolecules interact with surface alkyl amine through strong H-bonding and amidization/imidization in situ reactions under the chosen extrusion conditions. These in situ physical (H-bonding) and chemical (amidization/imidization) reactions can be schematically represented in the form of a structural model as follows [123] (Figure 4.3).

The results of several previous publications of other researchers can serve as model systems conforming the mechanism of nanostructural architecture formation described above. For example, Wouters et al. [278,279] studied ionomeric thermoplastic elastomers based on maleated ethylene/propylene copolymers, poly[(E-*co*-P)-*g*-MA]s, which exhibited good mechanical properties, but rather high melt viscosities due to very strong ionic interactions. Poly[(E-*co*-P)-*g*-MA] itself is unable to form hydrogen bonds, since the grafted MA groups do not possess an H-bond donor. An equimolar amount of a primary alkyl amine is used to open the anhydride rings, thereby forming an amide–acid structure [280], to introduce the possibility of intermolecular H-bonding to cross-link the rubber. Primary aliphatic amines were used because of their fast reaction in solution with anhydride groups at low temperatures, without the need for a catalyst. The reaction with amines has been used extensively for MA copolymers, such as poly(styrene-*co*-MA) [54,281,282] and poly(ethylene-*co*-MA) [54,283], usually to obtain maleimide copolymers. It is well known that the reaction of anhydride unit of MA copolymers with alkylamine proceeds via formation of amide-acid linkage, which then undergoes ring closure to form an imide structure [54]. Thus, Wermeesch et al. [281] reported the chemical modification of poly(styrene-alt-MA) with *N*-alkylamines by reactive extrusion. Sun et al. [284] thermoreversibly cross-linked the maleated ethylene/propylene copolymers [EPM-*g*-MA (2.1 wt.%)] via a reaction of grafted anhydride units with primary alkyl amines (C3–18) of different length. The effect of the type and amount of primary amine on the structure and morphology, studied by FTIR spectroscopy and small-angle x-ray scattering (SAXS), and the mechanical properties were investigated in order to obtain insight into the cross-linking mechanism. Authors showed that microphase-separated aggregates for both the starting graft copolymer and its alkylamide–acid and salts (complexes) act as thermor-eversible physical cross-links. Better tensile properties and elasticity were observed for the octadecylamine, which authors explained by the packing of the long alkyl tails in a crystalline-like order. Irreversible imide formation occurred for all amide–acids and amide-complexes at high temperatures, resulting in the disappearance of the aggregates and, hence, a dramatic decrease in mechanical properties. Taking into consideration a known fact that the organo-MMT silicates are able to swell in compounds with branched alkyl chain [26], it can be proposed that exfoliation of anhydride-containing i-PP chains between silicate galleries is most probable, and therefore their orientation and interaction with alkylamine will be easily realized in melt-swelling state under the chosen reactive extrusion conditions [123].

Lee et al. [285] examined the thermal behavior of organic (alkylammonium cation) modified MMT and its effect on the formation of PP/oligo(PP-*g*-MA) (2%)/Organo-MMT (8%) nanocomposite. They showed that the decrease in interlayer spacing at the processing temperature (around 180°C–200°C) was due to the release of surface organic ions by thermal decomposition. Xie et al. [286,287] demonstrated that the catalytic sites on the aluminosilicate layer of Organo-MMT reduced the thermal stability of a fraction of alkylammonium ions by an average of 15°C–25°C. Taking into consideration the possibility of the thermal degradation of Organo-MMT at the higher processing temperatures, synthesis of nanocomposites via in situ modification was carried out at relatively low temperatures around 160°C–175°C. Under the chosen reactive extrusion conditions, the acceleration of thermal degradation reactions of i-PP and poly(i-PP-*g*-MA) chains by the modified clay surface layers was not observed (from MFI and TGA measurements) contrary to the arguments proposed in the literature [8]. TGA-DTA analysis showed that the nanocomposites degrade via a one-step process and exhibit high thermal stability up to 420°C. The observed endo-peaks at around 162°C–165°C in DTA curves were associated with the melting transition of nanocrystalline structures, values of which are similar to those determined by the DSC method [123]. The internal structure of nanocomposites based on crystalline polymers on macroscopic, microscopic, and nanoscopic levels was described by Zhang et al. [283]. Finite element analysis in combination with homogenization theory based on asymptotic expansion utilized to predict the effective properties of the nanocomposites using a computer program written in FORTRAN language. Authors also studied the effect of the degree of crystallization, elastic moduli of crystal inclusions and nanoparticles, and volume fractions of nanoparticles on the properties of the nanocomposites.

The mechanism of formation of the nanostructural architectures in the synthesized nanocomposites may also be interpreted by using modified models according to the following schemes (Figures 4.4 and 4.5) [123].

As can be seen in Figure 4.4, in situ polymer/organosilicate reaction easily proceeds between two alkylamine-modified surfaces and MA-grafted oligomer in tetragonal structural units of silicate galleries. Strong H-bonding and formed amide/imide intermolecular linkages provide a favorable condition for the effective penetration and exfoliation processes during the formation of nanostructural architectures. The representative model (Figure 4.5) describes an important role of alkylamine conformation and geometry of the modified surface in the in situ modification and formation of nanostructural composites.

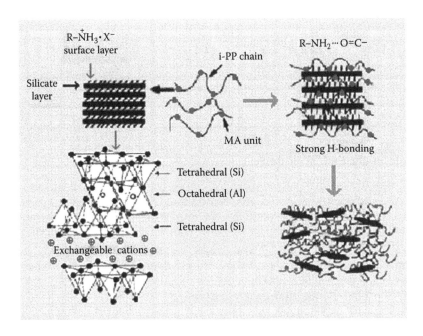

FIGURE 4.4
The proposed mechanism of formation of the nanostructural architectures in PP/oligo(PP-g-MA)/organo-MMT nanosystem. (Reproduced from Rzayev, Z.M.O. et al., *Adv. Polym. Technol.*, 26, 41, 2007.)

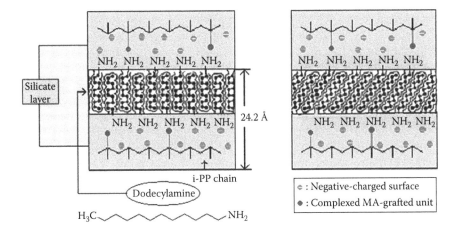

FIGURE 4.5
The proposed mechanism of the formation of nanostructural architectures in PP/oligo(PP-g-MA)/organo-MMT nanosystem. (Reproduced from Rzayev, Z.M.O. et al., *Adv. Polym. Technol.*, 26, 41, 2007.)

Recently, Moad et al. [288,289] designed and prepared novel copolymer intercalant/dispersant/exfoliant systems that are effective with unmodified clays at low levels (<20% with respect to clay), can be combined with commercial PP and clay in a conventional melt-mixing process, and do not require the use of additional compatibilizers. PP-clay nanocomposites prepared by direct melt mixing using unmodified MMT clays and a copolymer additive added at a level of only 1 wt.% with respect to PP for 5 wt.% clay. Authors investigated the following two classes of dispersants: (1) polyethylene oxide-based nonionic surfactants and (2) amphiphilic copolymers based on a long chain (octadecyl acrylate) and a more polar comonomer (MA, *N*-vinylpyrrolidone and methyl methacrylate). According to the authors, these copolymer additives provide substantially improved clay dispersion and cause partial exfoliation. Authors showed that nanocomposites possess a tensile modulus up to 40% greater than the precursor PP while elongation at break, impact strength, thermal and thermo-oxidative stability are significantly improved over organoclay-based nanocomposites.

4.4.4 Polyolefin Elastomers (Rubbers)

The most commonly used PO elastomers in nanotechnology include ethylene-propylene rubber (EPR), ethylene-propylene-diene terpolymer rubber (EPDM), styrene-ethylene-butadiene-styrene (SEBS), and polyethylene octane elastomer. A typical example of thermoplastic elastomer is thermoplastic PO, which generally consists of about 70 wt.% isotactic PP and 30 wt.% EPR. In this blend, PP comprises the major phase and acts as the matrix of PO. The morphology and mechanical behavior of thermoplastic PO elastomers reinforced with silicate clays have been studied by many research groups [247,278,280,290–306]. Tjong and Ruan [293] prepared a thermoplastic polyolefin (TPO) containing 70 wt.% poly(SEBS)-*g*-MA and 30 wt.% PP and its nanocomposites reinforced with 0.3–1.5 wt.% organoclay by melt mixing. They investigated the mechanical and fracture behaviors of the TPO/clay nanocomposites. Authors used the essential work of fracture (EWF) approach to evaluate the tensile fracture behavior of the nanocomposites toughened with elastomer. Tensile tests showed that the stiffness and tensile strength of TPO was enhanced by the addition of low loading levels of organically modified MMT. EWF measurements revealed that the fracture toughness of the TPO/clay nanocomposites increased with increasing clay content. Paul et al. [295] reported that the size of elastomer particles and the aspect ratio of the clay particles tended to decrease with increasing clay content. They found that the mechanical properties of PO-based nanocomposites depend greatly on the clay and elastomer contents; supertough nanocomposites with significantly high impact strengths can be achieved by the addition of 30%–40% rubber and 2–7 wt.% MMT clay to PP. Recently, Mishra et al. [302] prepared PO/clay nanocomposites in which the PO contained 25 wt.% PP and 75wt.% EPDM. According to the authors, the nanocomposites exhibited remarkable improvements in the tensile and storage moduli over their pristine PO blend. Austin and Kontopoulou [303] also incorporated organoclay into a rubber-rich PO (30% PP and 70% EPR-*g*-MA) to improve its stiffness and thermal stability. Some researchers used organoclay to reinforce EPDM and other types of rubbers, as low loading levels of clays are needed to achieve the desired properties [308,309]. Li et al. [310] prepared the EPDM/organo-MMT nanocomposites with an MA-grafted EPDM oligomer as a compatibilizer via melt intercalation. They found that the silicate layers of organo-MMT were exfoliated and dispersed uniformly as a few monolayers in nanocomposites. According to the authors, the nanocomposites exhibited great improvements in the tensile strength and tensile modulus, and the incorporation of organo-MMT gave rise to a considerable reduction of loss modulus/storage modulus ratio (Tan δ = Tg).

4.4.4.1 *Poly(Ethylene-co-Propylene)*

The layered silicates dispersed at the nanoscale are reported to provide an effective reinforcement to rubber materials [307–311]. However, there are very limited studies on the preparation and properties of rubber/clay nanocomposites, especially by means of a conventional rubber compounding process. Nasegawa and Usuki [245–247] studied the dispersion process of organo-MMT (organophilic clay galleries modified with stearyl ammonium ions) silicate layers in PO(PP or EPR)/clay nanocomposites based on MA-modified (around 0.09–4.5 wt.% of MA unit) POs. By using XRD measurements, SEM and TEM microscopy patterns, they observed the spontaneous full exfoliation of silicate layers for these nanocomposites, in which organo-MMT dispersed in the PP-*g*-MA matrixes with different compositions. They prepared E-P rubber-clay nanocomposites by melt-compounding (E-P)-*g*-MA with organophilic clay, and their properties were examined. TEM observation of authors confirmed the exfoliation and homogeneous dispersion of silicate layers in the nanocomposites [247]. It is thought that MA units (or maleic acid groups generated by hydrolysis) grafted to the E-P polymer chains selectively absorb to the dispersed silicate layers and form strong ionic interaction, because MA units have good affinity to ionic surfaces of silicate layers. According to the authors, the silicate layers, dispersing at the nanometer level, play a role of large pseudo-cross-link points and improve creep resistance of nanocomposites.

Nanocomposite technology using small amounts of silicate layers can lead to improved properties of thermoplastic elastomers with or without conventional fillers such as carbon black, talc, etc. Mallick et al. [305] investigated the effect of EPR-*g*-MA, nanoclay and a combination of the two on phase morphology and the properties of (70/30 w/w) nylon 6/EPR blends prepared by the melt-processing technique. They found that the number average domain diameter (Dn) of the dispersed EPR phase in the blend decreased in the presence of EPR-*g*-MA and clay. This observation indicated that nanoclay could be used as an effective compatibilizer in nylon 6/EPR blend. X-ray diffraction study and TEM analysis of the blend/clay nanocomposites revealed the delaminated clay morphology and preferential location of the exfoliated clay platelets in nylon 6 phase.

4.4.4.2 Poly(Ethylene-co-Octene)

Poly(ethylene-*co*-octene) (PEO) as a relatively new PO elastomer was manufactured by Dow Chemical Co. using a metallocene catalyst. It is known that PEO forms homogeneous and fine dispersion blend with elastomers, PA and PP [184,199–201,298,312,313]. Wahit et al. used PEO as an effective toughening agent for the preparation of rubber-toughened PP/PA6 nanocomposites [299,301,314–316]. They showed that the addition of PEO to the PA6/PP (PA6 (30/70) nanocomposites increased the toughness of the nanocomposites but with limited success due to its immiscibility with PA6. Recently, these authors prepared PEO elastomer and PEO-grafted MA (POE-*g*-MA) toughened nanocomposites of PP/PA6 containing 4 wt.% organophilic-modified MMT clay by melt compounding followed by injection molding. PP-*g*-MA was used by authors to compatibilize the blend system [316]. According to the authors, elastomer domains of POE-*g*-MA show a finer and more uniform dispersion than that of POE in the PP/PA6/organoclay matrix. They proposed that reduction in dispersed particle size and an increase in adhesion between the phases due to the reaction between the amino group of PA6 and the anhydride group of POE-*g*-MA during melt compounding and formation of PEO-*g*-PA6 copolymer are important factors for the toughness and morphology improvement of these blends.

4.4.4.3 Poly(Ethylene-co-Propylene-co-Diene)

Ethylene-propylene-diene terpolymer (EPDM) is an unsaturated PO rubber with wide applications. Several researchers [136,160,317] reported the results of investigations on grafting of MA onto EPDM in the molten phase and further utilization of EPDM-g-MA in thermoplastics as well as in the preparation of EPDM/organoclay nanocomposites [302,304–306,318–320]. Uzuki et al. [318] prepared the hybrid nanocomposites based on organophilic MMT and EPDM rubber by a melt compounding process. From the analysis by x-ray diffraction and TEM, the rubber molecules were found to be intercalated into the galleries of organoMMT and the silicate layers of organoMMT are uniformly dispersed as platelets of 50–80 nm thickness in the EPDM matrix. Dynamic mechanical studies reveal a strong rubber–filler interaction in the hybrid nanocomposite, which is manifested in the lowering of tan δ at the glass transition temperature. The hybrid nanocomposites exhibit great improvement in tensile and tear strength, and modulus, as well as elongation-at-break. Moreover, the permeability of oxygen for the hybrid nanocomposite was reduced remarkably.

According to Chang et al. [319], EPDM/organophilic MMT hybrid nanocomposites could successfully be prepared by a simple melt-compounding process. The organoclay prepared by treating sodium MMT with the octadecylammonium ion and subsequently with a low-molecular-weight EPDM oligomer. X-ray diffraction analysis indicates that matrix rubber molecules could be intercalated into the gallery of the organoclay effectively. Moreover TEM images confirmed that silicate layers of the organoclay are uniformly dispersed on the nanometer scale in the EPDM rubber matrix. The uniform dispersal of the clay nano-layers in the rubber matrix is considered to form a physical barrier against a growing crack, which leads to the increase in the resistance to tearing. These observations of authors indicated that the intercalation of rubber macromolecules into the galleries of organo-MMT can offer the hybrid nanocomposite effective enhancement in toughness, strength, and stiffness. In the EPDM hybrid filled with 10 phr of organo-MMT [dodecylammonium ion ($C_{18}H_{37}NH_3^+$) modified montmorillonite], the oxygen permeability was decreased to 60% relative to unfilled EPDM. This unique barrier property was explained by the authors in forming the dispersion of the impermeable clay with planar orientation in the rubber matrix, as observed by TEM micrographs.

Lee and Goettler [294] prepared the polymer blend nanocomposites, comprising nanoscale platelets derived from layered silicates treated with an organic modifier in thermoplastic vulcanizates (TPV) and polypropylene/EPDM blend, by direct melt intercalation. Authors showed that the interlayer spacing and dispersion of the organo-MMT are greatly affected by polar forces between the nanoclay and the polymeric matrix material. The structure–property relationship studies of authors in this class of nanocomposites indicate that with the increase of organoclay loading, the tensile modulus of TPV/clay nanocomposites increases by up to 170% at 8 wt.% organoclay loading, while tensile strength gradually decreases with the increase of organoclay loading. According to the authors, the dispersion of the nanoclay in physical polymer blends can be controlled using different addition sequences in the blending process, which can significantly affect the phase partitioning and mechanical properties.

Recently, Ahmadi et al. [320] prepared EPDM/clay nanocomposites with organoclay that was intercalated with MA-grafted EPDM (MA-g-EPDM) and EPDM-clay composites with pristine clay via indirect melt intercalation method. Authors characterized the dispersion of the silicate layers in the EPDM matrix by XRD and TEM analysis methods. They showed that the particles of organoclay were completely exfoliated in EPDM matrix, and the mechanical, thermal, and chemical properties of nanocomposites were significantly improved compared with conventional composites.

4.5 Polyolefin/Silica Hybrids

Silane-based coupling agents [321], most frequently γ-aminopropyltriethoxysilane, vinyltrialkoxy-silanes, tetraalkoxysilanes, etc., were used to improve the surface adhesion in various polymer composites, to surface modify PO films, and to prepare functional polymer/nanosilica hybrids [322]. Rzaev et al. [323–325] demonstrated that reaction of plasma-activated PO films [polyethylene (PE), isotactic polypropylene (i-PP), poly(E-*co*-P), and poly(E-*co*-P-*co*-butene-1)] with selective organosilicon compounds such as hexamethyldisiloxane, tetraethylsilane, γ-aminopropyltriethoxysilane, and olig o(MA-alt-vinyltriethoxysilane) [oligo(MA-alt-VTES)] led to a formation polysiloxane crosslinked hybrid structure in the surfaces [323]. The probable reaction steps for the plasmatreated PP with oligomer are as follows: (1) intermolecular esterification of anhydride units with PP surface hydroxyl groups, (2) intramolecular reaction between carboxyl and ethoxysilyl groups, and (3) polycondensation of $-Si(C_2H_5)_3$ fragments, as initiated by free $-COOH$, via the formation of cross-linked structure. The formation of surface PP/nanosilica hybrid structure can be presented schematically as follows (Figure 4.6) [323].

Melt compounding is most commonly utilized for the preparation of PO/silica nanocomposites. POs and their blends, such as PP [326–337], PE [338–343], ethylene-propylene copolymer [344–354], ethylene-octene copolymer [347], thermoplastic POs [348–350], PP/EPDM [351,352], and PP/liquid-crystalline polymer (LCP) [353–357] blends, have been used as the matrices in the preparation of PO/silica nanosystems and nanomaterials by twin-screw extrusion and injection molding or lab-scale single-screw extrusion and compression molding.

Yang and Kofinas [358] reported the synthesis and dielectric properties of sulfonated styrene-*b*-(ethylene-ran-butylene)-*b*-styrene (S-SEBS) block copolymer/TiO_2 nanoparticles/VTES crosslinker nanosystems. Authors incorporated VTES into the block copolymer matrices in order to decrease the dielectric loss of composites, improve the compatibility between polymer and silica domains through covalent bonding into domain interfaces. They showed that higher permittivity composites can thus be obtained with a significant decrease in loss tan δ (<0.001) when crosslinked with VTES. Bounor-Legare et al. [359] utilized alkoxysilane (coupling agents)-modified POs. They developed an original route to obtain organic–inorganic hybrid materials in molten state without the presence of solvent that can be integrated into processing operations of POs and other thermoplastic

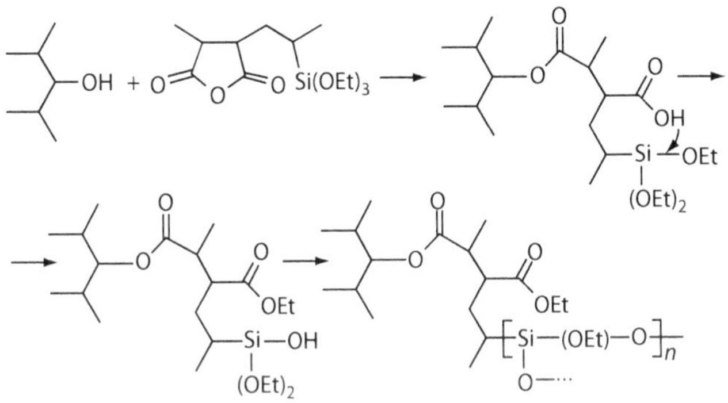

FIGURE 4.6
Plasma-treated PP and oligo(MA-alt-VTES) interaction through the formation of surface PP/nanosilica hybrid structure. (Akovali, G., et al., *J. Appl. Polym. Sci.*, 58, 645, 1996.)

polymers, such as extrusion including two successive steps: (1) the cross-linking of polymer, which contains pendant ester groups such as poly(ethylene-*co*-vinyl acetate), through ester-alkoxysilane interchange reaction in molten state in the presence of dibutyl-tin oxide as catalyst, and (2) the hydrolysis–condensation reactions of available alkoxysilane groups in the polymer network leading to the silica network grafted and confined in the organic network.

Liu and Kontopoulou [349] reported the morphology and physical properties of thermoplastic PO blend–based nanocomposites containing nanosilica. They found that with the addition of nanosilica into the PO blends, the size of the dispersed elastomer phase decreased and, therefore, impact properties of the PO composite improved. Zhou et al. [333,334] grafted the nanosized silica with poly(glycidyl methacrylate) (PGMA) and then melt-mixed together with PP and functionalized PP (PP-*g*-NH$_2$), which synthesized by reaction of PP-*g*-MA with hexamethylene diamine using a known method [336]. Thus, the authors grafted nano-SiO$_2$ particles onto PP matrix through the reaction of epoxy groups of PGMA with amine groups of PP-*g*-NH$_2$ during compounding. Chen et al. [337] reported an approach for the preparation of PP/SiO$_2$ nanocomposites by in situ reactive processing. They covalently bonded nanosilica to the matrix polymer via polyurethane (PU) elastomer and amine functionalized PP-*g*-NH$_2$. Taking advantages of rubber-type grafting PU and interfacial reactive compatibilization with PP-*g*-NH$_2$, authors observed a synergetic toughening effect for the PP nanocomposites. According to the authors, only very low concentrations of nano-SiO$_2$ (1.5–2.5 vol.%) and PU (<4 vol.%) were sufficient to greatly increase notched impact strength of PP. Similar PP/silica nanocomposites with improved thermal and mechanical properties synthesized by Reddy et al. using epoxide resin [329] and zinc-ion [339]-coated silica nanoparticles.

Yang et al. [351,352] reported two processing methods for the preparation of PP/EPDM/SiO$_2$ ternary composites: (1) one-step processing method, in which the elastomer and the filler directly melt blended with PP matrix, and (2) two-step processing method, in which the elastomer and the filler were mixed first, and then melt blended with pure PP. To control the interfacial interaction among the components, authors used two kinds of PP (virgin PP and grafted with MA, PP-*g*-MA) and SiO$_2$ (treated with or without coupling agent). They found that the formation of filler-network structure could be a key for a simultaneous enhancement of toughness and modulus of PP and its formation seemed to be dependent on the work of adhesion and processing method. According to the authors, in the two-step processing, a super toughened PP ternary composite with the Izod impact strength 2–3 times higher than PP/EPDM binary blend and 15–20 times higher than pure PP. Wu and Chu [348] prepared nanocomposites of blends of PP and dynamically vulcanized EPDM rubber (PP/EPDM)/SiO$_2$. The treated SiO$_2$ was melt blended with PP/EPDM blend in the presence of PP-*g*-MA, which acted as a functionalized compatibilizer. The strong interaction caused by the grafting reaction improved the dispersion of silica in the PO matrix. Similar results were obtained by Zhang et al. [328]. They observed the change of phase morphology and properties of immiscible PP/PS blends compatibilized with nano-SiO$_2$ particles. The compatibility of PP/PS blends was dramatically improved with the addition of nano-SiO$_2$ particles, which possess excellent hydrophobicity and contain a large number of alkyls on their surface. Authors observed a very homogeneous size distribution by introducing nano-SiO$_2$ particles in the blends at short mixing time. Hu et al. [353–357] reported novel approach to fibrillation of LCP in PP/LCP/nanosilica blends and effect of nanosilica filler on the structure, rheological, morphological, and mechanical properties of nanocomposites. Bikiaris et al. [326,327] prepared i-PP/SiO$_2$ (1–10 wt.%) nanocomposites with untreated and surface-treated silica nanoparticles by melt compounding using a co-rotating screw extruder. All nanocomposites were transparent as pure i-PP, indicating fine dispersion of the silica nanoparticles into i-PP matrix and the retention of their nano-sizes. According to the authors, when PP-*g*-MA copolymer was added as a compatibilizer [327], it resulted in a higher adhesion between i-PP matrix and SiO_2 nanoparticles, due to the interactions that took place between the reactive groups. They found that a decrease in the size of silica provided a further enhancement in the mechanical properties of i-PP/silica nanocomposite.

4.6 Summary and Outlook

This review summarizes the main advances published over the last 15 years outlining different synthesis routes of functionalized POs, reactive PO/thermoplastics and rubber (synthetic and natural) blends, PO silicate layered nanocomposites and PO/nanosilica hybrids through intercalation/exfoliation of POs and their functionalized grafts between silicate galleries in melt by reactive extrusion systems. Special attention is devoted to the mechanism of in situ grafting reactions and H-bonding effect in the reactive blend processing and nanostructure formation. One of the important concepts that has been a driving force behind rapid development in the polymer nanoscience and nanoengineering areas is the preparation of a new generation of layered organoclay nanostructures in situ melt compounding of the POs and various immiscible polymer blends at the nanoscale, as well as to prepare the polymer nanomaterials with unique physical and chemical structures and high performance engineering properties. Further development may open new ways of utilizing a combination of various functional monomers, especially donor–acceptor-type monomer systems, PO/other polymer blends, surfactants and functionalized organoclays, as well as PO reactive compatibilizers to prepare polymer nanomaterials with controllable structures and unique properties. The successful development of fundamental and applied research in this area calls for unified efforts of specialists in different fields of science and engineering, such as chemistry, physics, ecology, materials science, etc., to solve the problems of creating economically, ecologically, and technologically attractive polymer nanomaterials.

Undoubtedly, it can be expected that future development in this interesting branch of polymer chemistry and engineering will lead to (1) the discovery of new unique features of in situ chemical, physical, and engineering processes; (2) widening the use of functional monomers, synthetic, artificial, and natural polymer assortments, predominantly POs and functional copolymers of α-olefins; and (3) the designation of new apparatus and effective techniques for the preparation and testing of the nanoarchitecture formation, control of morphology at the nano scale, and their future scientific and industrial applications.

Acknowledgments

This review also contains the results of systematic investigations on functional copolymer/organo-silicate nanocomposites, which were carried out according to the Polymer Science and Technology Program of the Chemical Engineering Department of Hacettepe

University (HU). The support of the Turkish National Scientific and Technical Council (TÜBİTAK) through project TBAG-HD/249 and HU Scientific Research Foundation (BAB) grant funded by the Turkish Planning Committee (DPT) through the DPT-2007 K120930 project is kindly acknowledged.

References

1. Dubois, P., Alexandre, M., Hindryckx, F., and Jérôme, R. 1998. Homogeneous polyolefin-based composites. *Journal of Macromolecular Science: Reviews in Macromolecular Chemistry and Physics* C38:511–565.
2. Alexandre, M. and Dubois, P. 2000. Polymer-layered silicate nanocomposites: Preparation, properties and uses of a new class of materials. *Materials Science and Engineering R* 28:1–63.
3. Ray, S. S. and Okamoto, M. 2003. Polymer/layered silicate nanocomposites: A review from preparation to processing. *Progress in Polymer Science* 28:1539–1641.
4. Goettler, L. A., Lee, K. Y., and Thakkar, H. 2007. Layered silicate reinforced polymer nanocomposites: Development and applications. *Polymer Reviews* 47:291–317.
5. Manias, E., Touny, A., Wu, L., Strawhecker, K. B., Lu, B., and Chung, T. C. 2001. Polypropylene/montmorillonite nanocomposites. Review of the synthetic routes and materials properties. *Chemistry Materials* 13:3516–3523.
6. Adhikari, R. and Michler, G. H. 2009. Polymer nanocomposites characterization by microscopy. *Polymer Reviews* 49:141–180.
7. Komori, Y. and Kuroda, K. 2000. In *Polymer-Layered Silicate Nanocomposites*, ed. T. J. Pinnavaia, and G. W. Beall. New York: Wiley.
8. Newman, A. C. D. 1987. *Chemistry of Clays and Clay Materials, Mineralogical Society Monograph, No. 6.* New York: Wiley.
9. Schmidt, H. 1990. In *Polymer Based Molecular Composites*, ed. D. W. Schaefer, and J. E. Mark. Pittsburgh, PA: Material Research Society.
10. Lagalay, G. 1999. Introduction: From clay mineral–polymer interaction to clay–polymer nanocomposites. *Applied Clay Science* 15:1–9.
11. Okamoto, M., Nam, P. H., Maiti, P., Kotaka, T., Hasegawa, N., and Usuki, A. 2001. A house of cards structure in polypropylene/clay nanocomposites under elongational flow. *Nano Letters* 1:295–298.
12. Crosby, A. J. and Lee, J.-Y. 2007. Polymer nanocomposites: The "nano" effect on mechanical properties. *Polymer Reviews* 47:217–229.
13. Kawasumi, M., Hasewaga, N., Kato, M., Usuki, A., and Okada, A. 1997. Preparation and mechanical properties of polypropylene-clay hybrids. *Macromolecules* 30:6333–6338.
14. Okada, A. and Usuki, A. 1995. The chemistry of polymer-clay hybrids. *Materials Science and Engineering C* 3:109–115.
15. Osman, M. A., Mittal, V., Morbidelli, M., and Suter, U. W. 2003. Polyurethane adhesive nanocomposites as gas permeation barrier. *Macromolecules* 36:9851–9858.
16. Osman, M. A., Mittal, V., Morbidelli, M., and Suter, U. W. 2004. Epoxy-layered silicate nanocomposites and their gas permeation properties. *Macromolecules* 37:7250–7257.
17. Messersmith, P. B. and Gianelis, E. P. 1995. Synthesis and barrier properties of poly(ε-caprolactone)-layered silicate nanocomposites. *Journal of Polymer Science, Part A: Polymer Chemistry* 33:1047–1057.
18. Osman, M. A., Mittal, V., and Lusti, H. R. 2004. The aspect ratio and gas permeation in polymer-layered silicate nanocomposites. *Macromolecular Rapid Communications* 25:1145–1149.
19. Lan, T., Kaviratna, P. D., and Pinnavaia, T. J. 1994. On the nature of polyimide-clay hybrid composites. *Chemistry of Materials* 6:573–575.

20. Yeh, J. M., Liou, S. J., Lai, C. Y., Wu, P. C., and Tsai, T. Y. 2001. Enhancement of corrosion protection effect in polyaniline via the formation of polyaniline-clay nanocomposite materials. *Chemistry of Materials* 13:1131–1136.

21. Burnside, S. D. and Gianellis, E. P. 1995. Synthesis and properties of new poly(dimethylsiloxane) nanocomposites. *Chemistry of Materials* 7:1597–1600.

22. Kashiwagi, T., Morgan, A. B., Antonucci, J. M., van Landingham, M. R., Jr., Harris, R. H., Awad, W. H., and Shields, J. R. 2003. Thermal and flammability properties of a silica–poly(methyl methacrylate) nanocomposites. *Journal of Applied Polymer Science* 89:2072–2078.

23. Zhang, J. G. and Wilkie, C. A. 2001. Preparation and flammability properties of polyethylene–clay nanocomposites. *Polymer Degradation and Stability* 80:163–169.

24. Tang, Y., Hu, Y., Wang, S. F., Gui, Z., Chen, Z., and Fan, W. C. 2002. Preparation and flammability of ethylene-vinyl acetate copolymer/montmorillonite nanocomposites. *Polymer Degradation and Stability* 78:555–559.

25. Yeh, J. M., Liou, S. J., Lin, C. Y., Cheng, C. Y., Chang, Y. W., and Lee, K. B. 2002. Anticorrosively enhanced PMMA-clay nano-composite materials with quaternary alkylphosphonium salt as an intercalating agent. *Chemistry of Materials* 14:154–161.

26. Weiss, A. 1963. Organic derivatives of mica-type layer silicates. *Angewandte Chemie International Edition* 2:134–144.

27. Osman, M. A., Ernst, M., Meier, B. H., and Suter, U. W. 2002. Structure and molecular dynamics of alkane monolayers self-assembled on mica platelets. *Journal of Physical Chemistry B* 106:653–662.

28. Osman, M. A., Ploetze, M., and Suter, U. W. 2003. Surface treatment of clay minerals–thermal stability, basal-plane spacing and surface coverage. *Journal of Material Chemistry* 13:2359–2366.

29. Liao, B., Liang, H., and Pang, Y. 2001. Polymer-layered silicate nanocomposites. 1. A study of poly(ethylene oxide)/Na+—montmorillonite nanocomposites as polyelectrolytes and polyethylene-block-poly(ethylene glycol) copolymer/Na+—montmorillonite nanocomposites as fillers for reinforcement of polyethylene. *Polymer* 42:10007–10011.

30. Ding, C., Jia, D., He, H., Guo, B., and Hong, H. 2005. How organo-montmorillonite truly affects the structure and properties of polypropylene. *Polymer Testing* 24:94–100.

31. Gorras, G., Tortora, M., Vittoria, V., Kaempfer, D., and Mulhaupt, R. 2003. Transport properties of organic vapors in nanocomposites of organophilic layered silicate and syndiotactic polypropylene. *Polymer* 44:3679–3685.

32. Utracki, L. A. and Simha, R. 2004. Pressure–volume–temperature dependence of polypropylene/organoclay nanocomposites. *Macromolecules* 37:10123–10133.

33. Shin, S. Y. A., Simon, L. C., Soares, J. B. P., and Scholz, G. 2003. Polyethylene–clay hybrid nanocomposites: *In situ* polymerization using bifunctional organic modifiers. *Polymer* 44:5317–5321.

34. Zheng, L., Farris, R. J., and Coughlin, E. B. 2001. Novel polyolefin nanocomposites: Synthesis and characterizations of metallocene-catalyzed polyolefin polyhedral oligomeric silsesquioxane copolymers. *Macromolecules* 34:8034–8039.

35. Passaglia, E., Sulcis, R., Ciardelli, F., Malvaldi, M., and Narducci, P. 2005. Effect of functional groups of modified polyolefins on the structure and properties of their composites with lamellar silicates. *Polymer International* 54:1549–1556.

36. Giannelis, E. P. 1996. Polymer layered silicate nanocomposites. *Advance Materials* 8:29–35.

37. La Mantia, F. P., Dintcheva, N. D., Filippone, G., and Acierno, D. 2007. Structure and dynamics of polyethylene/clay films. *Journal of Applied Polymer Science* 102(5):4749–4758.

38. Beecroft, L. L. and Ober, C. K. 1997. Nanocomposite materials for optical applications. *Chemistry of Materials* 9:1302–1317.

39. LeBaron, P. C., Wang, Z., and Pinnavaia, T. J. 1999. Polymer-layered silicate nanocomposites: An overview. *Applied Clay Science* 15:11–29.

40. Vaia, R. A., Jandt, K. D., and Giannelis, E. P. 1996. Microstructural evolution of melt intercalated polymer-organically modified layered silicates nanocomposites. *Chemistry of Materials* 8:2628–2635.

41. Krishnamoorti, R., Vaia, R. A., and Giannelis, E. P. 1996. Structure and dynamics of polymer-layered silicate nanocomposites. *Chemistry of Materials* 8:1728–1734.
42. Wang, Z. and Pinnavaia, T. J. 1994. Clay-polymer nanocomposites formed from acidic derivatives of montmorillonite and an epoxy resin. *Chemistry of Materials* 6:468–474.
43. Kuila, T., Srivastava, S. K., Bhowmick, A. K., and Saxena, A. K., 2006. Thermoplastic polyolefin based polymer-blend-layered double hydroxide nanocomposites. *Journal of Applied Polymer Science* 99:3441–3450.
44. Zeng, Q.-H., Wang, D.-Z., and Lu, C.-Q. 2002. Synthesis of polymer–montmorillonite nanocomposites by *in situ* in intercalative polymerization. *Nanotechnology* 13:549–553.
45. Vaia, R. A., Jandt, K. D., Kramer, E. J., and Giannelis, E. P. 1995. Kinetics of polymer melt intercalation. *Macromolecules* 28:8080–8085.
46. Vaia, R. A., Isii, H., and Giannelis, E. P. 1993. Synthesis and properties of two-dimensional nanostructures by direct intercalation of polymer melts in layered silicates. *Chemistry of Materials* 5:1694–1696.
47. Hasegawa, N., Kawasumi, M., Kato, M., Usuki, A., and Okada, A. 1998. Preparation and mechanical properties of polypropylene–clay hybrids using a maleic anhydride-modified polypropylene oligomer. *Journal of Applied Polymer Science* 67:87–92.
48. Xu, W., Liang, G., Zhai, H., Tang, S., Hang, G., and Pan, W.-P. 2003. Preparation and crystallization behaviour of PP/PP-*g*-MAH/Org-MMT nanocomposite. *European Polymer Journal* 39:1467–1474.
49. Böhning, M., Goering, H., Fritz, A., Brzezinka, K.-W., Turky, G., Schönhals, A., and Schartel, B. 2005. Dielectric study of molecular mobility in poly(propylene-*graft*-maleic anhydride)/clay nanocomposites. *Macromolecules* 38:2764–2774.
50. Tang, Y., Hu, Y., Song, L., Zong, R., Gui, Z., Chen, Z., and Fan, W. 2003. Preparation and thermal stability of polypropylene/ montmorillonite nanocomposites. *Polymer Degradation Stability* 82:127–131.
51. Xu, L., Nakajima, H., Manias, E., and Krishnamoorti, R. 2009. Tailored nanocomposites of polypropylene with layered silicates. *Macromolecules* 42:3795–3803.
52. Kato, M., Usuki, A., and Okada, A. 1997. Synthesis of polypropylene oligomer-clay intercalation compounds. *Journal of Applied Polymer Science* 66:1781–1785.
53. Trivedi, B. C. and Culbertson, B. M. 1982. *Maleic Anhydride*. New York: Plenum Press.
54. Rzaev, Z. M. 1984. *Polymers and Copolymers of Maleic Anhydride*. Baku: Elm.
55. Culbertson, B. M. 1985. Maleic and fumaric polymers, in *Encyclopedia of Polymer Science and Engineering*, Vol. 9, New York: Wiley-Interscience.
56. Pekel, N., Şahiner, N., Güven, O., and Rzaev, Z. M. 2001. Synthesis and characterization of N-imidazole-ethyl methacrylate copolymers and determination of monomer reactivity ratios. *European Polymer Journal* 37:2443–2451.
57. Dinçer, S., Köşeli, V., Kesim, H., Rzaev, Z. M. O., and Pişkin, E. 2002. Radical copolymerization of N-isopropylacrylamide with anhydrides of maleic and citraconic acids. *European Polymer Journal* 38:2143–2151.
58. Denizli, B. K., Can, H. K., Rzaev, Z. M. O., and Güner, A. 2006. Synthesis of copolymers of *tert*-butyl vinyl ether with maleic and citraconic anhydrides. *Journal of Applied Polymer Science* 100:2453–2463.
59. Mazi, H., Kibarer, G., Emregül, E., and Rzaev, Z. M. O. 2006. Bioengineering functional copolymers. IX. Poly[(maleic anhydride-*co*-hexene-1)-*g*-poly(ethylene oxide)]. *Macromolecular Bioscience* 6:311–321.
60. Gülden, G. and Rzaev, Z. M. O. 2008. Synthesis and characterization of copolymers of N-vinyl-2-pyrrolidone with isostructural analogs of maleic anhydride. *Polymer Bulletin* 60:741–752.
61. Demircan, D., Kibarer, G., Güner, A., Rzaev, Z. M. O., and Ersoy, E. 2008. The synthesis of poly(MA-*alt*-NIPA) copolymer, spectroscopic characterization, and the investigation of solubility profile-viscosity behavior. *Carbohydrate Polymers* 72:682–694.
62. Cowie, G. J. M. 1985. *Alternating Copolymers*. New York: Plenum.

63. Rzaev, Z. M. O. 2000. Complex-radical alternating copolymerization. *Progress in Polymer Science* 25:163–217.
64. Rzaev, Z. M. O., Dinçer, S., and Pişkin, E. 2007. Functional copolymers of *N*-isopropylacrylamide for bioengineering applications. *Progress in Polymer Science* 32:534–595.
65. Rzaev, Z. M. O. 2002. Hyperbranched macromolecules through donor-acceptor type copolymerization of allyl-vinylene bifunctional monomers. *Polymer International* 51:998–1012.
66. Rzayev, Z. M., Bryksina, L. V., and Sadikh-zade, S. I. 1973. Charge transfer complexes of maleic anhydride in radical homo- and copolymerization. *Journal of Polymer Science Symposis Part 1* 42:519–529.
67. Rzayev, Z. M. and Sadikh-zade, S. I. 1973. Radical copolymerization of maleic anhydride with organotin acrylates. *Journal of Polymer Science Symposis Part 2* 42:541–552.
68. Mamedova, S. M., Rasulov, N. S., and Rzaev, Z. M. 1987. Alternating copolymerization of maleic anhydride with allyl chloroacetate. *Journal of Polymer Science, Part A: Polymer Chemistry* 25:711–717.
69. Rzaev, Z. M. O., Milli, H., and Akovalı, G. 1996. Complex-radical alternating copolymer of *trans*-stilbene with *N*-substituted maleimides. *Polymer International* 41:259–265.
70. Rzayev, Z. M., Salamova, U., and Akovalı, G. 1998. Complex-radical copolymerization of 2,4,4-trimethylpentene-1 with maleic anhydride. *European Polymer Journal* 34:981–985.
71. Rzayev, Z. M. O. and Güner, A. 2001. Synthesis and reactions of self-crosslinkable epoxy- and anhydride-containing macromolecules. *Polymer International* 50:185–193.
72. Şenel, S., Rzayev, Z. M. O., and Pişkin, E. 2003.Copolymerization of *N*-phenylmaleimide with 2-hydroxyethyl- and ethyl methacrylates. *Polymer International* 52:713–721.
73. Barsbay, M., Can, H. K., Rzayev, Z. M. O., and Güner, A. 2005. Design and properties of new functional water-soluble polymer of citraconic anhydride (CA) and related copolymers. *Polymer Bulletin* 53:305–314.
74. Chen, G.-Q., Wu, Z.-Q., Wu, J.-R., Li, Z.-C., and Li, F.-M. 2000. Synthesis of alternating copolymers of *N*-substituted maleimides with styrene via atom transfer radical polymerization. *Macromolecules* 33:232–234.
75. Benoit, D., Hawker, C. J., Huang, E. E., Lin, Z., and Russell, T. P. 2000. One step formation of functionalized block copolymers. *Macromolecules* 33:1505–1507.
76. Brouwer, H. De, Schellekens, M. A. J., Klumperman, B., Monteiro, M. J., and German, A. L. 2000. Controlled radical copolymerization of styrene and maleic anhydride and the synthesis of novel polyolefin-based block copolymers by reversible addition-fragmentation chain-transfer (RAFT) polymerization. *Journal of Polymer Science, Part A: Polymer Chemistry* 38:3596–3603.
77. Schmidt-Naake, G. and Butz, S. 1996. Living free radical donor-acceptor copolymerization of styrene and *N*-cyclohexyl maleimide and the synthesis of poly[styrene-*co*-(*N*-chlohexylmaleimide)]/polystyrene block copolymers. *Macromolecular Rapid Communications* 17:661–665.
78. Lokaj, J., Vlcek, P., and Kriz, J. 1999. Poly(styrene-*co*-*N*-butylmaleimide) macroinitiators by controlled autopolymerization and related block copolymers. *Journal of Applied Polymer Science* 74:2378–2385.
79. Lokaj, J., Holler, P., and Kriz, J. 2000. Copolymerization and addition of styrene and *N*-phenylmaleimide in the presence of nitroxide. *Journal of Applied Polymer Science* 76:1093–1099.
80. Butz, S., Baethge, H., and Schmidt-Naake, G. 2000. *N*-Oxyl mediated free radical donor-acceptor co- and terpolymerization of styrene, cyclic maleimide monomers and *n*-butyl methacrylates. *Macromolecular Chemistry and Physics* 201:2143–2151.
81. Shi, L., Wan, D., and Huang, J. 2000. Controllability of radical copolymerization of maleimide and ethyl α-(*n*-propyl)acrylate using 1,1,2,2-tetraphenyl-1,2-bis(trimethylsilyloxy) ethane as initiator. *Journal of Polymer Science, Part A: Polymer Chemistry* 38:2872–2878.
82. Park, E.-S. and Yoon, J.-S. 2003. Synthesis of polyethylene-*graft*-poly(styrene-*co*-maleic anhydride) and its compatibilizing effects on polyethylene/starch blends. *Journal of Applied Polymer Science* 88:2434–2438.

83. Deng, G. and Chen, Y. 2004. A novel way to synthesize star polymers in one pot by ATRP of *N*-[2-(2-bromoisobutyloxy)ethyl] maleimide and styrene. *Macromolecules* 37:18–26.

84. Zhao, Y.-Y., Zhang, J.-M., Jiang, J., Chen, C.-F., and Xi, F. 2002. Atom transfer radical copolymerization of *n*-hexylmaleimide and styrene in an ionic liquid. *Journal of Polymer Science, Part A: Polymer Chemistry* 40:3360–3366.

85. Chong, Y. K., Kristina, J., Le, T. P. T., Moad, G., Postma, A., Rizzardo, E., and Thang, S. H. 2003. Thiocarbonylthio compounds [S-C(Ph)S-R] in free radical polymerization with reversible addition-fragmentation chain transfer (RAFTpolymerization). Role of the free-radical leaving group (R). *Macromolecules* 36:2256–2272.

86. You, Y.-Z., Hong, C.-Y., and Pan, C.-Y. 2002. Controlled alternating copolymerization of St with MAh in the presence of DBTTC. *European Polymer Journal* 38:1289–1295.

87. Szablan, Z., Toy, A. A., Davis, T. P., Hao, X., Stenzel, M. H., and Barner-Kowollik, C. 2004. Reversible addition fragmentation chain transfer polymerization of sterically hindered monomers: Toward well-defined rod/coil architectures. *Journal of Polymer Science, Part A: Polymer Chemistry* 42:2432–2443.

88. Butler, G. 1992. *Cyclopolymerization and Cyclocopolymerization*. New York: Marcel Dekker.

89. Rzayev, Z. M., Mamedova, S. G., Yusifov, G. A., and Rustamov, F. B. 1988. Complex radical copolymerization of di-*n*-butyl stannyl butylstannyl dimethacrylate with maleic anhydride. *Journal of Polymer Science, Part A: Polymer Chemistry* 26:849–857.

90. Rzayev, Z. M., Medyakova, L. V., Kibarer, G., and Akovali, G. 1994. Complex-radical copolymerization of allyl cinnamate with styrene. *Macromolecules* 27:6292–6296.

91. Rzayev, Z. M., Akovali, G., and Medyakova, L. V. 1994. Alternating copolymerization of allyl *trans*-cinnamate and maleic anhydride. *Polymer* 35:5349–5354.

92. Rzaev, Z. M. O. and Salamova, U. 1997. Complex-radical cyclocopolymerization of allyl-α-(N-maleimido)-acetate with styrene and maleic anhydride. *Macromolecular Chemistry and Physics* 198:2475–2487.

93. Rzaev, Z. M. O., Akovalı, G., and Salamova, U. 1998. Effect of complex-formation and cyclization in radical copolymerization of allyl(metha)acrylates with maleic anhydride. *Journal of Polymer Science, Part A: Polymer Chemistry* 36:1501–1508.

94. Yürük, H., Rzaev, Z. M. O., and Akovalı, G. 1995. Complex-radical terpolymerization of allylglycidyl ether, maleic anhydride and methyl methacrylate. Design of self-crosslinking macromolecules. *Journal of Polymer Science, Part A: Polymer Chemistry* 33:1447–1454.

95. Rzaev, Z. M. O., Medyakova, L. V., Mamedova, M. A., and Akovali, G. 1995. Complex-radical terpolymerization of maleic anhydride, *trans*-stilbene and N-phenylmaleimide. *Macromolecular Chemistry and Physics* 196:1999–2009.

96. Baştürkmen, M., Rzaev, Z. M. O., Akovalı, G., and Kisakürek, D. 1995. Complex-radical terpolymerization of phenanthrene, maleic anhydride, and *trans*-stilbene. *Journal of Polymer Science, Part A: Polymer Chemistry* 33:7–13.

97. Rzaev, Z. M. O. 1999. Complex-radical terpolymerization of glycidyl(methyl) methacrylates, styrene and maleic anhydride. *Journal of Polymer Science, Part A: Polymer Chemistry* 37:1095–1102.

98. Medyakova, L. V., Rzaev, Z. M. O., Güner, A., and Kibarer, G. 2000. Complex-radical terpolymerization of acceptor-donor-acceptor systems: Maleic anhydride (*n*-butyl methacrylate)-styrene-acrylonitrile. *Journal of Polymer Science, Part A: Polymer Chemistry* 38:2652–2662.

99. Rzaev, Z. M. O., Güner, A., Kibarer, G., Can, H. K., and Aşici, A. 2002. Terpolymerization of maleic anhydride *trans*-stilbene and acrylic monomers. *European Polymer Journal* 38:1245–1254.

100. Can, H. K., Doğan, A. L., Rzaev, Z. M. O., Uner, A. H., and Güner, A. 2005. Synthesis and anti-tumor activity of poly(3,4-dihydro-2H-pyran-*co*-maleic anhydride-*co*-vinylacetate). *Journal of Applied Polymer Science* 96:2352–2359.

101. Devrim, Y., Rzaev, Z. M. O., and Pişkin, E. 2006. Synthesis and characterization of poly[(maleic anhydride-*alt*-styrene)-*co*-2- acrylamido-2-methyl-1-propanesulfonic acid]. *Macromolecular Chemistry and Physics* 207:111–121.

102. Can, H. K., Doğan, A. L., Rzaev, Z. M. O., Uner, A. H., and Güner, A. 2006. Synthesis, characterization, and antitumor activity of poly(maleic anhydride-*co*-vinylacetate-*co*-acrylic acid). *Journal of Applied Polymer Science* 100:3425–3432.

103. Devrim, Y., Rzaev, Z. M. O., and Pişkin, E. 2007. Physically and chemically cross-linked poly{[(maleic anhydride-*alt*-styrene)]-*co*-(2-acrylamido-2-methyl-1-propanesulfonic acid)}/poly(ethylene glycol) proton-exchange membranes. *Macromolecular Chemistry and Physics* 208:175–187.

104. Minoura, Y., Ueda, M., Mizunuma, S., and Oba, M. 1969. The reaction of polypropylene with maleic anhydride. *Journal of Applied Polymer Science* 13:1625–1640.

105. Nowak, R. M. and Jones, G. D. (Dow Chemical Co.). 1965. Extruding process for preparing functionalized polyolefin polymers. U.S. Patent 3,177,269.

106. Ide, F. and Hasegawa, A. 1974. Studies on polymer blend of nylon 6 and polypropylene or nylon 6 and polystyrene using the reaction of polymer. *Journal of Applied Polymer Science* 18:963–974.

107. Swiger, R. T. and Mango, L. A. 1977. Polyamide and modified polyethylene binary mixtures with high melt viscosity. DE Patent 2,722,270 A1.

108. Cha, J. and White, J. L. 2001. Maleic anhydride modification of polyolefin in an internal mixer and a twin-screw extruder: Experiment and kinetic model. *Polymer Engineering and Science* 41:1227–1237.

109. White, J. L. and Sasaki, A. 2003. Free radical graft polymerization. *Polym-Plastics Technology and Engineering* 42:711–735.

110. Ostenbrink, A. J., Borggreve, R. J. M., and Gaymans, R. J. 1989. *Integration of Fundamental Polymer Science and Technology*. New York: Elsevier Applied Science.

111. Sathe, S. N., Srinivasa, N., Rao, G. S., and Devi, S. 1994. Grafting of maleic anhydride onto polypropylene: Synthesis and characterization. *Journal of Applied Polymer Science* 53:239–245.

112. Martinez, J. M. G., Laguna, O., and Collar, E. P. 1998. Chemical modification of polypropylenes by maleic anhydride: Influence of stereospecificity and process conditions. *Journal of Applied Polymer Science* 68:83–95.

113. Gaylord, N. G. and Mishra, M. K. 1983. Nondegradative reaction of maleic anhydride and molten polypropylene in the presence of peroxides. *Journal of Polymer Science Letters* 21:23–30.

114. Gaylord, N. G. and Mehta, R. 1988. Peroxide-catalyzed grafting of maleic anhydride onto molten polyethylene in the presence of polar organic compounds. *Journal of Polymer Science, Part A: Polymer Chemistry* 26:1189–1198.

115. Simmons, A. and Baker, W. E. 1989. Basic functionalization of polyethylene in the melt. *Polymer Engineering and Science* 29:1117–1123.

116. Singh, R. P. 1992. Surface grafting onto polypropylene—A survey of recent developments. *Progress in Polymer Science* 17:251–281.

117. Bettini, S. H. P. and Agnelli, J. A. M. 1999. Grafting of maleic anhydride onto polypropylene by reactive processing. I. Effect of maleic anhydride and peroxide concentrations on the reaction. *Journal of Applied Polymer Science* 74:247–255.

118. Bettini, S. H. P. and Agnelli, J. A. M. 1999. Grafting of maleic anhydride onto polypropylene by reactive processing. II. Effect of rotor speed and reaction time. *Journal of Applied Polymer Science* 74:256–263.

119. Hagiwara, T., Saitoh, H., Tobe, A., Sasaki, D., Yano, S., and Sawaguchi, T. 2005. Functionalization and applications of telechelic oligopropylenes: Preparation of α,ω-dihydroxy- and diaminooligopropylenes. *Macromolecules* 38:10373–10378.

120. Li, C., Zhang, Y., and Zhang, Y. 2003. Melt grafting of maleic anhydride onto low-density polyethylene/polypropylene blends. *Polymer Testing* 22:191–195.

121. Pesetskii, S. S., Jurkowski, B., Krivogus, Y. M., Tomczyk, T., and Makarenko, O. A. 2006. PP/LDPE blends produced by reactive processing. I. Grafting efficiency and rheological and high-elastic properties of [PP/LDPE]-*g*-IA melts. *Journal of Applied Polymer Science* 102:5095–5104.

122. Güldoğan, Y., Eğri, S., Rzaev, Z. M. O., and Pişkin, E. 2004. Comparison of MA grafting onto powder and granular polypropylene in the melt by reactive extrusion. *Journal of Applied Polymer Science* 92:3675–3684.

123. Rzayev, Z. M. O., Yilmazbayhan, A., and Alper, E. 2007. A one step preparation of polypropylene-compatibilizer-clay nanocomposites by reactive extrusion. *Advance Polymer Technology* 26:41–55.

124. Ho, R. M., Su, A. C., Wu, C. H., and Chen, S. 1993. Functionalization of polypropylene via melt mixing. *Polymer* 34:3264–3269.

125. Heinen, W., Rozenmöller, C. H., Wenzel, C. B., de Croot, H. J. M., and van Duin, M. 1996. ^{13}C NMR study of the grafting of maleic anhydride onto polyethene, polypropene, and ethene–propene copolymers. *Macromolecules* 24:1151–1157.

126. Yong, L., Zhang, F., Endo, T., and Hirotsu, T. 2002. Structural characterization of maleic anhydride grafted polyethylene by ^{13}C NMR spectroscopy. *Polymer* 43:2591–2594.

127. Dorscht, B. M. and Tzoganakis, C. 2003. Reactive extrusion of polypropylene with supercritical carbon dioxide: Free radical grafting of maleic anhydride. *Journal of Applied Polymer Science* 87:1116–1122.

128. Xanthos, M. 1992. *Reactive Extrusion*. Munich, Germany: Hanser Publisher.

129. Nemeth, S., Jao, T.-C., and Fendler, J. H. 1996. Distribution of functional groups grafted onto an ethylene-propylene copolymer. *Journal of Polymer Science, Part B: Polymer Physics* 34:1723–1732.

130. Liu, N. C. and Baker, W. E. 1994. Basic functionalization of polypropylene and the role of interfacial chemical bonding in its toughening. *Polymer* 35:988–994.

131. Schellekens, M. A. J. and Klumperman, B. 2000. Synthesis of polyolefin block and graft copolymers. *Polymer Reviews* 40:167–192.

132. Moad, G. 1999. The synthesis of polyolefin graft copolymers by reactive extrusion. *Progress in Polymer Science* 24:81–142.

133. Liu, N. C., Baker, W. E., and Russell, K. E. 1990. Functionalization of polyethylenes and their use in reactive blending. *Journal of Applied Polymer Science* 41:2285–2300.

134. Coutinho, F. M. B. and Ferreira, M. I. P. 1994. Optimization of reaction conditions of bulk functionalization of EPDM rubbers with maleic anhydride. *European Polymer Journal* 30:911–918.

135. Jurkowski, B., Pesetskii, S. S., Olkhov, Y. A., Krivoguz, Y. M., and Kelar, K. 1999. Investigation of molecular structure of LDPE modified by itaconic acid grafting. *Journal of Applied Polymer Science* 71:1771–1779.

136. Pesetskii, S. S., Jurkowski, B., Krivoguz, Y. M., and Urbanowicz, R. 1997. Itaconic acid grafting on LDPE blended in molten state. *Journal of Applied Polymer Science* 65:1493–1502.

137. Krivoguz, Y. M., Pesetskii, S. S., Jurkowski, B., and Olkhov, Y. A. 2001. Solubility of additives: Grafting of itaconic acid onto LDPE by reactive extrusion. II. Effect of stabilizers. *Journal of Applied Polymer Science* 81:3439–3448.

138. Krivoguz, Y. M., Pesetskii, S. S., and Jurkowski, B. 2003. Grafting of itaconic acid onto LDPE by the reactive extrusion: Effect of neutralizing agents. *Journal of Applied Polymer Science* 89:828–836.

139. Pesetskii, S. S., Krivoguz, Y. M., and Jurkowski, B. 2004. Structure and properties of polyamide 6 blends with low-density polyethylene grafted by itaconic acid and with neutralized carboxyl groups. *Journal of Applied Polymer Science* 92:1702–1708.

140. Krivoguz, Y. M., Pesetskii, S. S., Jurkowski, B., and Tomczyk, T. 2006. Structure and properties of polypropylene/low density polyethylene blends grafted with itaconic acid in the course of reactive extrusion. *Journal of Applied Polymer Science* 102:1746–1754.

141. Devrim, Y. G., Eğri, S., Rzayev, Z. M. O., and Pişkin, E. 2004. Grafting of powder and granular polypropylene in melt using reactive extrusion, in Abstract *40th IUPAC World Polymer Congress*, Paris, France, July 4–9, p. 1975.

142. Devrim, Y. G., Eğri, S., Rzayev, Z. M. O., and Pişkin, E. 2004. Grafting and graft (co)polymerization of polypropylene, in Abstract *40th IUPAC World Polymer Congress*, Paris, France, July 4–9, p. 1977.

143. Güldoğan, Y., Rzayev, Z. M. O., and Pişkin, E. 2004. Polypropylene functionalization in solution and in melt by reactive extrusion, in Abstract *11th International Conference on Polymers and Organic Chemistry*, Praque, Czech., p. P42.

144. Devrim, Y., Rzaev, Z. M. O., and Pişkin, E. 2004. Anhydrido-carboxylation of polypropylene: Synthesis and characterization of poly(PP-*g*-MA), poly(PP-*g*-CA) and poly(PP-*g*-AA). *15th International Conference on Organic Synthesis*, in Abstract. JUPAC ICOS-15, August 1–6, Nagoya, Japan, p. 3F-01.

145. Devrim, Y., Rzaev, Z. M. O., and Pişkin, E. 2007. Functionalization of isotactic polypropylene with citraconic anhydride. *Polymer Bulletin* 59(4):447–456.

146. Dong, Q. and Liu, Y. 2003. Styrene-assisted free-radical graft copolymerization of maleic anhydride onto polypropylene in supercritical carbon dioxide. *Journal of Applied Polymer Science* 90:853–860.

147. Li, X., Xie, X. M., and Guo, B. H. 2001. Study on styrene-assisted melt free-radical grafting of maleic anhydride onto polypropylene. *Polymer* 42:3419–3425.

148. Cansarz, I. and Laskawski, W. 1979. Peroxide-initiated crosslinking of maleic anhydride-modified low-molecular-weight polybutadiene. I. Mechanism and kinetics of the reaction. *Journal of Polymer Science, Polymer Chemistry Edition* 17:683–692.

149. Cansarz, I. and Laskawski, W. 1979. Peroxide-initiated crosslinking of maleic anhydride-modified low-molecular-weight polybutadiene. II. Crosslinking degree and thermal degradation of cured polymers. *Journal of Polymer Science, Polymer Chemistry Edition* 17:1523–1529.

150. Gaylord, N. G., Oikawa, E., and Takahashi, A. 1971. Alternating graft copolymers. I. Grafting of alternating copolymers in the absence of complexing agents. *Journal of Polymer Science* B9:379–385.

151. Gaylord, N. G., Antropiusova, A., and Patnaik, K. 1971. Alternating graft copolymers. II. Grafting of alternating copolymers in the presence of complexing agents. *Journal of Polymer Science* B9:387–393.

152. Hu, G. H., Flat, J.-J., and Lambla, M. 1996. In *Reactive Modifiers for Polymers*, ed. S. Al-Malaika. London, U.K.: Chapman & Hall.

153. Abd El-Rehim, H. A., Hegazy, E. A., and El-Hag Ali, A. 2000. Selective removal of some heavy metal ions from aqueous solution using treated polyethylene-*g*-styrene/maleic anhydride membranes. *Reactive and Functional Polymers* 43:105–116.

154. Gabara, W. and Porejko, S. 1967. Grafting of maleic anhydride on polyethylene. I. Mechanism of grafting in a heterogeneous medium in the presence of radical initiators. *Journal of Polymer Science* 5:1547–1562.

155. Ranganathan, S., Baker, W. F., Russel, K. E., and Whitney, B. A. 1999. Peroxide initiated maleic anhydride grafting: Structural studies on an ester-containing copolymer and related substrates. *Journal of Polymer Science, Part A: Polymer Chemistry* 37:1609–1618.

156. Samay, G., Nagy, T., and White, J. L. 1995. Grafting maleic anhydride and comonomers onto polyethylene. *Journal of Applied Polymer Science* 56:1423–1433.

157. Jia, D., Luo, Y., Li, Y., Lu, H., Fu, W., and Cheung, W. L. 2000. Synthesis and characterization of solid-phase graft copolymer of polypropylene with styrene and maleic anhydride. *Journal of Applied Polymer Science* 78:2482–2487.

158. Liu, N. C. and Baker, W. E. 1992. Reactive polymers for blend compatibilization. *Advance Polymer Technology* 11:249–262.

159. Lorenzo, M. L. and Frigione, M. 1997. Compatibilization criteria and procedures for binary blends: A review. *Journal of Polymer Engineering* 17:429–437.

160. Datta, S. and Lohse, D. 1996. *Polymeric Compatibilizers*. New York/Munich, Germany: Hanser Publishers.

161. Covas, J. A., Machardo, A. V., and Van Dun, M. 2000. Rheology of PA-6/EPM/EPM-g-MA blends along a twin-screw extruder. *Advance Polymer Technology* 19:260–276.

162. Xanthos, M. and Dagli, S. S. 1991. Compatibilization of polymer blends by reactive processing. *Polymer Engineering Science* 31:929–935.

163. Xanthos, M. 1992. *Reactive Extrusion. Principles and Practice.* New York: Oxford University.
164. Natov, M., Mitova, V., and Vassileva, S. 2004. On the use of extruders as chemical reactors. *Journal of Applied Polymer Science* 92:871–877.
165. Cowie, J. M. G. 1993. *Polymers: Chemistry and Physics of Modern Materials.* London, U.K.: CRC Press.
166. Marechal, P., Coppens, G., Legras, R., and De-Koninck, J. M. 1995. Amine/anhydride reaction versus amide/anhydride reaction in polyamide/anhydride carriers. *Journal of Polymer Science, Part A: Polymer Chemistry* 33:757–766.
167. Galli, P. and Vecellio, G. 2004. Polyolefins: The most promising large-volume materials for the 21st century. *Journal of Polymer Science, Part A: Polymer Chemistry* 42:396–415.
168. Abacha, N. and Fellahi, S. 2002. Synthesis of PP-g-MAH and evaluation of its effect on the properties of glass fibre reinforced nylon 6/polypropylene blends. *Macromolecular Symposia* 178:131–138.
169. Abacha, N. and Fellahi, S. 2005. Synthesis of polypropylene-*graft*-maleic anhydride compatibilizer and evaluation of nylon 6/polypropylene blend properties. *Polymer International* 54:909–916.
170. Wu, S. 1985. Phase structure and adhesion in polymer blends: A criterion for rubber toughening. *Polymer* 26:1855–1863.
171. Okada, O., Keskkula, H., and Paul, D. R. 2004. Fracture toughness of nylon-6 blends with maleated rubbers. *Journal of Polymer Science, Part B: Polymer Physics* 42:1739–1758.
172. Dong, L., Xiong, C., Wang, T., Liu, D., Lu, S., and Wang. Y. 2004. Preparation and properties of compatibilized PVC/SMA-*g*-PA6 blends. *Journal of Applied Polymer Science* 94:432–439.
173. Tjong, S. C. and Xu, S. A. 1998. Impact and tensile properties of SEBS copolymer compatibilized PS/HDPE blends. *Journal of Applied Polymer Science* 68:1099–1108.
174. Wong, S. W. and Mai, Y. W. 1999. Effect of rubber functionality on microstructures and fracture toughness of impact-modified nylon 6,6/polypropylene blends: 1. Structure–property relationships. *Polymer* 40:1553–1566.
175. Wu, Y., Yang, Y., Li, B., and Han, Y. 2006. Reactive blending of modified polypropylene and polyamide 12: Effect of compatibilizer content on crystallization and blend morphology. *Journal of Applied Polymer Science* 100:187–192.
176. Tjong, S. C. and Meng, Y. Z. 1997. The effect of compatibilization of maleated polypropylene on a blend of polyamide-6 and liquid crystalline copolyester. *Polymer International* 42:209–217.
177. Meng, Y. Z. and Tjong, S. C. 1998. Effects of processing conditions on the mechanical performance of maleic anhydride compatibilized in-situ composites of polypropylene with liquid crystalline polymer. *Polymer Composites* 19:1–10.
178. Filippi, S., Yordanov, H., Minkova, L., Polacco, G., and Talarico, M. 2004. Reactive compatibilizer precursors for LDPE/PA6 blends. *Macromolecular Materials and Engineering* 289:512–523.
179. Chiono, V., Filippi, S., Yordanov, H., Minkova, L., and Magagnini, P. 2003. Reactive compatibilizer precursors for LDPE/PA6 blends. III: Ethylene–glycidylmethacrylate copolymer. *Polymer* 44:2423–2432.
180. Minkova, L., Yordanov, H., Filippi, S., and Grizzuti, N. 2003. Interfacial tension of compatibilized blends of LDPE and PA6: The breaking thread method. *Polymer* 44:7925–7932.
181. Chen, J. C. and Harrison, I. R. 1998. Modification of nylon-polyethylene blends via in situ fiber formation. *Polymer Engineering Science* 38:371–383.
182. Armat, R. and Moet, A. 1993. Morphological origin of toughness in polyethylene-nylon-6 blends. *Polymer* 34:977–985.
183. Lawson, D. F., Hergenrother, W. L., and Matlock, M. G. 1990. Influence of interfacial adhesion on toughening of polyethylene-octene elastomer/nylon 6 blends. *Journal of Applied Polymer Science* 39:2331–2352.
184. Margolina, A. and Wu, S. 1988. Percolation model for brittle-tough transition in nylon/rubber blends. *Polymer* 29:2170–2173.
185. Kayano, Y., Keskkula, H., and Paul, D. R. 1998. Fracture behaviour of some rubber-toughened nylon 6 blends. *Polymer* 39:2835–2845.

186. Borggreve, R. J. M. and Gaymans, R. J. 1988. Impact modification of poly(caprolactam) by copolymerization with a low molecular weight polybutadiene. *Polymer* 29:1441–1446.
187. Borggreve, R. J. M., Gaymans, R. J., Schuijer, J., and Inden Housz, J. F. 1987. Brittle-tough transition in nylon-rubber blends: Effect of rubber concentration and particle size. *Polymer* 28:1489–1496.
188. Holsti-Mettinen, R., Seppala, J., and Ikkala, O. T. 1992. Effects of compatibilizers on the properties of polyamide/polypropylene blends. *Polymer Engineering and Science* 32:868–877.
189. Kim, G. M., Michler, G. H., Gahleitner, M., and Mülhaupt, R. 1998. Influence of morphology on the toughening mechanisms of polypropylene modified with core-shell particles derived from thermoplastic elastomers. *Polymer Advance Technology* 9:709–715.
190. Kim, G. M., Michler, G. H., Rösch, J., and Mülhaupt, R. 1998. Micromechanical deformation processes in toughened PP/PA/SEBS-g-MA blends prepared by reactive processing. *Acta Polymerica* 49:88–95.
191. Wilkinson, A. N., Clemens, M. L., and Harding, V. M. 2004. The effects of SEBS-g-maleic anhydride reaction on the morphology and properties of polypropylene/PA6/SEBS ternary blends. *Polymer* 45:5239–5249.
192. Gonza'les-Montiel, A., Keskkula, H., and Paul, D. R. 1995. Impact-modified nylon 6/polypropylene blends: 1. Morphology-property relationships. *Polymer* 36:4587–4603.
193. Gonza'les-Montiel, A., Keskkula, H., and Paul, D. R. 1995. Impact-modified nylon 6/polypropylene blends: 2. Effect of reactive functionality on morphology and mechanical properties. *Polymer* 36:4605–4620.
194. Gonza'les-Montiel, A., Keskkula, H., and Paul, D. R. 1995. Impact-modified nylon 6/polypropylene blends: 3. Deformation mechanisms. *Polymer* 36:4621–4637.
195. Huang, J. J., Keskkula, H., and Paul, D. R. 2004. Rubber toughening of an amorphous polyamide by functionalized SEBS copolymer: Morphology and Izod impact behaviour. *Polymer* 45:4203–4215.
196. Lee, Y. and Char, K. 1994. Enhancement of interfacial adhesion between amorphous polyamide and polystyrene by in-situ copolymer formation at the interface. *Macromolecules* 27:2603–2606.
197. Dijkstra, K., Laak, J., and Gaymans, R. J. 1994. Nylon-6/rubber blends: 6. Notched tensile impact testing of nylon-6/(ethylene- propylene rubber) blends. *Polymer* 35:315–322.
198. Crespy, A., Caze, C., Coupe, D., Dupont, P., and Cavrot, J. P. 1992. Impact resistance performances of polyamide 6 blended with different rubber phases. *Polymer Engineering and Science* 32:273–276.
199. Yu, Z. Z., Ou, Y. C., and Hu, G. H. 1998. Influence of interfacial adhesion on toughening of polyethylene-octene elastomer/nylon 6 blends. *Journal of Applied Polymer Science* 69:1711–1718.
200. Li, Q. F., Kim, D. G., Wu, D. Z., Lu, K., and Jin, R. G. 2001. Effect of maleic anhydride graft ratio on mechanical properties and morphology on nylon 11/ethylene-octene copolymer blends. *Polymer Engineering Science* 41:2155–2161.
201. Chakrit, C. B., Sauvarop, L., and Jarunee, T. 2003. Effects of fillers, maleated ethylene propylene diene. Diene rubber, and maleated ethylene octene copolymer on phase morphology and oil resistance in natural rubber/nitrile rubber blends. *Journal of Applied Polymer Science* 89:1156–1162.
202. Zhao, R. and Dai, G. 2002. Mechanical property and morphology comparison between the two blends poly(propylene)/ethylene-propylene-diene monomer elastomer and poly(propylene)/maleic anhydride-g-ethylene-propylene-diene monomer. *Journal of Applied Polymer Science* 86:2486–2491.
203. Purnima, D., Maiti, S. N., and Gupta, A. K. 2006. Interfacial adhesion through maleic anhydride grafting of EPDM in PP/EPDM blend. *Journal of Applied Polymer Science* 102:5528–5532.
204. van Duin, M., Aussems, M., and Borggreve, R. J. M. 1998. Graft formation and chain scission in blends of polyamide-6 and -6.6 with maleic anhydride containing polymers. *Journal of Polymer Science, Part A: Polymer Chemistry* 36:179–188.

205. Lu, I., Wu, Z., and Negulescu, Q. 2004. II. Wood-fiber/high-density-polyethylene composites: Compounding process. *Journal of Applied Polymer Science* 93:2570–2578.
206. Mohanthy, A. K., Mishra, M., and Hinrichsen, G. 2000. Biofibres, biodegradable polymers and biocomposites: An overview. *Macromolecular Materials and Engineering* 276/277:1–24.
207. Paunikallio, T., Kasanen, J., Suvanto, M., and Pakkanen, T. T. 2003. Influence of maleated poly-propylene on mechanical properties of composite made of viscose fiber and polypropylene. *Journal of Applied Polymer Science* 87:1895–1900.
208. Aburto, J., Thiebaud, S., Alric, L., Borredon, E., Bikiaris, D., Prinos, J., and Panayiotou, C. 1997. Properties of octanoated starch and its blends with polyethylene. *Carbohydrate Polymers* 34:101–112.
209. Tasanatanachai, P., Singsat, N., Jamieson, A. M., and Magaraphan, R. 2008. Reathable film from reactive processing of LLDPE/NR blends with ENP and maleic anhydride. *International Polymer Processing* 23:146–151.
210. Aranberri-Askargorta, I., Lampke, T., and Bismarck, A. 2003. Wetting behavior of flax fibers as reinforcement for polypropylene. *Journal of Colloid and Interface Science* 263:580–589.
211. Qiu, W., Endo, T., and Hirotsu, T. 2004. Interfacial interactions of a novel mechanochemi-cal composite of cellulose with maleated polypropylene. *Journal of Applied Polymer Science* 93:1326–1335.
212. Beecroft, L. L. and Ober, C. K. 1997. Nanocomposite materials for optical applications. *Chemistry of Materials* 9:1302–1330.
213. Chrissopoulou, K., Altintzi, I., Anastasiadis, S. H., Giannelis, E. P., Pitsikalis, M., Hadjichristidis, N., and Theophilou, N. 2005. Controlling the miscibility of polyethylene/layered silicate nano-composites by altering the polymer/surface interactions. *Polymer* 46:12440–12451.
214. Durmus, A., Kasgoz, C., and Macosko, W. 2007. Linear low density polyethylene (LLDPE)/clay nanocomposites. Part I-Structural characterization and quantification of clay dispersion by melt rheology. *Polymer* 48:4492–4502.
215. Hotta, S. and Paul, D. R. 2004. Nanocomposites formed from linear low density polyethylene and organoclays. *Polymer* 45:7639–7654.
216. Lee, J., Kontopoulou, M., and Parent, J. S. 2004. Time and shear dependent rheology of male-ated polyethylene and its nanocomposites. *Polymer* 45:6595–6600.
217. Wang, K. H., Choi, M. H., Koo, C. M., Xu, I. J., Chung, M. C., Jang, S. W., Choi, H., and Song, H. 2002. Morphology and physical properties of polyethylene/silicate nanocomposite prepared by melt intercalation. *Journal of Polymer Science, Part B: Polymer Physics* 40:1454–1463.
218. Százdi, L., Abranyi, A., Pukánszky, B. Jr., Vansco, J. G., and Pukánszky, B. 2006. Morphology characterization of PP/clay nanocomposites across the length scales of the structural architec-ture. *Macromolecular Materials and Engineering* 291:858–868.
219. Vermogen, A., Masenelli-Varlot, K., Seguela, R., Duchet-Rumeau, J., Boucard, S., and Prele, P. 2005. Evaluation of the structure and dispersion in polymer-layered silicate nanocomposites. *Macromolecules* 38:9661–9669.
220. Wang, Z. M., Nakajima, H., Manias, E., and Chung, T. C. 2003. Exfoliated PP/clay nanocom-posites using ammonium-terminated PPas the organic modification for montmorillonite. *Macromolecules* 36:8919–8922.
221. Reichert, P., Nitz, H., Klinke, S., Brandsch, R., Thomann, R., and Mülhaupt, R. 2000. Poly(propylene)/organoclay nanocomposite formation: Influence of compatibilizer function-ality and organoclay modification. *Macromolecular Materials and Engineering* 275:8–17.
222. Százdi, L., Pukánszky, B. Jr., Földes, E., and Pukánszky, B. Jr. 2005. Possible mechanism of inter-action among the components in MAPP modified layered silicate PP nanocomposites. *Polymer* 46:8001–8010.
223. Kim, D. H., Fasulo, P. D., Rodgers, W. P., and Paul, D. R. 2007. Structure and properties of polypropylene-based nanocomposites: Effect of PP-g-MA to organoclay ratio. *Polymer* 48:5308–5323.
224. Lertwimolnun, W. and Vergnes, B. 2005. Influence of compatibilizer and processing conditions on the dispersion of nanoclay in a polypropylene matrix. *Polymer* 46:3462–3471.

225. Garcia-Lopez, D., Picazo, O., Merino, J. C., and Pastor, J. M. 2003. Polypropylene-clay nanocomposites: Effect of compatibilizing agents on clay dispersion. *European Polymer Journal* 39:945–950.

226. Galgali, G., Ramesh, C., and Lele, A. 2001. A rheological study on the kinetics of hybrid formation in polypropylene nanocomposites. *Macromolecules* 34:852–858.

227. Zhao, Z., Tang, T., Qin, Y., and Huang, B. 2003. Effects of surfactant soadings on the dispersion of clays in maleated polypropylene. *Langmuir* 19:7157–7159.

228. Morgan, A. B. and Harris, J. D. 2003. Effects of organoclay Soxhlet extraction on mechanical properties, flammability properties and organoclay dispersion of polypropylene nanocomposites. *Polymer* 44:2313–2320.

229. Solomon, M. J., Almusallam, A. S., Seefeldt, K. F., Somwangthanaroj, A., and Varadan, P. 2001. Rheology of polypropylene/clay hybrid materials. *Macromolecules* 34:1864–1872.

230. Gregoriou, V. G., Kandilioti, G., and Bollas, S. T. 2005. Chain conformational transformations in syndiotactic polypropylene/layered silicate nanocomposites during mechanical elongation and thermal treatment. *Polymer* 46:11340–11350.

231. Wang, K., Liang, S., Du, R., Zhang, Q., and Fu, Q. 2004. The interplay of thermodynamics and shear on the dispersion of polymer nanocomposite. *Polymer* 45:7953–7960.

232. Osman, M. A. and Rupp, J. E. P. 2005. Interfacial interactions and properties of polyethylene-layered silicate nanocomposites. *Macromolecular Rapid Communications* 26:880–884.

233. Durmus, A., Woo, M., Kasgoz, A., Macosko, C. W., and Tsapatsis, M. 2007. Intercalated linear low density polyethylene LLDPE)/clay nanocomposites prepared with oxidized polyethylene as a new type compatibilizer: Structural, mechanical and barrier properties. *European Polymer Journal* 43:3737–3749.

234. Durmus, A., Woo, M., Kasgoz, A., and Macosko, C. W. 2008. Mechanical properties of linear low density polyethylene (LLDPE)/clay nanocomposites: Estimation of aspect ratio and interfacial strength by composite models. *Journal of Macromolecular Science, Part B: Physics* 47:608–619.

235. Nguyen, Q. T. and Baird, D. G. 2006. Preparation of polymer-clay nanocomposites and their properties. *Advance Polymer Technology* 25:270–285.

236. Alexandre, M., Dubois, P., Jérôme, R., Garcia-Marti, M., Sun, T., Carces, J. M., Millar, D. M., and Kuperman, A. 1999. Polyolefin nanocomposites, WO Patent 9,947,598 A1.

237. Wang, K. H., Choi, M. H., Koo, C. M., Choi, Y. S., and Chung, I. J. 2001. Synthesis and characterization of maleated polyethylene/clay nanocomposites. *Polymer* 42:9819–9826.

238. Koo, C. M., Ham, H. T., Kim, S. O., Wang, K. H., and Chung, I. J. 2002. Morphology evolution and anisotropic phase formation of the maleated polyethylene-layered silicate nanocomposites. *Macromolecules* 35:5116–5122.

239. Wang, K. H., Chung, I. J., Jang, M. C., Keum, J. K., and Song, H. H. 2002. Deformation behavior of polyethylene/silicate nanocomposites as studied by real-time wide-angle x-ray scattering. *Macromolecules* 35:5529–5535.

240. Gopakumar, T. O., Lee, J. A., Kontopoulou, M., and Parent, J. S. 2002. Influence of clay exfoliation on the physical properties of montmorillonite/polyethylene composites. *Polymer* 43:5483–5491.

241. Zhou, Z., Zhai, H., Xu, W., Guo, H., Liu, C., and Pan, W. P. 2006. Preparation and characterization of polyethylene/-g-maleic anhydride–styrene/MMT nanocomposite. *Journal of Applied Polymer Science* 101:805–809.

242. Zhai, H., Li, Y., Guo, H., Liu, C., Zhou, Z., Song, Q., and Xu, W. 2005. Preparation and characterization of functional polyethylene and its nanocomposites. *Polymer Materials Science and Engineering* 21(2):252–255.

243. Sibolda, N., Dufourb, C., Gourbilleaub, F., Metznerb, M.-N., Lagrèvec, C., Le Pluarta, L., Madeca, P.-J., and Pham, T.-N. 2007. Montmorillonite for clay-polymer nanocomposites: Intercalation of tailored compounds based on succinic anhydride, acid and acid salt derivatives—A review. *Applied Clay Science* 38:130–138.

244. Masenelli-Varlot, K., Vigier, G., and Vermogen, A. 2007. Quantitative structural characterization of polymer-clay-nanocomposites and discussion of an "ideal" microstructure, leading to the highest mechanical reinforcement. *Journal of Polymer Science, Part B: Polymer Physics* 45:1243–12351.

245. Usuki, A., Kato, M., Okada, A., and Kurauchi, T. 1997. Synthesis of polypropylene-clay hybrid. *Journal of Applied Polymer Science* 63:137–138.

246. Nasegawa, H., Okamoto, H., Kawasumi, M., Hasegawa, N., and Usuki, A. 2001. A hierarchical structure and properties of intercalated polypropylene/clay nanocomposites. *Polymer* 42:9633–9640.

247. Nasegawa, H., Okamoto, H., and Usuki, A. 2004. Preparation and properties of ethylene-propylene rubber (EPR)-clay nancomposites based on maleic anhydride-modified EPR and organophilic clay. *Journal of Applied Polymer Science* 93:758–764.

248. Koo, C. M., Kim, M. J., Choi, M. N., Kim, S. O., and Chung, I. J. 2003. Mechanical and rheological properties of the maleated polypropylene-layered silicate nanocomposites with different morphology. *Journal of Applied Polymer Science* 88:1526–1535.

249. Liu, H., Lim, T., An, K. H., and Lee, S. J. 2007. Effect of ionomer on clay dispersions in polypropylene-layered silicate Nanocomposites. *Journal of Applied Polymer Science* 104:4024–4034.

250. Xu, B., Leisen J., Beckham, H. W., Abu-Zurayk, R., Harkin-Jones, E., and McNally, T. 2009. Evolution of clay morphology in polypropylene/montmorillonite nanocomposites upon equibiaxial stretching: A solid-state NMR and TEM approach. *Macromolecules* 42, DOI: 10.1021/ ma901754m (ASAP article).

251. La Mantia, F. P., Dintcheva, N. D., Scaffaro, R., and Marino, R. 2008. Morphology and properties of polyethylene/clay nanocomposite drawn fibers. *Macromolecular Materials and Engineering* 293:83–91.

252. Osman, M. A., Mittal, V., and Suter, U. W. 2007. Poly(propylene)-layered silicate nanocomposites: Gas permeation properties and clay exfoliation. *Macromolecular Chemistry and Physics* 208:68–75.

253. Xu, W., Liand, G., Wang, W., Tang, S., He, P., and Pan, W.-P. 2003. Poly(propylene)-poly(propylene)-grafted maleic anhydride-organic montmorillonite (PP-PP-g-MAH-Org-MMT) nanocomposites. II. Nonisothermal crystallization kinetics. *Journal of Applied Polymer Science* 88:3093–3099.

254. Ton-That, M.-T., Leelapornpisit, W., Utracki, L. A., Perrin-Sarazin, F., Denault, J., Cole, K. C., and Bureau, M. N. 2006. Effect of crystallization on intercalation of clay-polyolefin nanocomposites and their performance. *Polymer Engineering and Science* 46:1085–1093.

255. Li, J.-M. and Ton-That, M.-T. 2006. PP-based nanocomposites with various intercalant types and intercalant coverages. *Polymer Engineering and Science* 46:1061–1068.

256. Ratnayake, U. N. and Haworth, B. 2006. Polypropylene-clay nanocomposites: Influence of low molecular weight polar additives on intercalation and exfoliation behavior. *Polymer Engineering and Science* 46:1009–1015.

257. Peltola, P., Wälipakka, E., Vuorinen, J., Syrjälä, S., and Hanhi, K. 2006. Effect of rotation speed of twin screw extruder on the microstructure and rheological and mechanical properties of nanoclay-reinforces polypropylene nanocomposites. *Polymer Engineering and Science* 46:995–1000.

258. Kurokawa, Y., Yasuda, H., and Oya, A. 1996. Preparation of a nanocomposite of polypropylene and smectite. *Journal of Materials Science Letters* 15:1481–1483.

259. Kurokawa, Y., Yasuda, H., Kashiwagi, M., and Oya, A. 1997. Structure and properties of a montmorillonite/polypropylene nanocomposite. *Journal of Materials Science Letters* 16:1670–1672.

260. Oya, A., Kurokawa, Y., and Yasuda, H. 2000. Factors controlling mechanical properties of clay mineral/polypropylene nanocomposites. *Journal of Materials Science* 35:1045–1050.

261. Boucard, S., Duchet, J., Gérard, J. F., Prele, P., and Gonzalez, S. 2003. Processing of polypropylene–clay hybrids. *Macromolecular Symposia* 194:241–246.

262. Zhang, Q., Fu, Q., Jiang, J. X., and Lei, Y. 2000. Preparation and properties of polypropylene/ montmorillonite layered nanocomposites. *Polymer International* 49:1561–1564.

263. Parija, S., Nayak, S. K., Verma, S. K., and Tripathy, S. S. 2004. Studies on physico-mechanical properties and thermal characteristics of polypropylene/layered silicate nanocomposites. *Polymer Composites* 25:646–652.

264. Morgan, A. B. and Gilman, J. W. 2003. Characterization of polymer-layered silicate (clay) nano-composites by transmission electron microscopy and x-ray diffraction: A comparative study. *Journal of Applied Polymer Science* 87:1329–1338.
265. Wenyi, W., Xiaofei, Z., Guoquan, W., and Jianfeng, C. 2006. Preparation and properties of polypropylene filled with organo-montmorillonite nanocomposites. *Journal of Applied Polymer Science* 100: 2875–2880.
266. Tidjani, A., Wald, O., Pohl, M.-M., Henschel, M. P., and Schartel, B. 2003. Polypropylene–*graft*–maleic anhydride-nanocomposites: I. Characterization and thermal stability of nanocomposites produced under nitrogen and in air. *Polymer Degradation and Stability* 82:133–140.
267. Diagne, M., Gueye, M., Vidal, L., and Tidjani, A. 2005. Thermal stability and fire retardant performance of photo-oxidized nanocomposites of polypropylene-*graft*-maleic anhydride/clay. *Polymer Degradation and Stability* 89:418–426.
268. Ramos, F. F. G., Melo, T. G. A., Rabello, M. S., and Silva, S. M. L. 2005. Thermal stability of nanocomposites based on polypropylene and bentonite. *Polymer Degradation and Stability* 89:383–392.
269. Tjong, S. C. and Meng, Y. Z. 1997. Morphology and mechanical characteristics of compatibilized polyamide 6-liquid crystalline polymer composites. *Polymer* 38:4609–4615.
270. Tjong, S. C. and Meng, Y. Z. 1998. Rheology and morphology of compatibilized polyamide 6 blends containing liquid crystalline copolyesters. *Polymer* 39:99–106.
271. Tjong, S. C. and Meng, Y. Z. 1999. Microstructural and mechanical characteristics of compatibilized polypropylene hybrid composites containing potassium titanate whisker and liquid crystalline copolyester. *Polymer* 40:7275–7281.
272. Tjong, S. C., Meng, Y. Z., and Hay, A. S. 2002. Novel preparation and properties of polypropylene–vermiculite nanocomposites. *Chemistry of Materials* 14:44–51.
273. Toshniwal, L., Fan, Q.-J., and Ugbolue, S. C. 2007. Dyeable polypropylene fibers via nanotechnology. *Journal of Applied Polymer Science* 106:706–711.
274. Rzayev, Z. M. O. 2007. Nanotechnology methods in polymer engineering, in *Proceedings of Fourth Nanoscience and Nanotechnology Conference*, June 11–14, Bilkent, Ankara, Turkey, p. 154.
275. Rzayev, Z. M. O., Yilmazbayhan, A., and Alper, E. 2007. Synthesis and structural models of polypropylene/compatibilizer/organo-silicate nanocomposites by reactive extrusion in situ processing, in *Proceedings of Fourth Nanoscience and Nanotechnology Conference*, June 11–14, Bilkent, Ankara, Turkey, p. 153.
276. Nazli, K. O., Rzayev, Z. M. O., and Alper, E. 2007. Polypropylene/lamellar clay blends and their nanocomposite films, in *Proceedings of Fourth Nanoscience and Nanotechnology Conference*, June 11–14, Bilkent, Ankara, Turkey, p. 85.
277. Çaylak, N. and Rzayev, Z. M. O. 2007. X-ray diffraction study of isotactic polypropylene/organo-silicate nanostructure, in *Proceedings of Fourth Nanoscience and Nanotechnology Conference*, June 11–14, Bilkent, Ankara, Turkey, p. 211.
278. Wouters, M. E. L., Goossens, J. G. P., and Binsbergen, F. L. 2002. Morphology of neutralized low molecular weight maleated ethylene-propylene copolymers (MAn-*g*-EPM) as investigated by small-angle x-ray scattering. *Macromolecules* 35:208–216.
279. Wouters, M. E. L., Litvinov, V. M., Goossens, J. G. P., Binsbergen, F. L., van Duin, M., and Dikland, H. G. 2003. Morphology of ethylene-propylene copolymer based ionomers as studied by solid state NMR and small angle x-ray scattering in relation to some mechanical properties. *Macromolecules* 36:1147–1156.
280. Schmidt, U., Zchoche, S., and Werner, C. 2003. Modification of poly(octadecene-*alt*-maleic anhydride) films by reaction with functional amines. *Journal of Applied Polymer Science* 87:1255–1266.
281. Vermeesch, I. and Groneninclx, G. 1994. Chemical modification of poly(styrene-*co*-maleic anhydride) with primary *N*-alkyl-amine by reactive extrusion. *Journal of Applied Polymer Science* 53:1357–1363.
282. Schmidt-Naake, G., Becker, H. G., and Klak, M. 2001. Modification of polymers in the melt. *Macromolecular Symposia* 163:213–234.

283. Tsuwi, J., Appelhans, D., Zchoche, S., Friedel P., and Kremer, F. 2004. Molecular dynamics in poly(ethene-*alt*-*N*-alkylmaleimide)s as studied by broad band dielectric spectroscopy. *Macromolecules* 37:6050–6054.

284. Sun, C. X., van der Mee, A. J., Goossens, J. G. P., and van Duin, M. 2006. Thermoreversible cross-linking of maleated ethylene/ propylene copolymers using hydrogen-bonding and ionic interactions. *Macromolecules* 39:3441–3449.

285. Lee, W. L., Lim, Y. T., and Park, O. O. 2000. Thermal characteristics of organoclay and their effects upon the formation of polypropylene/organoclay nanocomposites. *Polymer Bulletin* 45(2):191–198.

286. Xie, W., Gao, Z., Pan, W.-P., Hunter, D., Singh, A., and Vaia, R. 2001. Thermal degradation chemistry of alkyl quaternary ammonium montmorillonite. *Chemistry of Materials* 13:2979–2990.

287. Xie, G.-I., Zhang, P., Gong, S.-G., Chao, W.-N., and An, X.-J. 2005. Multilevel model and prediction of effective properties of crystalline polymer nanocomposites. *Polymer Materials Science and Engineering* 21:23–27.

288. Moad, G., Dean, K., Edmond, L., Kukaleva, N., Li, G., Mayadunne, R. T. A., Pfaender, R., Schneider, A., Simon, G., and Wermter, H. 2005. Non-ionic, poly(ethylene oxide)-based surfactants as intercalants/dispersants/exfoliants for poly(propylene)-clay nanocomposites. *Macromolecular Materials and Engineering* 291:37–52.

289. Moad, G., Dean, K., Edmond, L., Kukaleva, N., Li, G., Mayadunne, R. T. A., Pfaender, R., Schneider, A., Simon, G., and Wermter, H. 2006. Novel copolymers as dispersants/intercalants/exfoliants for polypropylene-clay nanocomposites. *Macromolecular Symposia* 233:170–179.

290. Bandyopadhyay, A., Maiti, M., and Bhowmick, A. K. 2006. Synthesis, characterisation and properties of clay and silica based rubber nanocomposites. *Journal of Materials Science and Technology* 22:818–828.

291. Acharya, H., Srivastava, S. K., and Bhowmick, A. K. 2007. A solution blending route to ethylene propylene terpolymer/layered double hydroxide nanocomposites. *Nanoscale Research Letters* 2:1–5.

292. Chatterjee, K. and Naskar, K. 2007. Development of thermoplastic elastomers based on maleated ethylene propylene rubber and polypropylene by dynamic vulcanizaton. *e-Express Polymer Letters* 1:527–534.

293. Tjong, S. C. and Ruan, Y. H. 2008. Fracture behavior of thermoplastic polyolefin/clay nanocomposites. *Journal of Applied Polymer Science* 110:864–871.

294. Lee, K. Y. and Goettler, L. A. 2004. Structure-property relationships in polymer blend nanocomposites. *Journal of Polymer Engineering and Science* 49:1103–1111.

295. Lee, H., Fasulo, P. D., Rodgers, W. R., and Paul, D. R. 2005. TPO based nanocomposites. Part 1. Morphology and mechanical properties. *Polymer* 26:11673–11689.

296. Mehta, S., Mirabella, F. M., Rufener, K., and Bafna, A. 2004. Thermoplastic olefin/clay nanocomposites: Morphology and mechanical properties. *Journal of Applied Polymer Science* 92:928–936.

297. Kim, D. H., Fasulo, P. D., Rodgers, W. P., and Paul, D. R. 2007. Effect of the ratio of maleated polypropylene to organoclay on the structure and properties of TPO-based nanocomposites. Part I: Morphology and mechanical properties. *Polymer* 48:5960–5978.

298. Hassan, A., Wahit, M. U., and Chee, C. Y. 2005. Mechanical and morphological properties of PP/LLPDE/NR blends—Effects of polyoctenamer. *Polymer-Plastics Technology and Engineering* 44:1245–1256.

299. Wahit, M. U., Hassan, A., Mohd Ishak, Z. A., and Abu Bakar, A. 2005. The effect of polyethylene-octene elastomer on the morphological and mechanical properties of polyamide 6/polypropylene nanocomposites. *Polymers and Polymer Composites* 13:795–806.

300. Lim, J. W., Hassan, A., Wahit, M. U., and Rahmat, A. R. 2006. Rubber toughened polypropylene nanocomposite: Effect of polyethylene octene copolymer on mechanical properties and phase morphology. *Journal of Applied Polymer Science* 99:3441–3450.

301. Lim, J. W., Hassan, A., Rahmat, A. R., and Wahit, M. U. 2006. Morphology, thermal and mechanical behaviour of polypropylene nanocomposites toughened with poly(ethylene-*co*-octene). *Polymer International* 55:204–215.

302. Mishra, J. K., Hwang, K. J., and Ha, C. S. 2005. Preparation, mechanical and rheological properties of a thermoplastic polyolefin (TPO)/organoclay nanocomposite with reference to the effect of maleic anhydride modified polypropylene as a compatibilizer. *Polymer* 46:1995–2002.

303. Austin, J. R. and Kontopoulou, M. 2006. Effect of organoclay content on the rheology, morphology, and physical properties of polyolefin elastomers and their blends with polypropylene. *Journal of Polymer Engineering and Science* 46:1491–1501.

304. Zheng, H., Zhang, Y., Peng, Z., and Zhang, Y. 2004. Influence of clay modification on the structure and mechanical properties of EPDM/montmorillonite nanocomposites. *Polymer Testing* 23:217–223.

305. Mallick, S., Das, T., Das, C. K., and Khatua, B. B. 2009. Synergistic effect of nanoclay and EPR-g-MA on the properties of nylon/EPR blend. *Journal of Nanoscience and Nanotechnology* 9:3099–3105.

306. Li, W., Huang, Y. D., and Ahmadi, S. J. 2004. Preparation and properties of ethylene-propylene-diene rubber/organo-montmorillonite nanocomposites. *Journal of Applied Polymer Science* 94:440–445.

307. Lu, Y. L., Li, Z., Yu, Z. Z., Tian, M., Zhang, L. Q., and Mai, Y. M. 2007. Microstructure and properties of highly filled rubber/clay nanocomposites prepared by melt blending. *Composites Science and Technology* 67:2903–2913.

308. Kojima, Y., Fukumori, K., Usuki, A., Okada, A., and Kurauchi, T. 1993. Gas permeabilities in rubber-clay hybrid. *Journal of Materials Science Letters* 12:889–890.

309. Wang, Y., Zhang, L., Tang, C., and Yu, D. 2000. Preparation and characterization of rubber-clay nanocomposites. *Journal of Applied Polymer Science* 78:1879–1883.

310. Wang, S., Long, C., Wang, X., Li, Q., and Qi, Z. 1998. Synthesis and properties of silicone rubber/organomontmorillonite hybrid nanocomposites. *Journal of Applied Polymer Science* 69:1557–1561.

311. Wang, Z. and Pinnavaia, T. J. 1998. Nanolayer reinforcement of elastomeric polyurethane. *Chemistry of Materials* 10:3769–3771.

312. Bai, S.-L. and Wang, M. 2003. Plastic damage mechanisms of polypropylene/polyamide 6/polyethylene octane elastomer blends under cyclic tension. *Polymer* 44:6537–6547.

313. Bai, S.-L., Wang, G.-T., Hiver, J.-M., and G'Sell, C. 2004. Microstructures and mechanical properties of polypropylene/ polyamide 5/polyethelene-octane elastomer blends. *Polymer* 45:3063–3071.

314. Wahit, M. U., Hassan, A., Rahmat, A. R., Lim, J. W., and Mohd Ishak, Z. A. 2006. Effect of organoclay and ethylene-octene. *Plastics Composites* 25:933–954.

315. Wahit, M. U., Hassan, A., Mohd Ishak, Z. A., Rahmat, A. R., and Abu Bakar, A. 2006. Morphology, thermal, and mechanical behavior of ethylene octene copolymer toughened polyamide 6/polypropylene nanocomposites. *Journal of Thermoplastic Composite Materials* 19:545–567.

316. Wahit, M. U., Hassan, A., Mohd Ishak, Z. A., and Czigány, T. 2009. Ethylene-octene copolymer (POE) toughened polyamide 6/polypropylene nanocomposites: Effect of POE maleation. *e-EXPRESS Polymer Letters* 3:309–319.

317. Grigoryeva, O. P. and Karger-Kocsis, J. 2000. Melt grafting of maleic anhydride onto an ethylene–propylene–diene terpolymer (EPDM). *European Polymer Journal* 36:1419–1420.

318. Uzuki, A., Tukigase, A., and Kato, M. 2002. Preparation and properties of EPDM–clay hybrids. *Polymer* 43:2185–2189.

319. Chang, Y. W., Yang, Y., Ryu, S., and Nah, C. Preparation and properties of EEPDM/organo-montmorillonite hybrid nanocomposites. *Polymer International* 51:319.

320. Ahmedi, S. J., Huang, Y., and Li, W. 2005. Fabrication and physical properties of EDPM-organoclay nanocomposites. *Composites Science and Technology* 65:1069–1076.

321. Mittal, K. L. 1992. *Silanes and Other Coupling Agents.* AH Zeist, the Netherlands: VSB BV.

322. Jain, S., Goossens, H., Picchioni, F., Magusin, P., Mezari, B., and van Duin, M. 2005. Synthetic aspects and characterization of polypropylene-silica nanocomposites prepared via solid-state modification and sol-gel reactions. *Polymer* 46:6666–6681.

323. Akovali, G., Rzaev, Z. M. O., and Mamedov, D. G. 1996. Plasma surface modification of propylene-based polymers by silicon and tin-containing compounds. *Journal of Applied Polymer Science* 58:645–651.

324. Akovali, G., Rzaev, Z. M. O., and Mamedov, D. G. 1996. Plasma surface modification of poly-ethylene with organosilicon and organotin monomers. *European Polymer Journal* 32:375–383.

325. Akovali, G., Rzaev, Z. M. O., and Mamedov, D. G. 1995. Plasma polymerization of some silicon- and tin-containing monomers. *Polymer International* 37:119–128.

326. Bikiaris, D. N., Papageorgiou, G. Z., Pavlidou, E., Vouroutzis, N., Palatzoglou, P., and Karayannidis, G. P. 2006. Preparation by melt mixing and characterization of isotactic polypro-pylene/SiO$_2$ nanocomposites containing untreated and surface-treated nanoparticles. *Journal of Applied Polymer Science* 100:2684–2696.

327. Bikiaris, D. N., Vassiliou, A., Pavlidou, E., and Karayannidis, G. P. 2005. Compatibilisation effect of PP-g-MA copolymer on i-PP/SiO$_2$ nanocomposites prepared by melt mixing. *European Polymer Journal* 41:1965–1978.

328. Zhang, Q., Yang, H., and Fu, Q. 2004. Kinetics-controlled compatibilization of immiscible poly-propylene/polystyrene blends using nano-SiO$_2$ particles. *Polymer* 45:1913–1922.

329. Wu, C. L., Zhang, M. Q., Rong, M. Z., and Friedrich, K. 2005. Silica nanoparticles filled polypro-pylene: Effects of particle surface treatment, matrix ductility and particle species on mechanical performance of the composites. *Composites Science and Technology* 65:635–645.

330. Wu, C. L., Zhang, M. Q., Rong, M. Z., and Friedrich, K. 2002. Tensile performance improvement of low nanoparticles filled-polypropylene composites. *Composites Science and Technology* 62:1327–1340.

331. Reddy, C. S. and Das, C. K. 2006. Polypropylene-nanosilica-filled composites: Effects of epoxy-resin-grafted nanosilica on the structural, thermal, and dynamic mechanical properties. *Journal of Applied Polymer Science* 102:2117–2124.

332. Rong, M. Z., Zhang, M. Q., Zheng, Y. X., Zeng, H. M., Walter, R., and Friedrich, K. 2000. Irradiation graft polymerization on nano-inorganic particles: An effective means to design polymer-based nanocomposites. *Journal of Materials Science Letters* 19:1159–1161.

333. Zhou, H. J., Rong, M. Z., and Friedrich, K. 2006. Effects of reactive compatibilization on the performance of nano-silica filled polypropylene composites. *Journal of Materials Science Letters* 41:5767–5770.

334. Zhou, H. J., Rong, M. Z., Zhang, M. Q., Ruan, W. H., and Friedrich, K. 2007. Role of reactive compatibilization in preparation of nanosilica/polypropylene composites. *Polymer Engineering and Science* 47:499–509.

335. Ruan, W. H., Huang, X. B., Wang, X. H., Rong, M. Z., and Zhang, M. Q. 2006. Effect of draw-ing induced dispersion of nano-silica on performance improvement of poly(propylene)-based nanocomposites. *Macromolecular Rapid Communications* 27:581–585.

336. Lu, Q. W., Macosko, C. W., and Horrion, J. 2005. Melt amination of polypropylenes. *Journal of Polymer Science, Part A: Polymer Chemistry* 43:4217–4232.

337. Chen, J. H., Rong, M. Z., Ruan, W. H., and Zhang, M. Q. 2009. Interfacial enhancement of nano-SiO$_2$/polypropylene composites. *Composites Science and Technology* 69:252–259.

338. Zhang, M. Q., Rong, M. Z., Zhang, H. B., and Friedrich, K. 2003. Mechanical properties of low nano-silica filled high density polyethylene composites. *Polymer Engineering and Science* 43:490–500.

339. Reddy, C. S. and Das, C. K. 2006. Thermal and dynamic mechanical properties of low den-sity polyethylene-silica nanocomposites: Effect of zinc-ion coating on nanosilica. *Polymers and Polymer Composites* 14:281–290.

340. Huang, Y. Q., Jiang, S. L., Wu, L. B., and Hua, Y. Q. 2004. Characterization of LLDPE/nano-SiO$_2$ composites by solid-state dynamic mechanical spectroscopy. *Polymer Testing* 23:9–15.

341. Kontou, E. and Niaounakis, M. 2006. Thermo-mechanical properties of LLDPE/SiO$_2$ nanocom-posites. *Polymer* 47:1267–1280.

342. Gao, X. W., Hu, G. J., Qian, Z. Z., Ding, Y. F., Zhang, S. M., Wang, D. J., and Yang, M. S. 2007. Immobilization of antioxidant on nanosilica and the antioxidative behavior in low density polyethylene. *Polymer* 48:7309–7315.

343. Lee, J. A., Kontopoulou, M., and Parent, J. S. 2005. Synthesis and characterization of polyethylene-based ionomer nanocomposites. *Polymer* 46:5040–5049.
344. Reddy, C. S., Das, C. K., and Narkis, C. M. 2005. Propylene-ethylene copolymer nanocomposites: Epoxy resin grafted nanosilica as reinforcing filler. *Polymer Composites* 26:806–811.
345. Reddy, C. S. and Das, C. K. 2006. Propylene-ethylene copolymer filled nanocomposites: Influence of Zn-ion coating upon nano-SiO$_2$ on structural, thermal, and dynamic mechanical properties. *Polymer-Plastics Technology and Engineering* 45:815–820.
346. Reddy, C. S. and Das, C. K. 2006. Effects of epoxy resin-modified zinc-ion-coated nanosilica on structural, thermal and dynamic mechanical properties of propylene-ethylene copolymer. *Polymer International* 55:923–929.
347. Reddy, C. S., Patra, P. K., and Das, C. K. 2009. Ethylene-octene copolymer-nanosilica nanocomposites: Effects of epoxy resin functionalized nanosilica on structural, mechanical, dynamic mechanical and thermal properties. *Macromolecular Symposia* 277:119–129.
348. Wu, T. M. and Chu, M. S. 2005. Preparation and characterization of thermoplastic vulcanizate/silica nanocomposites. *Journal of Applied Polymer Science* 98:2058–2063.
349. Liu, Y. and Kontopoulou, M. 2006. The structure and physical properties of polypropylene and thermoplastic olefin nanocomposites containing nanosilica. *Polymer* 47:7731–7739.
350. Tanahashi, M., Hirose, M., Lee, J. C., and Takeda, K. 2006. Organic/inorganic nanocomposites prepared by mechanical smashing of agglomerated silica ultrafine particles in molten thermoplastic resin. *Polymer Advance Technology* 17:981–990.
351. Yang, H., Zhang, X., Qu, C., Li, B., Zhang, L., Zhang, Q., and Fu, Q. 2007. Largely improved toughness of PP/EPDM blends by adding nano-SiO$_2$ particles. *Polymer* 48:860–869.
352. Yang, H., Zhang, X. O., Guo, M., Wang, C., Du, R. N., and Fu, Q. 2006. Study on the phase structures and toughening mechanism in PP/EPDM/SiO$_2$ ternary composites. *Polymer* 47:2106–2115.
353. Lee, M. W., Hu, X., Yue, C. Y., Li, L., Tam, K. C., and Nakayama, K. 2002. Novel approach to fibrillation of LCP in an LCP/PP blends. *Journal of Applied Polymer Science* 86:2070–2078.
354. Zhang, L., Tam, K. C., Gan, L. H., Yue, C. Y., Lam, Y. C., and Hu, X. 2003. Effect of nano-silica filler on the rheological and morphological properties of polypropylene/liquid-crystalline polymer blends. *Journal of Applied Polymer Science* 87:1484–1492.
355. Lee, M. W., Hu, X., Li, L., Yue, C. Y., and Tam, K. C. 2003. Flow behaviour and microstructure evolution in novel SiO$_2$/PP/LCP ternary composites: Effects of filler properties and mixing sequence. *Polymer International* 52:276–284.
356. Lee, M. W., Hu, X., Li, L., Yue, C. Y., and Tam, K. C. 2003. Effect of fillers on the structure and mechanical properties of LCP/PP/SiO$_2$ in-situ hybrid nanocomposites. *Composites Science and Technology* 63:339–346.
357. Lee, M. W., Hu, X., Li, L., Yue, C. Y., Tam, K. C., and Cheong, L. Y. 2003. PP/LCP composites: Effects of shear flow, extensional flow and nanofillers. *Composites Science and Technology* 63:1921–1929.
358. Yang, T.-I. and Kofinas, P. 2007. Dielectric properties of polymer nanoparticle composites. *Polymer* 48:791–798.
359. Bounor-Legare, V., Angelloz, C., Blanc, P., Cassagnau, P., and Michel, A. 2004. A new route for organic–inorganic hybrid material synthesis through reactive processing without solvent. *Polymer* 45:1485–1493.

5

Preparation of Polyolefin Nanocomposites by In Situ Polymerization Using Clay-Supported Catalysts

Susannah L. Scott

CONTENTS

5.1 Introduction

5.1.1 The Evolution of Polyolefins: From Commodities to Specialties

About one-half of all synthetic resin produced worldwide is polyolefin.[1] The two principal compositions are polyethylene (mainly its high density and linear, low-density forms) and polypropylene (mainly its isotactic form). The widespread and enduring commercial success of olefins derives from a combination of favorable properties: they are inexpensive, highly processible, durable, and biocompatible. Polyolefins have even been described as "green polymers," based on their low density and high recyclability, as well as their chemical simplicity (halogen-free).[1]

The discoveries of the first-coordination catalysts capable of forming linear polyolefins by a non-radical mechanism at low temperatures and pressures are over half-a-century old. In 1951, Hogan and Banks at Phillips Petroleum observed that a silica-supported chromium catalyst converted ethylene and propylene to crystalline polyethylene and polypropylene, respectively.[2] In 1953, Ziegler found that soluble organotitanium complexes are capable of making linear polyethylene, and Natta used these catalysts in 1954 to achieve the stereospecific polymerization of propylene and styrene. The early polymerization catalysts did not immediately produce polyolefins with desirable properties: the polymers were brittle and had high residual catalyst contents. The development of high-activity δ-$MgCl_2$-supported Ziegler–Natta catalysts, and the use of α-olefins as comonomers were critical to the emergence of polyolefins as commodities in the 1970s.[1] Innovative process technologies (e.g., Unipol, Spheripol) were also critical. The polymers are typically highly linear, but may contain small amounts of short- and long-chain branches due to incorporation of oligomers. It is also quite polydisperse, with typical M_w/M_n values of 6 for Ziegler–Natta catalysts and up to 20 for Phillips' catalysts. Today, the majority of the world's polyolefins are still made using Phillips' and Ziegler–Natta catalysts, and it is not surprising that some catalyst systems for making polyolefin-clay nanocomposites by *in situ* polymerization have been formulated using components of these traditional catalysts (*vide infra*).

The "metallocene revolution"[3] that swept across the polymer industry in the 1980s made the development of polyolefins as specialty materials conceivable. Metallocene-based olefin polymerization catalysts are "single-site," and therefore inherently more tunable than the multi-sited heterogeneous Phillips' and Ziegler–Natta catalysts. Interaction with a cocatalyst is required to create the cationic active sites. The cocatalysts are Lewis acids, including alkylaluminums (e.g., Al^iBu_3) or boranes (e.g., $B(C_6F_5)_3$) that can abstract an alkyl group from the metallocene complex to give a cationic active site. Alkylaluminum-based cocatalysts also alkylate metallocene dichloride precatalysts and scavenge catalyst poisons such as adventitious water. The key discovery that made commercial development of metallocene catalysts feasible was Sinn and Kaminsky's treatment of trialkylaluminum cocatalysts with a limiting amount of water to form alkylaluminoxanes such as methylaluminoxane (MAO), $[MeAlO]_n$.[2] Metallocene polyolefins were first commercialized by Exxon Chemical in 1991 (polyethylene) and 1995 (polypropylene).[2] Chiral *ansa*-metallocenes as highly stereoselective catalysts for propylene polymerization were pioneered by Kaminsky and Brintzinger in the mid-1980s.[4] Metallocenes offer lower polydispersities than Ziegler–Natta catalysts and new mechanisms of stereocontrol, which has already led to new materials such as syndiotactic and hemi-isotactic polypropylene.

Broad patent coverage of the metallocene catalysts spurred many searches for non-metallocene single-site systems. Half-metallocenes such as Dow Chemical's constrained geometry catalyst are particularly effective at incorporating α-olefin comonomers into high molecular weight polyethylene to produce polyolefin elastomers.[5] Brookhart's discovery that ethylene oligomerization catalysts containing low-oxophilicity metals such as Ni and Pd can be converted to polymerization catalysts by increasing the ligand steric bulk established the possibility of olefin copolymerization with polar comonomers,[6] which has only recently been realized.[7] Other nontraditional metals such as Fe and Co have been incorporated in high-activity single-site catalysts by appropriate ligand design.[8] These late transition metal systems may be particularly compatible with clays due to their high tolerance for polar and protic environments.

Despite their high (albeit short-lived) activity in solution, it remains desirable to support single-site catalysts in order to reduce reactor fouling. The resulting polymers generally have increased bulk density and better particle morphology than those made by homogeneous polymerization.[9] Supported catalysts often show higher thermal stability than their homogeneous analogs, which allows the use of higher reactor temperatures leading to higher polymerization efficiencies. Clays have been widely investigated as support activators for metallocenes (*vide infra*), and a clay-supported catalyst is already in use by Japan Polychem for the production of a low-density polypropylene elastomer.[10]

The evolution of polyolefins toward specialty materials will be substantially broadened by enhancements in their stiffness, strength and low temperature toughness, reduced gas permeability and flammability, as well as higher thermal stability.[11] Such materials could be competitive replacements for less desirable commodity polymers (e.g., polyvinylchloride and polyurethane) as well as engineering thermoplastics (e.g., acrylonitrile-butadiene-styrene rubber).[12] Improvement in the target properties can be achieved at low cost by incorporation of mineral fillers, which increases tensile strength, modulus, and hardness. Strong interactions between the two components are crucial, since mixing a hydrophilic filler into a hydrophobic polyolefin as an inhomogeneous dispersion can actually degrade the mechanical properties of the polymer due to weak interfacial adhesion. Stress concentration results in dramatic decreases in elongation at break and impact strength (brittleness). Versatile methods of creating composites containing nanodispersed and strongly interacting clay fillers are needed, and are the subject of this chapter. Since polyolefin manufacturing by gas-phase and slurry polymerization is well-established, drop-in replacement catalysts must be solid, supported systems to be compatible with existing reactors to make nanocomposites or masterbatches for further blending. *In situ* polymerization by clay-supported catalysts is an obvious approach.

5.1.2 "Homogeneous" Polyolefin-Clay Composites by Polymerization Filling

In the early 1970s, groups at Imperial Chemical Industries, England, and at the Institute of Catalysis, Novosibirsk, demonstrated that single-component organometallic complexes with little or no homogeneous olefin polymerization activity become highly active for olefin polymerization when "fixed" or grafted, onto the surfaces of inorganic oxides, particularly those of silica and alumina.[13–15] These discoveries were quickly expanded to other activating supports, including clays. When the polymer is formed in this way from the surface of a clay mineral, and the deaggregated filler particles become encapsulated in the polymer, the resulting composite is called "homogeneous."[16] The clay content of the composite can be controlled by the polymerization time.

This approach was pioneered by researchers at Dupont Central Research and was first described in patents granted to Howard.[17–19] A team at the Institute of Synthetical Polymeric Materials in Moscow, led by Enikolopov, was also an early entrant in the field.[20,21] The technique, dubbed polymerization filling,[22,23] enabled the preparation of polymer-encapsulated filler (at high filler loadings) or uniformly filled polyolefins (at low filler loadings).[24,25] The composites showed higher strength and wear resistance, as well as better thermal stability and reduced flammability, compared to mechanical mixtures. In Russia, these filled polyolefins were called "Norplastics."[26] They included two classes of materials not accessible by either solution methods or conventional mechanical mixing due to either insolubility or extremely high viscosity: composites with very high filler content (up to 95%) and composites based on ultrahigh molecular weight polyethylene.[25] Norplastic concentrates, with high filler content (ca. 50 wt%), served as heat insulating materials with decreased brittleness and high specific impact ductility. The composite materials were also used as fillers for other thermoplastics. For example, Norplastics based on high molecular weight polyethylene were blended with low molecular weight polyethylene to achieve lower melt viscosity.

In polymerization filling, uptake of one or more monomers occurs on the surface of an inorganic filler, thereby creating the filled polymer directly during its synthesis. The method, which has been used for both radical and coordination polymerization, consists of three steps:[22] preparation of the filler, creation of active sites on the filler surface, and polymerization of the monomer on these active sites. Filler preparation involves the elimination of components that inhibit polymerization, such as water and acidic surface hydroxyl groups. This is usually achieved by thermal pretreatment. In the case of coordination polymerization, active sites are created by physical or chemical deposition of a transition metal catalyst or catalyst precursor on the filler surface. While physical deposition does not require the presence of functional groups on the filler surface, chemical deposition (grafting) involves formation of a covalent bond between the metal complex and the filler that persists during polymerization. The availability of appropriate functional groups to create these bonds limits the catalyst loading that can be achieved by grafting; however, additional grafting sites can be created by surface derivatization. Single-component catalysts such as transition metal alkyls, allyls, benzyls, borohydrides, etc., require no cocatalyst for activation. It is also possible to use metal halides in combination with an alkylating cocatalyst, typically an alkylaluminum compound. It may be desirable to deposit the cocatalyst on the filler first, prior to introduction of the catalyst, because the alkylaluminums are also very effective scavengers of catalyst poisons (such as water and hydroxyl groups associated with the filler surface). The state of aggregation of the filler can be greatly affected by deposition of the cocatalyst during active site synthesis. As polymerization proceeds, the filler particles become encapsulated by the growing polymer chains and often become further deaggregated as a result.

Researchers at Dupont reported the formation of high molecular weight polyethylene-clay composites by polymerization of ethylene over virgin kaolinite.[27] When calcined above 400°C, the clay became x-ray amorphous, although it retained some of its stacked structure. Howard et al. attributed its catalytic activity to naturally occurring Ti (0.6–2 wt%) in the clay,[27] although Ti-based catalysts are not known to spontaneously initiate ethylene polymerization in the absence of an external alkylating agent. It is possible that traces of Cr present in the clay were responsible for the polymerization activity, since Phillips' catalysts based on supported Cr oxides can generate active sites directly from ethylene, without a cocatalyst.[28] The distribution of clay in these polyethylene composites was described as grossly nonuniform.

Deposition of an alkylaluminum (Al^iBu_3 or Al^iBu_2Cl) prior to polymerization promoted extensive deaggregation of the clay in hydrocarbon solutions.[27] This role for alkylaluminum reagents, which are often used as cocatalysts for Ziegler–Natta and metallocene polymerization catalysts, has not always been recognized in more recent work, although it may be important in more recent efforts to create nanocomposites (*vide infra*). Ethylene uptake at 30°C–70°C and 100–200 psi over alkylaluminum-modified kaolinite resulted in homogeneous composites, with enhanced flexural moduli and elongation relative to unfilled polymer.[29] Homogeneity was judged by light transmission, hole development, density fractionation, and micronization fractionation. TEM images of the clay extracted from one of these composites show that the large clay particles were substantially disrupted during polymerization, as shown in Figure 5.1.

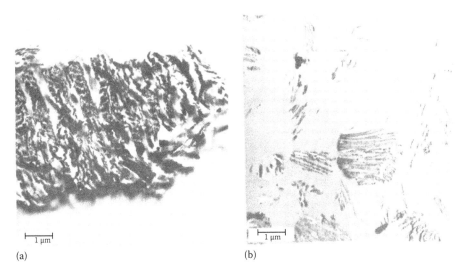

FIGURE 5.1
TEM images of a "homogeneous" kaolin-polyethylene composite prepared by polymerization filling (a), and the insoluble material remaining after extraction of polyethylene, showing fragmentation of the clay particles (b). Scale bars 1 μm. (Reproduced from Howard, E.G. et al., *Ind. Eng. Chem. Prod. Res. Dev.*, 20, 421, 1981. With permission.)

This early work was extended by the Russian group, who reported that natural kaolin containing traces of Ti, V, and Cr showed maximum activity after drying at 873–1273 K, and upon treatment with Al^iBu_3 with an initial rate of 0.22 mmol C_2H_4/(g clay·atm·min) at 353 K and 1.0 MPa.[30] The catalysts showed no induction period but activity declined gradually during the reaction. Extending the contact time between the alkylaluminum and the clay prior to polymerization led to loss of activity, likely due to overreduction of the transition metal active sites.

Addition of the Ziegler catalysts $TiCl_3$ or $TiCl_4$ directly to alkylaluminum-modified kaolinite resulted in the formation of brittle polyethylene, probably due to the formation of soluble active sites $Cl_xTiR_y^+$ which catalyzed simultaneous homogeneous polymerization.[27] However, hydrolysis/calcination of TiX_4- or CrX_3-modified clay (X = Cl, OR, OAc) ensured that no leaching occurred and produced homogeneous composites when activated by an alkylaluminum cocatalyst. The use of single-component catalysts activated by the clay was also successful. $Zr(CH_2Ph)_4$, $Cp_2Zr(CH_2Ph)_2$, Cp_2ZrCl_2, and $Zr(CH_2^tBu)_2(NMe_2)_2$ all showed high activity on alkylaluminum-modified clay; of these, $Zr(CH_2Ph)_4$ was the most active. The organometallic complexes MR_4 (M = Ti, Zr, Hf, Cr; R = CH_2Ph, CH_2CMe_2Ph, CH_2^tBu, CH_2SiMe_3) and R_nML_{4-n} where L is NMe_2, OR, acac, Cp, $OSiPh_3$, $N(SiR_3)_2$ also showed activity. In general, the polyethylenes were of ultrahigh molecular weights ($M_w > 1 \times 10^6$) with high polydispersities ($M_w/M_n > 6$), although accurate values were difficult to obtain because of their very low solubility.

In the early 1980s, Schöppel and Reichert reported ethylene polymerization using a Ziegler catalyst in the presence of a clay filler.[31] $TiCl_4$ was added to the clay (Icecap K or Burgess KE) and polymerization was initiated in presence of monomer by addition of the cocatalyst Al^iBu_3. Alternately, a preactivated supported Ziegler catalyst [$Mg(OEt)_2/TiCl_4/Al^iBu_3$] was mixed with either the clay or the $TiCl_4$-modified clay. Activities were higher for the latter systems. However, the best incorporation of the filler into the polyethylene was obtained when the intact Ziegler catalyst was used in combination with the $TiCl_4$-modified filler, as judged by SEM. At about the same time, Sohn and Park described reactions of low-pressure ethylene (300 Torr) over acid-treated (K10) montmorillonite that had been ion exchanged with various transition metal cations.[32] Selective ethylene dimerization to *n*-butenes was observed over Ni^{2+}-exchanged montmorillonite. In contrast, a Cr^{3+}-exchanged clay produced high-density polyethylene (0.968 g/cm³, melting point 142°C), as long as the catalyst was first calcined above 300°C in air. The highest activity was obtained after calcination at 500°C, which maximizes Lewis acidity in the clay. Unfortunately, the distribution of the clay in the polyethylene was not examined in this study.

Two particularly active catalysts supported on kaolin were derived from $Zr(BH_4)_4$ and $CrCp_2$ (chromocene).[33] Grafting of the former onto dehydrated kaolin led to the disappearance of the $\upsilon(OH)$ stretch of surface silanol groups according to Equation 5.1. The anchored catalyst becomes active upon heating above 200°C, which results in the formation of hydrides capable of inserting ethylene, Equation 5.2.

$$n(\equiv SiOH) + Zr(BH_4)_4 \rightarrow (SiO)_n Zr(BH_4)_{4-n} + \frac{n}{2} B_2H_6 + nH_2 \tag{5.1}$$

$$(\equiv SiO)_n Zr(BH_4)_{4-n} + \Delta \rightarrow (SiO)_n ZrH_{4-n} + \frac{4-n}{2} B_2H_6 \tag{5.2}$$

Anchored $CrCp_2$ required no additional heating, since catalysts of this type generate their active sites spontaneously upon exposure to ethylene.[34,35] As the calcination temperature of the clay increased, the amount of the organometallic complex taken up by the support decreased, however, the specific activity increased. Filled polyethylene prepared in this way was of very high molecular weight ($>1 \times 10^6$) and was difficult or impossible to process by injection molding or extrusion, although it can be pressure molded.[24] The presence of H_2 (16 vol% relative to C_2H_4) during polymerization over a kaolin-supported Ziegler catalyst system, $VOCl_3/AlEt_2Cl$, was effective in reducing the molecular weight by an order of magnitude.[36] Structural homogeneity of the composite was reported to be better when the clay was treated first with the transition metal complex followed by the alkylaluminum cocatalyst, rather than the reverse.

5.1.3 Toward Polyolefin-Clay Nanocomposites by *In Situ* Polymerization

Following these early explorations of "homogeneous" composites using the polymerization filling technique, as well as reports of dramatic changes in physical properties in clay-filled Nylon-6 by researchers at Toyota Central R&D,[37] many groups sought to further enhance the mechanical and thermal properties of polyolefins by achieving nanoscale dispersion of the clay filler. Exfoliation into individual clay sheets, each approximately 1 nm thick with aspect ratios greater than 100, in a hydrophobic polyolefin matrix has been a major challenge. Such clay dispersions are thermodynamically unstable except in the presence of compatibilizing agents,[38] some of which are of limited thermal stability[39] and most of which are expensive.

Preparation of polymer-clay nanocomposites by *in situ* polymerization (sometimes called intercalative polymerization or polymerization compounding) circumvents the enthalpic and entropic barriers that prohibit the intercalation of nonpolar polyolefins into polar clays. Since supported olefin polymerization catalysts are desirable anyway in high-volume polyolefin manufacturing (see Section 5.1.1), the design of clay-supported catalysts to prepare nanocomposites can achieve both goals at once. In the late 1990s, researchers explored the use of recently developed metallocene and post-metallocene catalysts with the potential for a high degree of control over molecular weight and molecular weight distribution, comonomer incorporation, and stereocontrol.

O'Hare et al. reported the first intercalation of a pre-activated polymerization catalyst into a layered silicate, in 1996.[40] A synthetic hectorite (LaponiteRD, $Na_{0.46}Mg_{5.42}Li_{0.46}Si_8(OH)_4O_{20}$) was first treated with methylaluminoxane (MAO) to remove accessible acidic protons causing no change in the interlayer spacing. Addition of the metallocenium ion $[Cp_2ZrMe(THF)^+]$ (**I**) resulted in displacement of essentially all of the interlayer Na^+ ions, accompanied by an increase in the basal spacing of 0.47–1.44 nm. The same catalyst was ion exchanged directly into anhydrous fluorotetrasilicic mica ($NaMg_{2.5}Si_4O_{10}F_2$), which lacks hydroxyl groups. Its basal spacing increased from 0.98 to 1.43 nm. Propylene polymerization at 40°C and 5 bar in the presence of added cocatalyst (MAO, Al/Zr = 1000) resulted in similar high activities for both intercalated catalysts, ca. 1000 kg PP/(mol Zr·h). However, the resulting polymers were of very low molecular weight (M_n ca. 2000; M_w/M_n ca. 2), and no investigation of the clay dispersion or the composite properties was reported.

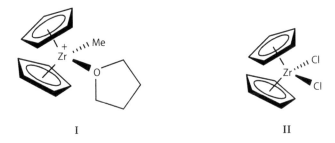

I   II

Suzuki and Suga reported the use of clays as solid acids to support and activate metallocene catalysts for olefin polymerization.[41] They were able to use much less alkylaluminum cocatalyst relative to solution polymerization conditions. The clays were slurried with $AlMe_3$ in toluene, then treated with a solution containing zirconocene dichloride, **II**, and $AlMe_3$. The metallocenium cation was presumed formed via abstraction of chloride and/or methyl ligands by acidic sites on the surface of the clay, and the low basicity of the clay surface was proposed to stabilize the coordinatively unsaturated cation. Propylene was copolymerized with 250 psi ethylene at 70°C. For acid-treated K10 montmorillonite, an activity of 3300×10^3 kg polymer/(g Zr·h) was obtained. Catalysts based on vermiculite, kaolin, and synthetic hectorite all showed lower but still appreciable activities. In this brief report, the Al/Zr ratio was not specified, and the clay dispersion was not reported.

Dubois et al. created polymer-clay composites by modifying kaolin with a conventional Ziegler–Natta catalyst.[42] The kaolin was first treated with $AlEt_3$ to induce its deaggregation. A support precursor, $Mg(^nBu)(^nOct)$, and a catalyst precursor, $Ti(O^nBu)_4$, were introduced, then $EtAlCl_2$ was added as a chlorinating/reducing agent to generate the supported catalyst $TiCl_3/MgCl_2$ in the presence of the clay. Ethylene was polymerized at 4 bar and 60°C in the presence of additional $AlEt_3$ as cocatalyst (Al/Ti = 160). The system showed activities up to 148 kg PE/(g Ti·h). With filler contents from 28% to 42%, the composites showed very high melt viscosities, as expected. The use of H_2 as a chain transfer agent to improve the processibility of the composites was explored. With 6 bar H_2, the activity increased significantly to 295 kg PE/(g Ti·h); however, the effect on the melt flow index was small. SEM images showed "heaps" of PE formed initially on the kaolin platelets, which later grew to cover the surface, Figure 5.2. The physical properties of the polymerization-filled composites were also evaluated.[43] The elongation at break and the impact resistance, while lower

(a)   (b)

(c)

FIGURE 5.2
SEM images of polyethylene grown on a kaolin surface, showing the clay prior to polymerization (a), as well as the material at later reaction times, when it contained 10 wt% polymer (b), and 20 wt% polymer (c). Scale bars represent 1 μm. (Reproduced from Hindryckz, F. et al., *J. Appl. Polym. Sci.*, 64, 423, 1997. With permission from John Wiley & Sons. Inc.)

than for pure PE, were far higher than for composites made by melt-blending (even with addition of an interfacial agent, or pretreatment of the clay with an aminosilane).

Parada et al. reported treating bentonite clay (predried at 350°C) sequentially with $AlEt_3$ and $TiCl_4$, and conducting ethylene polymerization at 10 psi and 25°C or 60°C in the presence of excess $AlEt_3$ as cocatalyst (Al/Ti = 50).[44] No shift in the d_{001} reflection of the clay was detected, therefore the catalyst was presumed dispersed on its external surface. High molecular weight, linear polyethylene was produced with activities up to 995 g PE/(g Ti · atm · h). Much higher activities were obtained when the bentonite was mixed with $MgCl_2$, presumably due to the formation of a conventional Ziegler–Natta catalyst. No analysis of the composite structure was undertaken.

Mülhaupt pioneered the *in situ* formation of polyethylene-clay composites using organo-clays.[45] The Na^+ form of bentonite (which is largely montmorillonite, $M^{x+Si}_8Al_{4-x}Mg_xO_{20}(OH)_4$) was made organophilic by ion exchange with (benzyl)(dimethyl)(stearyl)ammonium. The organoclay was dispersed in toluene, and ethylene polymerization was conducted at 40°C and 6 bar in the presence of an organometallic catalyst precursor (**II**, **III**, or **IV**) with MAO as the cocatalyst, or with the cationic catalyst **V** in the absence of a cocatalyst. The interaction between the catalysts and the clay, if any, was not established. The presence of the organoclay had no effect on the polymerization activities of the late transition metal catalysts **IV** and **V** (suggesting that the polymerization may have been largely homogeneous), but it suppressed the initial activity of **III**. In the resulting polyethylenes and as well as in ethylene/1-octene copolymers, the absence of the basal reflection of the organoclay at 1.96nm in the powder XRD was taken as evidence for exfoliation of the clay. Catalyst **II** produced polyethylene of moderate molecular weight and polydispersity (M_n 78,000; M_w/M_n 3.7) and high melting point (140°C) in the presence of a 6,000-fold excess of MAO.[46] With the same clay and using the same large excess of MAO, the chain-walking catalyst **IV** gave highly branched polyethylene with a melting point of 59°C. Catalyst **V** combined with synthetic hectorite gave only a viscous liquid.

| III | IV | V |

This preliminary study was expanded to include bentonite ion exchanged with (dimethyl)(distearyl)ammonium, as well as fluoromica and a synthetic hectorite (Optigel).[46] The clays were stirred in toluene in a high shear mixer (10,000 rpm) to enhance their dispersion prior to polymerization. The activity of catalyst **III** was affected less by the presence of the fluoromica and hectorite than by the organoclays. The clays had little effect on 1-octene incorporation, molecular weight, or branching frequency; however, they did suppress reactor fouling. The composites were visually homogeneous, and no basal reflection of the organoclay was detected by XRD. TEM images showed limited clay dispersion, although it was better in the composite made by *in situ* polymerization than in one made by melt compounding, as shown in Figure 5.3. The physical properties of the polyethylene composites, such as Young's modulus, stress at break and elongation at break, showed no improvement over the unfilled polymer or a melt-blended composite, but clay-filled ethylene/1-octene copolymers made by *in situ* polymerization were stiffer and stronger than the corresponding melt blends. The comonomer was suggested to improve the compatibility of polyethylene with clay.

The first claim to have made a true polyolefin-clay nanocomposite by *in situ* polymerization appeared in 1999. Coates et al. used a Brookhart catalyst in combination with synthetic fluorohectorite ($Li_{1.12}[Mg_{4.88}Li_{1.12}]Si_8O_{20}F_4$).[47] Olefin polymerization catalysts based on late transition metals are more tolerant of polar groups, including water, than are early transition metal catalysts (such as **I**) and are therefore potentially more compatible with clays. The interlayer Li^+ ions of the clay were first exchanged for 1-tetradecylammonium, expanding the galleries from 1.20 to 1.99nm. Addition of **VI** as a solution of its $[B(C_6H_3(CF_3)_2–3,5)_4^-]$ salt resulted in further increase in the basal spacing, to 2.76nm, but the irreversible catalyst uptake was apparently not a simple ion exchange, since both cation and anion were intercalated. Exposure of the clay-supported catalyst to ethylene at 80psi and 22°C for 12h resulted in highly branched, high molecular weight polyethylene (M_n 159,000; M_w/M_n 1.6). However, the activity was very low, with turnover frequencies of only 162h^{-1} for ethylene and 5.6h^{-1} for propylene. During the reaction, the basal reflection of the clay shifted to a lower angle and then disappeared, confirming the occurrence of polymerization in the galleries, as shown in Figure 5.4.

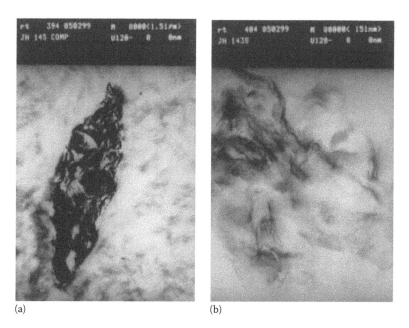

(a)   (b)

FIGURE 5.3

TEM images of composite materials containing an organically modified bentonite and high density polyethylene: (a) prepared by melt compounding and (b) prepared by *in situ* polymerization using catalyst III. (Reproduced from Heinemann, J. et al., *Macromol. Rapid Commun.*, 20, 423, 1999. With permission from Wiley-VCH Verlag GmbH.)

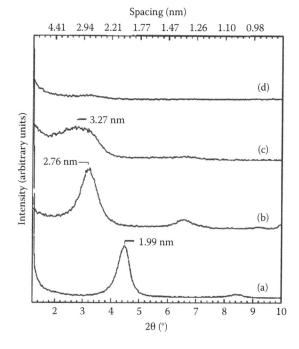

FIGURE 5.4

Powder x-ray diffraction of (a) 1-tetradecylammonium-modified fluorohectorite, (b) the organoclay after inter-calation of catalyst VI, (c) after exposure of the clay-supported catalyst to ethylene for 135 min, and (d) after exposure to ethylene for 24 h. (Reproduced from Bergman, J.S. et al., *Chem. Commun.*, 2179, 1999. With permission from The Royal Society of Chemistry.)

TEM images revealed an intercalated nanocomposite with layer spacings of 5–8 nm. No physical properties of the materials were reported.

VI

5.1.4 Scope and Goals of This Chapter

A viable process for manufacturing polyolefin-clay nanocomposites by *in situ* polymerization requires adequate catalytic activity, desirable polymer microstructure, and physical properties including processibility, a high level of clay exfoliation that remains stable under processing conditions and, preferably, inexpensive catalyst components. The work described in the previous two sections focused on achieving *in situ* polymerization with clay-supported transition metal complexes, and there was less emphasis on optimization of polymer properties and/or clay dispersion. Since 2000, many more comprehensive studies have been undertaken that attempt to characterize and optimize the entire system, from the supported catalyst to the nanocomposite material. The remainder of this chapter covers work published in the past decade on clay-polyolefin nanocomposites of ethylene and propylene homopolymers, as well as their copolymers, made by *in situ* polymerization. The emphasis is on the catalyst compositions and catalyst–clay interactions that determine the success of one-step methods to synthesize polyolefins with enhanced physical properties.

5.2 Homopolymerization of α-Olefins

5.2.1 Homopolymerization of Ethylene

5.2.1.1 Phillips' Catalysts for Ethylene Polymerization

Clay-supported analogs of the well-known Phillips' catalyst (Cr/SiO_2) were investigated by Yamamoto et al.[48] When Cr(III) was introduced into the interlayer spaces of a basic sodium montmorillonite (Kunipia F, pH 10) by ion exchange with $Cr(NO_3)_3$, the basal spacing of the clay, d_{001}, increased from 0.95 to 1.15 nm. Ethylene polymerization was conducted in isobutene at 1.5 MPa ethylene and 100°C, with an alkylmetal compound ($AlEt_3$, Mg^nBu_2, or $ZnEt_2$) added as the cocatalyst. An activator was required because the development of cocatalyst-independent activity by Phillips' catalysts requires very high temperature activation,[28,49] which destroys the layered structure of the clay. The cocatalyst serves to alkylate the chromium and to remove polar groups such as adsorbed water and hydroxyl groups associated with the clay that poison the active sites. The intercalated Cr catalyst was highly active with $AlEt_3$ as the cocatalyst, giving polymer at the rate of 2.11×10^3 kg PE/(mol Cr·h). Other cocatalysts were much less effective. The resulting polyethylene was characterized by a high molecular weight (M_w = 1.0 × 10^6) and a very broad molecular weight distribution (M_w/M_n = 28) suggesting that a variety of active sites were present (typical of Phillips' catalysts). In contrast, Cr(III) deposited on the external surfaces of the clay by grafting one of the molecular complexes Cr(acetate)$_3$ or Cr(acac)$_3$ on hydroxyl groups present at the clay edges gave catalysts with very poor activity. Good clay dispersion in the nanocomposites was observed by TEM, but only for those materials made with the intercalated Phillips' catalyst. No physical properties were reported for the filled polymers.

5.2.1.2 Ziegler–Natta Catalysts for Ethylene Polymerization

5.2.1.2.1 Hybrid MgCl$_2$-Clays as Ziegler–Natta Catalyst Supports

The MgCl$_2$ support is a crucial component of conventional Ziegler–Natta catalysts, including those based on MCl$_4$ (M = Ti, V) and VOCl$_3$. A hybrid Ziegler–Natta/clay catalyst support was prepared by Yang et al., who deposited microcrystallites of MgCl$_2$ in the interlayer regions of a sodium montmorillonite (Cloisite® Na$^+$).[50] The clay was first pretreated at 400°C, then dispersed in ^{n}BuOH, which also dissolves MgCl$_2$. After evaporation of the alcohol, the interlayer spacing had increased from 1.15 to 1.33 nm. A further increase to 1.58 nm occurred upon reaction with TiCl$_4$, which was presumed to react preferentially with the MgCl$_2$ phase rather than with the clay. Ethylene polymerization was conducted with added AlEt$_3$ (Al/Ti = 20) at 0.01 MPa and 50°C. Activities up to 748 kg PE/(mol Ti·h) were reported, similar to those of conventional Ziegler–Natta catalysts. The resulting polyethylenes had M_w values of ca. 1 × 10^6 and polydispersities (M_w/M_n) of ca. 4. However, over 50% of the polymer was not extractable by hot xylene; as is often the case with polyethylene-clay nanocomposites, solvent swelling results in formation of an insoluble gel. No basal reflections of the clay were detected by XRD, and a TEM image showed exfoliated clay. The nanocomposite exhibited sharply reduced polyethylene crystallinity, attributed to polymer chain confinement, and increased tensile strength and tensile elongation, caused by clay-induced rigidity of the amorphous polymer regions.

A similar approach to creating a supported Ziegler–Natta catalyst was used by the Woo group, who intercalated MgCl$_2$ in sodium montmorillonite (Kunipia F) as a support for VOCl$_3$.[51] Ethylene polymerization was conducted with added Al^{i}Bu$_3$ at room temperature and 1.2 atm. With activities up to 76 kg PE/(mol V·h), the supported catalyst was much more active than its homogeneous analog. No basal reflection was detected in the composite by XRD, and a TEM image revealed exfoliated clay. The filled polyethylenes were only partly extractable by hot decalin. They showed increased melting temperature and thermal decomposition temperature under N$_2$, attributed to restricted chain motion and the barrier properties of the clay, compared to unfilled polyethylene. Increases in tensile strength and tensile modulus were ascribed to the strong interactions between the clay and the polymer chains at the organic–inorganic interface.

The use of a Mg-rich clay to prepare Ziegler–Natta catalysts directly was pioneered by Rong et al.[52] Palygorskite is a naturally occurring magnesium silicate with the ideal formula Mg$_5$Si$_8$O$_{20}$(OH)$_2$(OH$_2$)$_4$. The presence of Al^{3+} (and to a lesser extent, Fe^{3+}) on some of the octahedral Mg^{2+} sites results in cation vacancies.[53] Its non-swelling, fibrillar structure arises due to the presence of ribbon-like octahedral sheets. The randomly oriented rods, with typical diameters from 10 to 50 nm and lengths from 1 to 3 μm, are aggregated into bundles, Figure 5.5.

The calcined clay was treated with TiCl$_4$. Attachment of Ti(IV) to the clay surface was suggested to occur at coordinatively unsaturated Mg ions (presumably formed by the removal of water during calcination) on the edges of the octahedral ribbons and to involve bridging chloride interactions, as in **VII**.[52] Upon activation by Al^{i}Bu$_3$ (Al/Ti = 15), ethylene polymerization was conducted in n-hexane at atmospheric pressure and 40°C.

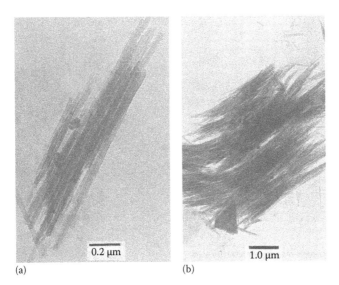

FIGURE 5.5
TEM images of palygorskite: (a) typical bundle of fibers with fine smectite flakes and (b) "bird's-nest" aggregate of fibers with entrapped grains of calcite and apatite. (Reproduced from Güven, N. et al., *Clays Clay Miner.*, 40, 457, 1992. With permission from The Clay Minerals Society.)

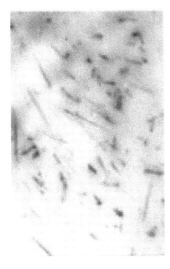

FIGURE 5.6
TEM image (magnification 20,000) of a polyethylene-palygorskite nanocomposite, prepared by *in situ* polymerization using the clay-supported Ziegler–Natta catalyst shown. (Reproduced from Rong, J. et al., *Macromol. Rapid. Commun.*, 22, 329, 2001. With permission from Wiley-VCH Verlag GmbH & Co. KGaA.)

TEM images showed high dispersion of the fibers in the polyethylene matrix, as shown in Figure 5.6. The nanocomposite was suggested to have a macromolecular comb (chain brush) structure, in which very long polymer chains were grown from several active sites on the mineral rod surface, as shown in Figure 5.7. The resulting chain entanglement resulted in good Izod impact strengths. At least one-half of the polyethylene was not extractable with decalin at 135°C. SEM images revealed relatively smooth fracture surfaces, indicating good adhesion between the polymer chains and the clay, in contrast to the coarse surfaces achieved by melt blending. The materials made by *in situ* polymerization showed improved tensile modulus, but inferior elongation at break. According to Ruckenstein et al., the palygorskite-supported catalyst caused the clay fibers to become encapsulated in polyethylene, suppressing the ability of the clay to act as a nucleation site, and thereby suppressing polymer crystallinity.[54]

VII

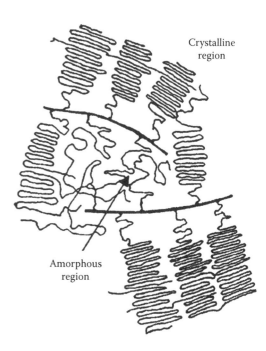

Crystalline
region

Amorphous
region

FIGURE 5.7
Proposed macromolecular comb structure of a polyethylene-palygorskite nanocomposite, prepared by *in situ* polymerization using a clay-supported Ziegler–Natta catalyst. (Reproduced from Rong, J. et al., *Polym. Comp.*, 23, 658, 2002. With permission from John Wiley & Sons, Inc.)

The effect of catalyst preparation conditions and reaction conditions on the ethylene polymerization activity of palygorskite-supported TiCl$_4$ was investigated by Li et al.[55] Despite a decrease in Ti uptake with increasing clay pretreatment temperature, maximum activity (1755 g PE/(g Ti·h) at 40°C and Al/Ti = 15) was observed when the clay was heated at 500°C. This was attributed to the creation of Lewis acidic surface sites that bind the catalyst. Above 500°C, surface area was lost rapidly. Polymerization activity was fairly constant from 30°C to 90°C, with a maximum at 60°C, and stable over time. The very high molecular weight (M_w ca. 5 × 10^6) of polyethylene was attributed to a decreased rate of chain termination compared to the homogeneous catalyst. Relatively, little cocatalyst was required for the heterogeneous system; the activity was a maximum for Al/Ti = 10. Al^{i}Bu$_3$ gave higher activity than either AlMe$_3$ or AlEt$_3$, presumably due to its weaker reducing ability that preserves the active Ti sites in their optimum oxidation state.

The low melt processibility of nanocomposites prepared using TiCl$_4$ supported on palygorskite was attributed to excessive entanglement of the chain brushes. This problem was overcome by Du et al. using a mixed catalyst system.[56] MgCl$_2$ was introduced into the reactor with the clay-supported catalyst and Al^{i}Bu$_3$ in *n*-hexane, forming the conventional Ziegler–Natta catalyst MgCl$_2$/TiCl$_4$/Al^{i}Bu$_3$ *in situ* (presumably via the transfer of soluble TiCl$_x$^{i}Bu$_y$ species from the clay-supported system). Ethylene was supplied at 1–6 atm, and the resulting unattached PE chains served to enhance the processibility of the macromolecular combs formed by the clay. The resulting materials had torsional moments less than 20 N·m. A similar enhancement was noted when the nanocomposite made without MgCl$_2$ was blended with unfilled polyethylene.[57] The compatibility of the palygorskite with the blended polymer was attributed to its encapsulation in polyethylene.

A Ziegler–Natta catalyst was created by supporting TiCl$_4$ directly on natural montmorillonite that was first dehydrated completely by calcination at 600°C.[58] The Ramazani group proposed that the catalyst binds via bridging chloride ligands to Mg^{2+} ions located in the interlayer regions, forming structures similar to **VII**. Ethylene polymerization was conducted in hexane with added AlR$_3$ (R = Et, ^{i}Bu) as the cocatalyst. A maximum activity of

193 g PE/(mmol Ti·h) was observed at 60°C and 7 bar, using Al^{i}Bu$_3$ at Al/Ti = 178; higher Al/Ti ratios led to overreduction of Ti. The XRD of the unprocessed powder showed no basal reflection; however, TEM micrographs were consistent with a combination of intercalated and exfoliated clay morphologies.

5.2.1.2.2 Organoclays as Ziegler–Natta Catalyst Supports

Intercalation of Ziegler–Natta (and other) catalysts may be facilitated by an increase in basal spacing, achieved by prior ion exchange of interlayer cations with long-chain alkyl-ammonium surfactants, as shown in Scheme 5.1. Ivanyuk et al. ion-exchanged sodium montmorillonite with dimethylbis(octadecyl)ammonium, then dried the organoclay at 353 K.[59] It was swollen with toluene, then treated with TiCl$_4$ (1 mol/g). Upon activation with Al^{i}Bu$_3$ (Al/Ti = 1), ethylene polymerization was conducted at 313 K and 0.3 MPa. The resulting nanocomposites, containing up to 24 wt% clay, showed no low-angle XRD reflections, suggesting complete exfoliation of the filler. Materials with 4–8 wt% clay filler had the best physical properties: an increase of up to 80% in elastic modulus with no decrease in tensile strength or specific elongation at rupture. Similarly, Li et al. reported a Ziegler–Natta catalyst prepared by addition of TiCl$_4$ to an alkylammonium-modified montmorillonite.[60] Polymerization of ethylene at 10 bar and 80°C in the presence of an unspecified alkylaluminum cocatalyst gave polyethylene with no clay diffraction peak from which delamination of the clay was inferred. Its presence in the nanocomposites resulted in improved thermal stability under N$_2$, increased tensile strength, and Young's modulus. These changes were optimum at a clay loading of 2.3 wt%.

Lomakin et al. prepared a supported catalyst by reaction of dehydrated Cloisite 20A (montmorillonite modified di(HT)dimethylammonium ions) with Al^{i}Bu$_3$ followed by the

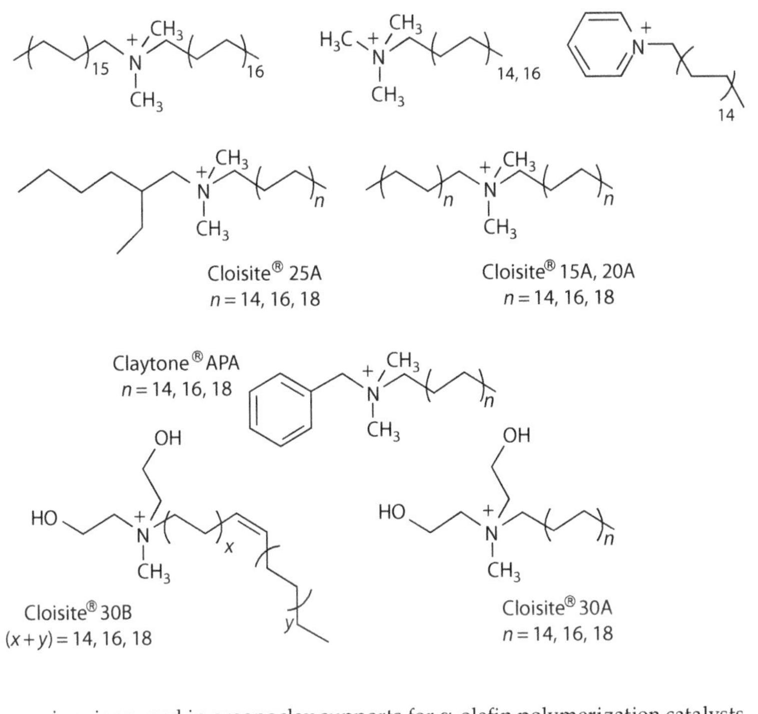

SCHEME 5.1
Quaternary ammonium ions used in organoclay supports for α-olefin polymerization catalysts.

Ziegler–Natta catalyst VCl$_4$.[61] In the surfactant, T (tallow) designates a mixture of long-chain saturated and unsaturated alkyls (65% C$_{18}$, 30% C$_{16}$, 5% C$_{14}$) derived from fatty acids, and HT indicates hydrogenated tallow. No details on ethylene polymerization conditions were given, but the basal reflection of the clay was absent from the XRD of nanocomposites containing up to 6.5 wt% clay. A TEM image shows an exfoliated morphology, with nanostacks distributed throughout the polymer matrix. Increased thermal stability in air was attributed to low-temperature intermolecular clay–induced radical cross-linking. The low O$_2$ permeability of the nanocomposite was credited with creating an oxygen-deficient environment that suppresses radical chain reactions leading to oxidative degradation.

Increased uptake of a Ziegler–Natta catalyst by the clay was achieved by Kwak et al. using Cloisite® 30B, a montmorillonite modified with bis(2-hydroxyethyl)(methyl)(T)ammonium ions.[62] The hydroxyl groups present in the surfactant provide additional grafting sites for the catalyst and cocatalyst. The organoclay was dried under vacuum, then dispersed in toluene prior to addition of TiCl$_4$ followed by AlEt$_3$. Three times more TiCl$_4$ was immobilized on the organoclay than on sodium montmorillonite. Ethylene polymerization was conducted at 30°C–50°C and 4 bar. Low activities, in the range 6–80 kg PE/(mol Ti·h), were attributed to fast deactivation of the grafted catalyst by unreacted hydroxyl groups of the surfactant. In fact, the catalyst based on sodium montmorillonite had comparable activity, and it produced a higher molecular weight polymer (M_w 200,000). However, complete exfoliation of the clay, as judged by the absence of a basal reflection by XRD, was confirmed only for material made by the organoclay-supported catalyst. This nanocomposite was dry blended with HDPE at 200°C via either compression or twin-screw extrusion. The dispersed clay was observed to reaggregate, leading to recovery of a basal reflection at $2\theta = 6°$. By TEM, 15 nm stacks comprised of ca. 10 layers were detected.

Bousmina et al. added TiCl$_4$ to Cloisite 30B, which had been dried in vacuum at 120°C and dispersed in toluene.[63] Slurry polymerization of ethylene was conducted in the presence of added AlEt$_3$. The basal reflection of the organoclay, at 1.8 nm, appeared at a lower diffraction angle in the resulting composite, suggesting intercalation of polyethylene (from which reaction of the catalyst with the hydroxyl groups of the interlayer cations was inferred). In contrast, the position of the basal reflection of sodium montmorillonite used in the same way remained unchanged at 1.17 nm. For the unmodified clay, the catalyst was presumed anchored on hydroxyl groups present on the edges of the aluminosilicate layers, resulting in polymerization in the bulk rather than in the clay galleries. Polyethylene was easily extracted with hot xylene from the composite containing unmodified clay, but not from composite containing the organoclay, attesting to the strong adsorption (designated *a*) of the polymer chains in PE-*a*-C30B. The ability of this adsorbed PE to act as a compatibilizer for HDPE was explored by melt blending at 180°C to form the composite HDPE/PE-*a*-C30B containing 3 wt% clay. Its low-angle XRD was featureless, suggesting an exfoliated structure; however, TEM images revealed that it is instead highly intercalated. The blend showed increased tensile modulus and yield strength, but decreased elongation at break and dynamic storage modulus.

Clay-supported Ziegler–Natta catalysts prepared using Cloisite® 20A and Cloisite® 30A (the hydrogenated version of Cloisite 30B) were compared by Kovaleva et al.[64] Each organoclay was stirred with AlMe$_3$ in toluene, followed by VCl$_4$. Polymerization was conducted at 20°C and 50 kPa in the presence of Al^{i}Bu$_3$ as cocatalyst. Surprisingly, the uptake of AlMe$_3$ and VCl$_4$ was higher for Cloisite 20A, although catalytic activity was higher for the catalyst made with Cloisite 30A (containing the hydroxylated surfactant). The polyethylenes had M_w values of 7×10^5 and polydispersity indices of ca. 7. Unlike the finding by Kwak et al.,[62] the material containing Cloisite 30A retained the basal reflection of the organoclay suggesting that the catalyst was immobilized only the external surfaces of the montmorillonite particles. In contrast, the material made with Cloisite 20A showed no d_{001} reflection by SAXS, and the catalyst was presumed immobilized in the clay galleries. The authors suggested that the higher organophilicity of Cloisite 20A led to better dispersion of the organoclay in toluene. However, the orders of addition of metal halide and alkylaluminum, which differ in the two reports,[62,64] may also be important for determining the locations of the active sites. Analysis of the nanocomposite containing Cloisite 20A by very cold neutron scattering gave the volume fraction of scattering particles as 86%. This material showed improved elastic modulus and decreased gas permeability.

5.2.1.2.3 Interlayer Cations as Ziegler–Natta Catalysts

A supported Ziegler–Natta catalyst was created by ion exchange of sodium montmorillonite with polyoxotitanium cations, formed by the slow hydrolysis of $Ti(OEt)_4$.[65] The clay took up 33.7 wt% Ti, while its Na^+ content declined from 0.89 to 0.0085 wt%. Woo et al. inferred intercalation of the catalyst from the increase in the basal spacing of the clay, from 1.19 to 5.86 nm. Ethylene polymerization was conducted in toluene, with Al^iBu_3 added as cocatalyst, at 60°C and variable pressure (<6.5 atm). The low activity, ca. 0.4 kg PE/(mol Ti·h·atm), was attributed to mass transfer resistance in the clay galleries. The molecular weight of the polyethylene was high, with M_n values from 390,000 to 700,000, and a large fraction was not extracted by decalin. The nanocomposites, containing 3.9–7.7 wt% clay, showed no d_{001} reflection by XRD and good clay exfoliation by TEM. Improved tensile strength, Young's modulus, and thermal stability were recorded. However, the improvements in thermal stability were less pronounced at higher clay loadings, due to clay-catalyzed polymer decomposition.

5.2.1.2.4 Acid-Treated Clays as Ziegler–Natta Catalyst Supports

No intercalation of the catalyst is necessary in order to achieve a high degree of clay dispersion during *in situ* polymerization if the stacking of the clay layers can be disrupted prior to reaction. Such disruption is induced by acid treatment. For example, montmorillonite refluxed in 30% H_2SO_4 undergoes a large increase in both accessible surface area (from ca. 40 m^2/g up to 150–200 m^2/g) and acidity.[66] The basal reflection disappears from the XRD after only 15 min acid treatment, although the lamellar structure is not entirely lost for 20 h. The structure of the clay changes as Mg^{2+} and Al^{3+} ions are leached from the octahedral layer, starting at the edges of the clay platelets.[67] The Al^{3+} ions that migrate to the interlayer region contribute to the Lewis and Brønsted acidity of the acid-treated clay. Strong Lewis and Brønsted acidity can also arise by isomorphic substitution of Si^{4+} by Al^{3+} in the tetrahedral sheets.[68,69]

In situ polymerization of ethylene was conducted with $Zr(CH_2Ph)_4$ supported on acid-treated montmorillonite by Scott et al.[70] While the organometallic complex shows very low activity as single-component Ziegler–Natta catalysts in solution, it can be activated simply by grafting onto inorganic oxides.[71] The clay was pretreated by stirring natural montmorillonite in a solution of 6 M H_2SO_4/1 M Li_2SO_4 for 5 h, followed by washing with deionized water and drying at 200°C for 2 h. The resulting material had a B.E.T. surface area of 190 m^2/g, and a strongly attenuated basal reflection at ca. 7°. Contact between the clay and a bright yellow solution of $Zr(CH_2Ph)_4$ in toluene resulted in a color change in both the clay and the solution, to dark green, indicating decomposition of the catalyst; the solid was inactive for polymerization. However, when the acid-treated clay was exposed

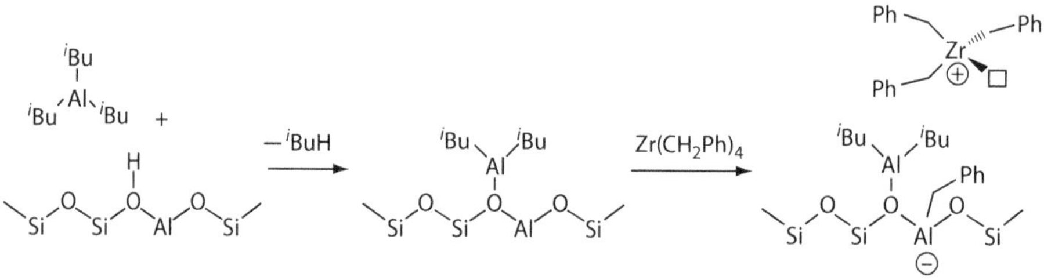

SCHEME 5.2
Possible activation mechanism by alkyl group abstraction for an organozirconium complex on acid-treated montmorillonite passivated with Al^{i}Bu$_3$.

to (CH$_3$)$_3$SiCl at room temperature to remove water and attenuate its Brønsted acidity by capping accessible hydroxyl groups followed by supporting Zr(CH$_2$Ph)$_4$, the resulting red catalyst showed an activity of 120 kg PE/(mol Zr·h) at 60°C and 100 psi.[72] The activity was higher, 280 kg PE/(mol Zr·h), when Al^{i}Bu$_3$ was used to dehydrate and dehydroxylate the clay. A possible activation mechanism involves the abstraction of a benzyl ligand by Lewis acid sites present in the interlayer region or on the clay surface (Scheme 5.2).

The higher activity with Al^{i}Bu$_3$ compared to (CH$_3$)$_3$SiCl may be due to the increased number of Lewis acid sites. Delayed exotherms of ca. 30°C were observed 15–30 min after initiation of the reaction, making temperature control difficult and suggesting that polymerization in the interior of the clay particles and subsequent fragmentation was causing exposure fresh active sites. With a prepolymerization step (15 min at 25°C), the exotherm was successfully suppressed. The distribution of clay in a resulting composite is shown in Figure 5.8. It was not possible to measure molecular weights, which were presumably very high, by GPC due to the insolubility of the polymers.

Ramazani et al. explored the use of acid-treated bentonite, prepared by refluxing the clay in hot H$_2$SO$_4$ for 2 h, as a Ziegler–Natta catalyst support.[73] The modified clay was dried at 200°C, then treated with AlR$_3$ (R = Et, ^{i}Bu) to create a supported aluminoxane

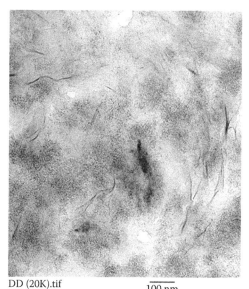

FIGURE 5.8
TEM micrograph (scale bar 100 nm) for a polyethylene nanocomposite containing acid-treated montmorillonite (5.4 wt%). The material was made by *in situ* polymerization using Zr(CH$_2$Ph)$_4$ supported on the Al^{i}Bu$_3$-treated clay, using a prepolymerization step.

DD (20K).tif
CYGA–SS
100 nm

by reaction of the alkylaluminum with surface hydroxyl groups. $TiCl_4$ was immobilized on the cocatalyst-modified clay. Ethylene polymerization was conducted in hexane with Al^iBu_3 as scavenger. A maximum activity of 90 g PE/(mmol Ti·h) was achieved at 60°C and 7 bar, with Al/Ti = 32. Both intercalated and exfoliated clay layers were visible by TEM, while no basal reflection for the clay was detected by XRD. The gas permeability of the nanocomposite was reduced by 75% at 9 wt% clay. The materials showed increased overall thermal stability in air, although the onset of thermal decomposition occurred at a much lower temperature compared to unfilled polyethylene.

5.2.1.3 Metallocene Catalysts for Ethylene Polymerization

Metallocenes produce more uniform polyethylene microstructures than do either Phillips' or Ziegler–Natta catalysts, giving random comonomer incorporation and narrow molecular weight distributions, often with much higher initial activities. Supported metallocene catalysts show lower tendencies for reactor fouling than their homogeneous analogs, require less cocatalyst to activate them, and produce polymer with better particle morphologies and bulk densities.[74] Methods of supporting metallocenes involve immobilization of either the catalyst or the cocatalyst. The latter approach is generally preferred, since it results in higher activity and produces polymer microstructures more like those achieved with homogeneous catalysts.[75] Consequently, and in contrast to the preparation of some clay-supported Ziegler–Natta catalysts, metallocenes are generally not added directly to clays. Instead, an alkylaluminum or alkylaluminoxane cocatalyst must be added first to passivate the clay surface.[76] The cocatalyst removes water and reacts with surface OH groups that poison the metallocene.

5.2.1.3.1 Organoclays as Metallocene Catalyst Supports

Tang et al. studied the effect of the order of addition of components on the catalytic activity of an organoclay-supported zirconocene.[77] Sodium montmorillonite (Kunipia-F) was calcined above 400°C, then ion exchanged with methylglycinium, causing the interlayer spacing to expand from 0.965 to 1.269 nm. Both unmodified clay and the organoclay were treated with MAO, followed by either Cp_2ZrCl_2 (**II**) or $[Cp_2ZrCl^+][BF_4^-]$. The active sites were described as homogeneous within a heterogeneous framework. The MAO-treated clays took up more catalyst than the untreated clays, and MAO-treated organoclay took up much more. Ethylene polymerization was conducted at 60°C and 1.3 atm, with added MAO as a catalyst. The most active catalyst was **II** supported on the MAO-modified organoclay. Homopolymerization activities up to 1200 kg PE/(mol Zr·h) were reported, giving polyethylene with M_w values up to 190,000 and good bulk density (up to 0.44 g/cm³).

The effect of varying the protonated amino acid (glycine, phenylalanine, glutamic acid, or arginine) or the corresponding methyl esters was further evaluated.[78] The interlayer spacing depended on the nature of the amino acid, but not on whether it was esterified. Organoclays containing a mixture of protonated amino acid and hexadecyltrimethylammonium were also prepared, resulting in further expansion of the clay galleries up to 1.572 nm. Subsequent intercalation of the catalyst made little difference to the clay spacing. Clays modified with the methyl esters gave higher activity than clays containing nonesterified amino acids, even though the latter took up more catalyst. This is reminiscent of the effects reported for organoclays with hydroxyl-containing quaternary ammonium surfactants in Ziegler–Natta systems (see Section 5.2.1.2.2). The highest metallocene uptake and catalyst activity were reported for organoclay containing protonated phenylalanine, and the effect was enhanced in the presence of quaternary ammonium ions. The highest recorded activity was 2985 kg PE/(mol Zr·h) at 25°C and 1.3 atm. The catalyst was inferred to remain associated with the clay during polymerization despite the addition of excess MAO because of the increased polymer bulk density relative to that of the product of homogeneous polymerization. A filtration experiment showed that excess MAO did not extract the zirconocenium cation into solution. A typical polyethylene product had $M_w = 1.5 \times 10^5$ and a polydispersity index between 3 and 4. The absence of low-angle XRD peaks indicated that the clay was exfoliated in the nanocomposite.

Chang et al. created a clay-supported catalyst using the *ansa*-metallocene *rac*-Et(Ind)$_2$ZrCl$_2$ (Brintzinger's catalyst, **VIII**) in combination with MAO to effect ethylene polymerization at room temperature and 20 psi.[79] The montmorillonite was first ion exchanged with cetylpyridinium, then dried at 120°C in vacuo. The higher activity of the catalyst system using the organoclay compared to the unmodified clay was attributed to deactivation by associated water. The activity was also higher when the organoclay was treated first with MAO, prior to addition of the metallocene. The XRD of the resulting nanocomposite containing no basal reflection, indicating highly dispersed clay. Measurement of proton relaxation times $T_{1\rho}$ by solid-state NMR showed that the fraction of mobile (amorphous) polymer increased with increasing clay content. The presence of the clay also slowed the rate of polyethylene crystallization; however, the effect was less pronounced at high filler loadings when the clay acted as a nucleating agent.

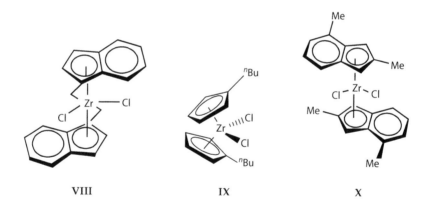

VIII   IX   X

Zapata et al. optimized catalytic activity for (*n*BuCp)$_2$ZrCl$_2$, **IX**, supported on an MAO-modified organoclay.[80] Sodium montmorillonite (Cloisite Na$^+$) was ion exchanged with octadecylammonium, followed by the metallocene. Ethylene polymerization was conducted at 60°C and 2 bar with added MAO. The observed activity of 4500 kg PE/(mol Zr·h·bar) was higher than for either the catalyst supported on the unmodified clay or the homogeneous catalyst. Curiously, pretreatment of the organoclay with MAO gave lower activity. The polymer produced by the supported catalyst showed increased molecular weight compared to the polymer produced by the homogeneous catalyst. For composites made by supporting the catalyst on the organoclay or on MAO-treated clays, complete clay exfoliation was inferred from the absence of a d_{001} reflection in the XRD, a conclusion supported by the TEM images.

The first report of bimodal PE from any supported metallocene involved a clay-supported catalyst.[81] Sodium montmorillonite (Kunipia-F) was dried at 500°C, then acid-treated with 6 N HCl. After drying again at 120°C, the clay was modified with (3-aminopropyl)triethoxysilane at 70°C. The XRD showed two characteristic interlayer spacings at 1.10 and 2.08 nm, possibly caused by occasional cross-linking of adjacent clay layers by the silane. Two larger interlayer spacings (1.33 and 2.45 nm) were observed when this clay was treated with Cp_2ZrCl_2 (**II**). Ethylene polymerization was conducted at 1 atm in the presence of added MAO for 30 min at 25°C. The catalyst had an activity of 2630 kg PE/(mol Zr·h) and produced bimodal polyethylene with $M_w = 6.6 \times 10^4$ and $M_w/M_n = 6$. Similar to the finding by Zapata et al.,[80] pretreatment of the clay with MAO reduced the amount of the zirconocene taken up, and its catalytic activity. The coexistence of different catalyst environments arising due to the siloxane pretreatment was suggested to be the origin of the unexpected molecular weight distribution. A physical explanation may be differences in rates of monomer diffusion into the two types of galleries (cross-linked or not).[82]

In some cases, the role of the surfactant in facilitating access of the catalyst to the interlayer regions may be indirect, exerting an effect primarily on the cocatalyst. Lee et al. found that treating the organoclay Cloisite® 25A with Al^iBu_3-modified MAO (MMAO, $(CH_3)_{0.7}(^iBu)_{0.3}[AlO]_n$) caused more than 90% of the modifier, (2-ethylhexyl)(dimethyl)(HT) ammonium to be lost from the interlayer spaces.[83] The MMAO-treated clay showed much decreased intensity in its basal reflection, which disappeared completely upon addition of Cp_2ZrCl_2 (**II**), suggesting extensive disruption of the layered structure. In contrast, sodium montmorillonite retained most of its layered structure when modified with MMAO followed by Cp_2ZrCl_2, although shifts in the position of the d_{001} reflection indicated sequential intercalation of both catalyst and cocatalyst. The supported catalyst was more active in the presence of the MMAO-modified organoclay than the MMAO-modified sodium montmorillonite, regardless of whether the zirconocene was added prior to polymerization or as a solution to the reactor. Slightly higher activities were observed when additional MMAO was present. High clay dispersion was inferred by XRD and TEM for the nanocomposite containing the organoclay. However, residual surfactant appeared to contribute to lower thermal stability of the material compared to a material made with sodium montmorillonite.

Tritto et al. also observed that most of the quaternary ammonium surfactant is displaced upon treatment of an organoclay with MAO. Cloisite® 15A (containing dimethyldi(HT) ammonium ions) was dried at 110°C for 24 h, then treated with $AlMe_3$-depleted MAO in toluene.[84] The interlayer spacing decreased and the stacking became less regular. Cp_2ZrCl_2 (**II**) was added to the MAO-modified clay in the presence of ethylene (40°C, 1 bar) to form the zirconocenium ion *in situ*. Its activity was higher in the presence of the clay than in its absence, but decreased with increasing clay content (attributed to reactions with impurities and/or deactivating interactions with the clay). The higher activities observed with longer MAO/clay precontact times may have been caused by more extensive disruption of the layered clay structure, as shown by TEM. The resulting composites had filler contents ranging from 9% to 34%; the polyethylenes had M_w values of ca. 10^5 and M_w/M_n ranging from 2.6 to 4.1. There was no low-angle XRD peak present in materials for which the polymerization was not quenched with acidified ethanol, but a broad peak corresponding to a basal spacing of 1.47–1.64 nm reappeared in the powder patterns of acid-quenched samples. The acid was suggested to react with intercalated MAO in regions of the clay not penetrated by the catalyst. The TEM images showed disordered clay aggregates; stacking did not appear to be induced by compression molding during sample preparation. A large decrease in thermal stability of the composite under N_2 was ascribed to the creation of strong Brønsted acid sites on the clay as a result

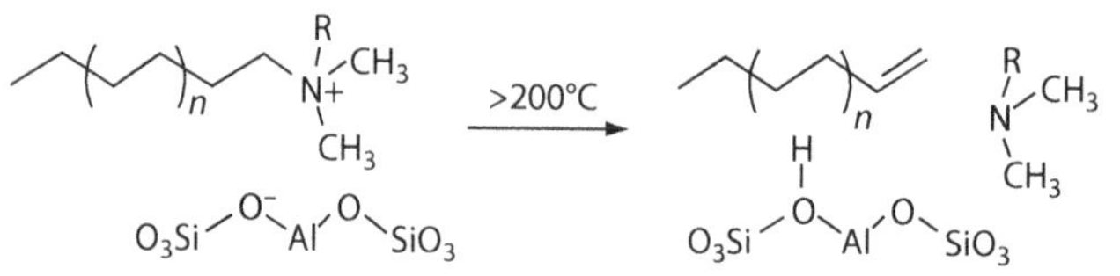

SCHEME 5.3
Proposed decomposition of tetraalkylammonium ions in organically modified clays by Hofmann β-elimination. Although the aluminosilicate oxygens are less basic than the amine, loss of the organic components by volatilization could lead to the formation of strongly acidic bridging hydroxyls.

of surfactant decomposition (Scheme 5.3). However, the composites showed appreciable stabilization under oxidizing conditions, presumably due to the diffusion barrier presented by the highly dispersed clay toward O_2.

5.2.1.3.2 Natural Clays as Metallocene Catalyst Supports

Clays without organic surfactants can be delaminated at least partially simply by treatment with an alkylaluminoxane,[27] allowing metallocenes to be supported on natural clays. For example, Wu et al. pretreated kaolin (B.E.T. surface area, $9\,m^2/g$) with MAO followed by Cp_2ZrCl_2 (**II**).[85] The grafted MAO was presumed to alkylate the metallocene and create the required cationic active sites. Ethylene polymerization was conducted at 80°C and 0.8 MPa. Much lower activities ($<100\,kg\ PE/(mol\ Zr\cdot h\cdot atm)$) were observed for the clay-supported catalyst compared to the homogeneous catalyst ($>2000\,kg\ PE\ (mol\cdot Zr\cdot h\cdot atm)$), although the molecular weight ($M_w = 2.5 \times 10^5$) was unchanged. It was not possible to extract all of the polyethylene from the composite using decalin, suggesting strong interfacial interactions between the polymer and the clay. This is consistent with SEM images of cryofractured and tensile-fractured surfaces and DMA measurements that showed increased glass transition temperatures T_g.

Wang et al. supported $(^nBuCp)_2ZrCl_2$ (**IX**) on MAO-treated sodium montmorillonite.[86] Ethylene polymerization was conducted at 50°C and 1 atm in the presence of additional MAO. The activity, initially very low, accelerated rapidly then declined. Such activity profiles are commonly observed for supported polymerization catalysts, where the induction period is associated with slow mass transfer of monomer into the pores, while the acceleration phase is attributed to fragmentation of the support, which exposes more active sites. XRD patterns recorded at different times showed the evolution of the layered clay structure, with expansion of the interlayer distance during the induction period as intercalated polyethylene was formed, followed by disappearance of the registry as the polymerization accelerated. It was suggested that the subsequent decline in activity was caused by polymer encapsulation of the catalyst, again restricting monomer access to active sites. Xu et al. reported different DSC peaks for the crystallization of intercalated and non-intercalated polymer, formed during the induction and acceleration phases, respectively.[87] The low melting point of the intercalated polyethylene was attributed to confinement of the polymer chains, which suppresses crystallinity. The melting point increased after delamination occurred, despite a decrease in the average molecular weight which may have been caused by greater exposure to the cocatalyst (since it can act as a chain-transfer agent).

The unbridged metallocene **X** was supported on MAO-modified montmorillonite in the presence of excess MAO, then centrifuged and washed to remove unsupported catalyst.[88] Polymerization of ethylene was conducted at 1 atm and 50°C in the presence of excess MAO. Clay-supported **X** showed lower activity ($0.4 \times 10^3\ kg\ PE/(mol\ Zr\cdot h)$) than its homogeneous analog, but the resulting polymer was higher in molecular weight and its molecular weight distribution was slightly broader.

Silica is a common and well-studied support for metallocene catalysts. A clay/metallocene catalyst system was prepared containing silica nanoparticles.[89] Sodium montmorillonite was ion exchanged with cetyltrimethylammonium, air-dried, then expanded further with dodecylamine to give an organoclay with a basal spacing of 5.2 nm. Exposure to $Si(OEt)_4$ resulted in hydrolysis of the silicate by intragallery water, giving a silica-containing clay with a basal spacing of 4.0 nm and silica nanoparticles with diameters <3.2 nm. The hybrid support was treated with MAO, then Cp_2ZrCl_2 (**II**), and used in the polymerization of ethylene with added MAO at 60°C and 1.2 atm. The extent of clay exfoliation was considerably greater when silica was present. Tang et al. attributed this effect to the presence of surface silanols on the silica nanoparticles where the cocatalyst was immobilized, resulting in a higher catalyst loading. The nanocomposites had higher tensile and storage moduli than either the unfilled polymer or the composite produced with clay but no silica. At 1665 MPa, the tensile modulus of the nanocomposite with 5.43 wt% filler may be high enough to make this material competitive with engineering plastics.

When titanocene dichloride (**XI**) was supported on palygorskite clay calcined at 800°C, then activated with 2000 equiv. MAO, and exposed to 0.1 MPa ethylene at 40°C in toluene, the supported catalyst was just as active in ethylene polymerization as the homogeneous catalyst.[90] However, it produced polymer with a higher molecular weight and melting point. The active site precursor was suggested to be a cationic titanocenium such as **XII**, formed by chloride transfer to Lewis acidic Mg^{2+} sites on the clay surface. This structure is analogous to those inferred by ^{13}C solid-state CP/MAS NMR spectroscopy for metallocenes $Cp'_2M(CH_3)_2$ (M is Zr, U, Th) supported on solid Lewis acids such as highly dehydroxylated alumina.[91]

XI   XII

Metallocenium active sites are probably associated with their grafted cocatalysts via ion-pairing, but are not covalently bonded to the clay. When a large excess of unsupported MAO is used as the cocatalyst, it may disrupt the close association between the metallocene and the clay, resulting in concurrent homogeneous and heterogeneous polymerization.[92] If this occurs, large regions of the composite will be devoid of clay. This problem was addressed by Novokshonova et al., who anchored AlR_3 (R = Me, ^iBu) onto wet sodium montmorillonite.[93] During the grafting onto surface hydroxyl sites, the alkylaluminum was partially hydrolyzed by interlayer water (9.6 wt%) associated with the clay. The resulting alkylaluminoxane-modified clays can be used to immobilize and activate metallocenes without the need for added MAO or other cocatalysts. Much of the work using AlR_3 to activate clay-supported catalysts probably involves this kind of *in situ* alkylaluminoxane generation on the clay surface. The productivities of zirconocene dichloride (**II**) and *rac*-Et(Ind)$_2$ZrCl$_2$ (**VIII**), supported in this way were comparable to those of their homogeneous analogs: while their initial activities were lower, deactivation was also less rapid.[93,94] At the same time, the supported catalysts produced higher molecular weight polyethylene with a broader molecular weight distribution, indicating less uniform active sites.[92] Although no information was provided on clay dispersion, the filler content was presumably very low due to the extremely high catalyst activity.

The Na^+ form of fluorotetrasilicic mica, $Na[(Mg_{2.5})Si_4O_{10}F_2]$ was cation exchanged with a variety of metal ions (Mg^{2+}, Fe^{3+}, Co^{2+}, Ni^{2+}, Cu^{2+}, Zn^{2+}), then calcined before its reaction with Cp_2ZrCl_2 (**II**) or $(^nBuCp)_2ZrCl_2$ (**IX**).[95] Ethylene polymerization was conducted with Al^iBu_3 in n-hexane at 60°C and 0.7 MPa. The activity was strongly dependent on the charge of the exchangeable cation, being essentially zero for Na^+ and highest for Fe^{3+}, at 870 kg PE/(mol Zr·h) for **II**. Polymerization activity was completely suppressed when the cation-exchanged clay was calcined above 250°C. At this temperature, cations migrate from the interlayer spaces into clay sheets, causing the mica to lose its ability to swell. The polyethylenes were characterized by M_w values of ca. 1×10^5 and $2 < M_w/M_n < 3$. The nature of the association between the catalyst and the clay is unclear, since fluorotetrasilicic mica contains no hydroxyl groups for anchoring either the catalyst or the cocatalyst. Washing the clay-supported catalysts led to >90% loss of activity, as the weakly bound zirconocene was removed. SEM images show that polymer forms initially at the edges of the clay sheets, as shown in Figure 5.9, supporting the hypothesis that the catalyst interacts with the clay. The presence of the catalyst did not affect the basal spacing, and no shift in the position of the d_{001} reflection was noted until after polymerization was initiated. The layered structure was destroyed later in the reaction, as the polymerization solvent became intercalated into the interlayer spaces. It was suggested that this results in access to more activating sites inside the clay. Although the evolution of catalytic activity over time was not reported, it presumably increases.

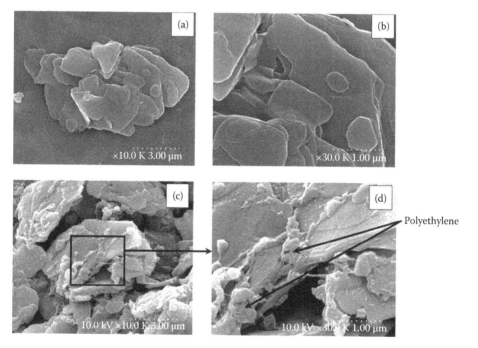

FIGURE 5.9
SEM images of Mg^{2+}-exchanged mica, dried at 200°C (frames a and b), and the same clay after supporting Cp_2ZrCl_2 and conducting polymerization of ethylene for 20 min at 40°C and 0.7 MPa in the presence of Al^iBu_3 (frames c and d). (Reproduced from Kurokawa, H. et al., *Appl. Catal. A: Gen.*, 360, 192, 2009. With permission from Elsevier B.V.)

5.2.1.3.3 Pillared Clay as Metallocene Catalyst Supports

Ferreira et al. produced a high-activity metallocene catalyst (1457 kg PE/(mol · Zr · h · atm)) by supporting Cp_2ZrCl_2 (**II**) on a pillared clay. Calcium montmorillonite was pillared with Keggin ions, $[Al_{13}O_4(OH)_{24} \cdot 12H_2O]^{7+}$, to increase its Lewis acidity.[96] The clay was pre-treated at 200°C to remove adsorbed water, then exposed to a toluene solution of MAO at 50°C to attenuate its Brønsted acidity. High activity was maintained over a period of 1 h at 50°C. The apparent absence of the usual rapid metallocenium deactivation processes was attributed to the suppression of bimolecular interactions in the supported catalyst.

5.2.1.3.4 Acid-Treated Clays as Metallocene Catalyst Supports

In solution, metallocene catalysts are activated by Lewis acids such as MAO or $B(C_6F_5)_3$, or Brønsted acids such as trialkylammoniums. Since acid-treated clays possess both Lewis and Brønsted acidity,[67] they can in principle act as "support activators" for metallocenes. Acid-treated montmorillonite and tetrafluorosilicic mica were claimed to give highly active MAO-free metallocene catalysts not only when combined (if necessary) with an alkylating agent (e.g., AlR_3),[41] but also in its absence for pre-alkylated metallocenes.[10] An activity of 490 kg PE/(mol · Zr · h) was reported at 85°C and 0.7 MPa for $(MeCp)_2ZrMe_2$ (**XIII**) supported on acid-treated montmorillonite, in the absence of any other cocatalyst.[10] Nakano et al. studied the effect of clay acidity by disrupting the layered structure of montmorillonite with H_2SO_4, then drying the acid-treated clay at 200°C.[97] It was treated with $AlEt_3$, then with various amounts of 2,6-dimethylpyridine to reduce its Brønsted acidity. The modified clays were stirred with a mixture of *rac*-$Me_2Si(2$-Me-4-Ph-1-Ind$)_2ZrCl_2$, **XV**, and Al^iBu_3. In the absence of 2,6-dimethylpyridine, the clay possesses strongly acidic sites (pK_a ca. 1.5), and the DRUV-vis spectrum showed a peak at ca. 550 nm associated with metallocenium ions. Clays whose acidity was attenuated did not show this peak and were inferred to contain only the neutral metallocene. Ethylene polymerization was conducted at 75°C and 0.8 MPaG. Polymerization activity was observed to decrease with decreasing acid strength of the clay.

XIII   XIV   XV

Supported metallocene catalysts were combined with K10 montmorillonite by Weiss et al.[98] During preparation of the clay, bentonite was treated with mineral acid, causing some of the octahedral alumina sheet to dissolve. This created dendritic silica, whose hydroxyl termination served to immobilize an alkylaluminum cocatalyst (either $AlMe_3$ or Al^iBu_3). The supported cocatalysts activated the simple metallocene dichlorides **II** and **XI**, by alkylation and alkyl abstraction, to create the metallocenium active sites. They were active for ethylene polymerization at 10 bar and 50°C without further addition of cocatalyst. Al^iBu_3 was more effective than $AlMe_3$, and activity increased with the Al/Zr,Ti ratio, up to 2000. The montmorillonite-supported zirconocenes were much more active than their homogeneous analogs, giving a maximum activity of 14,083 kg PE/(mol Zr·h). Unsubstituted titanocene dichloride **XI** suffered a large loss in activity when supported, but the substituted titanocenes **XVI–XIX** showed large activity increases, particularly dinuclear **XVIII**.

XVI   XVII   XVIII

XIX

The activation of pre-alkylated metallocene catalysts by homogeneous, non-coordinating Brønsted acids has a heterogeneous analog in their activation by sulfated zirconia, a strong solid acid cocatalyst.[99] Sun and Garcés reasoned that acidic clays might also function as effective heterogeneous metallocene activators, by delocalizing negative charge over the clay framework.[100] The low accessibility of these strong acid sites was overcome by exfoliating an unspecified clay via Brønsted acid treatment followed by freeze drying to give an acidic lamellar aerogel with high surface area and no basal reflection by XRD. The aerogel was treated with AlR_3 (R is Me, Et, ^nPr or ^iBu), then Cp_2ZrMe_2 (**XV**), and the resulting solid was active for ethylene polymerization (up to 157 kg PE/(mol Zr·h) at 10 psi C_2H_4 and 70°C in toluene); no activity was detected in the filtrate, confirming that the catalyst was heterogeneous. Despite the reaction of many of the acidic protons in the system with AlR_3, the clay itself was proposed to be the metallocene activator since the aerogel treated with $AlMe_3$ was less effective than the Al^nPr_3-treated material.

Woo et al. created supported MAO cocatalysts by hydrolyzing $AlMe_3$ in the presence of acid-treated clays, then used them to activate Cp_2ZrCl_2 (**II**).[101] The clays were hydrated K10 montmorillonite (4.9 wt% H_2O) and K10 dehydrated at 160°C. Each was treated with excess $AlMe_3$ until methane evolution ceased. The $AlMe_3$ reacted with all of the included water, but only some of the hydroxyl groups. Those present in the octahedral sheets were presumed inaccessible. Slurry polymerization of ethylene (1 atm) was initiated upon addition of the zirconocene at 50°C, with excess $AlMe_3$ remaining in solution. The catalyst made with hydrated K10 was more active than the catalyst made with dehydrated K10, although its high initial activity declined rapidly. Some unbound $AlMe_3$ was beneficial to the activity. It was suggested to scavenge hydroxyl groups newly exposed as the polymerization proceeded. Basic Kunipia clay (10.0 wt% H_2O) was not an effective cocatalyst when treated with $AlMe_3$, but was somewhat effective upon treatment with MAO. After acidification by intercalation of $ZrOCl_2 \cdot 8H_2O$, the Kunipia treated with $AlMe_3$ was a highly effective cocatalyst.

The reactivity of Cp_2ZrMe_2 was investigated by Scott et al. in combination with an acid-treated montmorillonite.[70] The metallocene-clay catalyst system was tested for ethylene polymerization activity from 20°C to 80°C at 100 psi in the absence of any other cocatalyst. The polymer yield was minimal. When the clay was modified with excess MAO and then washed, an activity of 125 kg PE/(mol·Zr·h) was recorded. However, the polymer showed little evidence of clay incorporation. It was suggested that, even without adding soluble MAO, methyl exchange could lead to leaching of the cocatalyst and thereby to polymerization that was largely homogeneous.

5.2.1.4 Early Transition Metal–Based Post-Metallocene Catalysts

The *in situ* synthesis of polyethylene-clay nanocomposites using Dow Chemical's constrained geometry catalyst, *ansa*-{η^5:η^1-(N)-1-[(*tert*-butylamido)dimethylsilyl]-2,3,4,5-tetramethylcyclopentadienyl}dimethyltitanium(IV) (**XX**), was explored by the Jérôme group.[102] To avoid direct contact between the very moisture-sensitive catalyst and the clay, they dried kaolin at 105°C and treated it with MAO prior to addition of **XX**. Ethylene slurry polymerization was conducted at 70°C and 9–20 bar in heptane. The use of $AlMe_3$-depleted MAO as the cocatalyst gave higher activity. The ultrahigh molecular weight polymers were insoluble in hot 1,2,4-trichlorobenzene. To reduce molecular weight and increase solubility, polymerization was conducted in the presence of 1.2 bar H_2; soluble polyethylene with a much lower molecular weight (M_n 10,200; M_w/M_n 10.9) was formed.

XX   XXI

The same group reported that supporting **XX** on sodium montmorillonite that had been treated with $AlMe_3$-depleted MAO gave an activity of 178 kg PE/(g·Ti·h) at 70°C and 10 bar.[103] The apparent enhancement of the activity even beyond that of the homogeneous catalyst was attributed to reduced reactor fouling. The same catalyst supported on lithium hectorite or kaolin was less active. The importance of pre-drying the clay, and of robust immobilization of MAO in the interlayer region (using a poor solvent such as heptane instead of toluene), was emphasized. The clay-supported catalysts produced ultrahigh molecular weight polyethylene that did not flow at 190°C. Very high melt viscosity provided kinetic stabilization of the filler dispersion. Extensive gel formation precluded molecular weight determination by size-exclusion chromatography. Although no low-angle peaks were evident in the XRD, suggesting exfoliation of the clay, there was also no evidence that the filler affected the mechanical properties of these composites. Apparent increases in the melting point (up to 145°C) were attributed to slow heat transfer through the materials.

The melt viscosity of the composites was reduced significantly by polymerization in the presence of H_2 (initially at 1.2 bar), which induces chain transfer. The H_2 also (unexpectedly) resulted in lower catalytic activity. The small-angle XRD showed no peaks, suggesting exfoliation of the clay, but a sloping baseline may indicate large and irregular distributions of interlayer spaces. TEM micrographs were consistent with exfoliated clay confined to discrete regions in the polymer matrix. This morphology suggests a combination of heterogeneous polymerization (giving rise to polymer-dispersed clay) and homogeneous polymerization (giving rise to polymer without clay). Compression molding resulted in the reappearance of a broad XRD signal at $2\theta = 4°$–$8°$, attributed to a polyethylene-intercalated structure. The intensity of the signal was even greater for melt-kneaded samples.[76] Improvements in tensile properties were evident for the lower molecular weight material, where they were attributed to the fine dispersion and preferential orientation of nanofiller.

The ability of the phenoxyketimine titanium complex **XXI**[104] to generate polyethylene-clay nanocomposites was investigated using MAO-treated montmorillonite as the activator.[70] Although this catalyst–cocatalyst combination showed very high activity, 4500 kg PE/(mol·Ti·h) at 20°C and 100 psi; it produced polyethylene that did not appear to contain any of clay (which was recovered separately at the bottom of the reactor) suggesting that the polymerization was predominantly homogeneous.

5.2.1.5 Late Transition Metal–Based Post-Metallocene Catalysts

Ethylene polymerization catalysts based on late transition metals may be intrinsically more compatible with clay supports than early transition metal catalysts, because of their higher tolerance for water and other polar impurities. Two families of late transition metal catalysts have been particularly successful in producing high molecular weight polyethylene: those based on bis(imino)pyridine ligands and those containing α-diimine or related ligands.

5.2.1.5.1 Bis(imino)pyridyl Complexes of Iron and Cobalt

When 2,6-bis[1-(2,6-diisopropylphenylimino)ethyl]pyridineiron(II) dichloride (**XXII**, L = Cl) was adsorbed on the organoclay Cloisite 20A (dried at 100°C then treated with MAO), its color changed from deep blue to brick red.[105] No change in the basal spacing of the clay was observed. Ethylene polymerization was conducted at 6 bar and 30°C for 1 h, in the presence of additional MAO. Compared to the homogeneous catalyst, the clay-supported catalyst showed two to three times lower activity (400–900 kg PE/(mol·Fe·h), depending on the Al/Fe ratio), but gave polyethylene with a similar bimodal molecular weight distribution, and increased molecular weight. However, the melting point was suppressed relative to that of polyethylene made without the clay. As the Al/Fe ratio increased, the molecular weight decreased due to an increasing contribution from the low molecular weight component to the bimodal distribution. As judged by XRD, the extent of clay exfoliation decreased with increasing clay content. TEM images showed a uniform distribution of clay particles.

The iron catalyst can also be assembled *in situ*. The Na^+ form of fluorotetrasilicic mica was ion exchanged with Fe^{3+}, then the iron-modified clay was treated with a solution of a bis(imino)pyridine ligand in 1-butanol.[106] Formation of the intercalated metal complexes was inferred by the appearance of the characteristic $\upsilon(C = N)$ mode at $1588\,cm^{-1}$ and an increase in the basal spacing of the clay from 1.0 to 1.6 nm for **XXIII** and **XXIV**. The bulkier **XXII** apparently did not form due to the inability of the ligand to penetrate the interlayer space.

XXII $R_1 = {}^iPr$, $R_2 = H$

XXIII $R_1 = Me$, $R_2 = H$

XXIV $R_1 = R_2 = Me$

Polymerization was conducted at 0.2–0.7 MPa ethylene and 60°C in the presence of a trialkylaluminum. The best activity, $1643\,kg\,PE/(mol\cdot Fe\cdot h)$, was obtained using catalyst **XXIII** with Al^iBu_3 as cocatalyst. The catalyst produced highly linear polyethylene with a melting point of 134°C and a broad molecular weight distribution ($M_w/M_n = 7.1$).

The bis(imino)pyridine complexes **XXII** and **XXIV** (L = Cl) were also pre-formed and then incorporated into the sodium forms of montmorillonite, fluorotetrasilicic mica, and saponite. The clay-supported catalysts all showed good activity in ethylene polymerization in the presence of Al^iBu_3.[107] However, much better activity was obtained when the clays were first ion exchanged with higher valent cations (Mg^{2+}, Zn^{2+}, or Fe^{3+}). **XXIV** was more active than **XXII**, and the montmorillonite- and saponite-supported catalysts were more active than the mica-supported catalysts. The mica lacks hydroxyl groups and therefore the ability to form a supported alkylaluminoxane cocatalyst. An activity of 22.6×10^3 kg PE/(mol·Fe·h) was observed for **XXIV** on Mg^{2+}-montmorillonite at 60°C with 0.4 MPa C_2H_4 and Al/Fe = 1000. The molecular weights were modest (M_n ca. 3×10^4) and the polydispersity indices high (5–20) indicating multiple active sites. The molecular weight unexpectedly increased with reaction temperature, probably due to changes in the distribution of active sites. No characterization of the clay incorporated into the composites was reported.

Tritto et al. examined the influence of steric hindrance in the *o*-positions on the activity of clay-supported iron catalysts.[108] Unmodified sodium montmorillonite (Dellite® HPS) and two organoclays ion exchanged with different amounts of dimethyldi(HT)ammonium (Dellite 72T and Dellite 67G) were dried at 110°C, then treated with Al^iBu_3-modified MAO (MMAO). The reaction of MMAO with hydroxyl groups present on the clay surface was accompanied by the formation of methane. MMAO also displaced more than 60% of the surfactant, causing the basal spacing of the organoclays to decrease. The blue complexes **XXII** and **XXIII** turned bright orange in the presence of the MMAO-modified clays (as is observed in solutions of MMAO), then slowly changed to green-brown over time. An increase in the basal spacing of the clay was taken to indicate catalyst intercalation.

Ethylene polymerization was conducted at atmospheric pressure at 40°C. Precontact between the catalyst and the clay led to lower activities than injection of the catalyst in the presence of ethylene, probably due to clay-induced catalyst decomposition. In contrast to their homogeneous behavior,[109] the productivity of the less sterically hindered catalyst **XXIII** and the M_w value of its polyethylene were not appreciably lower than that of the bulkier catalyst **XXII**. The broad, bimodal molecular weight distributions of the homogeneous catalysts were retained upon heterogenization, but with less of the lower mass component (indicating less chain transfer to aluminum). The resulting polymers showed two melting endotherms, at ca. 100°C and 130°C. TEM images revealed that better clay dispersion was obtained with a smaller amount of surfactant. The nanocomposites showed decreased polymer stability in an inert atmosphere, attributed to strong Brønsted acid sites on the clay formed by surfactant decomposition (Scheme 5.3). A remarkable increase in oxidative stability was attributed to the barrier effect of the clay on O_2 diffusion.

The ability of **XXII** to generate polyethylene-clay nanocomposites was investigated in the presence of acid-treated montmorillonite modified with MAO by Scott et al.[70] Although the catalyst was activated by the supported cocatalyst and showed high ethylene polymerization activity (5000 kg PE/(mol·Fe·h) at 30°C and 100 psi, it produced polyethylene that contained little clay, suggesting that most of the polymerization occurred homogeneously.

5.2.1.5.2 Nickel Catalysts Bearing α-Diimine Ligands

Wang and Liu treated sodium montmorillonite with triethylaluminoxane (TEAO), then added the precatalyst N,N'-bis(2,6-diisopropylphenyl)-1,4-diaza-2,3-dimethyl-1,3-butadienenickel dibromide, **XXV**, to create a supported catalyst containing 16 µmol Ni/g.[110] The basal spacing of the montmorillonite increased from 1.21 to 1.46 nm due to intercalation of the catalyst and cocatalyst. Ethylene polymerization was conducted at 1 atm and 0°C. The initial activity of the clay-supported catalyst was similar to that of its homogeneous analog, ca. 400 kg PE/(mol·Ni·h). However, ethylene uptake by the heterogeneous catalyst accelerated slowly, presumably due to an increased rate of mass transport of the monomer into the interlayer spaces as the clay was delaminated. The polydispersity index was low (1.7) at short reaction times (10 min), but rose to 11 after 20 min. The molecular weight distribution, analyzed by GPC and TREF, was bimodal. The first-formed polymer was highly branched with a low melting point, while at later times, a less branched, higher melting polymer was formed.

XXV

XXVI $R_1 = R_2 =$ Me
XXVII $R_1 =$ Me, $R_2 =$ H
XXVIII $R_1 = _iPr$, $R_2 =$ H

The Yang group created a supported catalyst using **XXV**, MAO, and fibrous palygorskite clay.[111] Calcination of the clay at 500°C prior to the reaction resulted in strong suppression of its main d_{110} reflection, which subsequently remained unchanged upon reaction with MAO and the nickel catalyst. When the clay was pretreated with MAO, its surface area increased while its pore volume and pore size decreased; uptake of the nickel catalyst increased by one-third. MAO was inferred to aid in dispersing the clay as nanoclusters. Polymerization was conducted in toluene with added MAO as a cocatalyst, using 6.87×10^5 Pa C_2H_4 and reaction temperatures from 0°C to 60°C. Activity was a maximum at 20°C, and with Al/Ni > 1000. The activated catalyst was suggested to be located near the clay surface, rather than interacting directly with it. Activity was lower in the presence of the clay, but this was largely due to the absence of the large initial surge in activity observed for the homogeneous catalyst. It was higher than for a silica-supported catalyst, attributed to the Lewis acid sites of the clay that participate in catalyst activation. Clay dispersion was not reported, but SEM images showed that the polymer particle morphology replicated that of the clay, as clusters of individual fibers (see Figure 5.5).

It is also possible to assemble the catalyst in the presence of the filler. Kurokawa et al. introduced Ni^{2+} into fluorotetrasilicic mica by ion exchange, then added an α-diimine ligand to form the catalyst precursors **XXV** and **XXVI–XXVIII** *in situ*.[112] The solvent used for addition of the ligand was important: toluene and 2-butanol were unable to deliver the ligand into the interlayer spaces, but acetonitrile worked, and resulted in an increase in basal spacing from 1.0 to 1.4 nm. Intercalation was more successful with the less sterically bulky ligands. The nickel complexes were strongly held by the clay and not extracted by CH_3CN, suggesting that they were probably cationic, with L = H_2O. Slurry polymerization of ethylene was conducted at 40°C–60°C with 0.4–1.0 MPa C_2H_4. There was no activity in the absence of a cocatalyst (MAO, $AlEt_3$, or Al^iBu_3, Al/Ni = 327), or without the α-diimine ligand. The highest activity, 2300 kg PE/(mol·Ni·h), was obtained at 50°C with MAO, but the highest molecular weight, M_n 1.2×10^5, was obtained with Al^iBu_3. The catalysts were deemed single-site because the polydispersity indices of the polyethylenes ranged from 2.6 to 3.5. A typical polyethylene, made with the catalyst derived from **XXVI**, contained 16 methyl branches per 1000 carbons, and had a melting point of 116°C. The particle morphology replicated that of the clay particles, as shown in Figure 5.10, and there was no reactor fouling. Exfoliation of the clay in the early stages of polymerization was assumed, but not investigated.

Polyethylene nanocomposites were prepared by reaction of bis(4,4'-methylene-bis-(2,6-diisopropylimino))acenapthenenickel dibromide, **XXIX**, with Claytone APA, an organoclay containing dimethyl(benzyl)(HT)ammonium ions.[113] The clay was first treated with $AlEt_3$. The pendant amine groups of the catalyst were proposed to interact with grafted ethylaluminum sites on the clay surface. Ethylene polymerization was conducted at 0.1 MPa and from 0°C to 70°C in toluene in the presence of added $AlEt_2Cl$ as cocatalyst. The highest polymerization activity was observed at 50°C and a cocatalyst/catalyst ratio (Al/Ni) of 600. Decreasing activity at higher temperatures was presumably caused by catalyst deactivation and/or decreased ethylene solubility in toluene. For filler contents of 5.05 and 8.31 wt%, the XRD patterns of the resulting nanocomposites showed no d_{001} reflection corresponding to the 1.87 nm gallery spacing in Claytone APA, indicating the presence of an exfoliated structure in the nanocomposite, although a broad peak centered at 6.0° was observed with 11.91 wt% clay. The position of the latter requires a *decrease* in interlayer spacing, possibly caused by a bilayer-to-monolayer surfactant transition leading to stacking recovery during compression molding (although an alternate explanation is that the alkylaluminum displaced much of the quaternary ammonium surfactant, as others have reported).[83,84,108] TEM micrographs confirmed that the material contains disordered, exfoliated clay sheets and that some reaggregation occurred during compression molding.

FIGURE 5.10
SEM images of Ni²⁺-exchanged mica, dried at 200°C (a), and a polyethylene particle obtained using this clay after its reaction with an α-diimine ligand (b), (c), showing morphology replication. (Reproduced from Fujii, K. et al., *Catal. Commun.*, 10, 183, 2008. With permission from Elsevier B.V.)

XXIX

XXX

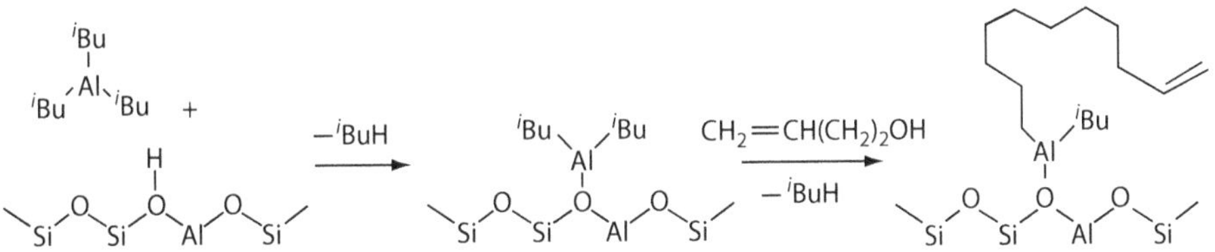

SCHEME 5.4
Tethering of the comonomer ω-undecylenyl alcohol to a clay surface.

In order to improve interfacial adhesion between the polymer and the clay, Simon et al. treated montmorillonite KSF first with HCl, then with AlR_3 (R = Me, ^{i}Bu), and finally with ω-undecylenyl alcohol to create pendant vinyl groups, as shown in Scheme 5.4.[114] The position of the basal reflection, d_{001}, decreased from $2\theta = 7.2°–3.2°$.[115] The catalyst (1,4-bis(2,6-diisopropylphenyl)-acenaphthenediimine)nickel dichloride, **XXX**, was intercalated and activated with MAO. Ethylene was polymerized at atmospheric pressure with an activity of 843 kg PE/(mol·Ni·h). The polymer had a low melting point due to extensive short chain branching, although the branching frequency was lower for the clay-supported catalyst than for its homogeneous analog. Copolymerization of ethylene with the tethered vinyl groups resulted in attachment of some chains to the clay surface, as evidenced by lower hot solvent extractability for the copolymer than for the homopolymer made by functionalizing the alkylaluminum-modified clay with undecyl alcohol. TEM images showed clay present in various states of dispersion. The attachment of polyethylene chains to clay platelets was confirmed by SEM and IR after solvent extraction.[115]

The thermal instability of quaternary ammonium surfactants (Scheme 5.3) imposes an upper limit on processing temperatures for polyethylene-organoclay nanocomposites. The Zhang group tackled this problem by creating a hybrid organic–inorganic surfactant based on a polyhedral oligosilsesquioxane (POSS) cube, terminated with eight propylammonium groups, **XXXI** (Oap-POSS).[116] It was intercalated into laponite, a synthetic, trioctahedral smectite comprised of octahedral magnesia sheets sandwiched between pairs of tetrahedral silica sheets. Li-substitution in the magnesia layer leads to a net negative charge on the clay layers, counterbalanced by Na^+ in the interlayer spaces, for an overall empirical formula $Na_{0.7}[(Si_8Mg_{5.5}Li_{0.3})O_{20}(OH)_4]$. Introduction of **XXXI** into laponite caused an expansion of the interlayer space from 1.48 to 1.86 nm. The modified clay was then treated with $AlEt_3$ (which reacts with the acidic protons of **XXXI**) followed by the nickel catalyst precursor **XXIX**, leading to a gallery space of 2.29 nm. Polymerization of ethylene at 0.1 MPa and 0°C, in the presence of $AlEt_2Cl$ as cocatalyst, gave exfoliated nanocomposites (3.92–9.75 wt% POSS-clay) showing no d_{001} reflection in their powder XRD patterns. TEM micrographs confirmed the high clay dispersion. ^{13}C NMR spectroscopy of the polyethylene showed that the chain branching frequency was much lower in the nanocomposite than in polyethylene made by the unsupported catalyst. Furthermore, methyl branches were observed exclusively in the filled polyethylene, while higher branches were abundant in the unfilled polyethylene. In this type of late transition metal catalyst system, extensive branching is largely due to facile chain-walking,[117] which is apparently suppressed in the confined interlayer space. The improvement in storage modulus E' reflects the combined reinforcement of the nanocomposite by both the clay and POSS components. Thermal stability in air was also improved, showing a 62°C increase in the temperature of 50% weight loss with 9.75 wt% POSS-clay.

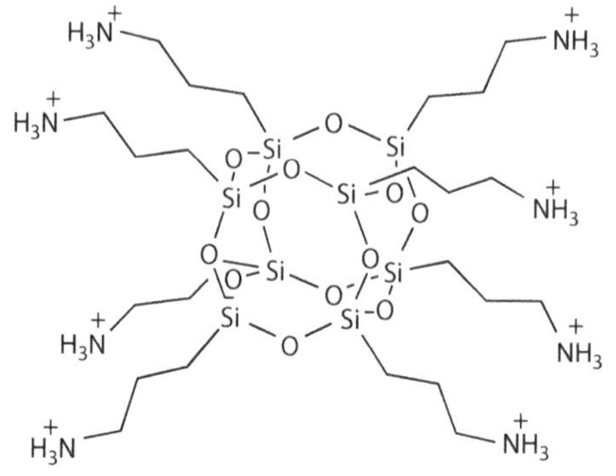

XXXI

The Zhang group also prepared polyethylene nanocomposites by intercalating the nickel α-diimine catalyst precursor **XXIX** into the interlayer space of a Zn–Al layered double hydroxide (LDH).[118] The chloride form of the LDH was first ion exchanged with dodecyl sulfate, causing the gallery spacing to expand from 0.87 to 2.65 nm, then treated with AlEt₃. Addition of a solution of the nickel complex resulted in further gallery expansion, to 3.31 nm. Ethylene polymerization was conducted at 0.1 MPa and 0°C in heptane in the presence of added AlEt₂Cl as cocatalyst. IR spectra confirmed the presence of the surfactant (and hence, the LDH) in the resulting polymer. The XRD of the resulting nanocomposites showed no d_{003} reflection (corresponding to adjacent LDH layers) for filler contents from 6.1–12.9 wt%, indicating loss of the layered structure during polymerization, while the polyethylene reflections also changed, suggesting restricted crystal growth due to confinement of the polymer chains. TEM micrographs showed the structure to be mainly intercalated, with some larger LDH aggregates and exfoliated material also present.

5.2.1.5.3 Nickel Catalysts Bearing α-Iminocarboxamidato Ligands

Activation of single-component nickel(II) catalysts by extensively acid-activated Li⁺ montmorillonite was explored by Scott et al.[70] The (α-iminocarboxamidato)(benzyl)nickel(II) complexes **XXXII–XXXIV** can be isolated in base-stabilized or base-free forms. Upon activation by a soluble Lewis acid such as B(C₆F₅)₃, the base-stabilized catalyst precursors undergo a spontaneous rearrangement of their supporting ligand, as shown in Scheme 5.5.[119] Although the base-free complex [N-(2,6-diisopropylphenyl)-2-(2,6-diisopropylphenylimino)

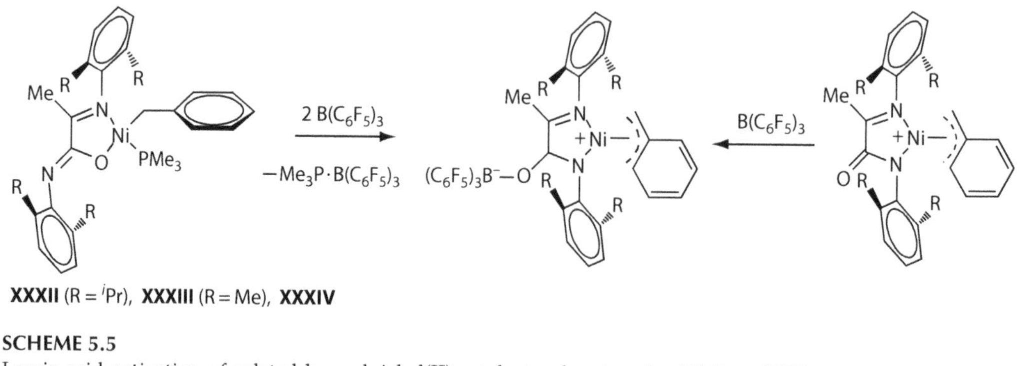

XXXII (R = ʲPr), **XXXIII** (R = Me), **XXXIV**

SCHEME 5.5
Lewis acid activation of related benzylnickel(II) catalysts, showing the (N,O) to (N,N) rearrangement of the α-iminocarboxamidato supporting ligand and the formation of the Zwitterionic active site (R = H, Me, ʲPr).[119,120]

propanamidato-κ^2-N,N]Ni(η^3-CH$_2$Ph) (**XXXIV**) does not rearrange, it is still inactive without a Lewis acid cocatalyst.[120] Activation involves formation of a Zwitterion (cationic at Ni) by coordination of the Lewis base at the carbonyl group. A reddish-yellow toluene solution of either **XXXII** or **XXXIV** (R = ^{i}Pr) was stirred with extensively acid-treated montmorillonite (pretreated with 17 wt% 6 M H$_2$SO$_4$/1 M Li$_2$SO$_4$ at 100°C for 6 h, then washed and dried at 200°C). The clay acquired a deep yellow color. The filtered, supported catalyst **XXXIV** polymerized ethylene with an activity of 965 kg PE/(mol·Ni·h) at 40°C and 100 psi in the absence of any cocatalyst. Thus, the clay functions as both support and activator for these nickel catalysts, as was proposed by researchers at Mitsubishi Chemical for metallocenes.[10,41] Polyethylene molecular weights were difficult to ascertain by high-temperature GPC because of extensive gel formation, but were estimated to be ca. 5×10^5.

The activation of the supported nickel complex was attributed to its interaction with Lewis acid sites present in the clay (likely coordinatively unsaturated Al cations), Scheme 5.6. Such Al sites can be found exposed at the edges of the clay platelets, and by isomorphic substitution into the silicate layer, as well as in the interlayer spaces where they migrate during the acid pretreatment. The base-stabilized complex **XXXII** showed somewhat lower activity, due to partial catalyst decomposition by loss of the α-iminocarboxamidato ligand (probably during dechelation involved in the clay-induced (N,O)-to-(N,N) rearrangement). A similar difference in activities was observed for catalyst **XXXIII**, which can be isolated in PMe$_3$-stabilized forms as both (N,O)- and (N,N)-ligated isomers.[120] The requirement for a carbonyl group to interact with a Lewis acid site in the clay was also evaluated. An analog containing a methylene instead of the carbonyl oxygen, Scheme 5.6, and R = ^{i}Pr showed even higher activity, 1340 kg PE/(mol Ni·h) at 40°C and 100 psi. Activation occurs via the methylene group, as was previously demonstrated for the homogeneous analog coordinated to B(C$_6$F$_5$)$_3$.[121]

Since the layered structure of the montmorillonite is largely disrupted by the acid treatment, its basal reflection is weak prior to polymerization, and it is completely absent in

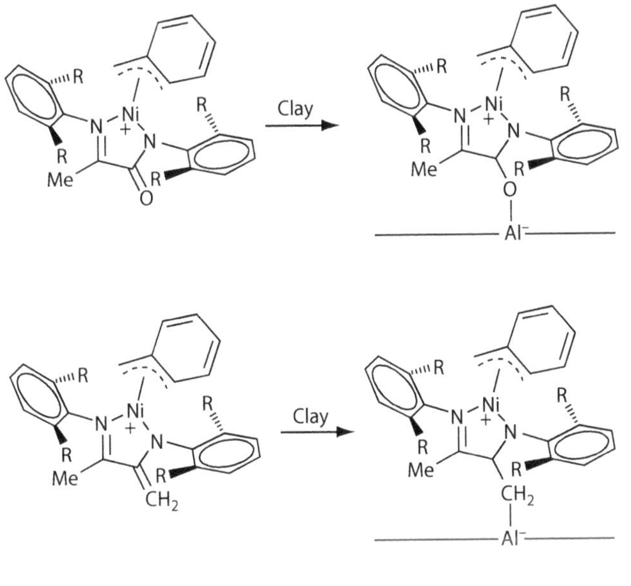

SCHEME 5.6
Proposed heterogeneous activation of (α-iminocarboxamidato)benzylnickel(II) catalysts by acid-activated montmorillonite.[70]

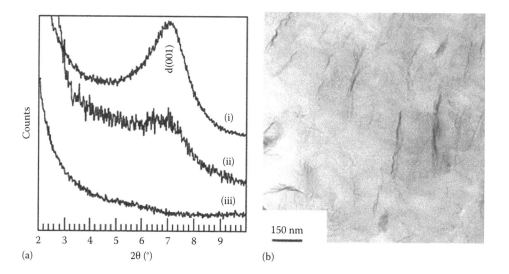

FIGURE 5.11
(a) Powder XRD of natural montmorillonite: (i) as raw clay, (ii) after treatment with H_2SO_4/Li_2SO_4, and (iii) incorporated into a 3 wt% clay-polyethylene nanocomposite generated by *in situ* polymerization using clay-activated **XXXIV**. (b) TEM image of the nanocomposite described in (a). (Reproduced from Scott, S.L. et al., *Chem. Commun.*, 4186, 2008. With permission from The Royal Society of Chemistry.)

nanocomposites made with clay-supported **XXXIV**, Figure 5.11. TEM images confirmed the high extent of clay dispersion and exfoliation into individual clay sheets. The material showed no evidence of restacking after annealing at 170°C for 30 min, presumably due to its high melt viscosity. Ethylene polymerization by each of the clay-supported catalysts resulted in significant increases in melting point and flexural modulus. However, these changes were not attributed directly to the presence of the clay, since the polymers had significantly less branching and higher molecular weights than the polyethylenes produced by the analogous, homogeneous catalyst systems $LNi(CH_2Ph)(PMe_3)/B(C_6F_5)_3$. Adsorption and activation of the precatalysts on the clay presumably suppresses the rates of chain termination and chain-walking, while accelerating chain propagation.

5.2.2 Homopolymerization of Propylene

In targeting polypropylene-clay nanocomposites by *in situ* polymerization, the need for both regio- and stereocontrol of the propylene insertion event is added to the requirements for sustained catalytic activity and high clay dispersion. Stereocontrol arises from asymmetry in the active site, due to either the ligand structure or the growing polymer chain. The active sites in Ziegler–Natta catalysts based on $TiCl_4/MgCl_2$ may become isospecific by adsorption of a Lewis base (the internal donor) near the active sites.[122] Chiral, stereorigid metallocenes with either C_2 or C_s symmetry can also be used to produce the desired highly isotactic polypropylenes.[123]

5.2.2.1 Ziegler–Natta Catalysts for Propylene Polymerization

5.2.2.1.1 Organoclays as Ziegler–Natta Catalyst Supports

The first polypropylene/clay nanocomposite made by intercalative polymerization was reported by Qi et al.[124] Sodium montmorillonite was ion exchanged with (hexadecyl/octadecyl)trimethylammonium, then vacuum-dried at 110°C. The organoclay was ground with MgCl$_2$, then mixed with toluene to make a slurry before addition of TiCl$_4$. The catalyst was activated with AlEt$_3$, and polymerization was conducted at 70°C–80°C. The clay loading in the composite was controlled by varying the polymerization time. No information about catalytic activity or polypropylene microstructure was reported. For a material containing 4.6 wt% clay, no basal reflection was detected by powder XRD above 2θ = 1.5°, implying a gallery size greater than 8.8 nm. With a higher clay loading (8.1 wt%), a small XRD feature indicated some clay aggregation. TEM images of injection-molded samples showed groups of 2–7 strongly aligned clay layers, with intercalated polymer and gallery spacings from 4 to 10 nm. The materials showed increased storage modulus and thermal stability under N$_2$.

Low-temperature (<200°C) decomposition of alkylammonium surfactants often used in organoclays, and leading to restacking of the clay layers, is an even more serious issue during melt processing of polypropylene compared to polyethylene. The problem was addressed by Han et al.,[125] who ion exchanged sodium montmorillonite with 1-hexadecylimidazolium (IMMT), Scheme 5.7. The *d*-spacing of the organoclay increased to 2.76 nm, and the temperature for onset of thermal decomposition increased to 270°C. The IMMT was added to a decane solution of MgCl$_2$ and 2-ethylhexan-1-ol containing phthalic anhydride as the internal donor. TiCl$_4$ was added to the slurry to give a material containing 0.98 wt% Ti. Polymerization was conducted in the presence of AlEt$_3$ as cocatalyst and diphenyldimethoxysilane as external donor. Unfortunately, the surfactant caused a significant reduction in catalytic activity.[126] The polypropylenes recovered from the reactor had M_w values in the range (4–8) × 10^5 and M_w/M_n values from 6 to 8. The isotacticity index (fraction of polymer insoluble in *n*-heptane) ranged from 90% to 97%, although it is not clear if the presence of the clay diminished the solubility of polypropylene. XRD of the nanocomposites showed no basal reflections for the clay, even after melt extrusion or injection molding at 200°C for 5 min. TEM images show a mixture of fully exfoliated clay and aggregates containing about 25 clay layers.

To improve catalytic activity while maintaining stability during subsequent high-temperature processing, the same group prepared montmorillonite ion exchanged with hexadecyltriphenylphosphonium ions (PMMT).[126] The surfactant shows onset of thermal decomposition above 300°C. The organoclay-supported Ziegler–Natta catalyst, made in

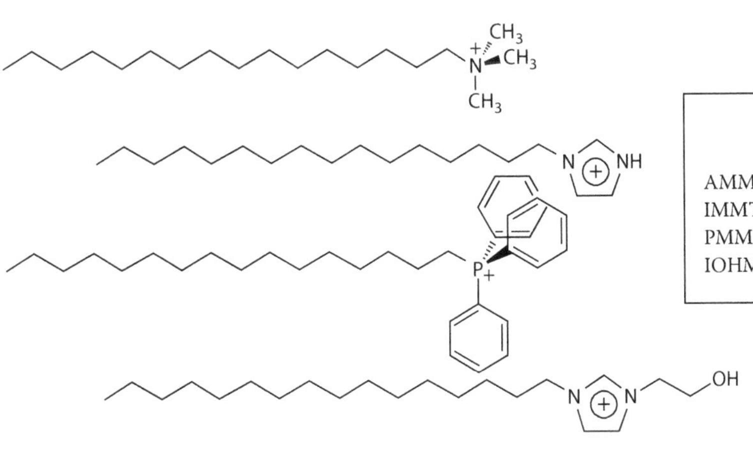

	$T_{0.01}$	$T_{0.10}$
AMMT	104	281
IMMT	267	410
PMMT	311	344
IOHMMT	250	400

SCHEME 5.7
Comparison of the thermal stability of organomontmorillonites containing various surfactants in a catalyst system for the *in situ* polymerization of propylene. The temperatures (°C) correspond to the indicated fractional weight loss measured by TGA under N$_2$.[126,127]

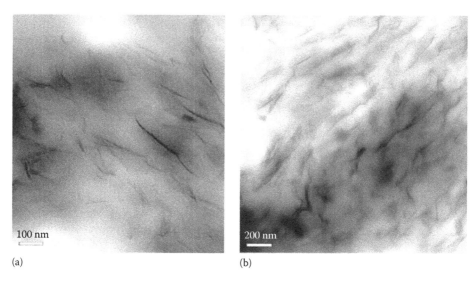

FIGURE 5.12
TEM images of polypropylene nanocomposites prepared with a Ziegler–Natta catalyst supported on 3-hexadecyl-1-(2-hydroxyethyl)-3*H*-imidazol-1-ium-exchanged montmorillonite: (a) containing 5 wt% clay and (b) containing 19 wt% clay. (Reproduced from Du, K. et al., *Macromol. Rapid Commun.*, 28, 2294, 2007. With permission from Wiley-VCJ Verlag GmbH & Co. KGaA.)

the same manner, contained 1.2 wt% Ti and produced isotactic polypropylene with similar molecular weight, polydispersity, and stereoregularity. Melting points and isotacticity indices of up to 163°C and 99%, respectively, were reported. However, catalytic activities were even higher than for the catalyst without clay, and much higher than for the catalyst on the alkylimidazolium-containing support, up to 1300 kg/(mol·Ti·h) at 70°C. XRD and TEM confirmed the dispersion and exfoliation of the clay in the nanocomposite. Even better clay exfoliation with up to 19 wt% clay loading was observed by TEM for polypropylene nanocomposites made using montmorillonite ion exchanged with 3-hexadecyl-1-(2-hydroxyethyl)-3*H*-imidazol-1-ium ions (IOHMMT), Figure 5.12.[127] The hydroxyl groups in the surfactant may interact with $MgCl_2$ and/or $TiCl_4$ to increase their association with the clay. The materials showed significant improvements in tensile strength and modulus, as well as elongation at break.

5.2.2.1.2 Hybrid MgCl$_2$-Clays as Ziegler–Natta Catalyst Supports

Zhang et al. created a hybrid catalyst support by treating sodium montmorillonite (dried at 400°C) with a solution of $MgCl_2$ in ethanol.[128] After filtering and evaporating the ethanol, $MgCl_2$ was deemed deposited in the interlayer spaces, since the galleries expanded from 1.15 to 1.39 nm. Further addition of $TiCl_4$ and ethyl benzoate (as internal donor) resulted in the adsorption of the catalyst onto the $MgCl_2$ component of the modified clay, with further gallery expansion to 1.71 nm. Propylene was polymerized at 70°C and 0.8 MPa in heptane with added $AlEt_3$ (Al/Ti = 20–40) as cocatalyst and $Ph_2Si(OMe)_2$ as external donor (Ti/Si = 4). The catalytic activity, ca. 3×10^5 g PP/(mol·Ti·h), and the resulting molecular weight of the polymer, M_w 1.5×10^5, were comparable to those for the $MgCl_2$-supported catalyst without clay; however, the isotacticity index was lower, at 85%. The nanocomposite showed no x-ray diffraction peak for the clay (2.4 wt%), which was observed to be partially exfoliated by TEM.

Ramazani et al. mixed calcined clay (untreated bentonite or sodium montmorillonite) with $Mg(OEt)_2$ in toluene, then added $TiCl_4$ with dibutyl phthalate as the internal donor.[129] The d_{001} reflection of the clay shifted to lower angle upon intercalation of the catalyst. Polymerization of propylene (1–4 bar) was conducted in hexane at 40°C–70°C with added cocatalyst ($AlEt_3$ or Al^iBu_3), dimethoxy(methyl)(cyclohexyl)silane as the external donor, and H_2. Maximum activity was observed at 60°C using montmorillonite, with Al^iBu_3 and Al/Ti = 55. The presence of 0.1 bar H_2 increased catalyst activity, up to 800 g PP/(mmol Ti·h), but caused the molecular weight to decrease from M_w = 260,000 to 100,000. The isotacticity index was 99%. The powder XRD pattern showed no d_{001} reflection; TEM confirmed a mostly random dispersion for the clay in the polymer. SEM images showed a smooth fractured surface without holes or knots, indicating good clay dispersion and good adhesion between the polymer and the clay. The apparent melting point of 167°C, higher than that of unfilled polypropylene (161°C), was attributed to slow heat transfer in the filled polymer. The gas permeability was reduced by 50% at a clay content of 1 wt%, and declined to 20% of the original value in the presence of 5 wt% clay. This is not far from the predicted value of 11% assuming highly exfoliated clay.

5.2.2.1.3 Acid-Treated Clays as Ziegler–Natta Catalyst Supports

The clay-supported catalyst systems comprised of $Zr(CH_2C_6H_5)_4$ and acid-treated montmorillonite passivated with $(CH_3)_3SiCl$ and/or Al^iBu_3 were described in Section 5.2.1.2.4. These systems are also capable of polymerizing propylene to form moderately isotactic polypropylene.[70] Catalytic activity toward propylene is lower than toward ethylene, with activities of only 20–30 kg PP/(mol Zr·h) at 50°C and 140 psi. Polypropylene composites with relatively high clay content (15–30 wt%) were recovered from the reactor; attempts to reduce the clay loading using longer reaction times were unsuccessful, suggesting that catalyst deactivation was severe. Visually, and by its melting point of ca. 150°C, the polymers appeared to be mixtures of atactic and isotactic forms, as has been reported for polypropylene made with the analogous homogeneous catalyst activated by $B(C_6F_5)_3$.[130] Further characterization of the polypropylene was severely hampered by the insolubility of the composite. However, TEM images confirmed partial exfoliation of the clay, Figure 5.13.

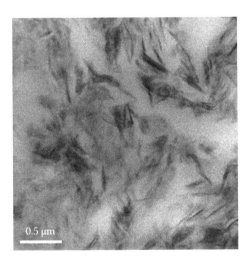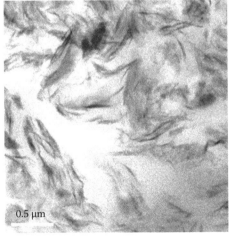

0.5 µm   0.5 µm

FIGURE 5.13
TEM images of a polypropylene nanocomposite containing 15 wt% clay, prepared by *in situ* polymerization with $Zr(CH_2C_6H_5)_4$ supported on acid-treated montmorillonite that had been passivated with $(CH_3)_3SiCl$.

5.2.2.2 Metallocene Catalysts for Propylene Polymerization

5.2.2.2.1 *Organoclays as Metallocene Catalyst Supports*

Researchers at Dow Chemical described the use of an organoclay containing an ammonium-based surfactant that activated a high isotacticity metallocene catalyst without the need for an external activator or cocatalyst.[131] Unfortunately, neither the surfactant nor the catalyst was specified. Sun and Garcés conducted the polymerization at 25°C, reporting a catalytic activity of 26×10^6 g PP/(mol catalyst·h). No polymerization occurred in the absence of the organoclay; therefore, the catalyst was deemed heterogeneous. The isotactic polypropylene had M_w of 181,000, M_w/M_n of 2.16, and 98.5% *mmm* triads. The powder XRD of the composite showed no basal reflection for the clay component. Although optical and low-magnification TEM images showed some large clay particles, high-resolution TEM images were consistent with considerable clay exfoliation into small stacks and single platelets with high aspect ratios (>100). The material showed a twofold improvement in Young's modulus (600 kpsi at 10.5 wt% clay). A reduced thermal expansion coefficient, improved melt flow strength, increased heat distortion temperature, and lowered gas diffusion coefficient were also claimed, although no details of these measurements were provided.

Hwu and Jiang used a montmorillonite ion exchanged with stearyltrimethylammonium to support MAO, followed by catalyst **VIII**.[132] Propylene polymerization was conducted in toluene at 20 psi and 20°C–80°C, without additional MAO. The presence of the organoclay resulted in an activity that was two orders of magnitude lower than that of the homogeneous catalyst, although the supported catalyst produced polymer with a higher content of isotactic pentads [*mmmm*]. For both supported and unsupported catalysts, the polymer melting point decreased precipitously when the reaction was conducted above 40°C. The XRD of the nanocomposite showed no basal reflection for the organoclay, although TEM images revealed only modest dispersion. The composites showed increased hardness and bulk density compared to unfilled polypropylene.

Reddy and Das made a silica-supported zirconocene catalyst *in situ*, by subliming $ZrCl_4$ onto a high surface area silica, then adding NaCp.[133] This catalyst was combined with an organically modified montmorillonite (Cloisite 20A), mixed with additional MAO (Al/Zr = 500). Propylene polymerization was conducted in the gas phase, at 8 bar and 70°C. Despite the absence of an obvious mechanism for interaction between the catalyst with the clay, or the presence of a solvent to swell the clay, the composite material was claimed to contain delaminated clay on the basis of decreased basal peak intensity in the XRD. The polypropylenes showed low melting points (132°C–134°C) consistent with large amorphous fractions.

To increase polymerization efficiency, Dong et al. prepared an organically modified montmorillonite using a mixture of two unfunctionalized and functionalized surfactants: hexadecyltrimethylammonium and (diethyl)(hexadecyl)(2-hydroxyethyl)ammonium, and then modified the organoclay with MAO.[134,135] The increased hydroxyl content of the clay galleries was proposed to reduce leaching of MAO, compared to organoclay functionalized with hexadecyltrimethylammonium alone. Intercalation of the isospecific, C_2-symmetric catalyst rac-Me₂Si(2-Me-4-Ph-Ind)₂ZrCl₂, **XXXV**, gave a very broad XRD pattern, which was interpreted in terms of nonuniform catalyst incorporation into the clay galleries. Propylene polymerization was conducted at 50°C and 0.5 MPa in the presence of additional MAO to give nanocomposites containing 1.0–6.7 wt% clay. The catalyst made with the surfactant mixture showed activities of $(5–30) \times 10^3$ kg PP/(mol·Zr·h), increasing with the amount of added MAO. Interestingly, the catalyst made with the unfunctionalized surfactant alone was more active, perhaps due to the absence of metallocene deactivation by the hydroxyl groups of the functionalized surfactant. The products were isotactic polypropylenes with melting points of ca. 157°C, M_n values of (1–2) × 10^5, and polydispersities between 3.2 and 4.4. The nanocomposites showed high clay dispersion, as judged by TEM, with no basal reflections visible by XRD.

XXXV   XXXVI   XXXVII

5.2.2.2.2 Unmodified Clay as a Metallocene Catalyst Support

Novokshonova et al. prepared clay-supported alkylaluminoxane cocatalysts by treating wet Cloisite Na$^+$ with AlR$_3$ (R = Me, ^{i}Bu), then used them in combination with a variety of *ansa*-zirconocenes to polymerize propylene.[94] The resulting polypropylene molecular weights and polydispersity indices were higher than those produced by the corresponding homogeneous catalyst systems. However, the isotacticity response was complex. Relative to the polymers made by the unsupported catalysts, the [*mmmm*] pentad content was lower for supported **XXXV**, higher for supported *rac*-Me$_2$Si(Ind)$_2$ZrCl$_2$ (**XXXVI**), and virtually unchanged for supported *rac*-[1-(9-η^5-Flu)-2-(5,6-Cp-2-Me-1-η^5-Ind)Et]ZrCl$_2$ (**XXXVII**). No physical properties of the nanocomposites were reported.

5.2.2.2.3 Acid-Treated Clays as Metallocene Catalyst Supports

Weiss et al. reported that Brintzinger's catalyst, **VIII**, supported on Al^{i}Bu$_3$-treated K10 montmorillonite was much more active toward propylene (60,300 kg PP/(mol·Zr·h) at 50°C) than the homogeneous catalyst (5,700 kg PP/(mol·Zr·h) at 40°C), although molecular weight and isotacticity were lower.[98] No physical properties of the composite material were reported.

Novokshonova et al. prepared clay-supported propylene polymerization catalysts using K10 montmorillonite (9.6 wt% H$_2$O).[136] The clay was first used to hydrolyze AlR$_3$ (R is Me, ^{i}Bu; AlR$_3$/H$_2$O = 1), then a zirconocene catalyst (**XXXV, XXXVI, or XXXVII**) or its hafnium analog was added (Al/Zr,Hf = 100–3700). Slurry polymerization was conducted in toluene at 30°C–65°C. Catalysts made with AlMe$_3$-modified clay showed poor activity, but performed well on Al^{i}Bu$_3$-modified clay. With M = Zr and R = ^{i}Bu, an activity of 23,280 kg PP/(mol·Zr·h·[C$_3$H$_6$]) was achieved at 50°C, higher than for the homogeneous system. The clay-supported hafnium analog of **XXXVI** was even more active than the zirconocene catalyst. The supported catalysts produced more stereoerrors than their homogeneous counterparts. Clay-supported **XXXVI** gave highly isotactic polymer (>80% [*mmmm*] pentads) even at low monomer concentration (0.13 M in toluene), in contrast to the behavior of the homogeneous catalyst (68%). Clay-supported **XXXV** was even more isospecific, giving >90% [*mmmm*] pentads. Clay-supported **XXXVII** made elastic polypropylene containing isolated stereoerrors. Addition of the borate cocatalyst [Ph$_3$C$^+$][B(C$_6$F$_5$)$_4$$^-$] gave higher polymerization rates at much lower Al/Zr, Hf ratios (100–500, compared to 2000–3000 without the borate). It also resulted in tremendously improved catalyst stability at 40°C–50°C and caused increases in molecular weight and isotacticity. No analysis of the clay distribution in the composite materials was reported.

5.3 Copolymerization

Polyolefin properties are modified by incorporation of comonomers during the polymerization. When made using a clay-supported catalyst, a copolymer nanocomposite can be obtained in one step. Incorporation of α-olefins (typically 1-hexene or 1-octene) into linear polyethylene creates branches that reduce brittleness and improve processability. Some catalysts are particularly well suited for copolymerization. For example, polyethylene-octene (POE) elastomers made using metallocene-based catalysts show uniform comonomer incorporation and narrow molecular weight distributions.[137] The constrained geometry catalyst **XX** is known for its superior ability to incorporate α-olefins. The incorporation of α-olefins bearing a polar functional group, or containing a precursor to such a functional group, into polyethylene or polypropylene can be used to improve the interfacial adhesion between the polyolefin and other materials.

Copolymers can be obtained from a single feed (e.g., ethylene) by concurrent tandem catalysis, using two catalysts. The first makes ethylene oligomers, while the second copolymerizes the oligomers with ethylene to give linear, low-density polyethylene (LLDPE).[138] This is illustrated in Scheme 5.8. While the need for mutual chemical compatibility limits possible catalyst combinations in solution, dispersed supported catalysts may be unable to interact as long as they remain immobilized, and may therefore tolerate each other better. However, it is necessary to match the catalytic activity of the two catalysts (via the catalyst ratio, as well as the reaction conditions) so that both contribute significantly to the overall reaction.

5.3.1 Ethylene/α-Olefin Copolymerization

5.3.1.1 By Comonomer Addition

The Jérôme group made clay nanocomposites containing ethylene/1-octene copolymers and ethylene/1-octene/1,9-decadiene terpolymers using the catalyst **XX**.[102] To avoid

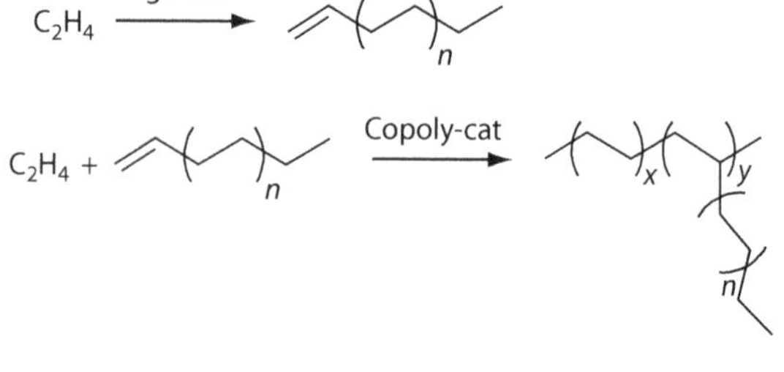

SCHEME 5.8
Concurrent tandem oligomerization/copolymerization of ethylene by a dual catalyst system.

contact between the very moisture-sensitive catalyst and the clay, they deposited **XX** on kaolin that had been dried at 105°C and treated with either MAO or $AlMe_3$-depleted MAO. Ethylene slurry polymerization was conducted in the presence of the comonomer(s) at 70°C and 10–20 bar in heptane. Strong melting point suppression was taken as evidence of comonomer incorporation. Catalytic activity was higher in the presence of H_2 (1.1–3 bar), although H_2 did not affect the incorporation of comonomers or the melting point of the copolymers.

Tang et al. reported ethylene/1-octene copolymerization using Cp_2ZrCl_2, **II**, supported on an MAO-modified organoclay.[77] The clay was made by ion exchange of sodium montmorillonite (Kunipia-F) with protonated amino acids and amino acid methyl esters. Catalytic activity for ethylene polymerization increased in the presence of 1-octene. This positive "comonomer effect" was attributed to an increased rate of ethylene diffusion in the less dense copolymer. The resulting nanocomposite material showed good bulk density[77] and 1-octene incorporation levels up to 5.7 mol%.[78] A clay-supported zirconocene catalyst prepared on montmorillonite pretreated with (3-aminopropyl)triethoxysilane followed by MAO showed a strong tendency for block copolymerization with ethylene/1-octene.[81]

Fan et al. reported copolymerization of ethylene with 1-hexene at 50°C and 1 atm using unbridged $(2,4-Me_2Ind)_2ZrCl_2$, **X**, supported on MAO-modified montmorillonite in the presence of excess MAO.[88] They observed slightly lower catalytic activity in the presence of the comonomer, ca. 200 kg P/(mol·Zr·h) and lower levels of comonomer incorporation (<10 mol%) than for the homogeneous catalyst.

Ziegler–Natta and metallocene catalysts are not capable of the direct copolymerization of ethylene with polar comonomers. To circumvent this problem, Dong et al. effected the copolymerization of ethylene with *p*-methylstyrene (MS) using an organoclay-supported metallocene catalyst to prepare a polymer nanocomposite containing reactive benzylic C–H bonds for subsequent functionalization.[139] Sodium montmorillonite ion exchanged with hexadecyltrimethylammonium ions was treated with MAO followed by *rac*-Et[Ind]$_2$ZrCl$_2$, **VIII**. The basal spacing of the organoclay increased from 2.0 to 3.8 nm due to intercalation of the catalyst. Copolymerization was conducted at 50°C and 0.5 MPa in the presence of additional MAO. The catalytic activity was higher in the presence of *p*-methylstyrene. The fractions of the polymer extracted by hot xylene had viscosity-average molecular weights $M\eta$ of (5–10) × 10^4. By ^1H NMR, their comonomer contents were judged to be 0.1–1 mol%, causing decreases in the melting point and crystallinity proportional to the extent of comonomer incorporation. Lower comonomer incorporation by the clay-supported catalyst compared to the corresponding homogeneous catalyst was attributed to hindered diffusion of the comonomer in the interlayer regions.

Following the methods of Lu and Chung,[140,141] the nanocomposites were derivatized in one of two ways. Treatment with benzoyl peroxide and maleic anhydride at 75°C led to free radical maleation of the copolymer (Scheme 5.9) without appreciably changing the molecular weight. The appearance of $\upsilon(C=O)$ bands at 1860 and 1780 cm^{-1} in the IR spectrum confirmed the formation of (PE-*co*-*p*-MS)-*g*-MA/OMMT, with maleic anhydride incorporated at levels proportional to the original *p*-methylstyrene content. Alternately, the copolymer was metallated with ^nBuLi/TMEDA at 45°C, followed by living anionic graft from polymerization of methyl methacrylate (MMA) to give (PE-*co*-*p*-MS)-*g*-PMMA/OMMT containing 10 mol% MMA. TEM images showed the clay to be exfoliated, and there was no reappearance of its basal reflection in the XRD, even after compression molding at 200°C.

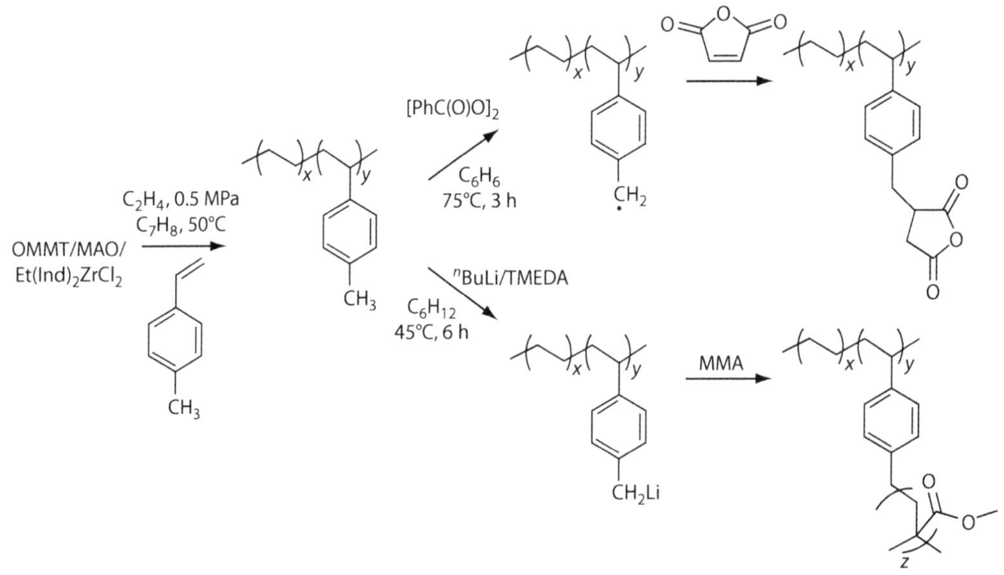

SCHEME 5.9
Functionalization of polyethylene/clay nanocomposites by intercalative copolymerization of ethylene with
p-methylstyrene and subsequent derivatization of the copolymer at its benzylic positions.[139]

5.3.1.2 By Tandem Concurrent Catalysis

One-pot, sequential oligomerization–copolymerization of ethylene and *in situ* nanocomposite formation was achieved by Wang et al. using two zirconium-based catalysts.[142] Zr(acac)$_2$Cl$_2$, **XXXVIII**, was intercalated into H$^+$-MMT, possibly becoming anchored through reaction with surface hydroxyl groups, and causing the interlayer spacing to increase from 1.38 to 1.86 nm. In the presence of AlEt$_2$Cl as cocatalyst at 50°C, ethylene (5 × 10^5 Pa) was converted to α-olefins in the clay galleries (Scheme 5.10a). The supported oligomerization catalyst showed greatly improved selectivity to C$_{6/8}$ α-olefins (90%, linear and branched), compared to its homogeneous analog (20%).[143] As has been reported for many supported ethylene polymerization catalysts, the oligomerization activity profile showed slower activation and much slower deactivation relative to the homogeneous catalyst. After 10 min of pre-oligomerization, the second catalyst, [*rac*-Et(Ind)$_2$ZrCl$_2$] (**VIII**), and its cocatalyst, MAO,

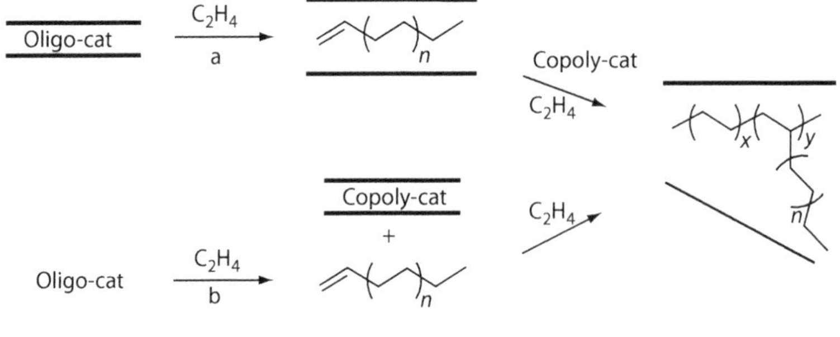

SCHEME 5.10
Two approaches to tandem formation of copolymer nanocomposites: (a) immobilization of the oligomerization catalyst, resulting in expansion of the clay layers during formation of α-olefins; and (b) immobilization of the copolymerization catalyst, resulting in exfoliation of the clay by migration of the α-olefins into the interlayer spaces.

were added to cause the formation of an ethylene/α-olefin copolymer, which showed the expected melting point depression for branched polyethylene. Copolymerization behavior was optimal at a catalyst ratio **XXXVIII/VIII** of 7.2; higher ratios led to the formation of highly branched products and much larger melting point depressions. High clay dispersion in the composites was assumed based on the absence of a basal reflection in the XRD.

XXXVIII

Hu et al. developed a dual catalyst system for the preparation of PE/MMT nanocomposites with an early–late transition metal combination.[144] The ethylene oligomerization catalyst was **XXXIX**. It was supported on MAO-treated montmorillonite and activated with additional MAO. After 10 min, pre-oligomerization with 1×10^5 Pa C_2H_4 at 50°C or 60°C, a solution containing $Me_2Si(Ind)_2ZrCl_2$ (**XXXVI**) and MAO was added to induce copolymerization and formation of an exfoliated nanocomposite. The clay dispersion was reported to be stable for months at room temperature. The presence of the ethylene oligomers led to a dramatic increase in polymerization activity and a decrease in the molecular weight; however, the resulting materials contained little clay and did not exhibit improved mechanical or thermal stability. By adjusting the catalyst ratio to obtain a branch frequency of 46 per 1000°C (and lower activity), better physical properties were observed, including a 50% increase in tensile strength and a 30°C increase in the onset of thermal decomposition.

XXXIX $R_1 = {}^iPr$, $R_2 = H$
XXXX $R_1 = Me$, $R_2 = OMe$
XXXXI $R_1 = R_2 = Me$

The same group used a similar approach with **XXXX** as the ethylene oligomerization catalyst and *rac*-Et(Ind)$_2$ZrCl$_2$ (**VIII**) as the copolymerization catalyst.[145] The oligomerization catalyst was again supported on MAO-treated montmorillonite, and additional MAO was used as the cocatalyst. During a pre-polymerization step at 90°C and 0.7 MPa C_2H_4, the formation of oligomers caused expansion of the interlayer spacing from 2.0 to 4.0 nm, allowing subsequent copolymerization to proceed in the galleries (assuming intercalation of the homogeneous zirconocene catalyst). The resulting nanocomposites containing LLDP showed clay exfoliation by XRD and TEM. Similar improvements in physical properties were reported.

The effect of clay pretreatment was investigated by supporting the oligomerization catalyst **XXXXI** on montmorillonite treated with MAO or AlMe$_3$, as well as on an organoclay (surfactant unspecified) treated with MAO.[146] After pre-oligomerization at 60°C and 0.7 MPa, ethylene copolymerization was initiated by addition of *rac*-Et(Ind)$_2$ZrCl$_2$, **VIII**. Almost all the α-olefins were incorporated, suggesting that this oligomerization catalyst produces lower molecular weight oligomers than either **XXXIX** or **XXXX**. The copolymers had melting points of ca. 120°C and M_n values from 98,000 to 143,000. XRD analysis showed more complete exfoliation (i.e., no d_{001} reflection) for the copolymer made with AlMe$_3$-modified montmorillonite, compared to those made with MAO-modified clays which showed mixed intercalated/exfoliated structures. AlMe$_3$ was suggested to intercalate more readily than MAO into the clay galleries and to form MAO *in situ* by partial hydrolysis with the interlayer water. The resulting stronger interactions with the clay were deemed responsible for better physical properties of the nanocomposite, such as impact strength. The tensile strength, Young's modulus, and flexural strength increased by factors of two, seven, and three, respectively, at 1.3 wt% clay loading, while the flexural modulus remained virtually unchanged relative to the unfilled copolymer, and elongation at break decreased.

An alternate strategy for concurrent tandem copolymerization is to add a homogeneous oligomerization catalyst to a slurry of the clay-supported copolymerization catalyst (Scheme 5.10b). A dual catalyst system was composed of the same bis(imino)pyridyliron catalyst (**XXXXI**) and *rac*-Et(Ind)$_2$ZrCl$_2$ (**VIII**) supported on MAO-modified montmorillonite.[147] Polymerization was conducted at 0.1 MPa C$_2$H$_4$ and 60°C with excess MAO. Slow initiation of the supported copolymerization catalyst was ascribed to the time required for the cocatalyst to diffuse into the interlayer spaces of the clay; during the induction period, oligomerization proceeded but then slowed as copolymerization was established. SEM images showed that better polymer particle morphology resulted when the polymerization catalyst was supported. There was no comonomer enhancement of the polymerization activity. At 0.1 MPa C$_2$H$_4$, the polymerization activity decreased slightly as the Fe/Zr ratio increased from 1:12 to 1:2; however, the effect was reversed when the pressure was increased to 0.7 MPa. At low Fe/Zr ratios, all oligomers were incorporated into the copolymer. The melting point, density, and crystallinity of the copolymer all decreased as the Fe/Zr ratio increased.

5.3.2 Ethylene/Polar Monomer Copolymerization

In order to enhance the interaction between polyethylene and the clay, and thereby to stabilize the clay dispersion in the nanocomposite, simultaneous *in situ* polymerization/functionalization of ethylene by copolymerization with functional group–containing comonomers was explored. Dong et al. created nanocomposites of organomontmorillonite (30.1 wt% hexadecyltrimethylammonium) and hydroxyl-functionalized polyethylene (PE-OH).[148] The catalyst, *rac*-Et[Ind]$_2$ZrCl$_2$ (**VIII**), was intercalated into the MAO-modified organoclay, resulting in a material containing 0.11 wt% Zr and 12.6 wt% Al, with a basal spacing of 3.8 nm. Ethylene copolymerization with 10-undecen-1-ol was conducted in the presence of added AlEt$_3$ (which reacts with and masks the alcohol) and MAO (to activate the metallocene) at 50°C and 0.5 MPa. As for the homogeneous catalyst, the clay-supported catalyst showed greater activity in the presence of the comonomer, which was incorporated into polyethylene up to 1.56 mol% based on ^{1}H NMR analysis. The presence of the comonomer caused its melting point to decrease from 132.3°C to 126.7°C.[149] The copolymer nanocomposites, containing 4–6 wt% clay, were almost XRD-silent in their as-prepared forms, and showed no recovery of their d_{001} signals upon molding at 200°C with as little as 0.16 mol% comonomer incorporation. In contrast, nanocomposites made in the same way but without the comonomer showed some stacking recovery accelerated by mechanical shear. It was manifested in the reappearance of the d_{001} peak after compression molding (5 min at 200°C), with greater recovery observed after injection molding at the same temperature. The shift of the basal reflection to 2θ values even larger than for the initial organoclay was suggested to arise from thermal decomposition of the surfactant during molding.

5.3.3 Propylene–Comonomer Copolymerization

An *in situ* strategy for preparing functionalized polypropylene/clay nanocomposites by copolymerization with a reactive monomer was developed by Dong et al.[150] An organoclay was created by ion exchange of sodium montmorillonite with hexadecyltrimethylammonium. An intercalated Ziegler catalyst was prepared by first dissolving anhydrous $MgCl_2$ in 2-ethylhexan-1-ol at 130°C, followed by addition of decane and phthalic anhydride at 60°C and the organoclay. This mixture was added to a large excess of $TiCl_4$ at –20°C. Dibutylphthalate was added as the internal donor. After heating at 120°C, filtering, washing, and drying, a solid containing 2.5 wt% Ti, 4.0 wt% Mg, 9.8 wt% ester, and 64.1 wt% organoclay was obtained. Its d_{001} spacing was 3.25 nm. Polymerization was conducted with $AlEt_3$ as cocatalyst and dimethoxydiphenylsilane as external donor in 0.7 MPa propylene at 60°C. Copolymers were prepared with 5-hexenyl-9-BBN (9-BBN = 9-borabicyclo[3.3.1]nonane, **XXXXII**) as comonomer. Upon borane removal with H_2O_2/NaOH in THF, polypropylene with anchored OH groups was obtained. The XRD reflection of the clay was absent in this material, and it reappeared to a lesser extent upon melting than for the nanocomposite made without the reactive comonomer.

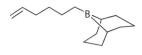

XXXXII

5.4 Future Research Directions

Understanding how polymerization catalysts interact with clays is an ongoing challenge, due to the complexity of heterogeneous support-cocatalyst-catalyst systems that include interlayer cations. The location of the catalyst and the robustness of its attachment to the clay (whether by covalent bonding or ion-pairing) under reaction conditions are probably important factors in determining the extent of dispersion and exfoliation of clay in the polymer product. The paramagnetism of natural clays complicates solid-state NMR studies. Characterization work with diamagnetic synthetic clays, coupled with computational modeling of clay-catalyst interactions, may shed light on how to optimize these interactions.

The preparation of polyolefin-clay nanocomposites with nanodispersions that remain stable during processing, even at elevated temperatures, remains an important goal.[12] For many potential applications, commercial success will preclude the use of large amounts of expensive compatibilizing agents. A promising approach is the *in situ* polymerization of ethylene with inexpensive polar comonomers, such as CO or acrylates. The incorporated functional groups will strengthen interfacial interactions between the polyolefin and the clay. Recent developments in the design of late transition metal catalysts capable of achieving such copolymerizations[17] should make the development of thermodynamically stable polyolefin-clay nanocomposites feasible.

The effect of stereochemistry and polydispersity on nanocomposite properties has yet to be explored fully. An intriguing result was obtained by Quijada et al. who performed melt mixing of PP-*g*-MA with octadecylamine-modified montmorillonite or hectorite, then blended these materials with different isotactic polypropylenes.[151] Nanocomposites with better clay exfoliation were obtained using metallocene polypropylene compared to Ziegler–Natta polypropylene, presumably because of its lower polydispersity. While stereoselective metallocene catalysts have been used successfully for *in situ* propylene polymerization in combination with clay supports (see Section 5.2.2.2), the materials properties of these nanocomposites have thus far received insufficient attention.

Acknowledgments

I am grateful to Fumihiko Shimizu and Toru Suzuki of the Mitsubishi Chemical Research Center, Yokohama, for many helpful discussions, and also to my UCSB coworkers Brian Peoples, Cathleen Yung, Mabel Caipa, Bryanna Kunkel, Jenny McCahill, and Philippe Perrotin for their many experimental contributions.

References

1. Galli, P. and Vecellio, G. 2004. Polyolefins: The most promising large-volume materials for the 21st century. *Journal of Polymer Science, Part A: Polymer Chemistry* 42:396–415.
2. Bowden, M. E. and Smith, J. K. 1994. *American Chemical Enterprise: A Perspective on 100 Years of Innovation*. Philadelphia, PA: Chemical Heritage Foundation.
3. Benedikt, G. M. and Goodall, B. L. 1998. *Metallocene-catalyzed polymers. Materials, Properties, Processing & Markets*. New York: Plastics Design Library.
4. Kaminsky, W., Külper, K., Brintzinger, H. H., and Wild, F. R. W. P. 1985. Polymerization of propene and butene with a chiral zirconocene and methylalumoxane as cocatalyst. *Angewandte Chemie International Edition* 24:507–508.
5. Stevens, J. C. 1996. Constrained geometry and other single site metallocene polyolefins catalysts: A revolution in olefin polymerization. *Studies in Surface Science and Catalysis* 101:11–20.
6. Ittel, S. D., Johnson, L. K., and Brookhart, M. 2000. Late-metal catalysts for ethylene homo- and copolymerization. *Chemical Reviews* 100:1169–1204.
7. Nakamura, A., Ito, S., and Nozaki, K. 2009. Coordination-insertion copolymerization of fundamental polar monomers. *Chemical Reviews* 109:5215–5244.
8. Britovsek, G. J. P., Gibson, V. C., Kimberley, B. S., Maddox, P. J., McTavish, S. J., Solan, G. A., White, A. J. P., and Williams, D. J. 1998. Novel olefin polymerization catalysts based on iron and cobalt. *Chemical Communications* 849–850.
9. Hlatky, G. G. 2000. Heterogeneous single-site catalysts for olefin polymerization. *Chemical Reviews* 100:1347–1376.
10. Takahashi, T., Nakano, H., Uchino, H., Tayano, T., and Sugano, T. 2002. Study of clay mineral "support-activator" in metallocene catalyst. *Polymer Preprints* 43:1259–1260.

11. Ciardelli, F., Coiai, S., Passaglia, E., Pucci, A., and Ruggeri, G. 2008. Nanocomposites based on polyolefins and functional thermoplastic materials. *Polymer International* 57:805–836.
12. Qin, Y.-W. and Dong, J.-Y. 2009. Preparation of nano-compounded polyolefin materials through *in situ* polymerization technique: Status quo and future prospects. *Chinese Science Bulletin* 54:38–45.
13. Ballard, D. G. H. 1973. Pi and sigma transition metal carbon compounds as catalysts for the polymerization of vinyl monomers and olefins. *Advances in Catalysis* 23:263–325.
14. Candlin, J. P. and Thomas, H. 1974. Supported organometallic catalysts. *Advances in Chemistry Series* 132:212–239.
15. Zakharov, V. A. and Yermakov, Y. I. 1979. Supported organometallic catalysts for olefin polymerization. *Catalysis Reviews—Science and Engineering* 19:67–103.
16. Rothon, R. N. 2003. *Particulate-Filled Polymer Composites*. Shrewsbury, U.K.: Rapra Technology Ltd.
17. Howard, E. G. 1975. Filled polyolefin compositions. DE 2,459,118. Priority application date December 13, 1974.
18. Howard, E. G. 1978. Low viscosity/inorganic filler. U.S. 4,104,243. Priority application date June 2, 1975.
19. Howard, E. G. 1978. Injection molding of homogeneous polyolefin compositions with high filler content. FR 2,355,855. Priority application date December 30, 1976.
20. Kostandov, L. A., Enikolopov, N. S., Dyachkovskii, F. S., Novokshonova, L. A., Gavrilov, Y. A., Kudinova, O. I., Maklakova, T. A., Akopyan, L. A., and Brikenshtein, K. A. 1980. Composite material. SU 763,379. Priority application date June 25, 1976.
21. Kostandov, L. A., Enikolopov, N. S., Dyachkovskii, F. S., Akopyan, L. A., Novokshonova, L. A., Gavrilov, Y. A., Kudinova, O. I., Brikenshtein, K. A., and Maklakova, T. A. 1980. Application of polymer coatings to solid materials. FR 72,437,416. Priority application date September 29, 1978.
22. D'yachkovskii, F. S. and Novokshonova, L. A. 1984. The synthesis and properties of polymerisation-filled polyalkenes. *Russian Chemical Reviews* 53:200–222.
23. Dubois, P., Alexandre, M., Hindryckx, F., and Jérôme, R. 1998. Polyolefin-based composites by polymerization-filling technique. *Polymer Reviews* 38:511–565.
24. Enikolopian, N. S., Fridman, A. A., Stalnova, I. O., and Popov, V. L. 1990. Filled polymers: Mechanical properties and processability. *Advances in Polymer Science* 96:1–67.
25. Dyachkovskii, F. S. 1993. New synthetic polyolefin composites. *Trends in Polymer Science* 1:274–280.
26. Enikolopian, N. S., Fridman, A. A., Popov, W. L., Stalnova, I. O., Brikenstein, A. A., Rudakov, W. M., Gherasina, N. P., and Tchalykh, A. E. 1986. Characteristics of polymerization-filled polyethylenes ("Norplastics") and compositions on the basis of norplastics. *Journal of Applied Polymer Science* 32:6107–6120.
27. Howard, E. G., Lipscomb, R. D., MacDonald, R. N., Glazar, B. L., Tullock, C. W., and Collette, J. W. 1981. Homogeneous composites of ultrahigh molecular weight polyethylene and minerals. 1. Synthesis. *Industrial & Engineering Chemistry Product Research and Development* 20:421–428.
28. McDaniel, M. P. 2008. Polymerization on Phillips-type catalysts. In *Handbook of Heterogeneous Catalysis*, G. Ertl, H. Knözinger, F. Schüth, and J. Weitkamp (eds.), pp. 3733–3792. Weinheim, Germany: Wiley-VCH.
29. Howard, E. G., Glazar, B. L., and Collette, J. W. 1981. Homogeneous composites of ultrahigh molecular weight polyethylene and minerals. 2. Properties. *Industrial & Engineering Chemistry Product Research and Development* 20:429–433.
30. Vlasova, N. N., Matkovskii, P. Y., Yenikolopyan, N. S., Papoyan, A. T., Vostorgov, B. Y., and Sergeyev, V. I. 1985. Effect of various factors on the kinetics of ethylene consumption during its polymerization on the surface of kaolin particles treated with organo-aluminum compounds. *Polymer Science USSR* 27:2380–2385.
31. Schöppel, W. and Reichert, K.-H. 1982. Polymerization of ethene by Ziegler catalysts in the presence of fillers. *Makromolekulare Chemie, Rapid Communications* 3:483–488.
32. Sohn, J. R. and Park, H. B. 1982. Cation exchanged silicate catalyst for ethylene polymerization. *Journal of Korean Chemical Society* 26:282–290.

33. Semikolenova, N. V., Nesterov, G. A., and Zakharov, V. A. 1986. Preparation of polymerization-filled polyethylene in the presence of catalysts based on organic and hydride compounds of Ti, Zr, Cr. *Polymer Science USSR* 28:186–192.

34. Karol, F. J., Karapinka, G. L., Wu, C.-S., Dow, A. W., Johnson, R. N., and Carrick, W. L. 1972. Chromocene catalysts for ethylene polymerization: Scope of the polymerization. *Journal of Polymer Science, Part A-1: Polymer Chemistry* 10:2621–2637.

35. Theopold, K. H. 1998. Homogeneous chromium catalysts for olefin polymerization. *European Journal of Inorganic Chemistry* 15–24.

36. Dubnikova, I. L., Petrosyan, A. I., Topolkarayev, V. A., Tovmasyan, Y. M., Meshkova, I. N., and D'yachkovskii, F. S. 1988. Effect of the initial component characteristics and the structural homogeneity of compositions on the plastic properties of polymerization-filled high density polyethylene. *Polymer Science USSR* 30:2509–2518.

37. Okada, A. and Usuki, A. Twenty years of polymer-clay nanocomposites. *Macromolecular Materials and Engineering* 291:1449–1476.

38. Usuki, A., Hasegawa, N., and Kato, M. 2005. Polymer-clay nanocomposites. *Advances in Polymer Science* 179:135–195.

39. Frankowski, D. J., Capracotta, M. D., Martin, J. D., Khan, S. A., and Spontak, R. J. 2007. Stability of organically modified montmorillonites and their polystyrene nanocomposites after prolonged thermal treatment. *Chemistry of Materials* 19:2757–2767.

40. Tudor, J., Willington, L., O'Hare, D., and Royan, B. 1996. Intercalation of catalytically active metal complexes in phyllosilicates and their application as propene polymerisation catalysts. *Chemical Communications* 2031–2032.

41. Suzuki, T. and Suga, Y. 1997. Solid acid supported metallocene catalysts for olefin polymerization. *Polymer Preprints* 38:207–208.

42. Hindryckx, F., Dubois, P., Jérôme, R., Teyssié, P., and Garcia Marti, M. 1997. Polymerization-filled composites prepared with highly active filler-supported Al/Ti/Mg catalysts. I. Synthesis of homogeneous polyethylene-based composites. *Journal of Applied Polymer Science* 64:423–438.

43. Hindryckx, F., Dubois, P., Jérôme, R., Teyssié, P., and Garcia Marti, M. 1997. Polymerization-filled composites prepared with highly active filler-supported Al/Ti/Mg catalysts. II. Properties of homogeneous polyethylene-based composites. *Journal of Applied Polymer Science* 64:439–454.

44. Parada, A., Bracho, L., Chirinos, J., and Rajmankina, A. M. T. 1998. Ethylene polymerization with catalysts supported on clay and blends of clay-$MgCl_2$. *Ciencia* 6:118–122.

45. Heinemann, J., Reichert, P., Thomann, R., and Mülhaupt, R. 1999. Transition metal catalyzed ethene homo- and copolymerization in the presence of exfoliated organophilic layered silicates and polyolefin nanocomposite formation. *Polymer Preprints* 40:788–789.

46. Heinemann, J., Reichert, P., Thomann, R., and Mülhaupt, R. 1999. Polyolefin nanocomposites formed by melt compounding and transition metal catalyzed ethene homo- and copolymerization in the presence of layered silicates. *Macromolecular Rapid Communications* 20:423–430.

47. Bergman, J. S., Chen, H., Giannelis, E. P., Thomas, M. G., and Coates, G. W. 1999. Synthesis and characterization of polyolefin-silicate nanocomposites: A catalyst intercalation and *in situ* polymerization approach. *Chemical Communications* 2179–2180.

48. Yamamoto, K., Ishihama, Y., Isobe, E., and Sugano, T. 2009. Ethylene polymerization behavior of Cr(III)-containing montmorillonite: Influence of chromium compounds. *Journal of Polymer Science, Part A: Polymer Chemistry* 47:2272–2280.

49. Demmelmaier, C. A., White, R. E., van Bokhoven, J. A., and Scott, S. L. 2008. Evidence for a chromasiloxane ring size effect in Phillips (Cr/SiO_2) polymerization catalysts. *Journal of Catalysis* 262:44–56.

50. Yang, F., Zhang, X.-Q., Zhao, H.-C., Chen, B., Huang, B.-T., and Feng, Z.-L. 2003. Preparation and properties of polyethylene-montmorillonite nanocomposites by *in situ* polymerization. *Journal of Applied Polymer Science* 89:3680–3684.

51. Cui, L. and Woo, S. I. 2008. Preparation and characterization of polyethylene (PE)/clay nanocomposites by *in situ* polymerization with vanadium-based intercalation catalyst. *Polymer Bulletin* 61:453–460.

52. Rong, J., Jing, Z., Li, H., and Sheng, M. 2001. A polyethylene nanocomposite prepared via *in-situ* polymerization. *Macromolecular Rapid Communications* 22:329–334.

53. Güven, N., D'Espinose de la Caillerie, J.-B., and Fripiat, J. J. 1992. The coordination of aluminum ions in the palygorskite structure. *Clays and Clay Minerals* 40:457–461.

54. Rong, J., Sheng, M., Li, H., and Ruckenstein, E. 2002. Polyethylene-palygorskite nanocomposite prepared via *in situ* coordinated polymerization. *Polymer Composites* 23:658–665.

55. Rong, J., Li, H., Jing, Z., Hong, X., and Sheng, M. 2001. Novel organic/inorganic nanocomposite of polyethylene. I. Preparation via *in situ* polymerization approach. *Journal of Applied Polymer Science* 82:1829–1837.

56. Du, Z., Zhang, W., Zhang, C., Jing, Z., and Li, H. 2002. A novel polyethylene/palygorskite nanocomposite prepared by in-situ coordination polymerization. *Polymer Bulletin* 49:151–158.

57. Du, Z., Rong, J., Zhang, W., Jing, Z., and Li, H. 2003. Polyethylene/palygorskite nanocomposites with macromolecular comb structure via *in situ* polymerization. *Journal of Materials Science* 38:4863–4868.

58. Ramazani, A. and Tavakolzadeh, F. 2008. Preparation of polyethylene/layered silicate nanocomposites using *in situ* polymerization approach. *Macromolecular Symposia* 274:65–71.

59. Ivanyuk, A. V., Gerasin, V. A., Rebrov, A. V., Pavelko, R. G., and Antipov, E. M. 2005. Exfoliated clay-polyethylene nanocomposites obtained by *in situ* polymerization. Synthesis, structure, properties. *Journal of Engineering Physics and Thermophysics* 78:926–931.

60. Zhang, F., Li, S., Karaki, T., and Adachi, M. 2005. Synthesis of polyethylene/montmorillonite nanocomposites by *in-situ* intercalative polymerization. *Japanese Journal of Applied Physics* 44:658–661.

61. Lomakin, S. M., Novokshonova, L. A., Brevnov, P. N., and Shchegolikhin, A. N. 2008. Thermal properties of polyethylene/montmorillonite nanocomposites prepared by intercalative polymerization. *Journal of Materials Science* 43:1340–1353.

62. Jin, Y.-H., Park, H.-J., Im, S.-S., Kwak, S.-Y., and Kwak, S. 2002. Polyethylene/clay nanocomposite by *in situ* exfoliation of montmorillonite during Ziegler-Natta polymerization of ethylene. *Macromolecular Rapid Communications* 23:135–140.

63. Gaboune, A., Sinha Ray, A., Ait-Kadi, A., Riedl, B., and Bousmina, M. 2006. Polyethylene/clay nanocomposites prepared by polymerization compounding method. *Journal of Nanoscience and Nanotechnology* 6:530–535.

64. Kovaleva, N. Y., Brevnov, P. N., Grinev, V. G., Kuznetsov, S. P., Pozdnyakova, I. V., Chvalun, S. N., Sinevich, E. A., and Novokshonova, L. A. 2004. Synthesis of polyethylene-layered silicate nanocomposites by intercalation polymerization. *Polymer Science Series A* 46:651–656.

65. Cui, L., Cho, H. Y., Shin, J.-W., Tarte, N. H., and Woo, S. I. 2007. Polyethylene-montmorillonite nanocomposites: Preparation, characterization and properties. *Macromolecular Symposia* 260:49–57.

66. Narayanan, S. and Deshpande, K. 1998. Acid activation of montmorillonite: Effect on structural and catalytic properties. *Studies in Surface Science and Catalysis* 113:773–778.

67. Rhodes, C. N. and Brown, D. R. 1995. Autotransformation and ageing of acid-treated montmorillonite catalysts: A solid-state ^{27}Al NMR study. *Journal of Chemical Society, Faraday Transactions* 91:1031–1035.

68. Briones-Jurado, C. and Agacino-Valdés, E. 2009. Brønsted sites on acid-treated montmorillonite: A theoretical study with probe molecules. *Journal of Physical Chemistry A* 113:8994–9001.

69. Haffad, D., Chambellan, A., and Lavalley, J. C. 1998. Characterisation of acid-treated bentonite. Reactivity, FTIR study and ^{27}Al MAS NMR. *Catalysis Letters* 54:227–233.

70. Scott, S. L., Peoples, B. C., Yung, C., Rojas, R. S., Khanna, V., Sano, H., Suzuki, T., and Shimizu, F. 2008. Highly dispersed clay-polyolefin nanocomposites free of compatibilizers, via the *in-situ* polymerization of α-olefins by clay-supported catalysts. *Chemical Communications* 4186–4188.

71. Ballard, D. G. H., Jones, E., Wyatt, R. J., Murray, R. T., and Robinson, P. A. 1974. Highly active polymerization catalysts of long life derived from σ- and π-bonded transition metal alkyl compounds. *Polymer* 15:169–174.

72. Peoples, B. C. 2008. *In situ* production of polyolefin-clay nanocomposites. PhD thesis, University of California, Santa Barbara.

73. Ramazani, A., Tavakolzadeh, F., and Baniasadi, H. 2009. *In situ* polymerization of polyethylene/ clay nanocomposites using a novel clay-supported Ziegler-Natta catalyst. *Polymer Composites* 30:1388–1393.

74. Jenny, C. and Maddox, P. 1998. Supported polyolefin catalysts. *Current Opinion in Solid State and Materials Science* 3:94–103.

75. Kaminsky, W. 1996. New polymers by metallocene catalysis. *Macromolecular Chemistry and Physics* 197:3907–3945.

76. Dubois, P., Alexandre, M., and Jérôme, R. 2003. Polymerization-filled composites and nano-composites by coordination catalysis. *Macromolecular Symposia* 194:13–26.

77. Liu, C., Tang, T., Zhao, Z., and Huang, B. 2002. Preparation of functionalized montmorillonites and their application in supported zirconocene catalysts for ethylene polymerization. *Journal of Polymer Science, Part A: Polymer Chemistry* 40:1892–1898.

78. Liu, C., Tang, T., Wang, D., and Huang, B. 2003. *In situ* ethylene homopolymerization and copolymerization catalyzed by zirconocene catalysts entrapped inside functionalized mont-morillonite. *Journal of Polymer Science, Part A: Polymer Chemistry* 41:2187–2196.

79. Kuo, S.-W., Huang, W.-J., Huang, S.-B., Kao, H.-C., and Chang, F.-C. 2003. Syntheses and characterizations of *in situ* blended metallocene polyethylene/clay nanocomposites. *Polymer* 44:7709–7719.

80. Zapata, P., Quijada, R., Covarrubias, C., Moncada, E., and Retuert, J. 2009. Catalytic activity during the preparation of PE/clay nanocomposites by *in situ* polymerization with metallocene catalysts. *Journal of Applied Polymer Science* 113:2368–2377.

81. Liu, C.-B., Tang, T., and Huang, B.-T. 2004. Zirconocene catalyst well spaced inside modified montmorillonite for ethylene polymerization: Role of pretreatment and modification of mont-morillonite in tailoring polymer properties. *Journal of Catalysis* 221:162–169.

82. Maneshi, A., Soares, J. B. P., and Simon, L. C. 2006. A single-gallery model for the *in situ* production of polyethylene nanocomposites. *Macromolecular Symposia* 243:277–286.

83. Lee, D.-H., Kim, H.-S., Yoon, K.-B., Min, K. M., Seo, K. H., and Noh, S. K. 2005. Polyethylene/ MMT nanocomposites prepared by *in situ* polymerization using supported catalyst systems. *Science and Technology of Advanced Materials* 6:457–462.

84. Leone, G., Bertini, F., Canetti, M., Boggioni, L., Stagnaro, P., and Tritto, I. 2008. *In situ* polym-erization of ethylene using metallocene catalysts: Effect of clay pretreatment on the prop-erties of highly filled polyethylene. *Journal of Polymer Science, Part A: Polymer Chemistry* 46:5390–5403.

85. Wang, X., Wu, Q., Dong, J.-Y., Hu, Y., and Qi, Z.-N. 2002. Characterization of polyethylene/ kaolin composites by polymerization filling with Cp_2ZrCl_2/MAO catalyst system. *Journal of Applied Polymer Science* 85:2913–2921.

86. Wang, Q., Zhou, Z.-Y., Song, L.-X., Xu, H., and Wang, L. 2004. Nanoscopic confinement effects on ethylene polymerization by intercalated silicate with metallocene catalyst. *Journal of Polymer Science, Part A: Polymer Chemistry* 42:38–43.

87. Xu, J.-T., Wang, Q., and Fan, Z.-Q. 2005. Non-isothermal crystallization kinetics of exfoliated and intercalated polyethylene/montmorillonite nanocomposites prepared by *in situ* polymer-ization. *European Polymer Journal* 41:3011–3017.

88. Wang, W., Fan, Z.-Q., and Feng, L.-X. 2005. Ethylene polymerization and ethylene/1-hexene copolymerization using homogeneous and heterogeneous unbridged bisindenyl zirconocene complexes. *European Polymer Journal* 41:2380–2387.

89. Wei, L., Tang, T., and Huang, B. 2004. Synthesis and characterization of polyethylene/clay-silica nanocomposites: A montmorillonite/silica-hybrid-supported catalyst and *in situ* polym-erization. *Journal of Polymer Science, Part A: Polymer Chemistry* 42:941–949.

90. Yan, X. W., Wang, J. D., Shan, Y. B., and Yang, Y. R. 2006. Characteristics of titanocene catalyst supported on palygorskite for ethylene polymerization. *Chinese Chemical Letters* 17:653–656.

91. Marks, T. J. 1992. Surface-bound metal hydrocarbyls. Organometallic connections between heterogeneous and homogeneous catalysis. *Accounts of Chemical Research* 25:57–65.
92. Novokshonova, L. A., Kovaleva, N., Meshkova, I., Ushakova, T. M., Krasheninnikov, V., Ladygina, T., Leipunskii, I., Zhigach, A., and Kuskov, M. 2004. Heterogenization of metalorganic catalysts of olefin polymerization and evaluation of active site non-uniformity. *Macromolecular Symposia* 213:147–155.
93. Novokshonova, L. A., Kovaleva, N. Y., Ushakova, T. M., Meshkova, I. N., Krasheninnikov, V. G., Ladygina, T. A., Zhigach, A. N., and Kuskov, M. L. 2005. Partially hydrolyzed alkylaluminums as the active heterogenized components of metallocene catalysts. *Kinetics and Catalysis* 46:853–860.
94. Ushakova, T. M., Lysova, M. V., Kudinova, O. I., Ladygina, T., Kiseleva, E. V., Novokshonova, L. A., Lyubimtsev, A. L., and Dybov, A. V. 2007. Olefin polymerization on immobilized zirconocene catalysts containing alkylaluminoxanes synthesized on the support surface. *Kinetics and Catalysis* 48:669–675.
95. Kurokawa, H., Morita, S., Matsuda, M., Suzuki, H., Ohshima, M., and Miura, H. 2009. Polymerization of ethylene using zirconocenes supported on swellable cation-exchanged fluorotetrasilicic mica. *Applied Catalysis A: General* 360:192–198.
96. Belleli, P. G., Eberhardt, A., Dos Santos, J. H. Z., Ferreira, M. L., and Damiani, D. E. 2003. Metallocene heterogenization on acid supports. *Recent Research Developments in Polymer Science* 7:223–245.
97. Nakano, H., Takahashi, T., Uchino, H., Tayano, T., and Sugano, T. 2006. Polymerization behavior with metallocene catalyst supported by clay mineral activator. In *Progress in Olefin Polymerization Catalysts and Polyolefin Materials*, T. Shiono, K. Nomura, and M. Terano (eds.), pp. 19–24, vol. 161. Amsterdam, the Netherlands: Elsevier.
98. Weiss, K., Wirth-Pfeifer, C., Hofmann, M., Botzenhardt, S., Lang, H., Brüning, K., and Meichel, E. 2002. Polymerisation of ethylene or propylene with heterogeneous metallocene catalysts on clay minerals. *Journal of Molecular Catalysis A: Chemical* 182–183:143–149.
99. Ahn, H., Nicholas, C. P., and Marks, T. J. 2002. Surface organozirconium electrophiles activated by chemisorption on "super acidic" sulfated zirconia as hydrogenation and polymerization catalysts. A synthetic, structural and mechanistic catalytic study. *Organometallics* 21:1788–1806.
100. Sun, T. and Garcés, J. M. 2003. Acidic lamellar aerogel nanoplate activated olefin polymerization with metallocene catalysts. *Catalysis Communications* 4:97–100.
101. Jeong, D. W., Hong, D. S., Cho, H. Y., and Woo, S. I. 2003. The effect of water and acidity of the clay for ethylene polymerization over Cp_2ZrCl_2 supported on TMA-modified clay materials. *Journal of Molecular Catalysis A: Chemical* 206:205–211.
102. Alexandre, M., Martin, E., Dubois, P., Garcia-Marti, M., and Jérôme, R. 2000. Use of metallocenes in the polymerization-filling technique with production of polyolefin-based composites. *Macromolecular Rapid Communications* 21:931–936.
103. Alexandre, M., Dubois, P., Sun, T., Garcés, J. M., and Jérôme, R. 2002. Polyethylene-layered silicate nanocomposites prepared by the polymerization-filling technique: Synthesis and mechanical properties. *Polymer* 43:2123–2132.
104. Mason, A. F. and Coates, G. W. 2004. New phenoxyketimine titanium complexes: Combining isotacticity and living behavior in propylene polymerization. *Journal of the American Chemical Society* 126:16326–16327.
105. Ray, S., Galgali, G., Lele, A., and Sivaram, S. 2005. *In situ* polymerization of ethylene with bis(imino)pyridine iron(II) catalysts supported on clay: The synthesis and characterization of polyethylene-clay nanocomposites. *Journal of Polymer Science, Part A: Polymer Chemistry* 43:304–318.
106. Kurokawa, H., Matsuda, M., Fujii, K., Ishihama, Y., Sakuragi, T., Ohshima, M., and Miura, H. 2007. Bis(imino)pyridine iron and cobalt complexes immobilized into interlayer space of fluorotetrasilicic mica: Highly active heterogeneous catalysts for polymerization of ethylene. *Chemistry Letters* 36:1004–1005.
107. Hiyama, Y., Kawada, Y., Ishihama, Y., Sakuragi, T., Ohshima, M., Kurokawa, H., and Miura, H. 2009. Catalytic behavior of bis(imino)pyridineiron(II) complex supported on clay minerals during slurry polymerization of ethylene. *Bulletin of Chemical Society of Japan* 82:624–626.

108. Leone, G., Bertini, F., Canetti, M., Boggioni, L., Conzatti, L., and Tritto, I. 2009. Long-lived layered silicates-immobilized 2,6-bis(imino)pyridyl iron(II) catalysts for hybrid polyethylene nanocomposites by *in situ* polymerization: Effect of aryl ligand and silicate modification. *Journal of Polymer Science, Part A: Polymer Chemistry* 47:548–564.

109. Small, B. L., Brookhart, M., and Bennett, A. M. 1998. Highly active iron and cobalt catalysts for the polymerization of ethylene. *Journal of the American Chemical Society* 120:4049–4050.

110. Wang, Q. and Liu, P. 2005. Dual bimodal polyethylene prepared by intercalated silicate with nickel diimine complex. *Journal of Polymer Science, Part A: Polymer Chemistry* 43:5506–5511.

111. Yan, X., Wang, J., Yang, Y., and Zhang, L. 2005. Ethylene polymerization with palygorskite supported nickel diimine catalyst. *Chinese Journal of Chemical Engineering* 13:361–366.

112. Fujii, K., Ishihama, Y., Sakuragi, T., Ohshima, M., Kurokawa, H., and Miura, H. 2008. Heterogeneous catalysts immobilizing α-diimine nickel complexes onto fluorotetrasilicic mica interlayers to prepare branched polyethylene from only ethylene. *Catalysis Communications* 10:183–186.

113. He, F.-A., Zhang, L.-M., Jiang, H.-L., Chen, L.-S., Wu, Q., and Wang, H.-H. 2007. A new strategy to prepare polyethylene nanocomposites by using a late-transition-metal catalyst supported on AlEt$_3$-activated organoclay. *Composites Science and Technology* 67:1727–1733.

114. Shin, S.-Y. A., Simon, L. C., Soares, J. B. P., and Scholz, G. 2003. Polyethylene-clay hybrid nanocomposites: *In situ* polymerization using bifunctional organic modifiers. *Polymer* 44:5317–5321.

115. Shin, S.-Y. A., Simon, L. C., and Soares, J. B. P. 2006. Morphology of hybrid polyethylene-clay nanocomposites prepared by *in-situ* polymerization. *Polymer Preprints* 47:31–32.

116. He, F.-A. and Zhang, L.-M. 2006. Using inorganic POSS-modified laponite clay to support a nickel α-diimine catalyst for *in situ* formation of high performance polyethylene composites. *Nanotechnology* 17:5941–5946.

117. Guan, Z. B., Cotts, P. M., McCord, E. F., and McLain, S. J. 1999. Chain-walking: A new strategy to control polymer topology. *Science* 283:2059–2061.

118. He, F.-A. and Zhang, L.-M. 2007. New polyethylene nanocomposites prepared by *in-situ* polymerization method using nickel α-diimine catalyst supported on organo-modified ZnAl layered double hydroxide. *Composites Science and Technology* 67:3226–3232.

119. Lee, B. Y., Bazan, G. C., Vela, J., Komon, Z. J. A., and Bu, X. 2001. α-Iminocarboxamidato-nickel(II) ethylene polymerization catalysts. *Journal of the American Chemical Society* 123:5352–5353.

120. Rojas, R. S., Wasilke, J.-C., Wu, G., Ziller, J. W., and Bazan, G. C. 2005. α-Iminocarboxamide nickel complexes: Synthesis and uses in ethylene polymerization. *Organometallics* 24:5644–5653.

121. Kim, Y. H., Kim, T. H., Lee, B. Y., Woodmansee, D., Bu, X., and Bazan, G. C. 2002. α-Iminoenamido ligands: A novel structure for transition-metal activation. *Organometallics* 21:3082–3084.

122. Garoff, T., Virkkunen, V., Jääskeläinen, P., and Vestberg, T. 2003. A qualitative model for polymerisation of propene with a MgCl$_2$-supported TiCl$_4$ Ziegler-Natta catalysts. *European Polymer Journal* 39:1679–1685.

123. Coates, G. W. 2000. Precised control of polyolefin stereochemistry using single-site metal catalysts. *Chemical Reviews* 100:1223–1252.

124. Ma, J.-S., Qi, Z.-N., and Hu, Y.-L. 2001. Synthesis and characterization of polypropylene/clay nanocomposites. *Journal of Applied Polymer Science* 82:3611–3617.

125. He, A. H., Hu, H., Huang, Y., Dong, J.-Y., and Han, C. C. 2004. Isotactic poly(propylene)/monoalkylimidazolium-modified montmorillonite nanocomposites: Preparation by intercalative polymerization and thermal stability study. *Macromolecular Rapid Communications* 25:2008–2013.

126. He, A. H., Wang, L., Li, J., Dong, J.-Y., and Han, C. C. 2006. Preparation of exfoliated isotactic polypropylene/alkyl-triphenylphosphonium-modified montmorillonite nanocomposites via *in situ* intercalative polymerization. *Polymer* 47:1767–1771.

127. Du, K., He, A. H., Liu, X., and Han, C. C. 2007. High-performance exfoliated poly(propylene)/clay nanocomposites by *in situ* polymerization with a novel Z-N/clay compound catalyst. *Macromolecular Rapid Communications* 28:2294–2299.

128. Zhao, H.-C., Zhang, Z.-Q., Yang, F., Chen, B., Jin, Y.-T., Li, G., Feng, Z.-L., and Huang, B.-T. 2003. Synthesis and characterization of polypropylene/montmorillonite nanocomposites via an *in-situ* polymerization approach. *Chinese Journal of Polymer Science* 21:413–418.
129. Ramazani, A. 2010. Synthesis of polypropylene/clay nanocomposites using bisupported Ziegler-Natta catalyst. *Journal of Applied Polymer Science* 115:308–314.
130. Pellechia, C., Proto, A., Longo, P., and Zambelli, A. 1992. Polymerization of ethylene and propylene in the presence of organometallic compounds of titanium and zirconium activated with tris(pentafluorophenyl)boron. *Makromolekulare Chemie, Rapid Communications* 13:277–281.
131. Sun, T. and Garcés, J. M. 2002. High-performance polypropylene-clay nanocomposites by *in-situ* polymerization with metallocene/clay catalysts. *Advanced Materials* 14:128–130.
132. Hwu, J.-M. and Jiang, G.-J. 2005. Preparation and characterization of polypropylene-montmorillonite nanocomposites generated by *in situ* metallocene catalyst polymerization. *Journal of Applied Polymer Science* 95:1228–1236.
133. Reddy, C. S. and Das, C. K. 2006. *In situ* polypropylene nanocomposites: Gas-phase polymerization of propylene in the presence of nanofillers using nanosilica-supported-zirconocene catalyst. *Journal of Macromolecular Science A: Pure and Applied Chemistry* 43:1365–1378.
134. Yang, K., Huang, Y., and Dong, J.-Y. 2007. Efficient preparation of isotactic polypropylene/montmorillonite nanocomposites by *in situ* polymerization technique via a combined use of functional surfactant and metallocene catalysis. *Polymer* 48:6254–6261.
135. Dong, J.-Y. and Hu, Y. 2009. Polymerized delamination of clay in polyolefins: From efficient intercalative polymerization to product-retrievable nanocomposite preparation. *Polymeric Materials Science and Engineering* 100:296–297.
136. Novokshonova, L. A., Ushakova, T. M., Lysova, M. V., Kudinova, O. I., Krasheninnikov, V. G., and Lyubimtsev, A. L. 2007. Supported Zr- and Hf-cene catalysts for propylene polymerization. *Macromolecular Symposia* 160:107–113.
137. Löfgren, B., Kokko, E., and Seppälä, J. 2004. Specific structures enabled by metallocene catalysis in polyethenes. In *Long Term Properties of Polyolefins*, A.-C. Albertsson (ed.), (Advances in Polymer Science), vol. 169, pp. 1–12. Berlin, Germany: Springer-Verlag.
138. Wasilke, J.-C., Obrey, S. J., Baker, R. T., and Bazan, G. C. 2005. Concurrent tandem catalysis. *Chemical Reviews* 105:1001–1020.
139. Huang, Y., Yang, K., and Dong, J.-Y. 2007. An *in situ* matrix functionalization approach to structure stability enhancement in polyethylene/montmorillonite nanocomposites prepared by intercalative polymerization. *Polymer* 48:4005–4014.
140. Lu, H. L. and Chung, T. C. 1999. Synthesis of PP graft copolymers via anionic living graft-from reactions of polypropylene containing reactive *p*-methylstyrene units. *Journal of Polymer Science, Part A: Polymer Chemistry* 37:4176–4183.
141. Lu, H. L. and Chung, T. C. 2000. Synthesis of maleic anhydride grafted polyethylene and polypropylene, with controlled molecular structure. *Journal of Polymer Science, Part A: Polymer Chemistry* 38:1337–1343.
142. Wang, J., Liu, Z.-Y., Guo, C.-Y., Chen, Y.-J., and Wang, D. 2001. Preparation of a PE/MT composite by copolymerization of ethylene with *in-situ* produced ethylene oligomers under a dual functional catalyst intercalated into MT layer. *Macromolecular Rapid Communications* 22:1422–1426.
143. Oouchi, K., Mitani, M., Hayakawa, M., Yamada, T., and Mukaiyama, T. 1996. Ethylene oligomerization catalyzed with dichlorobis(β-diketonato)zirconium/organoaluminum chloride systems. *Macromolecular Chemistry and Physics* 197:1545–1551.
144. Guo, C.-Y., Ma, Z., Zhang, M., He, A. H., Ke, Y., and Hu, Y. 2002. Preparation of PE/MMT nanocomposite by monomer intercalation and *in situ* copolymerization. *Chinese Science Bulletin* 47:1267–1270.
145. Qian, J., Guo, C.-Y., Wang, H., and Hu, Y. 2007. Fabrication and characterization of PE/MMT nanocomposites via copolymerization of ethylene and *in situ* formed α-olefins. *Journal of Materials Science* 42:4350–4355.

146. Guo, C.-Y., Ke, Y., Liu, Y., Mi, X., Zhang, M., and Hu, Y. 2009. Preparation and properties of polyethylene/montmorillonite nanocomposites formed via ethylene copolymerization. *Polymer International* 58:1319–1325.

147. Zhang, Z., Guo, C.-Y., Cui, N., Ke, Y., and Hu, Y. 2004. Preparation of linear low density polyethylene by *in situ* copolymerization of ethylene with Zr supported on montmorillonite/Fe/methylaluminoxane catalyst system. *Journal of Applied Polymer Science* 94:1690–1696.

148. Huang, Y., Yang, K., and Dong, J.-Y. 2006. Copolymerization of ethylene and 10-undecen-1-ol using a montmorillonite-intercalated metallocene catalyst: Synthesis of polyethylene/montmorillonite nanocomposites with enhanced structural stability. *Macromolecular Rapid Communications* 27:1278–1283.

149. Dong, J.-Y. and Hu, Y. 2007. Synthesis of functionalized polyolefin/montmorillonite nanocomposites. *Polymer Preprints* 48:205–206.

150. Yang, K., Huang, Y., and Dong, J.-Y. 2007. Preparation of polypropylene/montmorillonite nanocomposites by intercalative polymerization: Effect of *in situ* polymer matrix functionalization on the stability of the nanocomposite structure. *Chinese Science Bulletin* 52:181–187.

151. Moncada, E., Quijada, R., and Retuert, J. 2007. Comparative effect of metallocene and Ziegler-Natta polypropylene on the exfoliation of montmorillonite and hectorite clays to obtain nanocomposites. *Journal of Applied Polymer Science* 103:698–706.

6

Effect of Clay Treatment on the In Situ Generation of Polyolefin Nanocomposites

Fabio Bertini, Laura Boggioni, and Incoronata Tritto

CONTENTS

6.1 Introduction

The most promising approach to effectively disperse organoclays into polyolefins matrices at nanometric scale is the *in situ* polymerization. This approach can eliminate both entropic and enthalpic barriers associated with intercalating polar silicates with nonpolar polymers, which are drawbacks for achieving polyolefin nanocomposites with good dispersion through melt blending. The *in situ* polymerization involves the intercalation of the catalyst within the silicates and the growth of the polymer chains between the intercalated clay layers (Figure 6.1).[1] This chapter will concentrate on the effect of clay treatment on the *in situ* generation of polyolefin nanocomposites with two families of modern homogeneous catalytic systems, early transition metal catalysts, namely, metallocenes (Figure 6.2a) and late-transition metal catalysts (Figure 6.2b).

Brintzinger, Kaminsky, and Ewen's discovery of homogeneous *ansa*-metallocenes, that when activated with methylaluminoxane (MAO) are active for olefin polymerization, is considered a major breakthrough in polymer science.[2] This opened up a broad range of opportunities for the controlled synthesis of new polyolefin compositions. Changes to the metallocene ligand backbone provided access to novel polymer materials with controlled microstructure, molecular weight, chain branching, and branch distribution, which can be fine-tuned by varying metallocene ligands and modifying the substitution pattern.[3]

New late-transition metal–based catalysts give access to a range of materials from highly branched, amorphous polyolefins to linear semicrystalline high-density polyolefins.[4,5] Nickel- and palladium-based cationic and neutral catalysts, owing to their low oxophilic

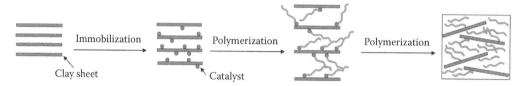

FIGURE 6.1
Catalyst immobilization within silicates and *in situ* olefin polymerization.

X = CH₂–CH₂, (CH₃)₂Si, (CH₃)₂C, (Ph)₂Si
$X = CH_2\text{–}CH_2$, $(CH_3)_2Si$, $(CH_3)_2C$, $(Ph)_2Si$
$R = CH_3$, ^tBu, benz
(a)

$R = Me$, iPr
$X = Cl$, Br
(b)

$Mt = Pd$, Ni

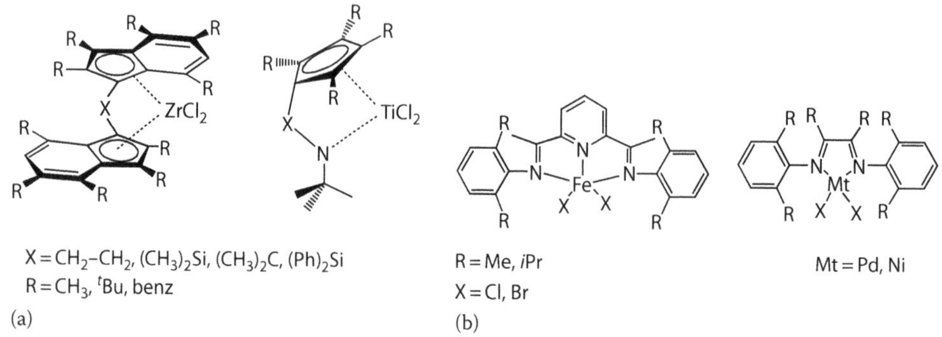

FIGURE 6.2
Schematic structure of metallocenes (a) and late-transition metal catalysts (b).

nature, can allow the incorporation of monomers bearing functional groups. In 1998, Brookhart[4b] and Gibson[5] independently discovered that iron catalysts bearing 2,6-bis(imino) pyridyl ligands, activated with MAO, are effective catalysts for ethylene oligomerization and polymerization. These catalysts show higher catalytic activities than metallocene catalysts but a shorter lifetime because of rapid deactivation during polymerization.

There is a great interest in immobilizing single-site catalysts on an inert carrier to utilize them in slurry or gas-phase processes for the industrial employment of both metallocenes and Brookhart catalysts. Their homogeneous nature prevents the use in industrial plants due to reactor fouling and extremely exothermic polymerization processes.[6] In this framework, inorganic support materials (such as silica, alumina, magnesium chloride) and organic supports (such as polystyrene) have been employed to heterogenize soluble catalysts.[7] Montmorillonite (MMT) has also been used as a support on which aluminoxane can be created by reacting aluminum alkyls with water present in the not dehydrated silicates or with the hydroxyl (OH) groups present on the surface or within the galleries.[8] The aluminoxane then reacts with the catalyst and forms the ion pair active for the olefin polymerization (Figure 6.3).[9]

The state of research on MMT, which among the family of layered aluminosilicates is the most common mineral used in polymer/nanoclay hybrid materials,[1] will be reviewed. MMT is constituted of stacks of hydrated aluminosilicate layers, whose crystal structure is composed of an octahedral alumina sheet sandwiched between two tetrahedral silica sheets (Figure 6.4).[10] The layers are separated by galleries where cations (e.g., Na⁺, K⁺) are present to balance the negative charge of the aluminosilicate sheets arising from isomorphic substitution of Al or Si with other metals.

The first most significant papers in the literature, which investigated the potential of layered silicates as possible host lattices of active catalysts for the synthesis of polyolefin nanocomposites, were reported by O'Hare, Mülhaupt, and Garcés. O'Hare et al.[11] used ion exchange to intercalate cationic metallocene species into clay interlayer galleries, followed

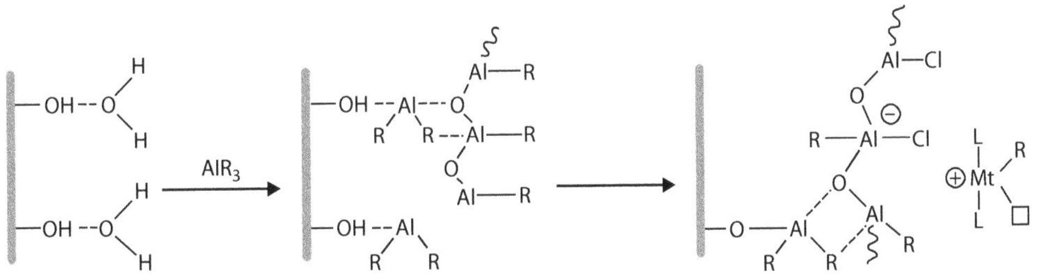

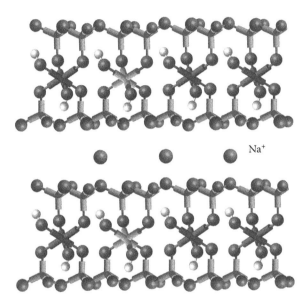

FIGURE 6.3
Formation of alkylaluminoxanes via the interaction of support-H_2O with AlR_3 and of heterogenized complexes active for olefin polymerization on the support surface.

FIGURE 6.4
Sodium montmorillonite structure. (Reproduced by permission of Southern Clay.)

by the addition of MAO activator for initiating propylene polymerization, resulting in low-molecular-weight propylene oligomers. Using an MAO activator and clay fillers rendered organophilic via ion exchange with various quaternary alkyl ammonium cations, Mülhaupt et al.[12] first prepared high-density polyethylene (PE)-MMT nanocomposites with high catalyst efficiency. They reported that the dispersion obtained by the *in situ* method was more effective in polyolefin nanocomposite formation, with respect to melt compounding methods. Sun and Garcés[13] reported the preparation of polypropylene (PP) nanocomposites by *in situ* polymerization with metallocene/clay catalysts. Organoclay precursors were prepared by ion-exchanging clays with selected amine complexes capable of activating metallocene catalysts for olefin polymerization. Thus, they did not use external activators, such as MAO, for initiating the olefin polymerization, but experimental details were not provided. TEM images revealed that the initial clay particles were exfoliated to some degree and an improvement in Young's modulus over two times was observed. Since then, a number of articles have appeared in the literature dealing with *in situ* olefin polymerization by previous intercalation of catalysts into the layers of clays.

According to the *in situ* polymerization method, the catalyst should be intercalated and fixed within the filler layers and the exfoliation degree should be dependent on the concentration of active sites formed between the silicate layers. To facilitate the intercalation of the catalyst, clays are generally treated in order to widen the interlayer spacing so that it is possible to accommodate the catalyst guest and to immobilize the catalyst to the layers. The layered silicates can be modified by reacting the OH groups on the surface edges and within the galleries with aluminum alkyls or aluminoxanes with a twofold purpose: (1) to have active catalysts, since especially the early transition metal catalysts are very sensitive to OH or polar groups and (2) to graft the cocatalyst, which will generate the ion pair and will attract the metal of the active site by electrostatic interactions. Moreover, to improve the dispersion of the phyllosilicates into the less polar polymer matrix, the alkaline cations are replaced by ionic exchange with suitable organic cations, such as alkyl-substituted ammonium or phosphonium ions. The presence of functional groups on the organic modifier can serve either to anchor the catalyst or as comonomer to graft the polymer chain to the clay layers. In order to understand the effect of clay treatment on the *in situ* generation of polyolefin nanocomposites, results on MMT pretreated with aluminoxane and on modified or functionalized MMT will be reviewed.

6.2 Effect of Aluminoxane Pretreatment of Clays on the *In Situ* Generation of Polyolefin Nanocomposites by Metallocenes

Jérôme et al.[14] prepared a series of nanocomposites consisting of PE and non-modified layered silicates (MMT, hectorite) by the so-called polymerization filling technique (PFT). This technique consists in the *in situ* polymerization after attaching the catalyst to the surface and within the interlayer of the clay (Figure 6.5).

Non-modified MMT (Na-MMT) and hectorite were first treated by trimethylaluminum-depleted MAO before being contacted with a titanium-based constrained geometry catalyst. Large amounts of MAO in heptane, a non-solvent for MAO, have been used to attach MAO to the filler in order to impart enough activity to the polymerization catalyst and to produce composites with a low filler content. The ethylene-based nanocomposites produced showed poor tensile properties due to the ultrahigh-molecular-weight. Hydrogen addition decreased the molecular weight of the polymer, resulting in ethylene-based nanocomposites with improved tensile and shear moduli. The origin of this improvement was related to the fine dispersion of nanoparticles in the PE matrix. The exfoliation of the layered silicates was confirmed by X-ray diffraction (XRD) and TEM analysis. This type of exfoliated nanocomposites was, however, thermodynamically unstable. The mechanical kneading of the molten nanocomposites resulted in the partial collapse of the exfoliated structure driven by the thermodynamic stability of the layered filler. Indeed, for sufficiently fluid PE matrices, the exfoliated structure collapsed in the melt into a non-regular intercalated structure.

A study on the effect of nanoscopic confinement on the ethylene polymerization process was undertaken by Wang et al.[15] The authors tried to immobilize the catalytic

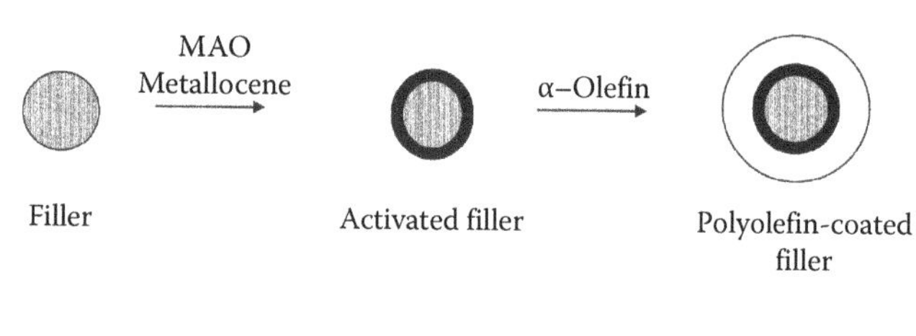

FIGURE 6.5
Schematic steps of PFT. (Adapted from Alexandre, M. et al., *Polymer*, 43, 2123, 2002; Dubois, P. et al., *Macromol. Symp.*, 194, 13, 2003.)

Filler

MAO
Metallocene
→

Activated filler

α–Olefin
→

Polyolefin-coated
filler

system (n-BuCp)$_2$ZrCl$_2$/MAO among galleries of pristine Na-MMT. The pristine clay was first intercalated by MAO and then by metallocene. After this treatment, the MMT gallery distance hardly changed although the regularity of the silicate was destroyed to a certain degree. The kinetic profile of the polymerization with this system was studied. The authors observed an induction period in the early stage of polymerization, the laminated structure of silicate restrained the mass transfer of monomer. As the polymerization continued, more and more PE was produced among the gallery of MMT, the gallery distance increased and the polymerization rate gradually increased. According to the authors, such enhancement of polymerization rate destroyed the laminated structure, the monomers could approach all active sites, and the polymerization rate reached to its maximum. Moreover, they observed that the nanoscopic confinement affected the molecular properties and crystallization behavior of PE produced in between the galleries.

Subsequently, different authors tried to modify the MMT by ion-exchange to replace internal sodium cations with ammonium salts having long alkyl chains. Hwu et al.[16] replaced sodium cations with positively charged stearyl trimethyl ammonium chloride to increase the *d*-spacing of the clay. They found that the *d*-spacing was higher when toluene was used as solvent. PP–MMT nanocomposites were prepared by using the *rac*-Et(Ind)$_2$ZrCl$_2$ catalyst supported on the modified MMT. TEM analysis showed that the clay was randomly dispersed into the PP matrix. The propylene-based nanocomposites had higher crystallinity, hardness, and improved thermal properties than pure PP.

Lee et al.[17] reported on *in situ* ethylene polymerization by unmodified Na-MMT and a commercial MMT modified with dimethyl-branched hexyl-hydrogenated tallow quaternary ammonium chloride (Cloisite®25A, C25A, *d*-spacing = 18.6 Å). Catalyst-supported clays with both the unmodified and modified clays were prepared by supporting first modified-MAO (MMAO) cocatalyst followed by Cp$_2$ZrCl$_2$ catalyst. They found that the *d*-spacing of C25A increased after MMAO and zirconocene incorporation while the *d*-spacing of Na-MMT was only slightly changed. In addition, they quantified the amount of supported MMAO and the content of residual modifier and found that most of the organic amine modifier was eliminated by MMAO. The most effective method for the preparation of nanocomposite was found to be the *in situ* polymerization of ethylene with the catalyst supported on C25A and additional MMAO cocatalyst. Based on wide-angle X-ray diffraction (WAXD) and TEM analysis, they claimed that a fully exfoliated nanocomposite was obtained with this *in situ* procedure. The decomposition temperature of the obtained nanocomposite increased up to 25°C, but the melting temperature was not improved much.

The research focus of Tritto et al.[18] was on getting an insight into the *in situ* polymerization technique, specifically on understanding the interactions between the clay and the aluminoxane required as cocatalyst in order to activate the metallocene precatalyst for polymerization. They tackled a thorough study on an *in situ* method for ethylene polymerization by Cp$_2$ZrCl$_2$ metallocene catalyst and MAO cocatalyst, using a commercial MMT modified with dimethyl dihydrogenated tallow quaternary ammonium chloride (Cloisite®15A, C15A, *d*-spacing = 32 Å). The structure of clay minerals presents, at the edges, bridging OH groups, which can act as Bronsted acid sites. Me–Al bonds of MAO cocatalyst can partially react with the OH groups leading to Al–O covalent bonding. In addition, since aluminum in MAO is either trivalent or tetravalent, interactions between MAO and the Lewis basic sites within the filler layers are possible, resulting in MAO supportation.[14] Since the clay pretreatment with the cocatalyst could be crucial for fixing the catalyst within the galleries of the clay and so for achieving nanocomposites by *in situ* polymerization, they investigated in detail the effect of the clay pretreatment. Particular attention was paid on (1) the evolution of the structure of the organoclay after the reactive pretreatment with MAO cocatalyst, (2) the influence of amount and structure of the pretreated organoclay, and (3) the influence of cocatalyst/catalyst ratio and polymerization time on the synthesis of a hybrid masterbatch with good filler dispersion within the continuous PE matrix.

To favor MAO fixation on the clay particles and to minimize the deactivation effects by the OH groups of the silicate surface on zirconocene catalyst activity in the subsequent polymerization step, the layered filler C15A was first stirred in toluene and then reacted with MAO for different contact times (C15A/MAO). The intermediates were isolated by filtration and characterized.

Elemental analysis, thermogravimetric analysis (TGA) (Figure 6.6), and WAXD revealed that the initial dispersion of the C15A in toluene removes only the excess of the unbound organic modifier and *d*-spacing is reduced to 28 Å. In contrast, the interactions between MAO and clay caused dramatic changes: most of the surfactant is removed and MAO was intercalated within the clay galleries. WAXD of C15/MAO intermediates did not show diffractions so the initial crystallographic order was destroyed (Figure 6.7). The absence of diffractions is one of the characteristic features of highly exfoliated clay structures. However, TEM micrographs showed platelets less defined and clay stacks within the aggregates more randomly oriented than those of pristine C15A (Figure 6.8). Thus, MAO was intercalated within C15A clay by replacing most of the organic surfactant within the

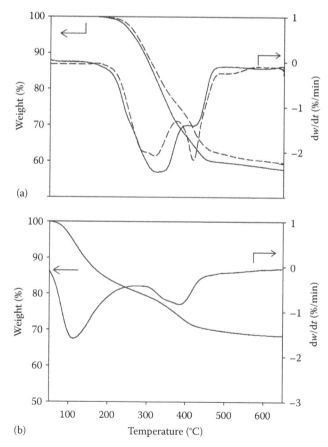

FIGURE 6.6
TGA and DTG curves of pristine C15A (straight line) and after the dispersion in toluene (dashed line) (a) and the intermediate C15A/MAO obtained after the organoclay pretreatment with MAO for 90 min (b). (Adapted from Leone, G. et al., *J. Polym. Sci., Part A: Polym. Chem.*, 46, 5390, 2008.)

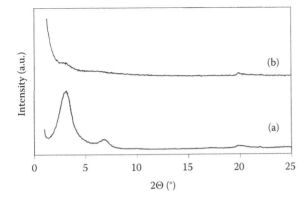

FIGURE 6.7
WAXD profiles of C15A after the dispersion in toluene (a) and C15A/MAO obtained after the organoclay pretreatment with MAO for 90 min (b). (Adapted from Leone, G. et al., *J. Polym. Sci., Part A: Polym. Chem.*, 46, 5390, 2008.)

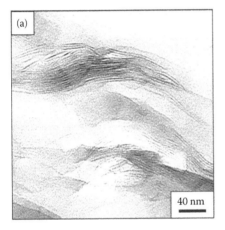
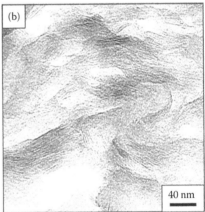

FIGURE 6.8
TEM micrographs of C15A after the dispersion in toluene (a) and the intermediate C15A/MAO obtained after the organoclay pretreatment with MAO for 90 min (b). (Adapted from Leone, G. et al., *J. Polym. Sci., Part A: Polym. Chem.*, 46, 5390, 2008.)

clay galleries. The various possible structures of the MAO or of the products of the MAO reaction with clay and organic modifier present in the layers originated disordered structures, which are not regular enough to give XRDs.

The *in situ* ethylene polymerization was carried out by mixing the metallocene to the toluene suspension of C15A/MAO. No further addition of MAO was needed. The metallocene reaction with MAO gives the alkyl zirconocene cation, that is, the species catalytically active for polymerization.[9] Thus, by electrostatic interactions, the metallocene was attracted into the clay galleries and the growth of polymer chains separated the silicate layers. The activity was generally higher in the presence of C15A than without any inorganic filler (Table 6.1). The polymerization activity and so the nanofiller content strongly depended on the contact time between MAO and clay. A higher MAO supportation time on C15A, which creates a higher disorder in the structure, promoted a larger formation of high active species in ethylene polymerization, which led to increased PE productivity and lower inorganic content in the nanocomposites. Highly filled (from 9 to 34 wt% of C15A content) hybrid PE nanocomposites with high molar masses were

TABLE 6.1

In Situ Ethylene Polymerization Using C15A/MAO and Cp_2ZrCl_2 Catalyst[a]

Entry	C15A Feed (mg)	t_{MAO}[b] (min)	Al/2M2HT[c] (mol/mol)	PE/C15A Yield (mg)	C15A Content[d] (wt%)	Activity[e]	$10^{-3} M_w$ (g/mol)	M_w/M_n	T_m (°C)
PE	—	—	—	781	—	328.2	nd	nd	138
1	150	90	50.7	1100	13.6	462.2	87	2.9	136
2	200	90	38.1	770	28.6	332.6	106	3.4	138
3	250	90	30.4	732	34.2	307.5	107	4.1	133
4	150	270	50.7	1310	11.5	548.1	nd	nd	136
5	200	270	38.1	1200	16.7	504.2	114	2.7	136
6	150	18 h	50.7	1670	9.0	695.9	103	2.8	136
7[f]	200	90	38.1	840	23.8	262.6	119	3.4	136
8[f]	200	270	38.1	1490	13.4	465.6	125	2.8	136
9[g]	200	90	38.1	1590	12.6	334.0	113	2.6	137

[a] Polymerization conditions: reaction temperature = 40°C, ethylene pressure = 1 bar, V = 50 mL (toluene as solvent), polymerization time = 1 h, [Al]/[Zr] = 4000 (Zr = 2.38 μmol, Al = $9.53 \cdot 10^{-3}$ mol). The aluminum content was calculated by assuming a general formula $-Al-O(Me)_x$.

[b] MAO contact time with clay.

[c] Dimethyl dihydrogenated tallow quaternary ammonium chloride (2M2HT).

[d] Calculated by the ratio of amount of the C15A in feed and PE/C15A yield.

[e] $kg_{PE}/[mol_{Zr} \cdot atm \cdot h]$.

[f] Al/Zr = 3000 (Zr = 3.20 μmol).

[g] Polymerization time = 2 h.

obtained in good yield. The presence of the C15A nanofiller improved the thermal stability of the polymer.

The WAXD patterns for the diffractograms of the hybrid composites displayed the presence of a broad diffraction peak in the low angular region, corresponding to a *d*-spacing of 14.7–16.4 Å, due to the clay structure (Figure 6.9). It was demonstrated that this peak resulted as a consequence of the final acid treatment required to terminate the polymerization since the acid reacts with the MAO present in the interlayer region of the clay not involved in the polymerization process.

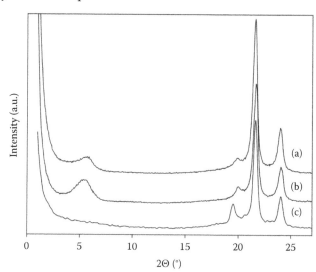

FIGURE 6.9
WAXD profiles of PE nanocomposites: sample 1 (13.6 wt% C15A) (a), sample 2 (28.6 wt % C15A) (b), and sample 1 without HCl treatment (c) of Table 6.2. (Adapted from Leone, G. et al., *J. Polym. Sci., Part A: Polym. Chem.*, 46, 5390, 2008.)

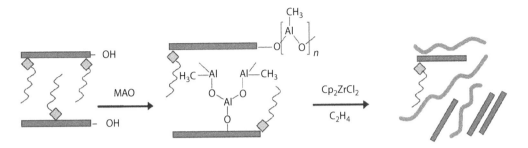

FIGURE 6.10
Interactions between an organo-modified MMT and MAO cocatalyst and *in situ* polymerization. (Adapted from Leone, G. et al., *J. Polym. Sci., Part A: Polym. Chem.*, 46, 5390, 2008.)

TEM analysis demonstrated the achievement of highly filled nanocomposites with a quite homogeneous distribution of the clay and a satisfactory dispersion. By varying MAO contact time in the pretreatment and experimental polymerization conditions, it was possible to tune the final morphology of the composites (Figure 6.10).

In conclusion, C15A was shown to be an effective inorganic support for MAO-activated zirconocene catalyst, enabling the *in situ* synthesis of organoclay-based highly filled nanocomposites, which will be easy to disperse with commercial polyolefins by melt blending. Other groups have reported that the intercalation of the catalyst was accomplished by treating the MMT with aluminum alkyls. Novokshonova et al.[8b] developed a method for studying the active centers of heterogeneous catalysts. This method is based on the mass-spectrometric study of the products of temperature-programmed desorption (MS-TPD) from the catalyst surface at the earliest stages of olefin polymerization. The formation of both alkylaluminoxanes and heterogenized complexes active in polymerization directly on the support surface was demonstrated. This method was used to introduce several fillers into high- and ultrahigh-molecular-weight polymer matrices and to prepare homogeneous ultrahigh-filled polymer composites containing as much as 90–95 wt% filler. Using this procedure of polymer filling, they synthesized a number of new composites with a unique combination of mechanical and performance characteristics. Moreover, new polyolefin nanocomposites with reduced gas permeability and inflammability and improved heat resistance, stiffness, and conductivity, were synthesized in the presence of supported and intercalated catalysts both Ziegler type and metallocenes.[8b,19]

6.3 Effect of Aluminoxane Pretreatment of Clays on the *In Situ* Generation of Polyolefin Nanocomposites by Late-Transition Metal Catalysts

Late-transition metal (Ni, Pd, Fe, and Co) olefin polymerization catalysts are particularly stable to Lewis bases and water in contrast to the vast majority of olefin polymerization catalysts.[4] Mülhaupt et al.[12] used both Pd- and Ni-based catalysts, [{iPr$_2$C$_6$H$_3$N=C(Me)C(Me)=NC$_6$H$_3$iPr$_2$-2,6}Pd(CH$_3$)(N≡CCH$_3$)]$^+$[BAr$_4$]$^-$ and *N,N*-bis(2,6-diisopropylphenyl)-1, 4-diaza-2,3-dimethyl-1,3-butadiene nickel dibromide, for preparing highly branched PE-clay nanocomposite. They found that they were much less sensitive to the addition of non-modified and modified layered silicates than metallocene catalysts.

Bergman[20] intercalated the clay with aliphatic 1-tetradecylammonium cations, to make an organically modified fluorohectorite, which was mixed in toluene with a Brookhart-type Pd catalyst [{2,6-iPr$_2$C$_6$H$_3$N=C(Me)C(Me)=NC$_6$H$_3$iPr$_2$-2,6}Pd(CH$_2$)$_3$CO$_2$Me]$^+$[B(C$_6$H$_3$(CF$_3$)$_2$)$_2$]$^-$. The suspension of organically modified fluorohectorite in a toluene solution of Pd catalyst yielded an orange–brown powder, the XRD of such powder revealed a structural change of the silicate with an increase in the basal spacing from 19.9 to 27.6 Å. Such *d*-spacing from molecular modeling calculation is sufficient for palladium catalyst entering in the interlayer galleries. The palladium complex could not be extracted with excess toluene, thus the reaction was irreversible, even though the driving force for this reaction was not known. After activation with perfluoroborates, this system polymerized ethylene to a rubbery, highly branched PE-clay nanocomposite.

Iron catalysts bearing 2,6-bis(imino)pyridyl ligands are effective catalysts for ethylene oligomerization and polymerization. They show higher catalytic activities than metallocene catalysts but they are rapidly deactivated during polymerization. The steric bulk of the substituents on the imino nitrogen donors plays a key role in controlling the rates of propagation and chain termination, thus influencing both activity and molecular mass of the polymer.

Tritto et al.[21] performed a study to understand the immobilization of iron catalysts within the layered silicates for achieving PE nanocomposites by intercalative *in situ* polymerization. The work focused on developing heterogeneous layered silicate-immobilized 2,6-bis(imino)pyridyl iron (II) dichloride catalysts in which the metallic complex precursor is intercalated into the layered host system. The heterogenization of iron single-site catalysts intercalated into layered silicate host was thought to provide a unique opportunity to increase catalyst lifetime and to prepare hybrid nanocomposite materials at the same time. The influence of steric hindrance of the immobilized catalyst ligands at the *ortho*-aryl ring positions in preparing PE with higher activities and in generating PE nanocomposites with intercalated/exfoliated morphologies was elucidated.

The iron complexes [((2,6-*i*PrPh)N=C(Me))$_2$C$_5$H$_3$N]Fe^{II}Cl$_2$ (**A**) and [((2,6-MePh)N=C(Me))$_2$C$_5$H$_3$N]Fe^{II}Cl$_2$ (**B**) were immobilized within the sodium- and organo-modified MMT sheets pretreated with MMAO activator. For these systems, MMAO is a better cocatalyst than MAO. The unmodified sodium-MMT Dellite®HPS (DNa), the organically modified, with a different amount of a quaternary ammonium having two methyl groups and two long alkyl tails (dimethyl dihydrogenated tallow, DMDHT), Dellite®72T (D72T), and Dellite®67G (D67G) were selected as catalyst supports.

The clay was pretreated with MMAO; MMAO suspension dried under vacuum thus obtaining a silicate-supported MMAO cocatalyst (DM). Then the Fe^{II} precatalyst suspended in toluene was added to the DM intermediate suspension to preform in situ a layered silicate-immobilized catalyst (DMA/DMB). Homogeneous **A** and **B** pristine iron complexes used are both intensely blue colored, after reaction with DM intermediates the color immediately changed to deep, bright orange, as after reaction with MMAO. The intermediates were thoroughly investigated by elemental analysis, TGA, TGA-FTIR, WAXD, and pyrolysis (Table 6.2).[22]

The characterization of the clay after the MMAO supportation (DM) demonstrated the replacement of most part of the organic modifier by MMAO. WAXD investigation of DM and DMA/DMB silicate-immobilized iron catalysts revealed a structural change of the silicate, an increase in the basal spacing due to intercalation of the Fe^{II} catalysts. The MMAO reaction with clay modified the original silicate crystallographic order and allowed for attracting the iron metallic complex within the galleries.

TABLE 6.2

Layered Silicate Pretreatment Characterization[a]

	Pristine Clay				MMAO-Supported Clay			Clay/MMAO/Fe^{II}-Supported Catalyst		
	N^b (%)	DMDHT^c (wt%)	d_{001}^{d} (Å)		N^b (%)	DMDHT^c (wt%)	d_{001}^{d} (Å)	Ar-*ortho* Ligand^e	N^b (%)	d_{001}^{d} (Å)
DNa	—	—	10	DM1	—	—	12	A	—	12
								B	—	12
D72T	0.94	40.2	26	DM2	0.30	12.8	27	A	0.33	35
								B	0.34	35
D67G	1.12	47.9	37	DM3	0.37	15.8	33	A	0.42	32
								B	0.40	33

[a] General conditions: 150 mg of clay was dispersed in toluene over 1 h at RT. The solvent was removed by vacuum and an MMAO heptane solution was added (MMAO = 22.5 mmol). The mixture was stirred over 90 min at RT and the solid clay/MMAO product was dried in vacuum. Then the Fe^{II} precatalyst (4.6 µmol) was added to preform a clay-immobilized iron-based catalyst.

[b] Determined by elemental analysis.

[c] The modifier DMDHT content was estimated from nitrogen amount obtained by elemental analysis.

[d] Determined by WAXD.

[e] **A**: *i*Pr; **B**: Me.

In addition, two synthesis strategies were studied (Figure 6.11). In route 1, the clay was first reacted with an MMAO heptane solution and then, after saturation with the monomer, the polymerization initiated by the addition of Fe^{II} precatalyst. Instead, in route 2, the precatalyst was activated by using the clay-immobilized cocatalyst and the polymerization initiated by monomer addition. The catalytic activities with the different montmorillonites and precatalysts used, following polymerization routes 1 and 2, are summarized in Figure 6.12.

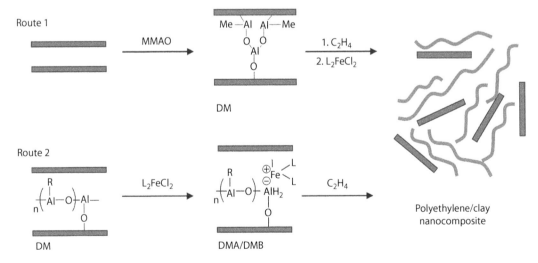

FIGURE 6.11

Schematic illustration of routes for the synthesis of MMAO-modified clay and clay-immobilized 2,6-bis(imino) pyridyl iron (II) catalysts for PE nanocomposites. (Adapted from Leone, G. et al., *J. Polym. Sci., Part A: Polym. Chem.*, 47, 548, 2009.)

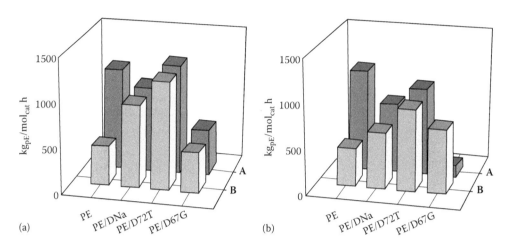

FIGURE 6.12
Ethylene polymerization activities with Fe^{II} precatalysts **A** and **B** by route 1 (a) and route 2 (b). (Adapted from Leone, G. et al., *J. Polym. Sci., Part A: Polym. Chem.*, 47, 548, 2009.)

Route 1 resulted to be a promising methodology to obtain ethylene polymerization with improved activity with respect to the homogeneous system, except in the case of using a highly organo-modified nanoclay (D67G). The initial difference in activity for the two iron precursors **A** and **B** vanished. This should arise from higher stability of active centers formed in the confined geometries of the layered structure.

Activities by route 2, where the metallic complexes were pretreated with inorganic support, were lower when compared to route 1 (Figure 6.12). The main reason was that the MMAO/Fe precontact reduced the catalytic activity also in the homogeneous polymerization. However, when immobilized within lamellae of nanoclays, both catalysts **A** and **B**, being protected from deactivation, have a longer lifetime. The catalytic activity was much higher with D72T than with D67G. This has to be related to the higher amount of organo-modifier, left in the inorganic catalyst support, which poisons the active polymerization centers.

A drawback of **A** and **B** homogeneous catalytic systems is their broad molecular mass distribution, coupled to a relevant amount of a fraction with a low molar mass. Very interestingly, the clay-immobilized catalysts displayed not only a much higher activity due to a maximum dispersion of active sites within the silicate host and a longer polymerization lifetime, but also an increase of high molar mass fraction polymers because of a decrease of chain transfer rates or a redistribution of the active center populations (Figure 6.13).

TEM investigations and TGA analysis under oxidative atmosphere evidenced that the PE-clay nanocomposites were characterized by an intercalated/exfoliated morphology and a higher onset decomposition temperature (Figure 6.14).

In conclusion, the MMAO reaction with clay modified the original silicate crystallographic order and allowed for attracting the iron complex within the galleries and thus for the growth of polymer chains within the filler galleries. The homogeneous and nanoscaled dispersion of the silicate in the PE matrix was demonstrated to be due to the capacity of these catalysts to be effectively nano-confined into the treated clay and to separate inorganic layers. Thus, by varying the nature of the inorganic support and of 2,6-bis(imino)pyridyl iron (II) catalytic systems, the clay pretreatment, and experimental

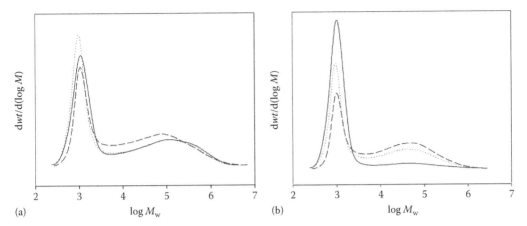

FIGURE 6.13
Size exclusion chromatography curves of pure PEs (straight line), and PE/D72T nanocomposites obtained with precatalyst **A**: route 1 (dashed line), route 2 (dashed-dotted line) (a). with precatalyst **B**: route 1 (dashed line), route 2 (dashed-dotted line) (b). (Adapted from Leone, G. et al., *J. Polym. Sci., Part A: Polym. Chem.*, 47, 548, 2009.)

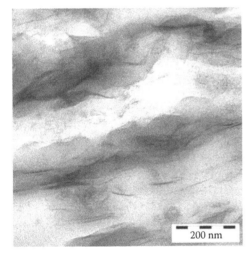

200 nm

FIGURE 6.14
TEM micrograph of nanocomposite obtained by *in situ* ethylene polymerization with precatalyst **B** immobilized on D72T following route 2. (Adapted from Leone, G. et al., *J. Polym. Sci., Part A: Polym. Chem.*, 47, 548, 2009.)

polymerization conditions, it is possible to tune the final properties of PE as well as the morphology of the nanocomposites.

Sivaram et al.[23] reported on a method for supporting the same late-transition metal **B** catalyst on a commercial organo-modified MMT (Cloisite®20 A). This clay is quite similar to D72T having the same organo-modifier and an average intergallery spacing of approximately 24 Å. The polymerization of ethylene was performed with clay-supported catalysts in the presence of MAO, with various amounts of clay and at various catalyst concentrations and catalyst/cocatalyst ratios. The catalyst activity for ethylene polymerization was independent of the Al/Fe ratio. They found that the clay dispersion is higher when the polymerization is performed at a higher Al/Fe ratio. Moreover, when ethylene was polymerized with a mixture of the homogeneous iron (II) catalyst and clay, the degree of dispersion was significantly lower than when the polymerization was performed with a clay-supported catalyst. This observation suggested that in the supported catalyst, at least some of the active centers resided within the galleries of the clay. A reduction in the molecular weight of the polymers as the concentration of clay increased was found.

For similar molecular weights, they observed that the storage modulus (G') and the loss modulus (G'') increase with an increase in the extent of clay dispersion.

6.4 Effect of Modification or Functionalization of Clays on the *In Situ* Generation of Polyolefin Nanocomposites

An alternative strategy consisting in introducing appropriate functional groups within the MMT galleries in order to chemically anchor the catalyst between the silicate layers was explored by various authors.

Kwak et al.[24] first reported the generation of a large number of OH groups as catalyst-anchoring sites in the interlayer galleries of MMT via the cation exchange reaction in the clay pretreatment step. They carried out ethylene polymerizations in the presence of commercial organophilically modified MMT (Cloisite®30B, C30B) containing methyl tallow bis(2-hydroxyethyl) quaternary ammonium between the silicate layers. The abundant OH groups of C30B as well as the wide interlayer spacing (18.8 Å) favored the fixation of $TiCl_4$ inside the interlayer space (Figure 6.15).

The polymerization of ethylene was conducted by injecting ethylene into the catalyst slurry. They found low catalyst efficiency. Thus, it appeared that the deactivation of titanium catalyst by the dangling OH groups present in the organo-modifier, which did not participate in the $TiCl_4$ fixation reaction, had a negative effect on the polymerization activity. Interestingly, the WAXD curve of dried reaction products displayed the complete disappearance of the basal peak. The PE-clay nanocomposites were then used as the masterbatch and blended with commercial high-density PE by melt extrusion. The WAXD analysis and TEM observation clearly indicated that, depending on the processing conditions, stacking recovery of montmorillonites occurred. A similar restacking was observed by Jérôme.[14]

Dong et al.[25] implanted OH groups into MMT interlayer galleries, via cation exchange reaction, to prepare highly exfoliated and dispersed isotactic PP-MMT nanocomposites. They used the (2-hydroxylethyl)hexadecyl diethylammonium iodine surfactant or a mixture of the OH-functionalized surfactant with conventional long alkyl chain ammonium salt (ca. 20%). During this first treatment step, a noticeable increase of the layer spacing was observed, from 9.8 Å for pristine MMT to 21.1 Å for organically modified MMT; this *d*-spacing is sufficient for the intercalation of MAO into organo-modified MMT. The OH-intercalated MMT was first treated with excess methylaluminoxane and then with *rac*-Me$_2$Si[2-Me-4-Ph-Ind]$_2$ZrCl$_2$ precatalyst to anchor by electrostatic interactions the metallocene in the interlayer galleries. It was noted that the modified MMT-based catalysts show

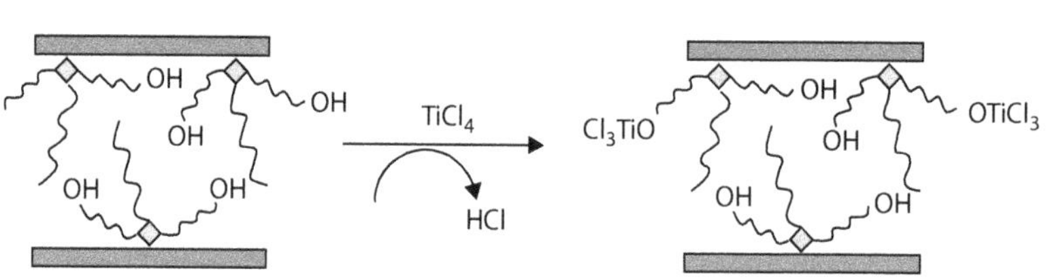

FIGURE 6.15
Mechanistic representation of the fixation of $TiCl_4$ between the silicate layers of MMT-containing OH groups (C30B). (Adapted from Jin, Y.H. et al., *Macromol. Rapid Commun.*, 23, 135, 2002.)

slightly broader WAXD diffraction peaks as compared with modified MMT. A series of isotactic propylene-based nanocomposites containing MMT at a loading range of 1.0–6.7 wt% were obtained in autoclave reactor. The use of a mixed surfactant in the cation-exchanged organic modification of pristine MMT provided high yields: anchoring of the catalyst facilitated the stabilization of catalyst species in MMT interlayer gallery without poisoning the intercalated metallocene. WAXD and small-angle X-ray scattering (SAXS) patterns evidenced no signals in the low angle diffraction region while TEM images showed well-dispersed nanoscopic layers, mostly lower than 50 Å in thickness. The TGA data indicated that the exfoliated nanocomposites generally possess better thermal stability, due to the laminated layers of MMT acting as physical barriers to the thermal transmission in the polymer matrix.

Highly dispersed clay–polyolefin nanocomposites via the *in situ* polymerization of ethylene by clay-supported catalysts were obtained by Scott et al.[26] Nickel-based catalysts with ligands, which can interact with Al sites in the clay surface were supported on acid-treated MMT without organic surfactants. They found that the clay could behave as a remote Lewis acid activator; moreover it also modified the relative rates of propagation and chain walking at the active site during polymerization. High activity was observed without addition of cocatalyst. No reflections of the clay were observed by WAXD of the nanococomposites containing 3–5 wt% of clay. TEM analysis revealed that the clay is both well distributed and highly exfoliated; no tactoids were visible.

A novel approach for preparing PE nanocomposites using dual nanofillers (MMT and silica nanoparticles) was proposed by Tang et al. (Figure 6.16).[27]

Pristine MMT was suspended in water and treated with cetyltrimethylammonium bromide. The resulting organo-modified MMT was mixed with dodecylamine, then tetraethoxysilane (TEOS) was added. The hydrolysis of TEOS by the intergallery water led to silica within the galleries. The suspension was centrifuged and the solid was dried to obtain MMT/silica hybrid support (MMT-Si). A common zirconocene catalyst Cp_2ZrCl_2/MAO was then immobilized on the MMT-Si surface by a simple method. XRD demonstrated that MMT solvated by dodecylamine gives a basal *d*-spacing of 52 Å along with several secondary reflections. The MMT-Si hybrid presents a distinct reflection (*d* = 40 Å), suggesting a long-range order of the clay layers in MMT-Si. XRD of the supported catalyst MMT-Si-cat did not show any reflection indicating the lack of the clay order. A supported catalyst without silica was prepared and used in the *in situ* polymerization as reference. Elementary analysis demonstrated a Zr loading of MMT-Si-cat about twofold than that of the MMT-cat without intragallery silica nanoparticles,

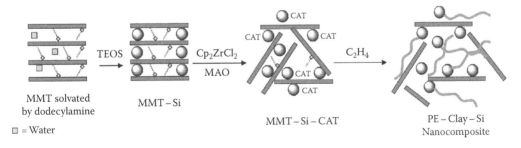

FIGURE 6.16
Schematic illustration of mechanism for the synthesis of PE nanocomposites using dual nanofillers (MMT and silica nanoparticles). (Adapted from Wei, L. et al., *J. Polym. Sci., Part A: Polym. Chem.*, 42, 941, 2004.)

suggesting that the intragallery silica nanoparticles played an important role in the fixation of catalyst.

Preparation of ethylene-based composites was carried out and neat PE was also synthesized for comparison. All supported catalysts showed lower activity than homogeneous catalyst. The nanocomposites with a low nanofiller loading (<10 wt%) exhibited good mechanical properties, that is, higher tensile moduli and storage moduli as compared with the homopolymer. The authors investigated the state of dispersion of the clay and silica nanoparticles in the nanocomposites by TEM. The micrographs evidenced that the aggregated clay particles had been exfoliated into very thin layers and homogeneously dispersed in the PE matrix. At higher magnification, many silica nanoparticles were observed around the exfoliated clay layer, the morphology of the silica nanoparticles was not uniform; most of them were 10–60 Å in diameter. Scanning electron microscopy (SEM) revealed that the nanocomposite powder produced with the supported catalyst had a granular morphology and a high bulk density, typical of a heterogeneous catalyst system.

Finally, a further strategy to obtain stable and exfoliated polyolefin nanocomposites involves the use of an organo-modifier containing a terminal double bond. The comonomer anchored within the silicate layers of MMT allows one to graft the polymer chains to the clay.

Simon et al.[28] first synthesized hybrid PE-clay nanocomposites using *in situ* polymerization with bifunctional organic modifiers. Pristine MMT was initially treated with an aluminum compound (trimethylaluminum or triisobutylaluminum) before being contacted with ω-undecylenyl alcohol. This intercalation procedure allowed to enhance the organophilic character of the MMT, introducing vinyl groups between the layers. After this intercalation process the transition metal catalyst 1,4-bis(2,6-diisopropylphenyl)-acenaphthenediimine nickel dichloride and MAO activator were introduced within the clay galleries and ethylene was added to synthesize PE-clay nanocomposites (Figure 6.17).

The yields obtained were comparable to that obtained in the absence of clay. The hybrid PE nanocomposites showed clay content in the range between 6 and 25 wt%. The melting temperature of the PE nanocomposites resulted to be higher than that of the homopolymer; this is due to the presence of the clay surface that decreases the rate of chain walking, leading to PE chains with fewer short branches. The authors noted both the homopolymerization of ethylene and the copolymerization between ethylene and the vinyl ends of the alcohol modifier connected to the surface during intercalation, which produced PE chains, chemically linked to the silicate surface. In fact, extraction in hot solvents leads to a partial extraction of the polymer chains from PE nanohybrids; SEM observation corroborated this

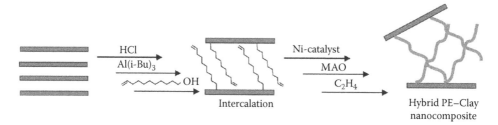

FIGURE 6.17
Synthetic approach using bifunctional organic modifier to produce PE chemically linked silicate layers prepared by *in situ* polymerization. (Adapted from Shin, S.Y. et al., *Polymer*, 44, 5317, 2003.)

evidence. TEM images evidenced that most of the silicate was homogeneously distributed over the PE matrix and the dimensions were in the order of tens of nanometers. However, the presence of clusters of silicate layers was evidenced.

In a recent work, PE-MMT nanocomposites with strong interfacial interactions were synthesized by a similar approach.[29] The authors carried out the polymerizations using a metallocene catalyst in the presence of modified MMT after sodium exchange with a polymerizable surfactant undecenylammonium cation. For the sake of comparison, dodecylammonium chloride–modified MMT was also used for *in situ* ethylene polymerization. The MMT content was 1 wt% in both PE nanocomposites. NMR spectra pointed out the success of linking the polymerizable surfactant to PE backbone. TEM images revealed that modified MMT with the polymerizable surfactant was better dispersed in the PE matrix than long alkyl chain–modified MMT.

Deformation behavior of PE-MMT nanocomposites was studied by means of morphology observation and XRD measurements. Nanocomposites with strong interfacial interactions between PE matrix and MMT platelets showed different deformation behavior during stretching with respect to neat PE and PE nanocomposites obtained without polymerizable surfactant. The existence of the strong interfacial interaction enabled the PE matrix and PE grafted to MMT to form a network-like structure. As a result, the mobility of MMT platelets is improved during the deformation of PE matrix. This results in an increase in the energy dissipated during elongation and thus greater fracture energy. The strong interfacial interaction is also the reason of restraining the cavity initiation and the cavity growth during stretching, which cause the break during elongation.

6.5 Conclusions

The literature results reviewed in this chapter proved that MMT can be exfoliated and give highly dispersed MMT-polyolefin nanocomposites via the *in situ* polymerization of α-olefins with an intercalated catalyst. A successful strategy relies on the organic modification of the clay via ion-exchange with a long-chain alkylammonium followed by intercalation of the precatalyst. The intercalation of the catalyst is possible provided that clay *d*-spacing is wide enough for the dimension of the catalyst. The metal precatalyst can be anchored to the silicate layer through reactions with functional groups on the organic modifier or on the clay. This method can be very efficient with robust precatalysts, but requires the synthesis of specific ligands and/or of specific modifiers. The pretreatment of the clay with aluminoxane or aluminum alkyls allows for the attraction of the catalyst within the galleries by electrostatic interactions between the positive metal and the negative charge of the counterions of the ion pair. In addition, aluminoxanes tend to destroy the crystallographic order of the clays. Further positive effects arise from the reaction between aluminoxane and the polar groups present on the clay, which may poison early transition metal catalysts, and from the displacement of the organo-modifier, which may have negative impact on the physical properties of the nanocomposite material. In this approach, care has to be taken in avoiding catalyst leaching. Interestingly, the synthesis of polyolefins chemically linked to the silicate layers seems to result in an increase in interfacial interactions between polyolefins and clay, which lead to restrain cavity initiation and growth during stretching, which cause the break during elongation. This methodology could also limit the layers' reaggregation after the processing of the nanocomposites at elevated temperatures, above the melting points of polyolefins, observed by some authors. In conclusion, even though the results reported above are very interesting, further in-depth investigations are needed to obtain more efficient systems, which allow independent control of activity and polymer molar masses, clay content and exfoliation, and particle morphology, in addition to polymer microstructures. The achievement of granular morphology, to avoid reactor fouling, is essential for the industrial development of high-performance polyolefin nanocomposites through *in situ* polymerization. The generation of stable and easy-to-disperse polyolefin-based nanocomposites from *in situ* polymerization would have a great industrial impact polyolefins being the highest volume commercial class of easy recycling synthetic polymers.

References

1. (a) Pinnavaia, T. J. and Beal, G. W. 2000. *Polymer-Clay Nanocomposites*. New York: Wiley. (b) Alexandre, M. and Dubois, P. 2000. Polymer-layered silicate nanocomposites: Preparation, properties and uses of a new class of materials. *Material Science and Engineering R: Reports* 28:1–63. (c) Ray, S. S. and Okamoto, M. 2003. Polymer/layered silicate nanocomposites: A review from preparation to processing. *Progress in Polymer Science* 28:1539–1641. (d) Chen, B., Evans, J. R. G., Greenwell, H. C., Boulet, P., Coveney, P. V., Bowden, A. A., and Whiting, A. 2008. A critical appraisal of polymer–clay nanocomposites. *Chemical Society Reviews* 37:568–594. (e) Pavlidou, S. and Papaspyrides, C. D. 2008. A review on polymer–layered silicate nanocomposites. *Progress in Polymer Science* 33:1119–1198. (f) Paul, D. R. and Robeson L. M. 2008. Polymer nanotechnology: Nanocomposites. *Polymer* 49:3187–3204. (g) Qin, Y. W. and Dong, J. Y. 2009. Preparation of nano-compounded polyolefin materials through in situ polymerization technique: Status quo and future prospects. *Chinese Polymer Bulletin* 54:38–45.
2. (a) Brintzinger, H. H., Fischer, D., Mülhaupt, R., Rieger, B., and Waymouth, R. M. 1995. Stereospecific olefin polymerization with chiral metallocene catalysts. *Angewandte Chemie International Edition English* 34:1143–1170. (b) Gladysz, J. A. (Ed.). 2000. Frontiers in metal-catalyzed polymerization. *Chemical Reviews* 100:1167–1682. (c) Scheirs, J. and Kaminsky W. (Eds.). 2000. *Metallocene-Based Polyolefins*. Chichester, U.K.: Wiley.
3. Resconi L., Cavallo L., Fait A., and Piemontesi, F. 2000. Selectivity in propene polymerization with metallocene catalysts. *Chemical Reviews* 100:1253–1345.
4. (a) Killian, C. M., Tempel, D. J., Johnson, L. K., and Brookhart, M. 1996. Living polymerization of α-olefins using Ni^{II}-α-diimine catalysts. Synthesis of new block polymers based on α-olefins. *Journal of the American Chemical Society* 118:11664–11665. (b) Small, B. L., Brookhart, M., and Bennett, A. M. A. 1998. Highly active iron and cobalt catalysts for the polymerization of ethylene. *Journal of the American Chemical Society* 120:4049–4050. (c) Ittel, S. D., Johnson, L. C., and Brookhart, M. 2000. Late-metal catalysts for ethylene homo- and copolymerization. *Chemical Reviews* 100:1169–1203.
5. Britovsek, G. P. J., Gibson, V. C., Kimberley, B. S., Maddox, P. J., MacTavish, S. J., Solan, G. A., White, A. J. P., and Williams, D. J. 1998. Novel olefin polymerization catalysts based on iron and cobalt. *Chemical Communications* 7:849–850.
6. (a) Hlatky, G. G. 2000. Heterogeneous single-site catalysts for olefin polymerization. *Chemical Reviews* 100:1347–1376. (b) Severn, J. R., Chadwick, J. C., Duchateau, R., and Friederichs, N. 2005. "Bound but not gagged"—Immobilizing single-site alpha-olefin polymerization catalysts. *Chemical Reviews* 105:4073–4147.
7. (a) Uozumi, T., Toneri, T., Soga, K., and Shiono, T. 1997. Copolymerization of ethylene and 1-octene with Cp*TiCl$_3$ as catalyst supported on 3-aminopropyltrimethoxysilane treated SiO$_2$. *Macromolecular Rapid Communications* 18:9–15. (b) Thune, P. C., Loos, J., Weingarten, F., Muller, F., Kretschmer, W., Kaminsky, W., Lemstra, P. J., and Niemantsverdriet, H. 2003. Validation of the

flat model catalyst approach to olefin polymerization catalysis: From catalyst heterogenization to polymer morphology. *Macromolecules* 36:1440–1445. (c) McKittrick, M. W. and Jones, C. W. 2004. Effect of site isolation on the preparation and performance of silica-immobilized Ti CGC-inspired ethylene polymerization catalysts. *Journal of Catalysis* 227:186–201. (d) Roscoe, S. B., Frechet, J. M. J., Walzer, J. F., and Dias, A. J. 1998. Polyolefin spheres from metallocenes supported on noninteracting polystyrene. *Science* 280:270–273. (e) Hong, S. C., Teranishi, T., and Soga, K. 1998. Investigation on the polymer particle growth in ethylene polymerization with PS beads supported *rac*-Ph₂Si(Ind)₂ZrCl₂ catalyst. *Polymer* 39:7153–7157.

8. (a) Weiss, K., Wirth-Pfeifer, C., Hofmann, M., Botzenhardt, S., Lang, H., Brüning, K., and Meichel, E. 2002. Polymerisation of ethylene or propylene with heterogeneous metallocene catalysts on clay minerals. *Journal of Molecular Catalysis A: Chemical* 182:143–149. (b) Novokshonova, L. A., Meshkova, I. N., Ushakova, T. M., Kudinova, O. I., and Krasheninnikov, V. G. 2008. Immobilized organometallic catalysts in the catalytic polymerization of olefins. *Polymer Science Series A* 50:1136–1150.

9. (a) Tritto, I., Donetti, R., Zannoni, G., and Sacchi, M. C. 1997. Dimethylzirconocene-methylaluminoxane catalyst for olefin polymerization: NMR study of reaction equilibria. *Macromolecules* 30:1247–1252. (b) Chen, E. Y. X. and Marks, T. J. 2000. Cocatalysts for metal-catalyzed olefin polymerization: Activators, activation processes, and structure-activity relationships. *Chemical Reviews* 100:1391–1434.

10. Brindley, G. W. and Brown, G. 1980. *Crystal Structure of Clay Minerals and Their X-Ray Identification*. London, U.K.: Mineralogical Society.

11. Tudor, J., Willington, L., O'Hare, D., and Royan, B. 1996. Intercalation of catalytically active metal complexes in phyllosilicates and their application as propene polymerization catalysts. *Chemical Communications* 2031–2032.

12. Heinemann, J., Reichert, P., Thomann, R., and Mülhaupt, R. 1999. Polyolefin nanocomposites formed by melt compounding and transition metal catalyzed ethene homo- and copolymerization in the presence of layered silicates. *Macromolecular Rapid Communications* 20:423–430.

13. Sun, T. and Garcés, J. M. 2002. High-performance polypropylene-clay nanocomposites by *in-situ* polymerization with metallocene/clay catalysts. *Advanced Materials* 14:128–130.

14. (a) Alexandre, M., Dubois, P., Sun, T., Garcés, J. M., and Jérôme, R. 2002. Polyethylene-layered silicate nanocomposites prepared by the polymerization-filling technique: Synthesis and mechanical properties. *Polymer* 43:2123–2132. (b) Dubois, P., Alexandre, M., and Jérôme, R. 2003. Polymerization-filled composites and nanocomposites by coordination catalysis. *Macromolecular Symposia* 194:13–26.

15. Wang, Q., Zhou, Z., Song, L., Xu, H., Wang, L., and Song, L. 2004. Nanoscopic confinement effects on ethylene polymerization by intercalated silicate with metallocene catalyst. *Journal of Polymer Science, Part A: Polymer Chemistry* 42:38–43.

16. Hwu, J. M. and Jiang, G. 2005. Preparation and characterization of polypropylene–montmorillonite nanocomposites generated by *in situ* metallocene catalyst polymerization. *Journal of Applied Polymer Science* 95:1228–1236.

17. Lee, D., Kim, H., Yoon, K., Min, K. E., Seo, K. H., and Noh, S. K. 2005. Polyethylene/MMT nanocomposites prepared by in situ polymerization using supported catalyst systems. *Science and Technology of Advanced Materials* 6:457–462.

18. Leone, G., Bertini, F., Canetti, M., Boggioni, L., Stagnaro, P., and Tritto, I. 2008. *In situ* polymerization of ethylene using metallocene catalysts: Effect of clay pretreatment on the properties of highly filled polyethylene nanocomposites. *Journal of Polymer Science, Part A: Polymer Chemistry* 46:5390–5403.

19. (a) Kovaleva, N. Y., Brevnov, P. N., Grinev, V. G., Kuznetsov, S. P., Pozdnyakova, I. V., Chvalun, S. N., Sinevich, E. A., and Novokshonova, L. A. 2004. Synthesis of polyethylene-layered silicate nanocomposites by intercalation polymerization. *Polymer Science Series A* 46:651–656. (b) Lomakin, S. M., Novokshonova, L. A., Brevnov, P. N., and Shchegolikhin A. N. 2008. Thermal properties of polyethylene/montmorillonite nanocomposites prepared by intercalative polymerization. *Journal of Materials Science* 43:1340–1353.

20. Bergman, J., Chen, H., Giannelis, E. P., Thomas, M. G., and Coates, G. W. 1999. Synthesis and characterization of polyolefin–silicate nanocomposites: A catalyst intercalation and *in situ* polymerization approach. *Chemical Communications* 2179–2180.
21. Leone, G., Bertini, F., Canetti, M., Boggioni, L., Conzatti, L., and Tritto, I. 2009. Long-lived layered silicates-immobilized 2,6-bis(imino)pyridyl iron (II) catalysts for hybrid polyethylene nanocomposites by *in situ* polymerization: Effect of aryl ligand and silicate modification. *Journal of Polymer Science, Part A: Polymer Chemistry* 47:548–564.
22. Bertini, F., Canetti, M., Leone, G., and Tritto, I. 2009. Thermal behavior and pyrolysis products of modified organo-layered silicates as intermediates for in situ polymerization. *Journal of Analytical and Applied Pyrolysis* 86:74–81.
23. Ray, S., Galgali, G., Lele, A., and Sivaram, S. 2004. *In situ* polymerization of ethylene with bis(imino)pyridine iron(II) catalysts supported on clay: The synthesis and characterization of polyethylene–clay nanocomposites. *Journal of Polymer Science, Part A: Polymer Chemistry* 43:304–318.
24. Jin, Y. H., Park, H. Y., Im, S. S., Kwak, S. Y., and Kwak, S. 2002. Polyethylene/clay nanocomposite by in-situ exfoliation of montmorillonite during Ziegler-Natta polymerization of ethylene. *Macromolecular Rapid Communications* 23:135–140.
25. Yang, K., Huang, Y., and Dong, J. Y. 2007. Efficient preparation of isotactic polypropylene/montmorillonite nanocomposites by in situ polymerization technique via a combined use of functional surfactant and metallocene catalysis. *Polymer* 48:6254–6261.
26. Scott, S. L., Peoples, B. C., Yung, C., Rojas, R. S., Khanna, V., Sano, H., Suzuki, T., and Shimizu, F. 2008. Highly dispersed clay–polyolefin nanocomposites free of compatibilizers, *via* the *in situ* polymerization of α-olefins by clay-supported catalysts. *Chemical Communications* 4186–4188.
27. Wei, L., Tang, T., and Huang, B. 2004. Synthesis and characterization of polyethylene/clay–silica nanocomposites: A montmorillonite/silica-hybrid-supported catalyst and *in situ* polymerization. *Journal of Polymer Science, Part A: Polymer Chemistry* 42:941–949.
28. Shin, S. Y., Simon, L. C., Soares, J. B. P., and Scholz, G. 2003. Polyethylene–clay hybrid nanocomposites: In situ polymerization using bifunctional organic modifiers. *Polymer* 44:5317–5321.
29. Ren, C., Jiang, Z., Du, X., Men, Y., and Tang, T. 2009. Microstructure and deformation behavior of polyethylene/montmorillonite nanocomposites with strong interfacial interaction. *The Journal of Physics Chemistry B* 113:14118–14127.

7

Modification of Inorganic Fillers and Interfacial Properties in Polyolefin Nanocomposites: Theory versus Experiment

Jie Feng and Hendrik Heinz

CONTENTS

7.1 Overview of Inorganic Fillers and Theoretical Approaches

7.1.1 Inorganic Fillers and Interfacial Properties

Polyolefin/nanofiller composites are in demand for many applications, including aerospace and automotive parts, tires, packaging, as well as electronic devices.[1–3] The most common polyolefins are polyethylene (PE), polypropylene (PP), and ethylene-propylene-diene terpolymer (EPDM). Filler materials include clay minerals, carbon nanotubes, fibers, and metal oxides. The structure of nanoscale fillers can be broadly categorized into 2D layers, 1D tubes, and 3D particles according to shape and aspect ratio.[3] In contrast to micrometer-size fillers, nanoscale fillers improve mechanical, optical, and catalytic properties of polyolefin nanocomposites with a much smaller amount of loading. The diversification of properties of neat polyolefins by nanoscale fillers depends significantly on interfacial features such as surface modification and the associated degree of dispersion within the polymer matrix. Therefore, a large ratio between interfacial area and volume of the filler material, a comparable dimension of the fillers and of the polymer chains, as well as short particle–particle spacing in comparison to microscale and macroscale composites

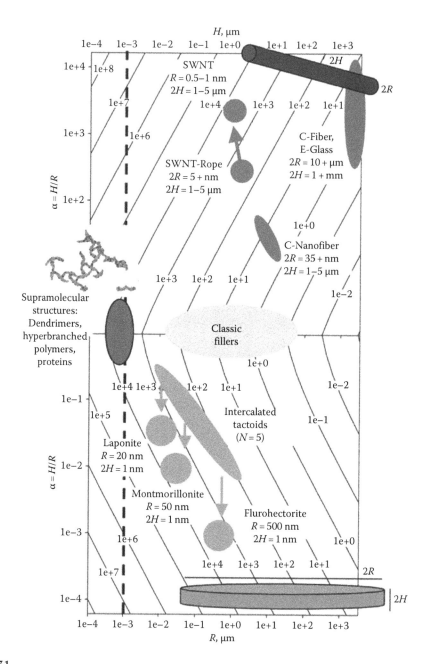

FIGURE 7.1
Logarithmic isolines of interfacial area/volume of particles in μm⁻¹ (=m²/mL) using the aspect ratio, α = H/R, and largest dimension of particle, R = radius, H = height, length, as parameters. Particles are approximated as cylinders with area/volume = 1/H + 1/R. Aspect ratios greater than 1 correspond to rods, length/diameter, and less than 1 to plates, height/diameter. Fully exfoliated and dispersed plates or rods with high aspect ratio, such as montmorillonite or SWCNTs, generate internal interfacial area comparable to that of macromolecular structures such as dendrimers and proteins, and two to three orders of magnitude more than classic fillers (calcite, silica, talc). Comparative plots such as this draw similarities between different fillers, including layered silicates (laponite, montmorillonite, fluorohectorite) with different degrees of exfoliation (N = number of layers per tactoid), roped SWCNTs, carbon nanofibers, and chopped glass fibers. (Reproduced from Vaia, R.A. and Wagner, H.D., *Mater. Today*, 7, 32, 2004.)

are desirable. The ratio of interfacial area to volume of various filler materials as a function of aspect ratio and largest dimension is illustrated in Figure 7.1.

7.1.2 Theoretical and Computational Methods

The complexity of nanoscale morphologies and molecular environments in filled polyolefin nanocomposites is challenging to understand using experimental techniques alone. Diffraction and imaging techniques, spectroscopy, thermal, electrical, and mechanical characterization often provide indirect information about structural and dynamic details which can be further explained and described by theory, modeling, and simulation. Theoretical models rely on a systematic interpretation of experimental findings leading to a new concept, or on relationships identified by modeling and simulation.

Computational techniques are available on molecular, mesoscale, and continuum length scales, as shown in Figure 7.2.[4] An advantage of such models is full control over composition, spatial arrangement of components, temperature, stress, applied electromagnetic fields, or external forces. The ever-decreasing cost of computational resources also helps advance computational and imaging methods. Simulation tools can be broadly classified into quantum mechanical (electronic structure) models, all-atom models, coarse-grain models, mesoscopic models, and continuum-level models. The features of these models are briefly described in the following.

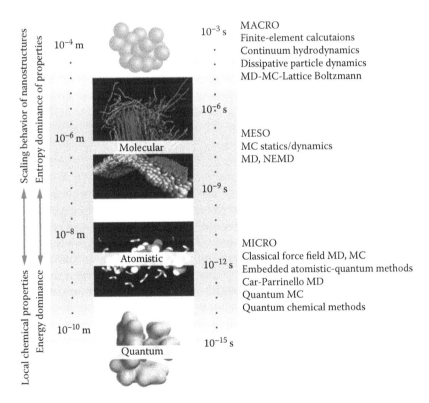

FIGURE 7.2
Illustration of the different length and timescales relevant for the simulation of nanocomposites. A selection of representative methods employed at each scale is shown. MC, Monte Carlo; MD, molecular dynamics; NEMD, nonequilibrium molecular dynamics. (Reprinted from Praprotnik, M. et al., *Annu. Rev. Phys. Chem.*, 59, 545, 2008.)

Electronic structure methods provide detailed insight into optic and electronic properties and are useful in multiscale approaches. They are currently less suited for the simulation of filler materials and nanocomposites as the maximum feasible number of atoms ($<10^3$) and timescales ($<10\,\mathrm{ps}$) are rather small. All-atom models on the basis of interatomic potentials (force fields) in combination with molecular dynamics (MD)[5] are a powerful tool that allows the simulation of systems of 10–100 nm size (up to 10^7 atoms) for periods approaching microseconds. MD simulations rely on Newton's classical equations of motion:

$$\nabla_{\vec{r}_i} E_{pot}(t) = \vec{F}_i(t) = m_i \frac{d^2 \vec{r}_i}{dt^2}. \qquad (7.1)$$

In Equation 7.1, $E_{pot}(t)$ represents the potential energy of the system at time t, $\vec{F}_i(t)$ the total force acting on i atom at time t, and m_i and $\vec{r}_i$ the mass and position of atom i at time t. The conservative potential energy E_{pot} is computed through interatomic potentials (force fields), which are derived from experimental data, known physical and chemical principles, and quantum mechanical calculations. The application of a small time step, e.g., 1 fs, allows the computation of changes in atomic coordinates using a thermodynamic ensemble such as NVT, NPT, NVE, or µVT including methods to control temperature, pressure, energy, or chemical potential. MD techniques are increasingly applied to filler materials and polymer nanocomposites.

Coarse-grained models reduce all-atomic detail and associated degrees of freedom, for example, by representation of three methylene groups of a PE chain by a single spherical bead. As a result, larger time and length scales by about a factor of 10^3 become accessible. MD or Monte Carlo (MC) methods are typically applied to these models. MC methods rely on specified stochastic motions of the coarse-grained particles such as translation, chain rotation, reptation of polymer chains, or bridging moves. These moves can occur in continuous 3D space (off-lattice models) or on-lattice models.[6] The acceptance probability of an attempted MC move is often the Metropolis criterion:

$$p_{i\rightarrow i+1} = \min\left[1, \exp\left(-\frac{\Delta E}{k_B T}\right)\right]. \qquad (7.2)$$

In Equation 7.2, $p_{i\rightarrow i+1}$ represents the probability of the system changing from current configuration i to a new configuration $i + 1$, ΔE the change in potential energy associated with the attempted move, k_B the Boltzmann constant, and T the temperature of the system. MC simulations are often performed in NVT and µVT ensembles, and widely applied to polymers as well as polymers in contact with filler particles. Brownian dynamics (BD)[7] and dissipative particle dynamics (DPD)[8] are further particle-based coarse-grained simulation methods similar to MD simulation. BD employs a continuum solvent model rather than explicit solvent molecules in MD and the total force is[7]

$$F_i(t) = \sum_{j\neq i} F_{ij}^C + R - \xi \frac{\partial r_i}{\partial t}. \qquad (7.3)$$

Hereby, the first term is the conservative force on particle i due to interaction with other particles and the additional terms represent interactions with the continuum medium. R is the Langevin random force and ξ is the friction coefficient. In DPD simulation, the polymer chain is represented by a coarse-grain model in which a coarse-grain particle (bead) represents a group of atoms. The total acting force on a DPD particle i is defined as[8]

$$F_i(t) = \sum_{j \neq i} F_{ij}^C + \sum_{j \neq i} F_{ij}^D + \sum_{j \neq i} F_{ij}^R \qquad (7.4)$$

where F_{ij}^C, F_{ij}^D, and F_{ij}^R are conservative, dissipative, and random forces between two DPD particles. Although chemical details are lost in coarse-grained methods, connective features of polymer chains as well as differential attraction between chemically different monomers and nanoparticle surfaces are retained.

Mesoscopic methods include several field-based approaches such as cell dynamical systems (CDS),[9] mesoscale density functional theory (DFT),[10] and self-consistent field (SCF)[11] theory. Most of these methods are related to the time-dependent Ginzburg–Landau equation (TDGL):

$$\frac{\partial \phi(r,t)}{\partial t} = M \nabla^2 \frac{\delta F[\phi(r,t)]}{\delta \phi(r,t)} + \zeta(r,t) \qquad (7.5)$$

where
$\phi(r,t)$ is an order parameter
M the mobility
F the free energy
$\zeta(r,t)$ a thermal noise parameter

Mesoscopic simulations have been applied to understand phase separation in polymer blends and in polymer/nanofiller mixtures.

7.2 Modification of Layered Silicates and Interfacial Properties in Polyolefin Nanocomposites

Interfacial interactions in clay/polymer nanocomposites have remained uncertain due to the difficulty in measuring local thermodynamic changes and cohesive forces. While this problem still persists, atomistic models for clay minerals can reproduce surface and interface properties in quantitative agreement with experiment[12,13] and provide detailed insight into the surface structure of organically modified layered silicates, interfacial regions with polyolefins, and dynamic processes associated with exfoliation.[14–19]

7.2.1 Thermodynamic Model for Dispersion

We can describe the dispersion of initially agglomerated layered silicates in nonpolar polymer matrices by a thermodynamic model, as shown in Figure 7.3.[16] During dispersion in the polymer matrix, the minerals (M) give up their cohesive free energy ΔG_M which equals the free energy of cleavage, the polymer (P) gives up locally its cohesive free energy per area ΔG_P, and the two components form a new interface (MP) associated with a free energy ΔG_{MP}. The overall change in free energy ΔG is

$$\Delta G = \Delta G_M + \Delta G_P + \Delta G_{MP}. \qquad (7.6)$$

The first two terms are positive, the third negative. The advantage of this free energy model is the focus on dispersion aspects and the full inclusion of entropic effects of the

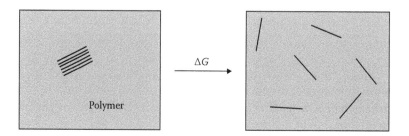

FIGURE 7.3
Exfoliation of mineral layers (M) in a polymer matrix (P) and the associated free energy change.[16] The difference between cleavage free energies ΔG_M, ΔG_P and surface tensions γ_M, γ_P can be more significant than the interface tension γ_{MP}. (Reproduced from Fu, Y.T. and Heinz, H., *Chem. Mater.*, 22, 1595, 2010.)

polymer in the second and third term.[16] Under the assumption that all surfaces would be well defined and do not reconstruct, Equation 7.6 can be rewritten per interfacial area using surface and interface tensions:

$$\gamma_{MP} = \gamma_M + \gamma_P + \Delta G_{MP} \tag{7.7}$$

γ_{MP} reflects the definition of the mineral–polymer interface tension and Equation 7.7 can be imported into Equation 7.6 to yield

$$\Delta G = \Delta G_M - \gamma_M + \Delta G_P - \gamma_P + \gamma_{MP} \tag{7.8}$$

Equation 7.8 states that the free energy of exfoliation can be lowered not only by reduction of the interface tension γ_{MP} but also by cleavage free energies lower than the corresponding surface tensions of the mineral and of the polymer. At the core of this model is reconstruction of the organically modified filler surfaces upon dispersion which can lead to significant differences between cleavage energies to separate the mineral surfaces and surface tensions of the cleaved surfaces.[13] This observation, supported by experimental data on interlayer densities and surface tensions, suggests that the choice of (solvent-free) organically modified minerals can control dispersion vs. agglomeration.[16]

7.2.2 Structure of Organically Modified Layered Silicates

The systematic study of confined hydrocarbon layers in the gallery space dates back to early work of Weiss and Lagaly[20–22] and subsequent experimental and theoretical validation.[12–17,23–47] The organization of interlayer structures and its relation to the cleavage energy is determined by the cation density on the clay mineral surface, interfacial interactions between surfactant head groups with the surface, packing of the hydrocarbon chains, and lateral rearrangements of the surfactants upon cleavage.

The interlayer structure of organically modified clay minerals has been described extensively at the molecular level by simulation techniques[12–17,26,28,30–33,35,36,40–42,44–46] in agreement with x-ray, IR, SFG, NMR, NEXAFS, DSC, TGA, and TEM data,[20–25,27,29,34,37,39,43] including head group–surface interactions, gallery spacing, interlayer density, chain conformation, and thermal transitions. The correspondence between simulation and experiment has often been quantitative such as within 0%–5% in basal plane spacing, justified percentages of gauche conformations in relation to IR, NMR, and DSC data, surfactant tilt angles in full agreement with NEFAXS data,[46] and transition temperatures of the surfactants (partial melting, lateral rearrangement on the surface) in agreement within 20 K or better.[12,31,33] A perspective view into representative interlayer structures of two different montmorillonites, $Na_{0.333}[Si_4O_8][Al_{1.667}Mg_{0.333}O_2(OH)_2]$ (CEC = 91 meq/100 g) and $Na_{0.533}[Si_4O_8][Al_{1.467}Mg_{0.533}O_2(OH)_2]$ (CEC = 143 meq/100 g), modified with a series of alkyl-ammonium surfactants is shown in Figures 7.4 and 7.5.[17]

The formation of flat-on alkyl layers in the interlayer space between the smectite layers of low CEC (Figure 7.4) and vermiculite layers of higher CEC (Figure 7.5) can be seen. The basal plane spacing increases in a plateau-and-step fashion as a function of chain length[14,21,23,34] and is associated with the formation of partially packed and densely packed alkyl layers (plateau) before an additional alkyl layer is formed, which pushes the gallery spacing up (step). In the same sequence, the interlayer density increases from a partially packed alkyl layer to a densely packed alkyl layer with increasing chain length (plateau in basal plane spacing) and decreases upon formation of a partial, additional alkyl layer (step in basal plane spacing).[14] In the same sequence, the percentage of gauche conformations increases from a partially packed alkyl layer to a densely packed alkyl layer (plateau in basal plane spacing) and decreases upon formation of a partial, additional alkyl layer (step in basal plane spacing). This is related to a trend toward more anti-conformations and less gauche conformations when the alkyl chains occupy more interlayer space (lower interlayer density). A superimposed influence on the percentage

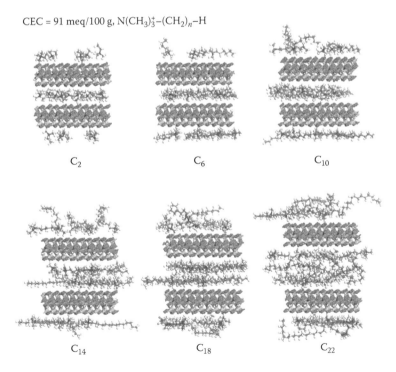

CEC = 91 meq/100 g, $N(CH_3)_3^+-(CH_2)_n-H$

C_2 C_6 C_{10}

C_{14} C_{18} C_{22}

FIGURE 7.4

Representative snapshots of montmorillonite of low CEC (91 meq/100 g) with trimethylalkylammonium surfactants $N(CH_3)_3^+-C_nH_{2n+1}$ at equilibrium distance. The formation of partially and densely packed alkyl layers can be seen. The quaternary ammonium head groups cannot form hydrogen bonds to the surface and remain shared between the two clay layers at equilibrium distance until a partial alkyl bilayer is formed. For a representation of this Figure in color, see ref. [17]. (Reproduced from Fu, Y.T. and Heinz, H., *Phil. Mag.*, 90, 2415, 2010.)

CEC 143 meq/100 g, NH$_3^+$–(CH$_2$)$_n$–H

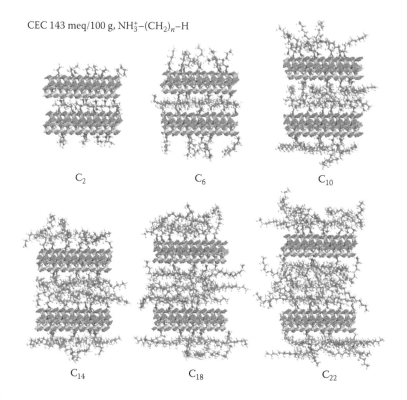

C$_2$ C$_6$ C$_{10}$

C$_{14}$ C$_{18}$ C$_{22}$

FIGURE 7.5
Representative snapshots of montmorillonite of high CEC (143 meq/100 g) with alkylammonium surfactants H$_3$N$^+$–C$_n$H$_{2n+1}$ at equilibrium distance. The formation of partially packed and densely packed alkyl monolayers, bilayers, and pseudo-trilayers can be seen. Head groups are positioned very close to the surface due to hydrogen bonds and, therefore, vertically separated between the two clay layers at equilibrium distance even for short chain length. (Reproduced from Fu, Y.T. and Heinz, H., *Phil. Mag.*, in press.)

of gauche conformations, however, arises through the binding mode of the head group. In case of quaternary ammonium surfactants (Figure 7.4), the head groups are rather space filling and do not vertically separate between the layers at equilibrium distance until at least an alkyl bilayer is formed. Therefore, the head groups have less preference for a specific orientation on the surface, leading to no additional gauche conformations in the alkyl backbone, and the barrier for lateral diffusion on the surface is comparatively low. In case of primary ammonium surfactants (Figure 7.5), however, up to three hydrogen bonds favor a tripod-like orientation of the head group on surface and the hydrogen-bonded head groups are located very close to one of the two surfaces (see Figure 7.5 versus Figure 7.4). This arrangement causes additional gauche conformations in the alkyl backbone near the N-terminal end, increasing the percentage of gauche conformations particularly for short alkyl chains, and the surfactants possess a higher barrier for lateral diffusion.[14]

7.2.3 Cleavage Energy and Potential for Exfoliation

The remainder of the section is dedicated to the miscibility of modified clay minerals with polyolefin matrices. The goal of homogeneous dispersion (Figure 7.3) depends largely on the interlayer environment into which the polymer enters during the initial stages of intercalation and the later stages of exfoliation (loss of orientation correlation between the clay mineral layers).[16] A sensitive property that can be explained on the basis of these observations is the cleavage energy ε_S, i.e., the energy needed to separate two layers of modified clay mineral:

$$\varepsilon_S = \frac{\Delta E_S - \Delta E_U}{2A} \tag{7.9}$$

The cleavage energy ε_S equals the energy difference per contact surface area $2A$ between a cleaved assembly of two layers at infinite separation ΔE_S and the united assembly of the two layers at equilibrium distance ΔE_U, assuming a slow cleavage process in thermodynamic equilibrium as shown in Figure 7.6.[13] Cleavage energies can be obtained by MD simulation with advanced equilibration techniques[16] and display a complex signature as a function of chain length, as shown in Figure 7.7. The chain lengths corresponding to the formation of densely packed alkyl monolayers and bilayers are indicated by numerals, and the cleavage energy exhibits a high sensitivity to head group binding and interlayer density.

In the case of primary ammonium surfactants, the head groups are already vertically separated between the two layers at equilibrium distance even for short C_2 alkyl chains (Figure 7.5). Therefore, the cleavage energy contains no significant Coulomb contribution

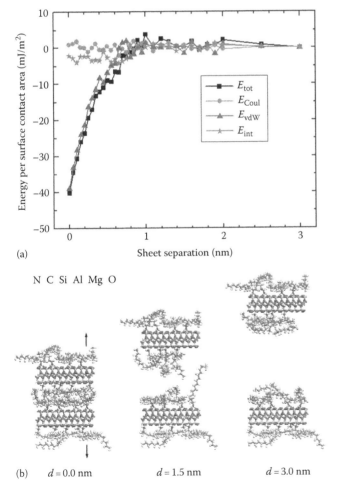

(a)

N C Si Al Mg O

(b) $d = 0.0$ nm $d = 1.5$ nm $d = 3.0$ nm

FIGURE 7.6
Illustration of the cleavage process. (a) Energy of interaction between two layers of octadecylammonium-montmorillonite as a function of sheet separation (CEC = 91 meq/100 g). (b) Snapshots of typical conformations, projected in the xz plane. Reconstruction of the surfaces upon cleavage as well as the proximity of the primary ammonium head groups to the montmorillonite surface due to hydrogen bonds can be seen. (Reproduced from Heinz, H. et al., *J. Chem. Phys.*, 124, 224713, 2006.)

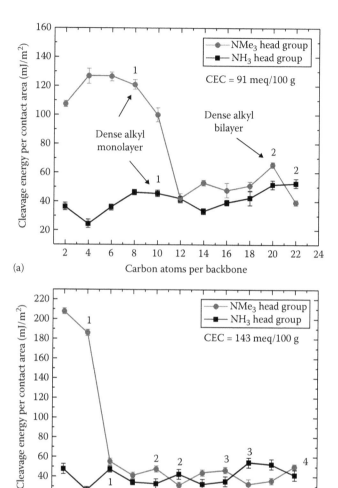

FIGURE 7.7
Cleavage energy of two montmorillonite layers modified with
C_n-alkylammonium ions as a function of chain length n and head group
structure (NH_3 and NMe_3): (a) low
CEC (91 meq/100 g) and (b) high CEC
(143 meq/100 g). The labels 1, 2, 3, and
4 indicate the chain length at which
a densely packed alkyl monolayer,
bilayer, pseudo-trilayer, and pseudo-
quadrilayer are formed in the interlayer
space. The formation of partially and
densely packed alkyl layers leads to
significant changes in cleavage energy
while corresponding surface tensions of
the cleaved mineral layers are in a narrow range of 40–45 mJ/m². Therefore,
modified clay minerals of low cleavage energy, 25–35 mJ/m² are expected
to better disperse in polymer matrices.
(Reprinted from Fu, Y.T. and Heinz, H.,
Chem. Mater., 22, 1595, 2010.)

and is determined mainly by the van der Waals contribution as a function of chain length
(Figure 7.7).[16] Maxima of the cleavage energy are found for densely packed alkyl multilayers
and minima for partially packed alkyl multilayers. In the case of quaternary ammonium
surfactants, the head groups are vertically shared between the two layers at equilibrium
distance for short alkyl chains (Figure 7.4), which leads to stronger cohesion (high cleavage
energy) due to dominant Coulomb contributions to the cleavage energy for shorter alkyl
chains (Figure 7.7).[16] Once a partial alkyl bilayer is formed, i.e., at chain length C_{12} at CEC
91 meq/100 g, the cleavage energy follows again the trend of van der Waals contributions
with maxima for densely packed alkyl layers and minima for partially packed alkyl layers
(Figure 7.7).

The observed trends are interesting as low cleavage energy supports the dispersion in
hydrophobic organic matrices for polymer nanocomposites with applications in automotive, aerospace, and packaging materials.[16] We find very low values near 25 mJ/m² for
n-butylammonium surfactants on montmorillonite of CEC 91 and 143 meq/100 g. Low
values near 30 mJ/m² are also observed for partial alkyl layers of primary ammonium

surfactants at low CEC and for partial alkyl layers of both primary and quaternary ammonium surfactants at high CEC (Figure 7.7). The highest cleavage energy for densely packed alkyl layers in the absence of strong Coulomb interactions is on the order of 50–60 mJ/m^2 (see C$_{20}$ in Figure 7.7a). When Coulomb contributions play a role, the higher CEC of 143 meq/100 g leads to a cleavage energy up to 210 mJ/m^2 (Figure 7.7b) and the lower CEC of 91 meq/100 g leads to a cleavage energy up to 127 mJ/m^2 (Figure 7.7a). The use of specifically designed organically modified montmorillonites with minimal cleavage energy as fillers in hydrophobic polymer matrices may increase the propensity of exfoliation relative to currently employed organoclays which rely on a narrow range of surfactants corresponding to comparatively high cleavage energies near 40 mJ/m^2.[16] Therefore, awareness about the widely tunable range of cleavage energies aids in the understanding and design of dispersion processes to prepare nanocomposites.

During the cleavage process, we also note the occurrence of surface reconstruction (Figure 7.6). The larger the distance of the cationic ammonium center from the surface, the lower is the barrier for lateral diffusion of the surfactants on the (nonhydrated) clay mineral surface.[14] Cleavage of the alkylammonium-modified clay minerals always involves lateral arrangements of head groups and changes in surfactant conformation, even if head group rearrangements may sometimes not be observed in the simulation due to limitations in timescale (1–100 ns).[13,16] Quantitative knowledge about the specific interlayer reconstruction processes, including the vertical separation of head groups between the two clay mineral layers, is required to understand the dispersion behavior in polymer matrices while the measurement of surface tensions[38,47] on already separated surfaces does not reveal such information.

In contrast to smectite and vermiculite systems of comparatively low packing density of the alkyl chains (<0.25), a higher CEC in micas and the use of ammonium or phosphonium surfactants with two, three, or four long alkyl chains can lead to high packing density (up to 1.0).[46] In these cases, the alkyl chains assume a lower tilt angle relative to the surface normal and resemble short polymer brushes.[31,33] Such brush-like systems exhibit less variety in interlayer environments as those shown in Figure 7.3 and cleavage energies typically range between 40 and 45 mJ/m^2 since they are essentially determined by van der Waals interactions at the interface between the two alkyl brushes.[13] Another characteristic of densely grafted surfactants is the possibility to form spatially separated homogeneous brush-like layers in the form of islands on parts of the surface of the clay mineral instead of a spatially uniform homogeneous brush-like layer across the entire surface.[33,48] Criteria for thermodynamic stability of one surface structure versus the other include the density of grafting sites, the degree of ion exchange, and the chain length.[33] Surface properties in the presence of island-like structures have not yet been explored. Likely, the presence of residual solvent in areas of excluded volume will play a role.

7.3 Modification of Carbon Nanotubes and Interfacial Properties in Polyolefin Nanocomposites

Carbon nanotubes (CNT) have attracted great attention due to very high mechanical modulus and strength, ultra-lightweight, and unique optical and electronic characteristics. The first two features make CNT a suitable nano-reinforcement for polymer-based nanocomposites. Interfacial properties of polymer/nanotube blends such as binding energies and chain conformations are important information to understand the reinforcement mechanism. Using all-atom MD simulation with the PCFF (polymer consistent force field), Lordi and Yao investigated interfacial interactions between polymers and single-walled carbon nanotubes (SWNCTs).[49] Their work indicates the ability of polymers including PPVs and PMMAs to form the helical structures around SWNCTs which influences interfacial adhesion. However, binding energies and sliding frictional stresses between SWCNTs and polymers play minor roles. By MD simulation with the COMPASS (condensed-phase optimized molecular potentials for atomistic simulation studies) force field, Zheng et al. compared the adhesion energies of various polymers including PE, PP, polystyrene (PS), and polyaniline (PANI) to SWCNTs, as shown in Figure 7.8.[50] The results demonstrate that the chirality of the nanotube affects adhesion energies between the SWCNTs and the chain molecules. The armchair nanotube (10, 10) was found to be the best choice for reinforcement in comparison to other SWCNTs in this study. Hereby, CNTs with (n, m) indices could be further classified using the rolling graphene model[50]:

$$\theta = \arctan\left(\frac{\sqrt{3}m}{2n+m}\right) \tag{7.10}$$

$$D_n = \frac{\sqrt{3}}{\pi}b\sqrt{(n^2+m^2+nm)} \tag{7.11}$$

where
  θ and D_n are chiral angle and diameter of an SWCNT
  b is the C–C bond length

Wei reported radius- and chirality-dependent conformations of PE on CNT interfaces by MD with the AMBER (assisted model building with energy refinement) force field.[51]

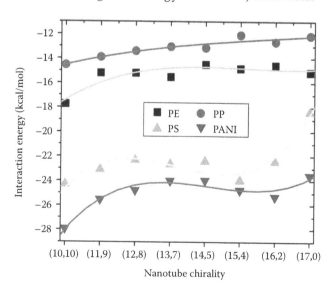

FIGURE 7.8
Adhesion energy between different chiral nanotubes and polymer chains. (Reproduced from Zheng, Q. et al., *J. Phys. Chem. C*, 111, 4628, 2007.)

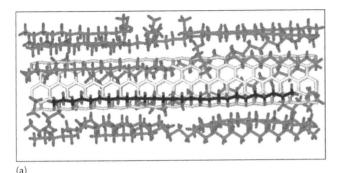

(a)

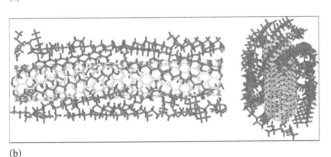

(b)

FIGURE 7.9

(a) Illustration of a snapshot of poly-ethylene molecules at the interface of CNT (5, 5) at $T = 50\,K$, only a portion of the nanotube is shown. The molecule shown in black is in registry with the CNT lattice. (b) Left: A snapshot of poly-ethylene molecules at the interface of CNT (10, 0) at $T = 50\,K$, only a portion of the nanotube is shown; Right: Tilted top view. The hydrogen atoms shown in the figure are not explicitly included in the simulation. (Reproduced from Wei, C., *Appl. Phys. Lett.*, 88, 093108, 2006.)

At a large radius of CNT (40, 40), the PE molecules wrapped on the tube in a similar way as on the graphite plane. For smaller radius CNT, PE wrapped on CNT (5, 5) at 0° and on CNT (10, 0) around 22° and –22° when temperature was low at 50 K, as shown in Figure 7.9. PE molecules were driven into extended conformations along the CNT axis when the temperature was increased to 280 K. Wei also found that the interfacial shear stress of PE/CNT composite increases linearly with increasing applied tensile strain along the axis direction of the CNT in the small strain regime due to changes in van der Waals binding energies.[52] However, interfacial bonds between polymer chains and CNTs are broken at further increase of the applied strain. Simulation results show that significant improvements of mechanical properties such as a 200% increase in Young's modulus can be achieved via addition of short CNTs. Liu et al. simulated PE/CNT composites in a temperature range from 300 to 600 K using MD simulation with the COMPASS force field.[53] In the simulation of single PE molecules (degree of polymerization $N = 50$–80) on CNTs, the chains wrapped closely onto the CNT at higher temperature $T = 600\,K$, but remained in linear conformations at lower temperature $T = 300\,K$. In systems with many polymer chains, PE chains much longer than the diameter of the CNTs were aligned along the CNT axis and formed a shell around the tube at $T = 300\,K$. Using MD simulation with the AMBER force field for the study of alkane mixture/CNT composites, Wei observed that CNTs prefer longer chains to shorter chains due to more binding sites and less edge effects for the longer molecules.[54] Wei also proposed a master equation combined with MD to investigate the adsorption–desorption kinetics of alkane molecules on CNT surfaces. The kinetic study revealed that fully adsorbed chains are more stable than partially adsorbed chains on the CNT surface. Al-Ostaz et al. investigated interfaces of PE/SWCNT composites by MD simulation with the COMPASS force field.[55] The interfacial shear strength was obtained via simulation of the pullout of a single SWCNT:

$$E_{pull\text{-}out} = \int_0^L 2\pi a(L-x)\tau_i dx = \pi a\tau_i L^2 \qquad (7.12)$$

then

$$\tau_i = \frac{E_{pull\text{-}out}}{\pi a L^2}. \qquad (7.13)$$

where

$E_{pull\text{-}out}$ is the energy difference between the completely packed SWCNT and the complete pullout state

a and L are the radius and the length of the SWCNT, respectively

x is the displacement of the SWCNT

At a coarse-grained level, polyolefin molecules and nanotubes have been modeled as bead-connected chains and smooth cylindrical surfaces, respectively. Milchev and Binder employed off-lattice bead-spring type model and coarse-grained MC simulation to study structures of polymer droplets on ultrathin cylindrical surfaces.[56] Transitions between a "barrel" morphology, i.e., polymer chains wrapped around the cylinder symmetrically as a barrel, and "clamshell" morphology, i.e., polymer chains adsorbed on one side of the cylinder, occur on changing the droplet size, the tube radius, and the strength of the adsorption potential.[56] Srebnik's group studied adsorption of polymer chains on nanotubes by off-lattice coarse-grained MC simulations.[57,58] The results indicate that the chain stiffness and curvature of CNTs play crucial roles in the wrapping behavior of chains on the tubes. Tubes of excessively small or large diameter disfavor wrapping of polymer chains. Weak van der Waals interactions lead to adsorption of full flexible chains on larger tubes as "clouds" which involve a rich conformational space due to dominance of entropy effects. If flexible chains were replaced by semiflexible or rigid chains, the van der Waals interactions between the chain beads could push the chains into the conformations of helices. With increasing chain concentration, more chains have been adsorbed and the adsorbed layer transformed from a uniform monolayer to thicker and nonuniform layers. Moreover, the adsorption of block copolymers containing hydrophobic (H) and polar (P) blocks was studied on CNT surfaces in aqueous solution.[57,58] In HPH triblock copolymer system, an interesting loop conformation of hydrophobic blocks wrapping around the tube was observed while the polar block extended toward the bulk. This conformation is helpful to reinforce the interface between the nanotube and polymers. By using DPD simulations, Wescott et al. reported the formation of percolating networks of CNTs in thin films of block copolymer melts and topologically interesting patterns.[59] The information on morphology and composition was converted into concentration profiles and projected onto finite-element modeling (FEM) units in order to obtain conductivities of the nanocomposites.

Using field-based models, it is more difficult to provide information about the chain conformation on the surface; however, attempts have been made to understand phase separation and mechanical properties of composites. Shou et al. combined SCF/DFT techniques with lattice spring model (LSM) to study the effects of the spatial distribution and aspect ratio of particles (rods and spheres) on the mechanical properties of the composite.[60] Buxton and Balazs combined TDGL theory for polymer blends with BD for nanorods in the simulations of nanocomposites.[61] A percolating network of nanorods was identified in the minority phase of a bicontinuous structure. Clancy and Gates developed a hybrid model for CNTs in a bulk poly(ethylene vinyl acetate) matrix.[62] Molecular structures of functionalized CNTs were obtained by MD simulation and used to compute interfacial thermal resistance and thermal conductivity by an analytical approach.

7.4 Modification of Spherical Particles and Interfacial Properties in Polyolefin Nanocomposites

Vacatello applied coarse-grain models and MC techniques to analyze the effects of chain stiffness on the interfacial properties of polymer–solid surfaces.[63] The polymer chains prefer parallel conformation over perpendicular alignment to the solid surface. Increasing chain stiffness improves the density and ordering of polymer chains near the solid surface. On the solid surface, the contact blocks can be classified into three categories: (1) trains, in which all segments of the block contact the surface, (2) loops, in which the head and end beads of the block contact the surface, but others stretch out of surface, and (3) tails, in which only one terminal bead in the block contacts the surface. With increasing chain stiffness, the average lengths of loop and tail blocks increase significantly, but much less change occurs at the train block. Vacatello also identified physical cross-link tendencies of spherical nanofillers in the polymer matrix, in which polymer chains bridge neighbor nanoparticles and in which each particle may be wrapped densely as an ordered shell.[64] The particles packed by polymers decrease the overall mobility of polymer chains in comparison to neat polymers without nanofillers.

Marla and Meredith investigated the mean force between two nanoparticles grafted with polymers by off-lattice coarse-grained MC.[65] There are two regions: when the grafting density and chain length are above a threshold value, the attractive mean force between the two grafted nanoparticles is generated via attractions between polymer chains and bridging of polymers between the particles. In contrast, the mean force between the two modified nanoparticles becomes repulsive when the grafting density and chain length are below a threshold value since the dominating factors of entropy and excluded volume effects drive the particles into dispersion.

Smith et al. studied matrix-induced interaction between nanoparticles by coarse-grained MD simulation.[66] Strong attraction between polymers and nanoparticles lead to dense adsorption of polymer chains onto the particle surfaces and dispersion of the nanoparticles into polymer matrix. In contrast, weak nanoparticle–polymer interaction lead to nanoparticle aggregation and the polymer bead density between neighbor nanoparticles decreased sharply due to depletion effects. By application of coarse-grained MD simulation and the polymer reference interaction site model (PRISM) theory, Hooper et al. studied effective forces between two particles embedded in the polymer matrix.[67] In the semi-dilute polymer solution, the effective force between the two nanoparticles shows a strong repulsive–attractive oscillation as a function of the particle–particle distance. In contrast, a strong attraction followed by a strong repulsion was observed upon separation of two nanoparticles in a polymer melt. By changing the pair potentials between the polymer beads and nanoparticles, contact aggregation, steric stabilization, and local bridging phenomena were identified for a range of interface modifications.

By using coarse-grained MD, the physical aging rate of polymer nanocomposites was studied at interfacial region via measuring the changes in the average pair energy per polymer particle.[68] The physical aging rate could be suppressed by attractive interactions between the polymer beads and nanoparticles in contrast to bulk regions. However, repulsive interactions resulted into an increase of the physical aging rate.

References

1. Kickelbick, G. 2003. Concepts for the incorporation of inorganic building blocks into organic polymers on a nanoscale. *Progress in Polymer Science* 28:83–114.
2. Nwabunma, D. and Kyu, T. (eds.). 2008. *Polyolefin Composites*. Hoboken, NJ: John Wiley & Sons Inc.
3. Vaia, R. A. and Wagner, H. D. 2004. Framework for nanocomposites. *Materials Today* 7:32–37.
4. Praprotnik, M., Delle Site, L., and Kremer, K. 2008. Multiscale simulation of soft matter: From scale bridging to adaptive resolution. *Annual Review of Physical Chemistry* 59:545–571.
5. Frenkel, D. and Smit, B. 2002. *Understanding Molecular Simulation: From Algorithms to Applications*, 2nd ed. San Diego, CA: Academic Press.
6. Metropolis, N., Rosenbluth, A. W., Marshall, N., Rosenbluth, M. N., and Teller, A. T. 1953. Equation of state calculations by fast computing machines. *Journal of Chemical Physics* 21:2819–2823.
7. Heyes, D. M. and Mitchell, P. J. 1994. Self-diffusion and viscoelasticity of dense hard-sphere colloids. *Journal of Chemical Society, Faraday Transactions* 90:1931–194.
8. Hoogerbrugge, P. J. and Koelman, J. M. V. A. 1992. Simulating microscopic hydrodynamic phenomena with dissipative particle dynamics. *Europhysics Letters* 19:155–160.
9. Oono, Y. and Puri, S. 1988. Study of phase-separation dynamics by use of cell dynamics systems. I. Modeling. *Physical Reviews A* 38:434–453.
10. Altevogt, P., Ever, O. A., Fraaije, J. G. E. M., Maurits, N. M., and van Vlimmeren, B. A. C. 1999. The MesoDyn project: Software for mesoscale chemical engineering. *Journal of Molecular Structure* 463:139–143.
11. Balazs, A. C., Singh, C., and Zhulina, E. 1998. Modeling the interactions between polymers and clay surfaces through Self-Consistent Field Theory. *Macromolecules* 31:8370–8381.
12. Heinz, H., Koerner, H., Anderson, K. L., Vaia, R. A., and Farmer, B. L. 2005. Force field for mica-type silicates and dynamics of octadecylammonium chains grafted to montmorillonite. *Chemistry of Materials* 17:5658–5669.
13. Heinz, H., Vaia, R. A., and Farmer, B. L. 2006. Interaction energy and surface reconstruction between sheets of layered silicates. *Journal of Chemical Physics* 124:224713.
14. Heinz, H., Vaia, R. A., Krishnamoorti, R., and Farmer, B. L. 2007. Self-assembly of alkylammonium chains on montmorillonite: Effect of chain length, headgroup structure, and cation exchange capacity. *Chemistry of Materials* 19:59–68.
15. Heinz, H., Vaia, R. A., Koerner, H., and Farmer, B. L. 2008. Photoisomerization of azobenzene grafted to montmorillonite: Simulation and experimental challenges. *Chemistry of Materials* 20:6444–6456.
16. Fu, Y. T. and Heinz, H. 2010. Cleavage energy of alkylammonium-modified montmorillonite and the relation to exfoliation in nanocomposites: Influence of cation density, head group structure, and chain length. *Chemistry of Materials* 22:1595–1605.
17. Fu, Y. T. and Heinz, H. 2010. Structure and cleavage energy of surfactant-modified clay minerals: Influence of CEC, head group, and chain length. *Philosophical Magazine* 90:2415–2424.
18. Zartman, G. D., Liu, H., Akdim, B., Pachter, R., and Heinz, H. 2010. Nanoscale tensile, shear, and failure properties of layered silicates as a function of cation density and stress. *Journal of Physical Chemistry C* 114:1763–1772.
19. Chen, B. and Evans, J. R. G. 2006. Elastic moduli of clay platelets. *Scripta Materialia* 54:1581–1585.
20. Lagaly, G. and Weiss, A. 1970. Arrangement and orientation of cationic tensides on silicate surfaces. 2. Paraffin-like structures in alkylammonium layer silicates with a high layer charge (mica). *Kolloid Zeitschrift & Zeitschrift für Polymere* 37:364–368.
21. Lagaly, G. and Weiss, A. 1971. Arrangement and orientation of cationic tensides on silicate surfaces. 4. Arrangement of alkylammonium ions in low-charged silicates in films. *Kolloid Zeitschrift & Zeitschrift für Polymere* 243:48–55.

22. Lagaly, G. 1976. Kink-block and gauche-block structures of bimolecular films. *Angewandte Chemie International Edition* 15:575–586.
23. Vaia, R. A., Teukolsky, R. K., and Giannelis, E. P. 1994. Interlayer structure and molecular environment of alkylammonium layered silicates. *Chemistry of Materials* 6:1017–1022.
24. Vaia, R. A. and Giannelis, E. P. 1997. Lattice model of polymer melt intercalation in organically-modified layered silicates. *Macromolecules* 30:7990–7999.
25. Vaia, R. A. and Giannelis, E. P. 1997. Polymer melt intercalation in organically-modified layered silicates: Model predictions and experiment. *Macromolecules* 30:8000–8009.
26. Hackett, E., Manias, E., and Giannelis, E. P. 1998. Molecular dynamics simulations of organically modified layered silicates. *Journal of Chemical Physics* 108:7410–7415.
27. Osman, M. A., Seyfang, G., and Suter, U. W. 2000. Two-dimensional melting of alkane monolayers ionically bonded to mica. *Journal of Physical Chemistry B* 104:4433–4439.
28. Pospisil, M., Capkova, P., Merinska, D., Malac, Z., and Simonik, J. 2001. Structure analysis of montmorillonite intercalated with cetylpyridinium and cetyltrimethylammonium: Molecular simulations and XRD analysis. *Journal of Colloid Interface and Science* 236:127–131.
29. Osman, M. A., Ernst, M., Meier, B. H., and Suter, U. W. 2002. Structure and molecular dynamics of alkane monolayers self-assembled on mica platelets. *Journal of Physical Chemistry B* 106:653–662.
30. Kuppa, V. and Manias, E. 2002. Computer simulation of PEO/layered silicate nanocomposites: 2. Lithium dynamics in PEO/Li$^+$ montmorillonite intercalates. *Chemistry of Materials* 14:2171–2175.
31. Heinz, H., Castelijns, H. J., and Suter, U. W. 2003. Structure and phase transitions of alkyl chains on mica. *Journal of American Chemical Society* 125:9500–9510.
32. Pospisil, M., Kalendova, A., Capkova, P., Simonik, J., and Valaskova, M. 2004. Structure analysis of intercalated layer silicates: Combination of molecular simulations and experiment. *Journal of Colloid Interface and Science* 277:154–161.
33. Heinz, H. and Suter, U. W. 2004. Surface structure of organoclays. *Angewandte Chemie International Edition* 43:2239–2243.
34. Osman, M. A., Ploetze, M., and Skrabal, P. J. 2004. Structure and properties of alkylammonium monolayers self-assembled on montmorillonite platelets. *Journal of Physical Chemistry B* 108:2580–2588.
35. Heinz, H., Paul, W., Binder, K., and Suter, U. W. 2004. Analysis of the phase transitions in alkyl-mica by density and pressure profiles. *Journal of Chemical Physics* 120:3847–3854.
36. Zeng, Q. H., Yu, A. B., Lu, G. Q., and Standish, R. K. 2004. Molecular dynamics simulation of the structural and dynamic properties of dioctadecyldimethyl ammoniums in organoclays. *Journal of Physical Chemistry B* 108:10025–10033.
37. Zhu, J. X., He, H. P., Zhu, L. Z., Wen, X. Y., and Deng, F. J. 2005. Characterization of organic phases in the interlayer of montmorillonite using FTIR and ^{13}C NMR. *Journal of Colloid Interface and Science* 286:239–244.
38. Lewin, M., Mey-Marom, A., and Frank, R. 2005. Surface free energies of polymeric materials, additives, and minerals. *Polymer Advance Technology* 16:429–441.
39. Osman, M. A., Rupp, J. E. P., and Suter, U. W. 2005. Gas permeation properties of polyethylene-layered silicate nanocomposites. *Journal of Material Chemistry* 15:1298–1304.
40. Pandey, R. B., Anderson, K. L., Heinz, H., and Farmer, B. L. 2005. Conformation and dynamics of a self-avoiding sheet: Bond-fluctuation computer simulation. *Journal of Polymer Science B* 43:1041–1046.
41. Greenwell, H. C., Harvey, M. J., Boulet, P., Bowden, A. A., Coveney, P. V., and Whiting, A. 2005. Interlayer structure and bonding in nonswelling primary amine intercalated clays. *Macromolecules* 38:6189–6200.
42. He, H. P., Galy, J., and Gerard, J. F. 2005. Molecular simulation of the interlayer structure and the mobility of alkyl chains in HDTMA$^+$/montmorillonite hybrids. *Journal of Physical Chemistry B* 109:13301–13306.

43. Jacobs, J. D., Koerner, H., Heinz, H., Farmer, B. L., Mirau, P., Garrett, P. H., and Vaia, R. A. 2006. Dynamics of alkyl ammonium intercalants within organically modified montmorillonite: Dielectric relaxation and ionic conductivity. *Journal of Physical Chemistry B* 110:20143–20157.

44. Fermeglia, M. and Pricl, S. 2007. Multiscale modeling for polymer systems of industrial interest. *Progress in Organic Coatings* 58:187–199.

45. Scocchi, G., Posocco, P., Fermeglia, M., and Pricl, S. 2007. Polymer-clay nanocomposites: A multiscale molecular modeling approach. *Journal of Physical Chemistry B* 111:2143–2151.

46. Heinz, H., Vaia, R. A., and Farmer, B. L. 2008. Relation between packing density and thermal transitions of alkyl chains on layered silicate and metal surfaces. *Langmuir* 24:3727–3733.

47. Kamal, M. R., Calderon, J. U., and Lennox, R. B. 2009. Surface energy of modified nanoclays and its effect on polymer/clay nanocomposites. *Journal of Adhesion Science and Technology* 23:663–688.

48. Hayes, W. A. and Schwartz, D. K. 1998. Two-stage growth of octadecyltrimethyl-ammonium bromide monolayers at mica from aqueous solution below the Krafft point. *Langmuir* 14:5913–5917.

49. Lordi, V. and Yao, N. 2000. Molecular mechanics of binding in carbon-nanotube-polymer composites. *Journal of Materials Research* 15:2770–2779.

50. Zheng, Q., Xue, Q., Yan, K., Hao, L., Li, Q., and Gao, X. 2007. Investigation of molecular interactions between SWNT and polyethylene/polypropylene/polystyrene/polyaniline molecules. *Journal of Physical Chemistry C* 111:4628–4635.

51. Wei, C. 2006. Radius and chirality dependent conformation of polymer molecule at nanotube interface. *Nano Letters* 6:1627–1631.

52. Wei, C. 2006. Adhesion and reinforcement in carbon nanotube polymer composite. *Applied Physics Letters* 88:093108.

53. Liu, J., Wang, X., Zhao, L., Zhang, G., Lu, Z., and Li, Z. 2007. The absorption and diffusion of polyethylene chains on the carbon nanotube: The molecular dynamics study. *Journal of Polymer Science: Part B: Polymer Physics* 46:272–280.

54. Wei, C. 2009. Adsorption of an alkane mixture on carbon nanotubes: Selectivity and kinetics. *Physical Review B* 80:085409-7.

55. Al-Ostaz, A., Pal, G., Mantena, P. R., and Cheng, A. 2008. Molecular dynamics simulation of SWCNT–polymer nanocomposite and its constituents. *Journal of Materials Science* 43:164–173.

56. Milchev, A. and Binder, K. 2002. Polymer nanodroplets adsorbed on nanocylinders: A Monte Carlo study. *Journal of Chemical Physics* 117:6852–6862.

57. Kusner, I. and Srebnik, S. 2006. Conformational behavior of semi-flexible polymers confined to a cylindrical surface. *Chemical Physics Letters* 430:84–88.

58. Gurevitch, I. and Srebnik, S. 2008. Conformational behavior of polymers adsorbed on nanotubes. *Journal of Chemical Physics* 128:144901.

59. Wescott, J. T., Kung, P., and Maiti, A. 2007. Conductivity of carbon nanotube polymer composites. *Applied Physics Letters* 90:033116.

60. Shou, Z., Buxton, G. A., and Balazs, A. C. 2003. Predicting the self-assembled morphology and mechanical properties of mixtures of diblocks and rod-like nanoparticles. *Composite Interfaces* 10:343–368.

61. Buxton, G. A. and Balazs, A. C. 2004. Predicting the mechanical and electrical properties of nanocomposites formed from polymer blends and nanorods. *Molecular Simulation* 30:249–257.

62. Clancy, T. C. and Gates, T. S. 2006. Modeling of interfacial modification effects on thermal conductivity of carbon nanotube composites. *Polymer* 47:5990–5996.

63. Vacatello, M. 2001. Monte Carlo simulations of the interface between polymer melts and solids. Effects of chain stiffness. *Macromolecular Theory and Simulations* 10:187–195.

64. Vacatello, M. 2001. Monte Carlo simulations of polymer melts filled with solid nanoparticles. *Macromolecules* 34:1946–1952.

65. Marla, K. T. and Meredith, J. C. 2006. Simulation of interaction forces between nanoparticles: End-grafted polymer modifiers. *Journal of Chemical Theory and Computation* 2:1624–1631.

66. Smith, J. S., Bedrov, D., and Smith, G. D. 2003. A molecular dynamics simulation study of nanoparticle interactions in a model polymer-nanoparticle composite. *Composites Science and Technology* 63:1599–1605.
67. Hooper, J. B., Schweizer, K. S., Desai, T. G., Koshy, R., and Keblinski, P. 2004. Structure, surface excess and effective interactions in polymer nanocomposite melts and concentrated solutions. *Journal of Chemical Physics* 121:6986–6997.
68. Liu, A. Y. H. and Rottler, J. 2009. Physical aging and structural relaxation in polymer nanocomposites. *Journal of Polymer Science, Part B: Polymer Physics* 47:1789–1798.

8

Polyolefin Nanocomposites with Layered Double Hydroxides

DeYi Wang, Francis Reny Costa, Andreas Leuteritz, Purv J. Purohit,
Andreas Schoenhals, Anastasia Vyalikh, Ulrich Scheler,
Udo Wagenknecht, Burak Kutlu, and Gert Heinrich

CONTENTS

8.1 Introduction

Layered double hydroxides (LDHs) are host–guest materials consisting of positively charged metal hydroxide sheets with intercalated anions and water molecules, called anionic clay minerals [1]. They can be represented by a general formula, $[M^{2+}_{1-x}M^{3+}_x(OH)_2]^{x+} A^{n-}_{x/n} \cdot yH_2O$, where M^{2+} and M^{3+} are divalent and trivalent metal cations, such as Mg^{2+}, Zn^{2+}, Ca^{2+}, etc., and Al^{3+}, Co^{3+}, Fe^{3+}, respectively, A^{n-} are interlayer anions, such as CO_3^{2-}, Cl^-, and NO_3^-. The anions occupy the interlayer region of these layered crystalline materials. The most commonly known naturally occurring LDH clay is hydrotalcite (Mg-Al type) with the chemical formula $Mg_6Al_2(OH)_{16}CO_30.4H_2O$, showing a structure represented in Figure 8.1.

With respect to easy exchangeability of anions in the interlayer of LDH, there are wide potential application fields for LDH research, such as catalysts, medical materials, gene delivery, etc. LDH materials have generated serious research activities in the area of their synthesis and modifications. One of the most recent applications investigated extensively is the use of LDH as nanofiller for a polymer matrix to improve material properties. In this chapter, polyolefin/LDH nanocomposites are described in detail including synthesis and characterization of modified LDH, preparation of polyolefin/LDH nanocomposites, and their special properties.

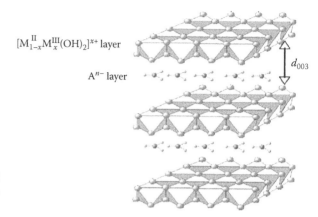

$[M_{1-x}^{II}M_x^{III}(OH)_2]^{x+}$ layer

A^{n-} layer

d_{003}

FIGURE 8.1
Schematic representation of LDH crystal structure. (From Costa, F.R. et al., *Polymer,* 46, 4447, 2005.)

8.2 Synthesis, Modification, and Characterization of LDH

8.2.1 Synthesis and Modification

To obtain good dispersion in polyolefin matrices, the organic modification of LDH is necessary before using it as nanofiller. Generally, there are several modification methods for preparing organic modified LDH; they are as follows:

Regeneration—Many LDH materials show a unique phenomenon called "memory effect," which involves the regeneration of the layered crystalline structure from their calcinated form, when the latter is dispersed in an aqueous solution containing suitable anions [2]. This property is often used to synthesize and modify LDHs with different types of intercalating anions. Typically, LDHs containing carbonate anion are heated to a temperature in the range of 350°C–800°C for several hours and the resultant, mixed metal oxide (more precisely a solid solution of the two metal oxides), is then dispersed in an aqueous solution of the desired anionic species. The dispersion is stirred mechanically overnight at room temperature to ensure the completion of the regeneration process. The regeneration property shown by LDHs is extensively reported in numerous literatures [3–5].

Anion exchange—This method takes the advantage of exchangeable interlayer anions present in LDHs by other anionic species. Based on this property, the LDHs containing one type of intercalating anionic species can be synthesized from the LDHs containing another type of intercalating anion. Usually, the original LDH is dispersed in an aqueous solution of the desired anionic species and the dispersion is stirred at room temperature for several hours. However, some anionic species show more affinity to the inter-gallery region of LDH than the other [6].

One step synthesis—It develops an effective and easy method to prepare organo-LDH. No additional measures other than controlling the pH of the medium were required to obtain a high degree of intercalation by the surfactant retaining the high crystallinity independent on the presence of other anions, like NO_3^-, CO_3^{2-}, and Cl^- [7]. The typical synthesis was carried out by the slow addition of a mixed metal salt solution (either metal nitrate salts or metal chloride salts as original source) to a modifier solution (like common surfactant, sodium dodecylbenzenesulfonate (DBS)) under continuous stirring maintaining the reaction temperature at 50°C. During the synthesis, the pH value was kept at a constant value by adding suitable amount of base solution. After the addition of the mixed metal salt solution, the resulting slurry was continuously stirred at the same temperature for 0.5 h and then was allowed to age in heater at some temperature for several hours. The final products were filtered, washed several times, and then dried in oven till the constant weight.

8.2.2 Characterization for LDH

Several typical technologies for characterization-modified LDH are introduced in this section and organic LDH modified by DBS as a model example. In detail, the technologies include wide angle x-ray scattering (WAXS), Fourier transform infrared spectra (FTIR), thermogravimetric analysis (TGA), elemental analysis (EA), and ^{27}Al NMR and ^{1}H MAS NMR. A comparison between modified and unmodified LDH are also shown.

WAXS—The WAXS patterns (Figure 8.2) of the LDH and organo-LDH synthesized by one-step synthesis method revealed that DBS anion can be efficiently intercalated within the LDH layers by this method [7].

In the unmodified LDH, the first basal reflection (003) at $2\theta = 11.8°$ corresponds to an interlayer distance of 0.77 nm. The absence of any distinguishable reflection at this position in the modified LDH synthesized by the one-step method indicates that no unmodified LDH phase was formed. The reflections up to several orders in the WAXS patterns point to high crystalline order of the organo-LDH, indicating that the presence of DBS does not affect the formation of layered structure during co-precipitation of the metal ions. The WAXS patterns of organo-LDH obtained in the presence of different starting anion species, that is, Cl^-, NO_3^-, and CO_3^{2-}, show the same interlayer distances proving the efficiency of the one-step method regardless of the starting material used.

The interlayer distance in the organo-LDH is about 2.98 nm, which resembles the value observed for DBS-modified LDH synthesized by other method [8]. The proposed scheme suggests a monolayer arrangement of surfactant anions in the interlayer region of the modified LDH by the one-step synthesis. The interlayer distance can be calculated

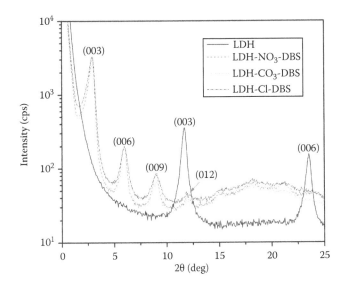

FIGURE 8.2
WAXS pattern of LDH and modified LDH by one-step synthesis. (From Wang, D.Y., *Chem. Mater.*, 21, 4490, 2009.)

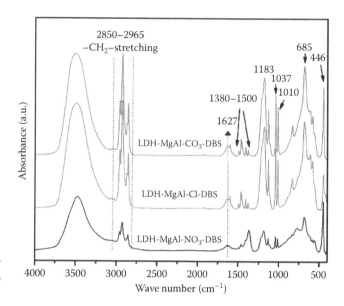

FIGURE 8.3
The FTIR spectra of LDH modified by one-step synthesis. (From Wang, D.Y., *Chem. Mater.*, 21, 4490, 2009.)

as $d_{calculated} = d_{layer} + d_{inter}$, where d_{layer} represents the thickness of the hydrotalcite brucite-like LDH sheet being 0.49 nm, and d_{inter} includes the length of intercalated species and absorptive water in the interlayer [7]. It has been reported that in LDH-DBS, the loss of the absorbed interlayer water causes 0.32 nm contraction in basal spacing [9]. Since the orientation of the molecular chain of DBS anion is proposed to be perpendicular to the layer and has a dimension of 2.14 nm, the interlayer distance is calculated and found to be 2.95 nm, which is in agreement with the experimental result (d = 2.98 nm).

FTIR—The FTIR spectra of the organo-LDHs reveal the presence of DBS anion in the materials (Figure 8.3) [7]. The characteristic vibration bands were detected for the SO_3^- group (symmetric stretching at 1037 cm^{-1} and asymmetric at 1182 cm^{-1}), the benzene group (C–C stretching at 1496 and 1602 cm^{-1}, C–H in plane bending at 1011 and 1131 cm^{-1}), and alkyl group (asymmetric stretching of CH_3 and CH_2 at 2958 and 2926 cm^{-1}, respectively, symmetric stretching of CH_3 and CH_2 at 2872 and 2855 cm^{-1}, respectively, and CH_2 scissoring at 1466 cm^{-1}). The bands recorded below 800 cm^{-1}, especially the sharp and strong characteristic band around 450 cm^{-1}, arise due to the vibration of metal–oxygen bond in the brucite-like lattice. These results indicate the presence of DBS molecules in all the modified LDHs and were consistent with the results of WAXS.

TGA and EA—The elemental analysis of organo-LDH obtained by the one-step method along with their thermograms is shown in Figure 8.4 [7]. The results exhibit a 2:1 Mg:Al ratio and very high (nearly 100%) intercalation of the DBS anions within LDH layers. The similar contents of C, H, and S in organo-LDH obtained from different starting materials indicate their similar chemical composition. These findings obviously establish the fact that the presence of nitrogen atmosphere is not always a necessary condition to incorporate DBS anion within LDH layer. The thermograms presented in Figure 8.4 reveal the similar thermal decomposition behavior for the organo-LDH materials synthesized by the one-step method that results from similarity of their chemical compositions identified by the elemental analysis. The characteristic features of these thermograms corresponding to the low-temperature weight loss (roughly below 250°C) and the thermal decomposition indicate similar amount of interlayer water and the same Mg:Al ratio in

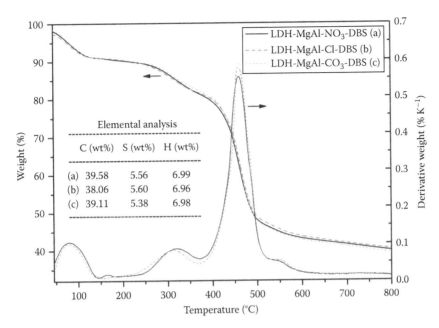

FIGURE 8.4
TGA and element analysis of modified LDH by one-step synthesis. (From Wang, D.Y., *Chem. Mater.*, 21, 4490, 2009.)

all samples under study, respectively. A quite similar conclusion also could be observed in the study of other kinds of LDH synthesized by the same method, such as organo Co–Al LDH [10].

Nuclear magnetic resonance (NMR)—Unlike other structural techniques, such as powder and single-crystal x-ray and neutron diffraction, which characterize the "long-range" order, giving an average view of a structure, solid-state NMR probes the local environment of a particular nucleus and, therefore, is highly suited to study amorphous or disordered materials, such as modified LDH. An extensive review of NMR studies related to both the structure and dynamics in LDH materials was reported by Rocha [11]. Herein, we concentrate on site-specific information available from the ^{1}H and ^{27}Al solid-state NMR.

The solid-state ^{1}H NMR spectrum of unmodified LDH is composed of ^{1}H signals from interlayer water and metal hydroxide groups [12]. Strong residual dipolar interactions and diversity of the local environment of individual hydroxyl sites within an LDH layer result in significant line broadening, which can be reduced by using magic angle sample spinning (MAS). It has been successfully used to identify the organic species introduced into the interlayer space of LDH, and therefore, can be applied for a control of modification extent and a quality of the modified products. Figure 8.5 shows the ^{1}H MAS NMR spectra of two DBS-modified LDHs produced by the regeneration and the one-step synthesis [7]. The peak positions in both spectra are identical, while linewidths in the case of regenerated LDH-DBS are significantly broader than those in the spectrum of one-step produced LDH-DBS. Based on ^{1}H CRAMPS (combined rotation and multiple pulse spectroscopy) experiment, which reduces strong ^{1}H–^{1}H dipolar interactions, this broadening has been attributed to the structural disorder in the regeneration-produced LDH that can originate from calcination products, for example, tetrahedral aluminum, which is clearly visible in the ^{27}Al NMR spectrum (see below).

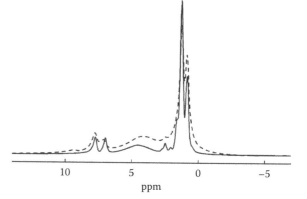

FIGURE 8.5
¹H MAS NMR spectrum of LDH-DBS produced by the regeneration method (dashed line) and by one-step route (solid line). (From Wang, D.Y., *Chem. Mater.*, 21, 4490, 2009.)

Solid-state ^{27}Al NMR has been widely applied to provide quantitative structural information on crystalline and amorphous aluminum-containing solids [13,14], because the coordination number of aluminum is manifested in characteristic ranges of the ^{27}Al chemical shifts [15]. In Figure 8.6a, the ^{27}Al solid-state NMR spectrum of unmodified LDH [7] is shown demonstrating an intense signal at 10 ppm, which is attributed to octahedral AlO_6 aluminum. The spectra of organo-modified LDH-DBS produced by one-step route and the regeneration (Figure 8.6b and c, respectively) are also dominated by the AlO_6 signals evidencing that LDH-DBS formed by either route is composed of octahedral aluminum. However, in the spectrum of regenerated LDH, an additional signal at 70 ppm is observed, which is a signature of tetrahedral AlO_4 aluminum. A quantitative analysis reveals 15% AlO_4 in the regenerated LDH, which is thought to be the product of calcination involved in the regeneration procedure. Partial transformation of AlO_6 to AlO_4 has been recently observed in BEHP-modified LDH also prepared by the regeneration method [16].

Additional information can be extracted analyzing the line shape of the ^{27}Al spectrum, because a nonzero ^{27}Al electric quadrupole moment gives rise to quadrupolar interactions, which are very sensitive to the local charge distribution. Thus, the lower quadrupolar

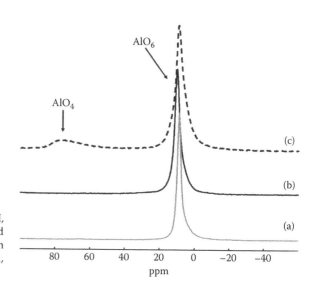

FIGURE 8.6
^{27}Al MAS NMR spectra of (a) unmodified LDH, (b) LDH-DBS produced by one-step route, and (c) LDH-DBS produced by the regeneration method. (From Wang, D.Y., *Chem. Mater.*, 21, 4490, 2009.)

coupling constant in LDH-DBS prepared by the one-step route in contrast to the regenerated LDH-DBS translates more ordered and symmetric octahedral Al positions similar to unmodified LDH. Thus, ^{27}Al solid-state NMR provides a complementary evidence of the formation of more completed and single-crystalline structure in "one-step" produced LDH-DBS.

The ^{27}Al triple-quantum (3Q)MAS spectrum, [17] that enables supplementary resolution by separating the intensity in a two-dimensional spectrum, allows distinguishing two nonequivalent AlO_6 sites in BEHP-modified LDH, which are characterized by essentially different quadrupolar parameters with rather similar chemical shifts [16]. These two crystallographically distinct Al sites with different local symmetry, found in all pristine and modified LDH materials regardless of the processing route, may originate from geometric isomers due to different configurations of the OH groups or due to the arrangement of magnesium cations in the LDH network [12]. It is worth to note that the coexistence of geometric isomers has also been observed in gibbsite, [18] whose structure is a single-cation natural analogue of synthetic Al-based LDH.

Thus, it has been shown in this paragraph, that high-resolution ^{1}H and ^{27}Al solid-state NMR is a powerful tool for the characterization of a quality of the synthesized organo-LDH, in terms of degree of order, intercalation behavior, and preservation of the local environment in the metal-hydroxide network after chemical modification.

8.3 Polyolefin/LDH Nanocomposite

8.3.1 Preparation and Characterization

Beside of the usages of LDHs in different fields, like catalysis, controlled chemical release, etc., their potential as nanofillers in preparing polymer nanocomposites is very popular in recent years. In this regard, unlike conventional layered silicates, which are widely used so far as nanofillers in polymer, LDHs possess certain inherent advantages. For example, being mostly of synthetic origin, the presence of impurities in LDHs is far less and also a wide range of chemical compositions can be obtained by changing the type and molar ratio of the metal ion pairs during the synthesis process. However, the major disadvantage of LDHs is the high charge density of layers, which firmly holds the metal hydroxide layers in the crystalline stacks and makes the intercalation of the polymeric materials into the interlayer region difficult. Principally, the methods used for intercalation of monomer or small oligomeric organic molecules into LDH are also applicable for high-molecular-weight species. But, due to a very small interlayer distance and high charge density of the LDH layers, the direct intercalation of large polymeric chains is more difficult. Two main methods for the preparation of polyolefin/LDH nanocomposites have been employed as follows:

Solution intercalation—To prepare polymer nanocomposites, typically the organically modified LDH is dispersed in a solution containing polyolefin. The resultant dispersion is then stirred or aged under nitrogen atmosphere to accomplish the polyolefin intercalation. For example, Qu et al. [19,20] prepared and characterized polyethylene/LDH by this method using dodecylsulfate (DS)-modified LDH (LDH-DS). The nanocomposites were obtained by refluxing the mixture of LDH-DS and the polyolefin solution in xylene.

Melt-compounding method—This method has definite technological advantage over the solution method as it can be easily adopted for industrial product manufacture using conventional polymer-processing equipments. For the preparation of polyolefin/LDH nanocomposite by melt compounding method, it becomes more difficult due to high thermodynamic incompatibility between the nonpolar polymer matrix and the polar LDH. However, in spite of being one of the most challenging and promising methods, melt-compounding method is always used to prepare polyolefin/LDH nanocomposite. For examples, Qu et al. [21] prepared polypropylene/Zn-Al–LDH and polyethylene/Mg-Al–LDH nanocomposites using organically modified LDH. Costa et al. [1,22] and Wilkie et al. [23,24] also reported the preparation polyolefin/LDH nanocomposite by this method. They observed that a high degree of exfoliation of the LDH particles can be obtained using this method.

Characterization—There are several typical methods for characterizing polyolefin/LDH nanocomposites as well as other polymer nanocomposites, such as FTIR, SEM, XRD, TEM, etc. Herein, TEM and solid-state NMR as powerful characterization technologies for polyolefin/LDH nanocomposite are mainly described.

TEM—TEM morphology analysis is used as a very useful method to describe the extent of intercalation and exfoliation of the nanofiller having layered structure. When the high-magnification TEM images are analyzed, both the nature of these primary particles and the matrix regions become more distinguishable. Figure 8.7 reveals the magnified view of one such particle and also the surrounding PE matrix region [25].

It can be noticed that the primary LDH particles are highly distorted and show ample evidences of crystal layer delamination/exfoliation from their surface. Such particles always show poor long-range crystalline symmetry. The morphological features observed in Figure 8.8 provide a clear understanding of the mechanism of exfoliation of the LDH-DBS particles during melt mixing [25]. At first, polymer chains penetrate within the interlayer region of the LDH-DBS particles and push apart the metal hydroxide sheets. With the course of time within the mixing channel of extruder, an increasing number polymer chains enter interlayer region and shearing of the screws facilitates the delamination of the surface layers one by one from a primary particle. As a result, not only the exfoliated particle fragments are formed, but also the size of the original primary particles are reduced.

These morphological features also prove that the melt-mixing method is certainly a promising method to obtain partially exfoliated particle morphology of LDH nanofillers in polyethylene. However, obtaining a high degree of LDH particle exfoliation in nonpolar matrix is still a challenging task. The strong electrostatic attractive force between the layers due to high charge density on the LDH layers makes the exfoliation process difficult, especially a nonpolar matrix, like polyethylene.

Solid-state NMR—The structural changes in LDH, modified to prepare the LDH-based polymer nanocomposites, have been studied by a combination of the ^{27}Al MAS and 3QMAS NMR. The ^{27}Al NMR data presented here (Figure 8.9) demonstrate that the chemical modification of LDH materials by the regeneration method significantly affects the local aluminum environment [26]. The ^{27}Al NMR spectra of parent and regenerated LDH (LDH and LDH-R in Figure 8.9) represent a single peak arising from octahedral Al, while in the spectra of calcined (CLDH) and surfactant intercalated LDH (BEHP-LDH and DBS-LDH), both AlO_6- and AlO_4-signals are clearly visible. The characteristic regeneration property of LDH after calcination and posterior rehydration (LDH-R)—"memory effect"—has been demonstrated by restoring the ^{27}Al NMR spectrum of the parent compound. However, a small trace of AlO_4 in LDH-R (ca. 5% of total spectrum intensity) indicates that regeneration is

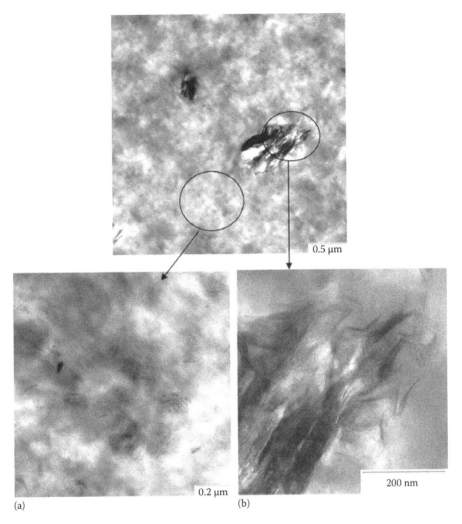

FIGURE 8.7
TEM micrographs showing that exfoliated LDH particles exist both in (a) the bulk matrix and (b) in the vicinity of the originating bigger particles. (From Costa, F.R., *Polym. Degrad. Stabil.*, 92, 1813, 2007.)

only 95% reversible in our experiment. Incorporation of surfactants into LDH by means of regeneration results in the changes in local aluminum structure indicating that ca. 29% of aluminum is converted from octahedral to tetrahedral sites in the case of BEHP intercalation, while DBS modification leads to the formation of only ca. 17% tetrahedral Al sites. The presence of tetrahedral aluminum in LDH compounds is attributed to calcination products, which are not converted after the rehydration process. Differences in the relative contents of AlO_4 in both surfactant-modified LDHs can be related to the presence of interlayer water molecules in DBS-LDH that is visible in the ^{1}H MAS NMR spectrum [26].

Finally, the DBS-LDH was added to low-density polyethylene (LDPE) to prepare polymer nanocomposite (LDH-PE) according the procedure described in [1]. Comparing the spectra of DBS-LDH and LDH-PE, one can see that after incorporation of surfactant-modified LDH into polymer matrix, the local Al structure has been preserved in the final polymer hybrid material.

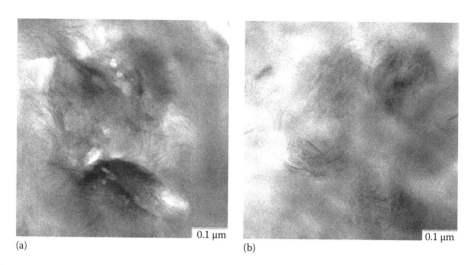

(a)   (b)

FIGURE 8.8
High-magnification TEM micrograph showing that LDH particles also undergo exfoliation into single layers: (a) PE-LDH4 and (b) PE-LDH5. (From Costa, F.R., *Polym. Degrad. Stabil.*, 92, 1813, 2007.)

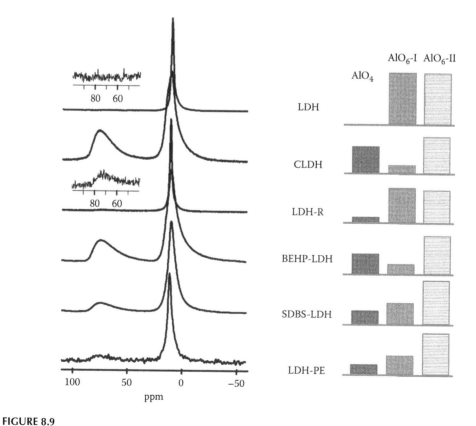

FIGURE 8.9
^{27}Al MAS NMR spectra of the pristine and modified LDH. The magnified amplitudes (×50) show the presence of tetrahedral Al in LDH-R as compared to parent LDH. On the right-hand side, the corresponding quantification diagrams are presented. (From Vyalikh, A., et al., *J. Phys. Chem. C*, 113, 21308, 2009.)

8.3.2 Properties of Polyolefin/LDH Nanocomposite

In general, when compared with the conventional polymer composites, polymer nanocomposites exhibit significant improvements in different properties at relatively much lower concentration of filler. The efficiency of various additives in polymer composites can be increased manyfold when dispersed in the nanoscale. This becomes more noteworthy when the additive is used to address any specific property of the final composite such as mechanical properties, conductivity, fire retardancy, thermal stability, etc. In case of polyolefin/LDH nanocomposites, similar improvements are also observed in many occasions. For example, the thermal properties of PE/LDH showed that even a small amount of LDH improves the thermal stability and onset decomposition temperature in comparison with the unfilled PE [22]; its mechanical properties revealed increasing LDH concentration brought about steady increase in modulus and also a sharp decrease in the elongation at break [25]. While in this section, fire-retardant properties and electric properties of polyolefin/LDH nanocomposite were described in detail.

8.3.2.1 Fire Retardancy

Polyolefin as one of the synthetic polymers is widely applied for engineering materials, electronic cases, interior decoration, and so on because of low cost, ease of fabrication, and processing. Due to its high flammability, in many of these applications, there are severe risks of fire-related causalities and loss of valuable properties. Therefore, it is often needed to make these products more flame retardant by using suitable flame retardants in the original compound.

The cone calorimeter investigation is a very popular and standard method for ranking and comparing the flammability properties of polymeric materials. In contrary, sustained combustion of the test samples in the test methods, like LOI, UL94, etc., depend on material's heat of combustion. Costa et al. [25] reported fire-retardant properties of the nanocomposite based on LDPE- and Mg-Al-based layered double hydroxide (Mg-Al–LDH). The results from the cone calorimeter investigations of the LDPE/LDH nanocomposites are summarized in Figure 8.10. Heat release rate (HRR) is the most important variable, which controls how fast a fire could reach an uncontrollable stage. It can be seen in Figure 8.10a that the increasing concentration of LDH in LDPE/LDH nanocomposites not only significantly reduces the peak heat release rate (PHRR), but also makes HRR curve increasingly flattened. This obviously results in slow burning rate of the materials at higher LDH concentration. In case of the unfilled polyethylene, cone calorimeter investigation under similar external heat flux (about 30 kW m^{-2}) gives a PHRR value over 800 kW m^{-2} [27]. The addition of small amount of LDH causes significant reduction in PHRR (below 600 kW m^{-2}) in LDPE. At higher LDH concentration (PE-LDH4, PE-LDH5 and PE-LDH6), the PHRR is lowered about 300 kW m^{-2}. The ignition of the unfilled polyethylene is followed by the formation of molten surface layer on which the flame floats, whereas a char layer floats on the molten layer in case of nanocomposites. The ignition time (T_{ig}), defined as the time at which the test samples catch fire, is also significantly increased with increasing LDH content. The unfilled LDPE has a T_{ig} below 100 s and that increases to above 120 s with the addition of 16.20 wt% LDH (PE-LDH6) (Figure 8.10b). The total heat release (THR) is a parameter that determines how big a fire is. Once the ignition takes place, THR steadily increases with burning time and attains a steady state before the flameout occurs. As can be seen in Figure 8.10c, THR is progressively reduced with increasing LDH content. At 10 min after the application of the external heat flux

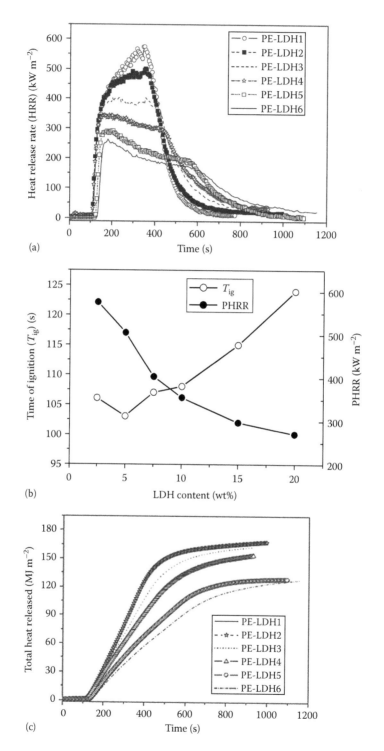

FIGURE 8.10
Cone-calorimeter investigation results showing (a) variation of HRR with time, (b) variation of time of ignition (T_{ig}) and PHRR with LDH content, and (c) THR with time in LDPE/LDH nanocomposites. (From Costa, F.R., *Polym. Degrad. Stabil.*, 92, 1813, 2007.)

TABLE 8.1

Cone Calorimetric Data of Various PP, PP/IFR, and PP/IFR/LDH Samples

Samples	PP	PP/IFR	PP/IFR/ MgAl-LDH	PP/IFR/ CaAl-LDH	PP/IFR/ ZnAl-LDH	PP/IFR/ CuAl-LDH
	────────▶		Increasing molecular weight per unit cell		────────▶	
IT (s)	60	48	55	52	55	54
pk-HRR (kW/m^2)	1275	506	330	351	318	424
pk-MLR (g/m^2 s)	0.254	0.115	0.086	0.089	0.081	0.110
pk-EHC (MJ/kg)	96.3	71.8	64.9	64.3	61.8	65.4

Source: Data from Zhang, M. et al., *Polym. Compos.*, 1000, 2009.

during cone calorimeter test, THR is reduced by about 17% and 44% in the samples PE-LDH4 and PE-LDH6, respectively. THR is often taken as the measure of the propensity to sustain a fire for long duration. An efficient flame retardant should reduce THR effectively when incorporated into a polymer. In this respect, LDH fillers have the definite advantage of improving the fire retardancy of polyolefin because LDH undergoes decomposition through endothermic reaction, which act as a heat sink reducing the total heat generated during combustion.

At LDH concentration above about 10 wt%, the nanocomposites not only burn at extremely slow rates, but also show low dripping tendency. However, LDH alone, at these concentrations, is not sufficient to obtain high LOI value or V-0 rating in UL94 testing.

In recent years, in order to meet the most demanding fire-retardancy requirements, a combination of nanofillers with conventional fire retardants is used to prepare combinative fire retardants. Some results show that the use of classical fire retardants in nanocomposites could reduce the amount of additive required [28–30].

Zhang et al. [31] studied fire-retardant properties of PP/IFR/LDH nanocomposites, which were comprised by a typical intumescent flame retardant (IFR) system and LDHs with different bivalent metal cations. The results of pk-HRR, pk-MLR, pk-EHC, and ignition time (IT) obtained from the cone calorimeter tests of various samples were listed in Table 8.1 [31]. It could been seen from Table 8.1 that the IT values of PP/IFR/LDH samples with different divalent metal cations of Zn, Mg, Cu, and Ca were 55, 55, 54, and 52 s, respectively, which are longer than 48 s of the PP/IFR sample without LDH. These data showed that the PP/IFR samples with LDHs are obviously harder to ignite than the sample only with the IFR.

8.3.2.2 Electric Properties

8.3.2.2.1 Electrodischarge Properties

The necessity to modify the pristine LDH with organic compounds finally leads to the idea possibly adding another function to the layered structure besides being reinforcing, flame retardance, or diffusion barrier in polymeric materials. One of the problems in cable industry is the burning behavior of cable insulation materials. Due to European regulations, the addition of halogenated flame retardants is very restricted. Present alternatives based on aluminum-trihydrate (ATH) or magnesium hydroxide lack on the necessary high amount from 55% to 65% filler in formulations to be effective as flame retardant. In combination with layered additives in the nanometer range based either on montmorillonite (cationic clay) or on LDH (anionic clay), the level of filler loading could be significantly reduced, thus allowing better processing of the material. For safety reasons, high-voltage cables have to have a certain conductivity to dissipate the electrostatic charge achieved during use in lifeline. Therefore, in maintenance long down times have to be considered to ensure the discharge of the cable mantle. The new type of filler should therefore provide an effect as flame retardant and introduce conductivity to the material. This approach is described in more detail in the following section [32,33].

Due to its structure and constitution, LDH can be considered as an intrinsic flame retardant. The organic modification of the LDH for a good dispersion in polymeric matrices should introduce conductivity to the filler. To achieve this goal, an organic conductive substance has to be intercalated in between the layers, providing an increase of interlayer distance, reduction of attractive forces, and turning the surface of the LDH hydrophobic. Among conductive organic materials, polyaniline attracts increasing attention especially after the discovery of its conductive features by MacDiarmid et al. in 1985 [34]. To introduce one of the named forms into the gallery of LDH, two different approaches were applied in order to achieve tunable electric properties. These routes are discussed in the following.

Route 1 is principally the diffusion of the monomers within the LDH and subsequent polymerization of the monomers inside the host material. These trials were carried out for LDH previously modified with aniline sulfonic acid (ANIS) and sodium dodecylbenzenesulfonate (DBS) by regeneration method. The chosen polymerization method was emulsion polymerization of aniline since it is one of the most promising methods to improve the processability of PANI [35]. The polymerization method was performed from the work of Liu [36], which was done by using oleic acid as dopant and surfactant. Organo-modified LDH was dispersed in water and followed by addition of monomer, aniline and dopant, acetic acid (AA). It was found out that in all cases, conductivity values were rather low with this approach.

In the second route, the polymerization was carried out first and then the intercalation of the PANI inside the layers was performed. The chosen polymerization technique was different than the previous one in order to maintain superior organosolubility of the polyaniline. Polymerization was performed based on the work of Xie et al. [37]. Aniline was polymerized in the presence of a mixture including oxidizing agent ammonium peroxodisulfate and doping agent dodecylbenzene sulfonic acid (DBSA) in 150 ml acetone and 350 ml water at a temperature of 0°C for 12 h. Then, PANI/DBSA products were converted into the emiraldine base (EB) form by treatment with 3% ammonia solution and washing with water. The resulting emeraldine base is soluble in 1-methyl-2-pyrrolidine (NMP), which has also advantage in comparison with other solvents in the manner of being nontoxic. In the following overnight stirring of EB in NMP and DBS-modified LDH was dispersed into the system to carry out the insertion of the PANI. The NMP swells the organo-LDH to a high extent. Due to the high internal surface, PANI is adsorbed from LDH caused by a concentration gradient and partition coefficient between the solid and liquid phases. To acquire conductive properties, EB was redoped with camphor sulfonic acid. Additionally, after stirring for 3 h to obtain salt, distilled water was added to the system to increase the separation yield. Water is totally miscible with NMP but shows no solubility with DBS-modified LDH or polyaniline. Thus, the layered structure collapses again, facilitating the separation of NMP and PANI-DBS-LDH.

In case of PANI-modified LDH also, a strong dependency on the conductivity could be found. In this case, additionally the diffusion properties and the partition coefficient between the organo-modified LDH and the solvent used during redoping have to be considered. It was found that redoping with camphorsulfonic acid (HCSA) works best in case of PANI-modified LDH. Further, the acid used during the original polymerization process led to the interesting result that although acetic acid used during synthesis gave highest conductivity in case of redoping with HCSA, the sample prepared with DBSA gave better results, as shown in Figure 8.11.

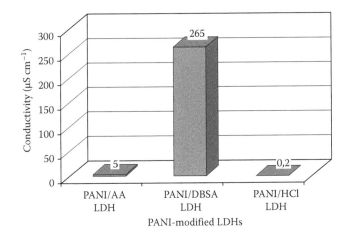

8.3.2.2.2 Dielectric Properties

Broadband dielectric spectroscopy is a powerful tool to investigate polymeric systems (see [38]) including polymer-based nanocomposites with different nanofillers like silica [39], polyhedral oligomeric silsesquioxane (POSS) [40–42], and layered silica systems [43–47] just to mention a few. Recently, this method was applied to study the behavior of nanocomposites based on polyethylene and Al-Mg LDH (AlMg-LDH) [48]. The properties of nanocomposites are related to the small size of the filler and its dispersion on the nanometer scale. Besides this, the interfacial area between the nanoparticles and the matrix is crucial for the properties of nanocomposites. Because of the high surface-to-volume ratio of the nanoparticles, the volume fraction of the interfacial area is high. For polyolefin systems, this interfacial area might be accessible by dielectric spectroscopy because polyolefins are nonpolar and, therefore, the polymeric matrix is dielectrically invisible [48].

For polymers, dielectric spectroscopy is sensitive to fluctuations of dipoles, which are related to the molecular mobility of groups, segments, or the polymer chain as well [38]. The molecular mobility is taken as a probe for structure. The basic quantity is the complex dielectric function $\varepsilon^*(f) = \varepsilon'(f) - i\varepsilon''(f)$ as a function of the frequency f and the temperature T. $\varepsilon'(f)$ is the real whereas $\varepsilon''(f)$ is the loss part ($i = \sqrt{-1}$). A relaxation process is indicated by a step-like decrease of $\varepsilon'(f)$ with increasing frequency and a peak in $\varepsilon''(f)$. From the maximum position of the peak a mean relaxation rate f_p can be deduced, which corresponds to the relaxation time of the fluctuation of the dipole moment of a given structural unit. For details see reference [49]. All shown measurements were carried out isothermally in the frequency range from 10^{-1} to 10^7 Hz by an ALPHA analyzer (Novocontrol®). The temperature of the sample is controlled by a Quatro Novocontrol system with stability better than 0.1 K.

Here the dielectric behavior of nanocomposites of polyethylene (LDPE; density ρ = 0.9225 g cm^{-3}; Melt flow index, MFI 3.52 g 10 min^{-1}) and AlMg layered double hydroxide (AlMg-LDH) is discussed. The inter-gallery anions were exchanged by DBS (for details see reference [8]). The samples were prepared by a masterbatch approach. First, the compatibilizer maleic anhydride–grafted polyethylene (MAH-g-PE; ρ = 0.926 g cm^{-3}, MFI 32.0 g 10 min^{-1}, MHA concentration 1.0 wt%) was mixed with the AlMg-LDH (ratio 2:1; weight) as the masterbatch in a twin-screw extruder (27 mm screw diameter, L/D ratio 36, 160°C–200°C temperature profile from the feed to the extruder barrel, 200 rpm screw speed and 6 kg h^{-1} feed rate). Second, the masterbatch was added in the selected concentration to LDPE and compounded using the same conditions as for the preparation of the masterbatch. The nanocomposites have a predominately exfoliated morphology [8].

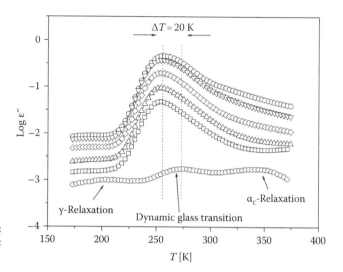

FIGURE 8.12

ε'' vs. T at f = 1 kHz: ○—MAH-g-PE;
□—2.43 wt%; △—4.72 wt%; ◇—8.94 wt%;
▽—12.73 wt%; ⊙—16.2 wt% LDH.

The monomer unit of polyethylene consists only of CH_2 units and is symmetric. No dipole moment is present for PE and no dielectric process is expected. The small losses measured in reality are due to the presence of impurities. Moreover, by oxidation, a few polar carbonyl groups can be formed [38,50]. Figure 8.12 compares the isochronal spectra (loss vs. temperature at fixed frequency) for MAH-g-PE with that of the nanocomposites. Because MAH-g-PE carries the polar anhydride group, several relaxation processes indicated by peaks can be detected [48,51]: At low temperatures, the γ-relaxation is observed corresponding to localized fluctuations. At higher temperatures (ca. 275 K) the dynamic glass transition (β-relaxation, segmental dynamics) takes place. At even higher temperatures, the α_c relaxation becomes active, which is assigned to the crystalline lamella.

The dielectric loss of the β-relaxation process increases with the concentration of the nanofiller (see Figure 8.12). The only polar component in the system, which increases with the concentration of LDH is the anion DBS. It is assumed that the alkyl tail of DBS is desorbed from the surface of the layers and forms a phase with the polymer segments. Therefore, these polar molecules are fluctuating together with the apolar PE segments and probe the molecular mobility of the latter. Therefore, with increasing LDH concentration (increasing DBS concentration), the dielectric loss of the dynamic glass transition increases. This is a probe technique [51,52] where the localization of the probe is known, namely, in the interfacial area between the nanoparticle and the matrix. Therefore, here the dielectric spectroscopy monitors selectively the molecular dynamics in an interfacial region close to the LDH sheets because the dielectric loss of PE is low and so the matrix of the nanocomposite is nearly dielectrically invisible. In addition to the increase of the β-relaxation, it shifts to lower temperatures by 20 K (see Figure 8.12). This leads to the conclusion that the molecular mobility in the interfacial region is higher than that in the neat PE.

Figure 8.13 gives ε'' vs. frequency and temperature for the nanocomposite with 16.2 wt% of AlMg-LDH. Besides the dynamic glass transition (β-relaxation), additional dielectric processes like melting or the interfacial process can be detected. For a discussion see [48]. With increasing temperature, the β-relaxation shifts to higher frequencies as expected. An analysis by the model function of Havriliak/Negami (HN) [53] shows that it consists of two processes (see inset Figure 8.14) for all concentrations of AlMg-LDH. These relaxation processes are assigned to two different regions of molecular mobility of the PE segments depending on the distance from the surface of the LDH. The low-frequency process

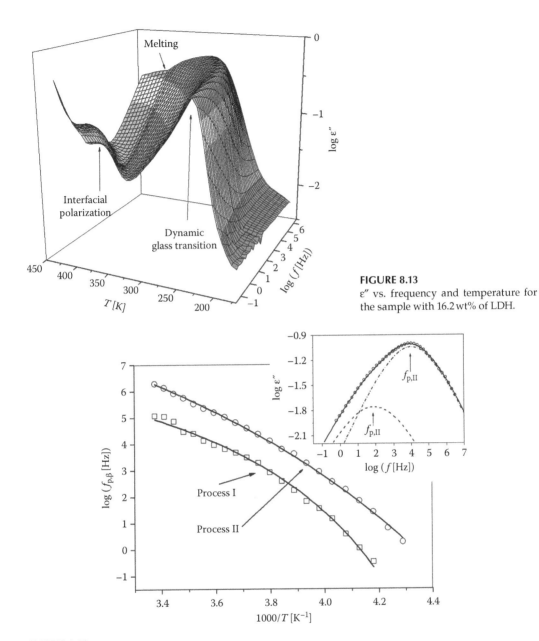

FIGURE 8.13
ε″ vs. frequency and temperature for the sample with 16.2 wt% of LDH.

FIGURE 8.14
Relaxation rates $f_{p,\beta}$ vs. $1/T$ for the both relaxation processes of the β-relaxation for the nanocomposite with 16.2 wt% LDH. Lines are fits of the VFT equation. The inset gives ε″ vs. frequency for the sample with 4.72 wt% LDH at $T = 263.2$ K. The solid line is a fit of two HN functions to the data. The dashed lines are the contribution of the individual process.

(process I) is assigned to PE segments and alkyl segments of DBS in close proximity to the LDH layers. A similar result is found for poly(ethylene oxide)/laponite nanocomposites by NMR [54]. The high-frequency process (process II) is assigned to the PE segments and parts of the alkyl tail of DBS in a greater distance from the LDH sheets because in this case, the alkyl tail with the bulky methyl group of the DBS molecules play the role of plasticizer for the PE segments in the region farther from the LDH sheets.

From this analysis, the relaxation rate f_p (see inset Figure 8.13) can be estimated and plotted vs. inverse temperatures (see Figure 8.13). For both processes, this trace is curved and can be described by the Vogel–Fulcher–Tammann (VFT) equation [55], which indicates that both processes are related to a glass transition but at different distances from the LDH layers. To have a signature as a glass transition, the extent of these regions should be in the order of 1–3 nm (see for instance [56–58]). This means that the extension of the interfacial region into the bulk matrix is about the same length. For all concentrations, the same temperature dependence for f_p of both processes is found. For detailed discussions see [48].

As discussed above for both processes, the temperature dependence of the relaxation rates follows the VFT equation. From the estimated parameters, a dielectric glass transition temperature can be calculated by $T_g^{Diel} = T(f_p = 10^{-2} \text{ Hz})$. An analysis using all the concentrations of LDH shows that the average difference in the glass transition temperature of both processes is ca. 10 K [48].

The analysis by the HN function delivers also the dielectric strength $\Delta\varepsilon$, which is proportional to the number of dipoles involved. In the inset of Figure 8.4, the sum of the dielectric strengths for the two processes ($\Delta\varepsilon_\beta$) is plotted vs. inverse temperature. As expected, for a glass transition [38] $\Delta\varepsilon_\beta$ decreases with temperature. More importantly the dielectric strength of the dynamic glass transition increases with the concentration of LDH. $\Delta\varepsilon_\beta$ is plotted at a temperature of $T_{Comp} = 263.1$ K vs. the concentration of LDH in Figure 8.15. For low concentrations, $\Delta\varepsilon_\beta$ varies linearly with c_{LDH}. This proves that $\Delta\varepsilon_\beta$ is due to the concentration of DBS molecules and supports a well-exfoliated state of the layers in the matrix. For higher values of c_{LDH}, $\Delta\varepsilon_\beta$ increases stronger than expected (see Figure 8.15). This point to a change in the structure of the nanocomposite. For higher concentration

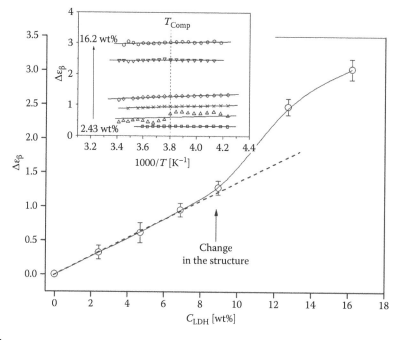

FIGURE 8.15
$\Delta\varepsilon_\beta$ vs. the concentration of LDH c_{LDH} at $T_{Comp} = 263.1$ K. The solid line is a guide to the eyes. The dashed lines indicates the linear behavior $\Delta\varepsilon_\beta \sim c_{LDH}$. The inset gives $\Delta\varepsilon_\beta$ for the β-relaxation vs. $1/T$: □—2.43 wt%; △—4.72 wt%; ×—6.89 wt%; ◇—8.94 wt%; ▽—12.73 wt%; and ○—16.2 wt% LDH. The solid lines are linear regression to the data.

of the nanofiller, the exfoliated LDH sheets cannot be arranged independently from each other. This leads to a correlation or orientation of the LDH sheets giving rise to a change in the concentration dependence of $\Delta\varepsilon_\beta$.

8.4 Conclusions and Future Outlook

To make the unmodified LDH clay a suitable precursor for the preparation of polymer nanocomposite, modification by anionic surfactants is necessary. In fact, a number of such surfactants were used to modify Mg–Al–LDH in order to enlarge the interlayer distance and to render it more organophilic, while study on the multifunctionalities modifier for LDH could be more interesting in the future. The characterization of the organically modified LDH can be carried out using various analytical techniques like XRD, FTIR, TGA, SEM, NMR, etc., while exact quantitative analysis for LDH should be further improved. To prepare nanocomposites based on polyolefin matrix using melt-compounding method, a conventional masterbatch technique is still an effective method. This is an obvious choice rather than an arbitrary one as a highly nonpolar polyolefin matrix hardly intercalates itself within the interlayer space of inorganic clays. Therefore, a functionalized polymer (PE-g-MAH) was used as compatibilizer between the two. The improvement of LDH particle dispersion in polymer matrix is certainly desired for further improvement in properties, especially mechanical properties. Such improvements will certainly be helpful to develop flame retardancy. However, concerning the low flame-retardant efficiency of LDH, combination with conventional active flame-retardants, like phosphates, might be very useful for improving the flame retardancy of polyolefin/LDH nanocomposites. This synergistic approach would also have a tremendous potential in reducing the total flame retardant concentration in polymer composites. We also believe that dielectric spectroscopy is a powerful technique to characterize the structure–property relationships of polymer-based nanocomposites. For polyolefin with incorporated LDH nanoparticles, it can be used to obtain detailed information about the interface between the exfoliated LDH layers and the matrix because the latter is nearly dielectric invisible. Besides further studies on PE/LDH nanocomposite, further work could concern other matrix polymers like polypropylene or polylactide as well as a broad variation of the LDH materials.

References

1. Costa, F. R., Goad, M. A., Wagenknecht, U., and Heinrich. G. 2005. *Polymer* 46:4447.
2. Miyata, S. 1980. *Clays and Clay Minerals* 28:50.
3. Cavani, F., Trifiro, F., and Vaccari, A. 1991. *Catalysis Today* 11:173.
4. Meyn, M., Beneke, K., and Legaly, G. 1990. *Inorganic Chemistry* 29:5201.
5. Stanimirova, T. S., Kirov, G., and Dinolova, E. 2001. *Journal of Material Science Letters* 20:453.
6. Khan, A. I. and O'Hare, D. 2002. *Journal of Materials Chemistry* 12:3191.
7. Wang, D. Y., Costa, F. R., Vyalikh, A., Leuteritz, A., Scheler, U., Jehnichen, D., Wagenknecht, U., Häussler, L., and Heinrich, G. 2009. *Chemistry of Materials* 21:4490.

8. Costa, F. R., Saphiannikova, M., Wagenknecht, U., and Heinrich, G. 2008. *Advances in Polymer Science* 210:101.
9. Costa, F. R., Leuteritz, A., Wagenknecht, U., Jehnichen, D., Häussler, L., and Heinrich, G. 2008. *Applied Clay Science* 38:153.
10. Wang, D. Y., Leuteritz, A., Wagenknecht, U., and Heinrich, G. 2009. *Transactions of the Nonferrous Metals Society of China* 19:1479.
11. Rocha, J. 2001. Solid state NMR and EPR studies of hydrotalcites. pp. 217–240, In *Layered Double Hydroxides: Present and Future*, ed., V. Rives. New York: Nova Science.
12. Sideris, P. J., Nielsen, U. G., Gan, Z., and Grey, C. P. 2008. *Science* 321:113.
13. Massiot, D., Bessada, C., Coutures, J. P., and Taulelle, F. 1990. *Journal of Magnetic Resonance* 90:31.
14. Müller, D., Gessner, W., Behrens, H. J., and Scheler, G. 1981. *Chemical Physics Letters* 79:59.
15. McKenzie, K. J. D. and Smith, M. E. 2002. *Multinuclear Solid-State NMR of Inorganic Materials.* Oxford, U.K.: Pergamon.
16. Vyalikh, A., Massiot, D., and Scheler, U. 2009. *Solid State NMR* 36:19.
17. Frydman, L. and Harwood, J. S. 1995. *Journal of the American Chemical Society* 117:5367.
18. Damodaran, K., Rajamohanan, R. P., Chakrabarty, D., Racherla, U. S., Manohar, V., Fernandez, C., Amoureux, J. P., and Ganapathy, S. 2002. *Journal of the American Chemical Society* 124:3200.
19. Chen, W. and Qu, B. 2003. *Chemistry of Materials* 15:3208.
20. Chen, W., Feng, L., and Qu, B. 2004. *Chemistry of Materials* 16:368.
21. Ding, P. and Qu, B. 2006. *Polymer Engineering and Science* 10:1153.
22. Costa, F. R., Satapathy, B. K., Wagenknecht, U., Weidisch, R., and Heinrich, G. 2006. *European Polymer Journal* 42:2140.
23. Manzi-Nshuti, C., Songtipya, P., Manias, E., Jimenez-Gasco, M. M., Hossenlopp, J. M., and Wilkie, C. A. 2009. *Polymer Degradation and Stability* 94:2042.
24. Manzi-Nshuti, C., Songtipya, P., Manias, E., Jimenez-Gasco, M. M., Hossenlopp, J. M., and Wilkie, C. A. 2009. *Polymer* 50:3564.
25. Costa, F. R., Wagenknecht, U., and Heinrich, G. 2007. *Polymer Degradation and Stability* 92:1813.
26. Vyalikh, A., Costa, F. R., Wagenknecht, U., Heinrich, G., Massiot, D., and Scheler, U. 2009. *Journal of Physical Chemistry C*, 113:21308.
27. Scudamore, M. J., Briggs, P. J., and Prager, F. H. 1991. *Fire Materials* 15:65.
28. Zanetti, M., Camino, G., Canavese, D., Morgan, A. B., Lamelas, F. J., and Wilkie, C. A. 2002. *Chemistry of Materials* 14:189.
29. Wang, D. Y., Song, Y. P., Wang, J. S., Ge, X. G., Wang, Y. Z., Stec, A. A., and Hull, T. R. 2009. *Nanoscale Research Letters* 4:303.
30. Wang, D. Y., Liu, X. Q., Wang, J. S., Wang, Y. Z., Stec, A. A., and Hull, T. R. 2009. *Polymer Degradation and Stability* 94:544.
31. Zhang, M., Ding, P., and Qu, B. J. 2009. *Polymer Composition* 30:1000.
32. Tronto, J., Leroux, F., Crepaldi, E. L., Naal, Z., Klein, S. I., Valim, J. B., 2006. *Journal of Physics and Chemistry of Solids* 67:968.
33. Kutlu, B., Leuteritz, A., Wagenknecht U., and Heinrich, G. 2010. Integration of antistatic properties into LDH-polyolefine nanocomposites, *Proceedings of the Polymer Processing Society*, 26th Annual Meeting (PPS-26) July 4–8, Banff, Canada.
34. Macdiarmid, A. G., Chiang, J., Halpern, M., Huang, W., Mu, S., Nanaxakkara, L. D., Wu, S. W., and Yaniger, S. I. 1985. *Molecular Crystals and Liquid Crystals* 121:173.
35. Blinova, N. V., Stejskal, J., Trchova, M., Prokes, J., and Omastova, M. 2007. *European Polymer Journal* 43:2331.
36. Liu, P. 2009. *Synthetic Metals* 159:148.
37. Xie, H., Ma, Y., and Feng, D. 2000. *European Polymer Journal* 36:2201.
38. Schönhals, A. 2002. Molecular dynamics in polymer model systems. In: *Broadband Dielectric Spectroscopy*, ed. F. Kremer, A. Schönhals, p. 225. Berlin, Germany: Springer.
39. Fragiadakis, D., Pissis, P., and Bokobza, L. 2005. *Polymer* 46:6001.
40. Bian, Y., Pejanovic, S., Kenny, J., and Mijovic, J. 2007. *Macromolecules* 40:6239.
41. Hao, N., Böhning, M., Goering, H., and Schönhals, A. 2007. *Macromolecules* 40:2955.

42. Hao, N., Böhning, M., and Schönhals, A. 2007. *Macromolecules* 40:9672.
43. Anastasiadis, S. H., Karatasos, K., Vlachos, G., Manias, E., and Gianneis, E. P. 2000. *Physical Review Letters* 84:915.
44. Schwartz, G. A., Bergman, R., and Swenson, J. 2004. *Journal of Chemical Physics* 120:5736.
45. Mijovic, J., Lee, H. K., Kenny, J., and Mays, J. 2006. *Macromolecules* 39:2172.
46. Elmahdy, M. M., Chrissopoulou, K., Afratis, A., Floudas, G., and Anastasiadis, S. H. 2006. *Macromolecules* 39:5170.
47. Böhning, M., Goering, H., Fritz, A., Brzezinka, K. W., Turky, G., Schönhals, A., and Schartel, B. 2005. *Macromolecules* 38:2764.
48. Schönhals, A., Goering, H., Costa, R. F., Wagenknecht, U., and Heinrich, G. 2009. *Macromolecules* 42:4165.
49. Schönhals, A. and Kremer, F. 2002. Theory of dielectric relaxation spectra in broadband dielectric spectroscopy In *Broadband Dielectric Spectroscopy*, ed. F. Kremer and A. Schönhals, p. 1. Berlin, Germany: Springer.
50. McCrum, N. G., Read, B. E., and Williams, G. 1967. *Anelastic and Dielectric Effects in Polymeric Solids*. New York: Wiley (Reprinted by Dover 1991).
51. van den Berg, O., Sengers, W. G. F., Jager, W. F., Picken, S. J., and Wübbenhorst, M. 2004. *Macromolecules* 37:2460.
52. van den Berg, O., Wübbenhorst, M., Picken, S. J., and Jager, W. F. 2005. *Journal of Non-Crystalline Solids* 351:2694.
53. Schönhals, A. and Kremer, F. 2002. Analysis of dielectric spectra. In *Broadband Dielectric Spectroscopy*, ed. F. Kremer and A. Schönhals, p. 59. Berlin, Germany: Springer.
54. Lorthioir, C., Lauprêtre, F., Soulestin, J., and Lefebvre, J. M. 2009. *Macromolecules* 42:218.
55. Vogel, H. 1921. *Physikalische Zeitschrift* 22:645; Fulcher, G. S. 1925. *Journal of the American Ceramic Society* 8:339; Tammann, G. and Hesse, W. 1926. *Zeitschrift fur Anorganische und Allgemeine Chemie* 156:245.
56. Donth, E. 1982. *Journal of Non-Crystalline Solids* 53:325.
57. Cangialosi, D., Alegria, A., and Colmenero, J. 2007. *Physical Reviews E* 76:011514.
58. Berthier, L., Biroli, G., Bouchaud, J. P., Cipelletti, L., El Masri, D., L'Hote, D., Ladieu, F., and Pierno, M. 2005. *Science* 310:1797.

9

Microstructure and Properties of Polypropylene/ Carbon Nanotube Nanocomposites

Suat Hong Goh

CONTENTS

9.1 Introduction

Among the many potential applications of carbon nanotubes (CNTs), their use as reinforcing fillers for the fabrication of polymer nanocomposites has received considerable attention [1–4]. Both single-walled and multiwalled carbon nanotubes (SWCNTs and MWCNTs, respectively) are noted for their outstanding thermal, electrical, and mechanical properties. Polypropylene (PP) is a widely used thermoplastic because of its low cost, good processability, and well-balanced physical and mechanical properties. Products of PP take the forms of fibers, films, and molded articles. This chapter highlights the microstructure and properties of PP/CNT nanocomposites. Since most studies dealt with isotactic polypropylene, the term "PP" in this chapter refers to isotactic polypropylene unless otherwise stated.

9.2 Fabrication of Nanocomposites

Three methods are generally used to fabricate PP/CNT nanocomposites: solution mixing, in situ polymerization, and melt mixing. For solution mixing, CNTs are suspended in a solvent by mechanical stirring or ultrasonication. The suspension is then mixed with polymer solution. The nanocomposites are obtained by evaporating off the solvent or by precipitation in a non-solvent. In view of the poor solubility of PP in common organic solvents, the use of high-boiling solvents at elevated temperatures is needed, for example, xylene at 120°C [5], tetrahydronaphthalene at 140°C [6], and decalin at 70°C [7]. The need to remove solvents makes this method not suitable for large-scale production of PP/CNT nanocomposites.

Funck and Kaminsky [8] reported the polymerization of propylene with a metallocene/methylaluminoxane (MAO) catalyst and in situ coating. The hydroxyl and carboxyl groups of acidified MWCNTs reacted with MAO to form a heterogeneous catalyst. The MAO anchored on MWCNTs surface was still able to form an active complex with the metallocene, enabling polymer chains to grow directly from the surface. Koval'chuk et al. [9,10] fabricated PP/MWCNT nanocomposites by in situ polymerization using zirconocenes activated by MAO in liquid propylene medium. The use of liquid propylene medium led to high yield and also high molecular weight of PP. Isotactic and syndiotactic PP were obtained by using isospecific C_2- and C_S-symmetry metallocene catalysts, respectively.

Melt mixing offers a simple and convenient means to fabricate PP/CNT nanocomposites using common processing equipment such as extruder, internal mixer, and injection molding machine. This method is suitable for large-scale productions of nanocomposites. Highly oriented PP/CNT nanocomposites can be produced using a dynamic packing injection molding (DPIM) process [11,12]. Nanocomposites were first prepared using a twin-screw extruder. The pellets were then subjected to DPIM, in which the melt was forced to move repeatedly in a chamber by two pistons that moved reversibly with the same frequency. The melt was thus subjected to a repeated shear force during the solidification process, leading to the formation of highly oriented PP chains and CNTs.

To facilitate the dispersion of CNTs in PP matrix, PP grafted with a small amount of maleic anhydride (PP-g-MA) is commonly added to serve as a compatibilizer [12–18]. Functionalized CNTs [15,18,19] and PP-grafted MWCNTs [20,21] have also been used to achieve a more uniform dispersion of nanotubes. Li et al. [20] allowed MA to react with hydroxyl groups of acidified CNTs in ethyl acetate under constant sonication. MA-grafted CNTs were dispersed in hot xylene, followed by the addition of benzoyl peroxide. The reaction between PP and MA-grafted CNTs led to the attachment of PP chains onto CNTs. Yang et al. [21] used a reactive blending method to graft PP onto MWCNTs. PP-g-MA was melt blended with amine-functionalized MWCNTs. The reaction between anhydride and amine groups led to the grafting of PP onto MWCNTs.

9.3 Microstructure of Nanocomposites

PP is a semicrystalline polymer; it exhibits three crystalline structures: α (monoclinic), β (hexagonal), and γ (triclinic). PP adopts the α form when prepared under usual industrial and laboratory conditions. The β form can be induced using a β-nucleating agent such as calcium suberate [22] or under a certain processing condition [6,23]. The γ form of PP is rarely observed when crystallized under normal processing conditions; it can be formed under a very slow cooling [24] or under a high pressure [25]. The crystallization/melting behavior and microstructure of PP in nanocomposites have been studied mainly using differential scanning calorimetry (DSC) and wide-angle x-ray diffraction (WAXD).

It is clear that CNTs increase the crystallization temperature (T_c) of PP. In other words, CNTs enable PP to crystallize at a higher temperature upon cooling from its melt, demonstrating the nucleation effect of CNTs. The nucleating effect of CNTs is also evidenced from polarized optical microscopy [5,14,15,26–29]. The incorporation of CNTs leads to the formation of a larger number but smaller spherulites, and the spherulites become more irregular with increasing amount of CNTs added. SWCNTs are more effective nucleating agents than MWCNTs as they produce larger increases in the T_c value [30]. On the other hand, the effect of CNTs on the melting temperature (T_m) is less clear. The T_m of PP has been reported to increase [21,28,31], decrease [13,14], or remain virtually unchanged [18,23,32] upon the addition of CNTs. The conflicting results are likely due to the fact that the reported T_m is not the equilibrium melting temperature (T_m°). T_m is dependent on lamellar thickness, and T_m° represents the melting temperature of crystals with infinite lamellar thickness. Several studies [29,33] reported the measurements of T_m° of PP in nanocomposites using the Hoffman–Weeks method. Those studies showed that the T_m° of PP decreased upon the addition of CNTs, indicating that the PP crystals in the nanocomposites were less perfect than those in pure PP. Causin et al. [33] found that the T_m° of PP decreased from 214°C to 197°C–199°C for PP/PP-*g*-MWCNT nanocomposites and to 191°C–193°C for PP/MWCNT nanocomposites. This is indicative of the less disrupting role of PP-*g*-MWCNTs, with respect to pristine MWCNTs, on the crystallization of PP. On the other hand, Zhou et al. [29] found that silane-functionalized MWCNTs produced a larger depression of T_m° than pristine MWCNTs. Unlike PP-*g*-MWCNTs on which the nanotubes are covered by PP, the silane moieties are distinctly different from PP and thus produce less perfect crystals.

PP and its nanocomposites exhibit double-melting behavior when crystallized under some specific conditions [5,7,23,29,30,34]. Grady et al. [7] attributed the lower melting peak to the β form and the higher melting peak to the α form. However, they did not provide WAXD results to support their claim. Other researchers found that for samples showing double-melting peaks, their WAXD patterns did not show the presence of the β form. Therefore, the double-melting phenomenon arises from a melting–recrystallization–melting mechanism. The lower melting peak is the melting of crystals formed during the prior non-isothermal crystallization process, and the higher melting peak is the melting of recrystallized crystals. Leelapornpisit et al. [23] reported that for PP prepared by slow cooling, it showed a double-melting peak, and its WAXD showed the presence of β form as evidenced by the appearance of a peak at 2θ = 16°. On the other hand, for PP/SWCNT nanocomposites prepared under the same cooling condition, only a single melting peak was observed, and WAXD did not reveal the presence of the β form. They proposed that SWCNTs acted as a nucleating agent specifically for the α form of PP and limited the β crystal formation. Yin et al. [34] recently reported an unusual finding. The PP sample exhibited a single melting peak and WAXD showed the absence of the β form. However, for a PP/MWCNTs nanocomposite containing 0.1 wt% of pristine MWCNTs, it exhibited a triple melting peak, and its WAXD pattern showed the presence of the β form. Interestingly, for PP/MWCNT nanocomposites containing 0.5 wt% or more of pristine MWCNTs, a single melting peak was observed. Their WAXD patterns did not show the presence of the β form.

In general, WAXD studies show the existence of the α form in PP and various nanocomposites. Figure 9.1 shows the WAXD patterns of PP and various PP/PP-*g*-MWCNT nanocomposites [21]. All the peaks in the WAXD patterns of PP and the nanocomposites arose from the various planes of the α form, and there was no evidence of the presence of β form (peak around 2θ = 16°). However, there was a change in the relative intensities of the 110 and the 040 peaks. A decrease in the I_{110}/I_{040} ratio suggested that PP-*g*-MWCNTs promoted the orientation along the *b* crystallographic axis. Pristine MWCNTs also produced a similar orientation effect [33,35]. Fereidoon et al. [31] found that when PP/SWCNT nanocomposites were prepared using a very slow cooling process (cooling of the melt from 180°C to room temperature over a period of 4h under a pressure of 10 MPa), the presence of the

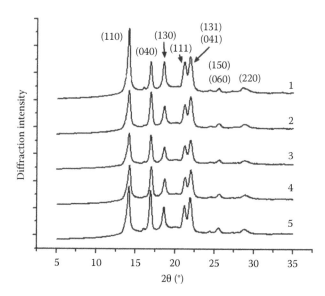

FIGURE 9.1
WAXD patterns of (1) PP and (2–5) PP/
PP-*g*-MWCNT composites with effective MWCNT contents of 0.5, 1.0, 1.5, and
2.0 wt%, respectively. (Reprinted from
Yang, B.X. et al., *Compos. Sci. Technol.*, 68,
2490, 2008.)

γ form was detected as shown by a broad peak at $2\theta = 19.8°$. Chang et al. [6] examined the WAXD patterns of PP fiber (with a draw down ratio of 110) and PP/SWCNTs fiber under various strains. For PP fiber without strain, the WAXD pattern showed the typical α form, and no changes were observed at 0.5% strain. However, when the PP fiber was stretched further (1% and 2%), WAXD patterns showed the presence of the β form as evidenced by a double peak near 16°. The same result was observed for the PP/SWCNTs fiber.

Small-angle x-ray scattering (SAXS) provides information on packing and lamellar thickness [5,33]. The scattering maxima of the SAXS intensity profile of nanocomposites were found to be less defined than that of pure PP, indicating that the lamellar packing of PP became less regular in the presence of CNTs. Furthermore, CNTs led to an increase of lamellar thickness. The lamellar thickness of PP increased from 14.3 to 17.5–18.7 nm in the presence of PP-*g*-MWCNTs [33].

In summary, CNTs serve as nucleating agents, enabling PP to crystallize at a higher temperature upon cooling from the melt. The spherulites become smaller in size and less regular with increasing CNTs content in the nanocomposite. CNTs lead to thicker but less regular lamellae. Under normal processing conditions, PP retains its α form in the presence of CNTs.

9.4 Properties of Nanocomposites

9.4.1 Flame Retardancy

CNTs improve the flame retardancy of PP [36–38]. Figure 9.2 shows the heat release rates of PP and two nanocomposites. The peak heat release rate (PHRR) of PP was reduced by 73% by the addition of 1 vol% of MWCNTs. Although all samples burned nearly completely, the two nanocomposites burned much slower than PP. Further improvement on flame retardancy was achieved by the functionalization of CNTs with intumescent flame retardant (IFR) [38]. At the same CNTs content of 1 wt%, the PHRR of PP was reduced by 68% using pristine CNTs, and by 75% using IFR-functionalized CNTs. It was suggested that CNTs

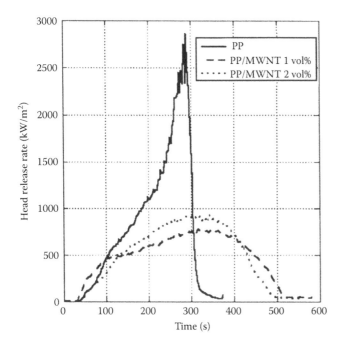

FIGURE 9.2
Heat release rates of PP and PP/MWCNT nanocomposites. (Reprinted from Kashiwagi, T. et al., *Macromol. Rapid Commun.*, 23, 761, 2002.)

formed a relatively uniform network-structured floccule layer to cover the entire sample surface. The layer reemitted much of the incident radiation back into the gas phase and reduced the transmitted flux to the PP layers below it, slowing the PP pyrolysis rate [37].

9.4.2 Thermal Stability

The thermal stability of PP in air and in nitrogen atmosphere is greatly improved by CNTs [18,28,31,39–43]. For example, the onset degradation temperature of PP in nitrogen increased from 278°C to 312°C, 320°C, and 352°C upon the addition of 1, 2, and 5 wt% of MWCNTs, respectively [40]. It was also observed that the scission of PP chains took a longer time in the presence of MWCNTs. The thermal stability of PP also depends on how well CNTs are dispersed in the matrix [44]. Figure 9.3 shows the thermogravimetric analysis curves of PP and three nanocomposites prepared by different processes: melt mixing (MM), solid-state shear pulverization (SSSP), and SSSP + MM. The dispersion improved in the order MM < SSSP < SSSP + MM. The nanocomposite with the best dispersion of CNTs exhibited the best thermal stability.

   The thermo-oxidative stability of PP/acidified MWCNT nanocomposites has been examined in detail [28]. The oxidative degradation takes place in two stages. In the first stage up to 230°C, MWCNTs accelerate the oxidation, while at higher temperatures, the trend is reversed. PP undergoes random chain scission in the first stage, and this is accelerated by the carboxylic acid groups on the MWCNTs. In the second stage, MWCNTs provide a shielding effect to hinder the removal of gases produced during decomposition and thus improve the oxidative stability of PP.

9.4.3 Electrical Properties

The electrical insulating properties of polymers are often regarded as one of their major advantages. However, there are applications that require polymers to conduct electric

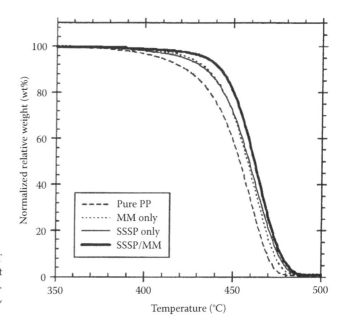

FIGURE 9.3
TGA curves of PP and PP/MWCNT nanocomposites fabricated by different methods. (Reprinted from Masuda, J. and Torkelson, J.M., *Macromolecules*, 41, 5974, 2008.)

current and to shield against electrostatic or electromagnetic fields. To impart conducting properties, fillers such as carbon black or metal fibers are often added. The electrical properties of a polymer composite experience abrupt changes when the filler content exceeds a certain quantity, the so-called percolation threshold. Below the threshold, fillers are dispersed as isolated clusters. Above the threshold, fillers tend to link together to form conductive networks [45].

As shown in Figure 9.4, the volume resistivity (resistance of a material to flow of electrons) of PP dropped from 10^9 to 10^3 Ω cm for a nanocomposite containing 2 wt% of MWCNTs [46]. Similarly, the dielectric constant and conductivity of PP/MWCNTs nanocomposite increased sharply above the percolation threshold, as shown in Figures 9.5 and 9.6 [45]. Moreover, the percolation threshold depended on how the nanocomposites were fabricated. The percolation threshold was lower for nanocomposites prepared using a Haake

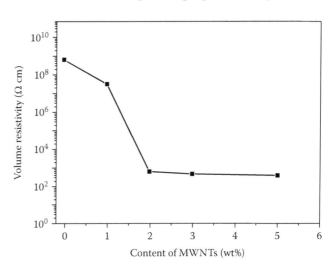

FIGURE 9.4
Electrical volume resistivity of PP and PP/MWCNT nanocomposites. (Reprinted from Seo, M.K. and Park, S.J., *Chem. Phys. Lett.*, 395, 44, 2004.)

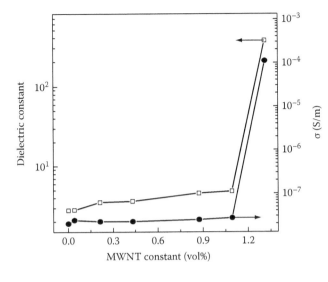

FIGURE 9.5
Dielectric constants and conductivities of PP and PP/MWCNT nanocomposites prepared using a Brabender mixer. (Reprinted from Tjong, S.C. et al., *Scr. Mater.*, 57, 461, 2007.)

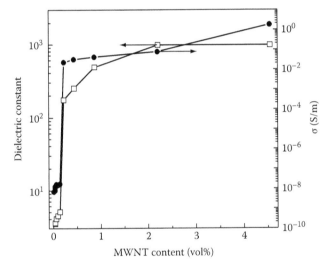

FIGURE 9.6
Dielectric constants and conductivities of PP and PP/MWCNT nanocomposites prepared using a Hakke mixer. (Reprinted from Tjong, S.C. et al., *Scr. Mater.*, 57, 461, 2007.)

mixer (rotation speed of 200 rev min^{-1}) than those prepared using a Brabender mixer (rotation speed of 60 rev min^{-1}). The high shear rate associated with the Haake mixer led to a better dispersion of MWCNTs as shown by transmission electron microscopy and optical microscopy. As a result, nanocomposites prepared using the Haake mixer exhibited a lower percolation threshold.

There were conflicting reports on the effect of compatibilizer on the electrical properties of PP nanocomposites [16,47]. Lee et al. [16] reported that the electrical conductivity of PP/MWCNTs nanocomposite increased by the addition of PP-*g*-MA. They considered that PP-*g*-MA improved the interfacial adhesion between PP and nanotubes, and generated additional electrical pathway. However, their later work [47] showed that the electrical conductivity of PP/MWCNTs nanocomposite decreased when PP-*g*-MA was added, and the conductivity remained almost constant irrespective of the PP-*g*-MA content. Moreover, the percolation threshold of PP/MWCNT nanocomposites (between 0.5 and 2 wt%) was shifted toward a higher value (between 1 and 2 wt%) when PP-*g*-MA was added [47].

MWCNTs are more effective than carbon black in improving the electrical properties of PP [47]. The addition of 5 wt% of carbon black hardly improved the electrical properties of PP whereas 2 wt% of MWCNTs increased the conductivity by 5–6 orders of magnitude. Moreover, the percolation threshold of carbon black was much higher (10–20 wt%) as compared to CNTs [48].

PP-grafted CNTs are better than pristine CNTs in improving the electrical properties of PP. The incorporation of 3 wt% of MWCNTs reduced the volume resistivity of PP from 10^{10} to $10^{8.5}$ Ω cm [49]. However, the volume resistivity of a nanocomposite containing 3 wt% of PP-grafted MWCNTs was 10^6 Ω cm [49].

9.4.4 Mechanical Properties

Table 9.1 summarizes the changes of various mechanical properties of PP brought by the incorporation of CNTs. The mechanical properties of PP used by various workers vary significantly, particularly the ultimate strain. It is therefore more meaningful to compare the percentages of improvement or reduction on the mechanical properties. A glance of Table 9.1 leads to the following observations. First, the Young's modulus (E) and the storage modulus (E') of PP are greatly improved by CNTs. Second, the improvement of tensile strength (σ) is less dramatic. Third, the ultimate strain (ε_b) of PP is often reduced, and in some cases the reduction is extremely large. Fourth, the mechanical properties are affected by the aspect ratio, alignment and dispersion of CNTs, and also the efficiency of stress transfer from the matrix to CNTs.

Masuda and Torkelson [44] prepared two series of nanocomposites using MWCNTs with different aspect ratios, one around 400 and the other around 2000. As shown in Table 9.1, MWCNTs with a larger aspect ratio were marginally better in improving the Young's modulus and tensile strength of PP. Both types of MWCNTs produced about similar reduction in ultimate strain. They also observed that improvement in mechanical properties depended on how well the nanotubes were dispersed.

The effect of alignment on the mechanical properties of PP fibers was studied by Moore et al. [50]. Two grades of PP were used, one with a low melt-flow rate (LMFR) and one with a high melt-flow rate (HMFR). The mechanical properties of the two batches of PP fibers were substantially different. The tenacity and the elongation at break of as-spun LMFR PP fibers were 1.19 g/denier and 615%, respectively, and those of HMFR PP fibers were 1.03 g/denier and 1320%, respectively. The incorporation of 1 wt% SWCNTs improved the tenacity of LMFR PP fibers to 1.55 g/denier and the elongation at break to 818%. In contrast, the tenacity and the elongation at break of HMFR PP/SWCNTs fibers were reduced to 0.22 g/denier and 11%, respectively. The incorporation of SWCNTs made the HMFR PP fibers very brittle. The alignment of fibers through post-drawing improved the tenacity. The tenacities of post-drawn LMFR and HMFR PP fibers were 9.0 and 6.88 g/denier, respectively. The stress-strain curves of post-drawn PP fibers are shown in Figures 9.7 and 9.8. As shown in Figure 9.7, the addition of 0.5 wt% or 1.0 wt% SWCNTs improved the stiffness and strength of LMFR PP fibers. However, at higher SWCNTs loadings of 1.5 and 2.0 wt%, the composite fibers were weaker than the neat fibers. SWCNTs produced detrimental effects on post-drawn HMFR PP fibers, as shown in Figure 9.8. HMFR PP fibers with higher SWCNTs contents could not be post-drawn as the fibers were too brittle. Therefore, alignment improved the stiffness and strength of fibers provided that the PP had a sufficiently LMFR and the SWCNTs content was not too high.

Zhao et al. [12] prepared PP/MWCNT nanocomposites with a highly oriented structure using DPIM. Comparison was made to similar nanocomposites prepared using a static

TABLE 9.1

Changes in Mechanical Properties of PP upon the Incorporation of CNTs

E	σ	ε_b	E'	Notes
+30%	—	−99%	—	iPP; 1.5 wt% MWNCTs [9]
+16%	−35%	−65%	—	sPP; 1.5 wt% MWCNTs [9]
+33%	+34%	+4300%	—	0.3 wt% MWCNTs; oriented sample [12]
+7%	+3%	+20%	—	0.3 wt% MWNCTs; unoriented sample [12]
+42%	+16%	−2%	—	1 wt% MWCNTs [13]
+23%	+4%	−20%	+48% at −100°C	1 wt% SWCNTs + 5 wt% PP-g-MA [14]
+30%	+4%	−14%	+80% at −100°C	1 wt% MWCNTs + 5 wt% PP-g-MA [14]
—	+10%	−5%	—	0.3 wt% MWCNTs; high orientation [15]
—	+18%	+100%	—	0.3 wt% MWCNTs; low orientation [15]
+26%	+18%	−17%	—	1 wt% MWCNTs [17]
+30%	+48%	−28%	—	1 wt% MWCNTs and 2 wt% PP-g-MA [17]
+34%	+24%	−42%	—	1 wt% MWCNTs [21]
+108%	+141%	+49%	+80% at −40°C	PP-g-MWCNTs 1.5 wt% [21]
+30%	—	—	+32% at 25°C	1 wt% MWCNTs [27]
+122%	+8%	−99%	—	2.5 wt% acidified MWCNTs [28]
+82%	—	−80%	—	1 wt% SWCNTs [31]
+16%	+3%	−95%	—	1 wt% CNTs + PP-g-MA [38]
+43%	+16%	−95%	—	1 wt% IFR-g-CNTs [38]
+270%	+400%	—	—	PP fiber + 1 wt% MWCNTs [42]
+250%	+1580%	—	—	Oriented PP fiber + 1 wt% MWCNTs [42]
+50%	−3%	−16%	—	0.93 wt% MWCNTs, small aspect ratio [44]
+57%	+3%	−15%	—	0.92 wt% MWCNTs; large aspect ratio [44]
+113%	+110%	—	—	3 wt% PP-g-MWCNTs [49]
+50%	+18%	—	—	3 wt% MWCNTs [49]
—	+30%	+33%	+27% at −50°C	As-spun PP fibers + 1 wt%
			+129% at 25°C	SWCNTs [50]
—	+44%	−2%	+42% at −50°C	Post-drawn PP fibers + 1 wt%
			+8% at 25°C	SWCNTs [50]
+27%	+10%	—	+33% at −40°C	1 wt% SWCNTs [51]
−3%	−6%	−63%	—	2.5 wt% SWCNTs [58]
+32%	+48%	−54%	—	2.5 wt% F-SWCNTs [58]

E = Young's modulus; σ = tensile strength; ε_b = ultimate strain (elongation at break); *E'* = storage modulus; reference number in brackets.

packing injection molding. Nanocomposites with a high degree of orientation exhibited more enhanced mechanical properties. The ultimate strain of the oriented sample was about 40 times better than the matrix polymer. The unusual increase in ductility was attributed to the increased mobility of both PP chains and MWCNTs because they were oriented along the tensile deformation direction, and also the bridging effect of the oriented MWCNTs on the crack development during tensile failure.

In another study, Jose et al. [42] found that the strength and modulus of PP fiber were improved from 25 MPa and 1.0 GPa, respectively, to 125 MPa and 3.7 GPa, by the addition of 1 wt% MWCNTs. When the composite fiber was subjected to a draw ratio of 4, the strength was improved further to 420 MPa, but the modulus dropped slightly to 3.5 GPa. The modulus of drawn PP fibers increased by three times when 1 wt% of SWCNTs was added [6].

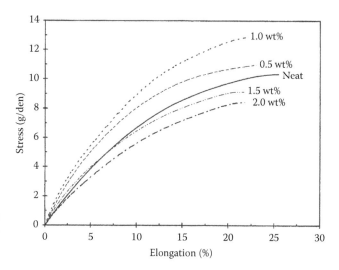

FIGURE 9.7
Stress–strain curves for post-drawn LMFR PP/SWCNT fibers. (Reprinted from Moore, E.M. et al., *J. Appl. Polym. Sci.*, 93, 2926, 2004.)

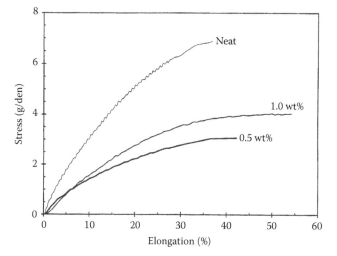

FIGURE 9.8
Stress–strain curves for post-drawn HMFR PP/SWCNT fibers. (Reprinted from Moore, E.M. et al., *J. Appl. Polym. Sci.*, 93, 2926, 2004.)

It is recognized that a uniform dispersion of CNTs and an efficient stress transfer from matrix to CNTs are essential to achieve effective mechanical enhancement. PP-*g*-MA is often added as a compatibilizer to improve the dispersion of CNTs in PP matrix. Prashantha et al. [17] found that the addition of PP-*g*-MA enabled MWCNTs to be more uniformly dispersed in PP, resulting in better improvements in tensile modulus and strength. The modulus and strength of PP were improved by 26% and 18%, respectively, by the addition of 1 wt% of MWCNTs. In comparison, the improvements were 30% and 48% when additional 2 wt% of PP-*g*-MA was added [17]. Lee et al. [14] suggested that there were hydrogen-bonding interactions between hydroxyl groups of acidified MWCNTs and MA groups of PP-*g*-MA. It was assumed that the hydrogen-bonding interactions led to the wrapping of MWCNTs by PP-*g*-MA to achieve a stronger adhesion between the matrix and the nanotubes.

Lopez-Manchado et al. [51] reported that SWCNTs were more effective fillers than carbon black. The Young's modulus and storage modulus at −40°C of PP were both improved by 13% upon the addition of 1 wt% of carbon black. In comparison, the Young's modulus and storage modulus were improved by 27% and 33%, respectively, by 1 wt% of SWCNTs. They also observed decreases in Young's modulus, tensile strength, and storage modulus when the SWCNTs content was increased from 0.75 to 1 wt%.

Lee et al. [14] found that improvements in tensile properties of PP fibers brought by SWCNTs and MWCNTs were about the same. PP-*g*-MA was added to help disperse the nanotubes. At a nanotube content of 1 wt%, the Young's modulus and tensile strength of PP fiber were increased by 23% and 4%, respectively, by SWCNTs, and by 30% and 11%, respectively, by MWCNTs. However, the storage modulus of PP fibers was improved more significantly by MWCNTs (about 80%) than by SWCNTs (about 48%) at −100°C.

Recent studies [21,49,52–57] have shown that polymer-grafted CNTs are highly effective in enhancing the mechanical properties of polymers. The polymer chains grafted onto CNTs wrap around the nanotubes, and the tendency of the nanotubes to aggregate is reduced. If the polymer grafted onto CNTs is the same or is miscible with the matrix polymer, the stress can be efficiently transferred from the matrix to CNTs through the grafted polymer chains. Yang et al. [21] studied the enhancement of the mechanical properties of PP using PP-*g*-MWCNTs. Figure 9.9 shows the stress–strain curves of PP and several nanocomposites. The addition of pristine MWCNTs increased the Young's modulus and tensile strength of PP, but the ultimate strain was significantly reduced (curve 6). PP-*g*-MWCNTs produced significant increases in Young's modulus, tensile strength, and ultimate strain of PP (curves 2–5). When the effective MWCNTs content was 1.5 wt%, the addition of PP-*g*-MWCNTs improved the Young's modulus, tensile strength, and ultimate strain by 108%, 141%, and 49%, respectively, as compared to PP. However, there was a downturn in the mechanical properties when the effective MWCNTs content was further increased to 2.0 wt%. The storage modulus of nanocomposite also showed a similar downturn at an effective MWCNTs content of 2.0 wt%, as shown in Figure 9.10. The downturn in mechanical properties is associated with increasing difficulty in dispersing a larger amount of PP-*g*-MWCNTs in the PP matrix. Figure 9.11 shows the scanning electron microscopy (SEM) micrographs of fracture surfaces after tensile testing of several nanocomposites. For the PP/PP-*g*-MWCNTs nanocomposite with an effective MWCNTs content of 1.5 wt%, the nanotubes were dispersed individually in the matrix (Figure 9.11a). Moreover, the nanotubes were broken instead of being pulled out, indicating strong interfacial adhesion. In

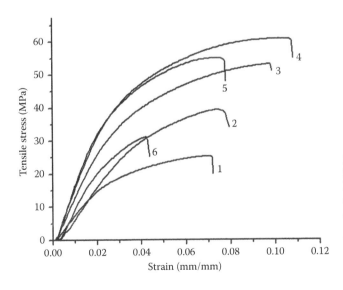

FIGURE 9.9
Stress–strain curves of PP (curve 1), PP/MWCNT nanocomposite containing 1.0 wt% pristine MWCNTs (curve 6), and PP/PP-*g*-MCWNT nanocomposites (curves 2–5; with effective MWCNT contents of 0.5, 1.0, 1.5 and 2.0 wt%, respectively). (Reprinted from Yang, B.X. et al., *Compos. Sci. Technol.*, 68, 2490, 2008.)

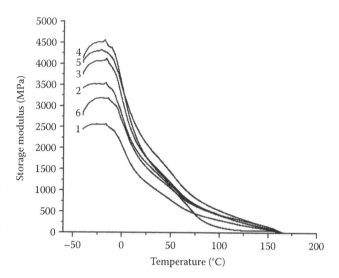

FIGURE 9.10
Storage modulus-temperature curves of PP (curve 1), PP/MWCNT nanocomposite containing 1.0 wt% pristine MWCNTs (curve 6), and PP/PP-*g*-MWCNT nanocomposites (curves 2–5; with effective MWCNT contents of 0.5, 1.0, 1.5, and 2.0 wt%, respectively). (Reprinted from Yang, B.X. et al., *Compos. Sci. Technol.*, 68, 2490, 2008.)

comparison, big bundles of MWCNTs were present in the PP/MWCNTs nanocomposite containing 1.0 wt% of pristine MWCNTs (Figure 9.11c), and the fracture surface showed a large number of unbroken nanotubes. For the PP/PP-*g*-MWCNTs nanocomposite with an effective MWCNTs content of 2.0 wt%, in addition to individually dispersed nanotubes, small bundles of nanotubes could be seen (Figure 9.11b). Therefore, the grafting of PP onto MWCNTs helps disperse the nanotubes in the matrix, but it is still difficult to completely disperse PP-*g*-MWCNTs at higher PP-*g*-MWCNTs contents. Similar downturns in mechanical properties at higher CNTs contents were also observed for other polymer/polymer-grafted CNT nanocomposites [52–55,57]. As mentioned earlier, the addition of PP-*g*-MWCNTs increased the lamellar thickness. Causin et al. [33] recently found that the changes in mechanical properties of PP/PP-*g*-MWCNT nanocomposites were closely related to the lamellar thickness, as shown in Figure 9.12. It is of interest to note that there was also a decrease in lamellar thickness when the effective MWCNTs content was increased from 1.5 to 2.0 wt%. Therefore, in addition to dispersion and interfacial adhesion, lamellar morphology can also play an important role in determining the mechanical properties of semicrystalline polymer/CNT nanocomposites.

Another recent study also showed that significant improvements in mechanical properties of PP were achieved using PP-*g*-MWCNTs [49]. The tensile strength and the Young's modulus of PP were improved by 18% and 50%, respectively, by the addition of 3 wt% of pristine MWCNTs. In comparison, the addition of 3 wt% of PP-*g*-MWCNTs led to increases in Young's modulus and tensile strength by 113% and 110%, respectively [49].

McIntosh et al. [58] found that fluorinated SWCNTs (F-SWCNTs) were more effective than SWCNTs in enhancing the mechanical properties of PP. The Young's modulus, tensile strength, and ultimate strain of PP were reduced by 3%, 6%, and 63%, respectively, by the incorporation of 2.5 wt% SWCNTs. However, the Young's modulus and the tensile strength of PP were improved by 32% and 48%, respectively, by the addition of 2.5 wt% of F-SWCNTs, although the ultimate strain was reduced by 54%. They observed that there was partial defluorination of F-SWCNTs when they were melt mixed with PP. They suggested that defluorination arose from in situ covalent bonding between the nanotubes and PP during melt mixing, leading to the grafting of PP onto SWCNTs.

The impact behavior of PP is affected by CNTs. Seo et al. [32] found that the notched Izod impact strength of PP was improved by about 20% upon the addition of 1 wt% of

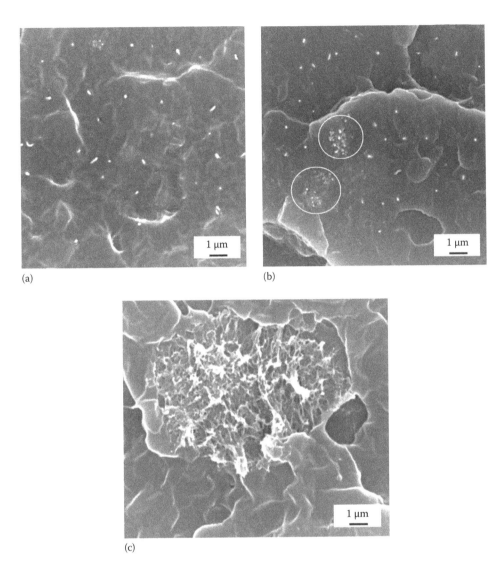

FIGURE 9.11
SEM micrographs of nanocomposites. (a) PP/PP-*g*-MWCNTs nanocomposite with an effective MWCNTs content of 1.5 wt%, (b) PP/PP-*g*-MWCNT nanocomposite with an effective MWCNTs content of 2.0 wt%, and (c) nanocomposite containing 1.0 wt% pristine MWCNTs. (Reprinted from Yang, B.X. et al., *Compos. Sci. Technol.*, 68, 2490, 2008.)

MWCNTs. Zhang and Zhang [26] made a detailed study on the impact behavior of PP/MWCNT nanocomposites. The notched Charpy impact strength of the nanocomposite depended strongly on testing temperatures. Toughening effect was not observed at temperatures below the glass transition temperature of PP. In contrast, the impact strength of PP was greatly enhanced at high temperatures, and longer nanotubes produced a slightly better toughening effect. For example, at 80°C, the addition of 1 vol% of short (1–2 μm) and long (5–15 μm) MWCNTs increased the impact strength of PP by 156% and 167%, respectively. They found that the smaller spherulite size induced by nanotubes was beneficial to impact resistance. Large spherulites resulted in higher level of stress concentration

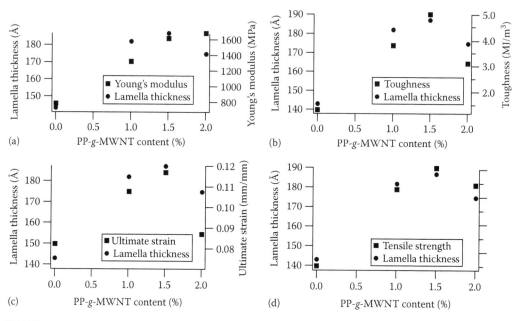

FIGURE 9.12
Variation of lamellar thickness and mechanical property as a function of PP-*g*-MWCNTs content. (a) Young's modulus, (b) toughness, (c) ultimate strain, and (d) tensile strength. (Reprinted from Causin, V. et al., *Eur. Polym. J.*, 45, 2155, 2009.)

and cracks propagated easily at the interfaces between larger spherulites. Prashantha et al. [17] also found that the notched Charpy impact strength of PP was improved by MWCNTs, and the addition of PP-*g*-MA further improved the impact strength. At a nanotube's content of 1 wt%, the impact strength was improved by about 30%. However, the improvement was about 50% when additional 2 wt% of PP-*g*-MA was added. However, they found that the un-notched Charpy impact strength of PP was reduced by the incorporation of MWCNTs [17,59]. Prashantha et al. [59] suggested that nanotubes limited the crack propagation, but eased the crack initiation. Better dispersion of nanotubes could avoid the crack initiation, and thus providing high strength and ductility to the nanocomposites [59]. Improvement in impact strength can also be achieved through orientation of PP and MWCNTs [11]. The notched Izod impact strength of PP fibers prepared by a static method was 3.9 kJ/m², and its impact strength was hardly improved by the addition of MWCNTs. For PP fibers prepared by a dynamic packing injection molding (DPIM) process, the notched Izod impact strength was 17 kJ/m². For a PP/MWCNTs nanocomposite containing 0.6 wt% of MWCNTs prepared by DPIM, the impact strength was improved further to 26 kJ/m² [11].

The area under the stress–strain curve of a polymer is the energy required to break the polymer, and hence it is sometimes used as a measurement of toughness. In this respect, the toughness of PP is often reduced significantly by CNTs due to the markedly reduced ultimate strain. For example, the toughness of PP was lowered by 36% upon the addition of 1 wt% pristine MWCNTs [21]. However, the toughness of PP was increased by 192% when 1 wt% of PP-*g*-MWCNTs was added [21]. Similar to other mechanical properties, the toughness of PP/PP-*g*-MWCNTs nanocomposites also varies in the same manner as the lamellar thickness, as shown in Figure 9.12b [33].

9.5 Conclusions

The thermal, electrical, and mechanical properties of PP can be significantly improved by the incorporation of CNTs. The main challenges in the development of high-performance PP/CNT nanocomposites are to disperse the nanotubes in the PP matrix homogeneously and to increase the interfacial adhesion between CNTs and PP. PP-grafted CNTs are particularly useful to enhance the mechanical properties of PP.

References

1. Moniruzzaman, M. and Winey, K. I. 2006. Polymer nanocomposites containing carbon nanotubes. *Macromolecules* 39:5194–5205.
2. Coleman, J. N., Khan, U., Blau, W. J., and Gun'ko, Y. K. 2006. Small but strong: A review of the mechanical properties of carbon nanotube-polymer composites. *Carbon* 44:1624–1652.
3. Coleman, J. N., Khan, U., and Gun'ko, Y. K. 2006. Mechanical reinforcement of polymers using carbon nanotubes. *Advanced Materials* 18:689–706.
4. Du, J. H. and Cheng, H. M. 2007. The present status and key problems of carbon nanotube based composites. *Express Polymer Letters* 1:253–273.
5. Avila-Orta, C. A., Medellin-Rodriguez, J., Davila-Rodriguez, M. V., Aguirre-Figueroa, Y. A., Yoon, K., and Hsiao, B. S. 2007. Morphological features and melting behavior of nanocomposites based on isotactic polypropylene and multiwalled carbon nanotubes. *Journal of Applied Polymer Science* 106:2640–2647.
6. Chang, T. E., Jensen, L. R., Kisliuk, A., Pipes, R. B., Pyrz, R., and Sokolov, A. P. 2005. Microscopic mechanism of reinforcement in single-wall carbon nanotube/polypropylene nanocomposites. *Polymer* 46:439–444.
7. Grady, B. P., Pompeo, F., Shambaugh, R. L., and Resasco, D. E. 2002. Nucleation of polypropylene crystallization by single-walled carbon nanotubes. *Journal of Physical Chemistry B* 106:5852–5858.
8. Funck, A. and Kaminsky, W. 2007. Polypropylene carbon nanotube composites by in situ polymerization. *Composites Science and Technology* 67:906–915.
9. Koval'chuk, A. A., Shevchenko, V. G., Shchegolikhin, A. N., Nedorezova, P. M., Klyamkina, A. N., and Aladyshev, A. M. 2008. Effect of carbon nanotube functionalization on the structural and mechanical properties of polypropylene/MWCNT composites. *Macromolecules* 41:7536–7542.
10. Koval'chuk, A. A., Shchegolikhin, A. N., Shevchenko, V. G., Nedorezova, P. M., Klyamkina, A. N., and Aladyshev, A. M. 2008. Synthesis and properties of polypropylene/multiwall carbon nanotube composites. *Macromolecules* 41:3149–3156.
11. Xiao, Y., Zhang, X., Cao, W., Wang, K., Tan, H., Zhang, Q. et al. 2007. Dispersion and mechanical properties of polypropylene/multiwall carbon nanotubes composites obtained via dynamic packing injection molding. *Journal of Applied Polymer Science* 104:1880–1886.
12. Zhao, P., Wang, K., Yang, H., Zhang, Q., Du, R., and Fu, Q. 2007. Excellent tensile ductility in highly oriented injection-molded bars of polypropylene/carbon nanotubes composites. *Polymer* 48:5688–5695.
13. Rahmatpour, A. and Aalaie, J. 2008. Steady shear rheological behavior, mechanical properties, and morphology of the polypropylene/carbon nanotube nanocomposites. *Journal of Macromolecular Science, Part B: Physics* 47:929–941.
14. Lee, G. W., Janannathan, S., Chae, H. G., Minus, M. L., and Kumar, S. 2008. Carbon nanotube dispersion and exfoliation in polypropylene and structure and properties of the resulting composites. *Polymer* 49:1831–1840.

15. Hou, Z., Wang, K., Zhao, P., Zhang, Q., Yang, C., Chen, D. et al. 2008. Structural orientation and tensile behavior in the extrusion-stretched sheets of polypropylene/multi-walled carbon nanotubes composite. *Polymer* 49:3582–3589.

16. Lee, S. H., Kim, M. W., Kim, S. H., and Youn, J. R. 2008. Rheological and electrical properties of polypropylene/MWCNT composites prepared with MWCNT masterbatch chips. *European Polymer Journal* 44:1620–1630.

17. Prashantha, K., Soulestin, J., Lacrampe, M. F., Claes, M., Dupin, G., and Krawczak, P. 2008. Multi-walled carbon nanotube filled polypropylene nanocomposites based on masterbatch route: Improvement of dispersion and mechanical properties through PP-*g*-MA addition. *Express Polymer Letters* 2:735–745.

18. Jin, S. H., Kang, C. H., Yoon, K. H., Bang, D. S., and Park, Y. B. 2009. Effect of compatibilizer on morphology, thermal, and rheological properties of polypropylene/functinalized mult-walled carbon nanotubes composite. *Journal of Applied Polymer Science* 111:1028–1033.

19. Zhou, Z., Wang, S., Lu, L., Zhang, Y., and Zhang, Y. 2008. Functionalization of multi-wall carbon nanotubes with silane and its reinforcement on polypropylene composites. *Composites Science and Technology* 68:1727–1733.

20. Li, W. H., Chen, X. H., Li, S. N., Xu, L. S., and Yang, Z. 2007. Synthesis of polypropylene wrapped carbon nanotubes composite via in situ graft method with maleic anhydride. *Materials Science and Technology* 23:1181–1185.

21. Yang, B. X., Shi, J. H., Pramoda, K. P., and Goh, S. H. 2008. Enhancement of the mechanical properties of polypropylene using polypropylene-grafted multiwalled carbon nanotubes. *Composites Science and Technology* 68:2490–2497.

22. Juhasz, P., Varga, J., Belina, K., and Marand, H. 2003. Determination of the equilibrium melting point of the β–form of polypropylene. *Journal of Thermal Analysis and Calorimetry* 69:561–574.

23. Leelapornpisit, W., Ton-That, M. T., Perrin-Sarazin, F., Cole, K. C., Denault, J., and Sigmard, B. 2005. Effect of carbon nanotubes on the crystallization and properties of polypropylene. *Journal of Polymer Science, Part B: Polymer Physics* 43:2445–2453.

24. Bruckner, S. and Meille, S. V. 1989. Non-parallel chains in crystalline γ-isotactic polypropylene. *Nature* 340:455–457.

25. Kardos, J. L., Christiansen, A. W., and Baer, E. 1966. Structure of pressure crystallized polypropylene. *Journal of Polymer Science, Part B: Polymer Physics* 4:777–778.

26. Zhang, H. and Zhang, Z. 2007. Impact behaviour of polypropylene filled with multi-walled carbon nanotubes. *European Polymer Journal* 43:3197–3207.

27. Bao, S. P. and Tjong, S. C. 2008. Mechanical behaviors of polypropylene/carbon nanotube nanocomposites: The effects of loading rate and temperature. *Materials Science and Engineering A* 485:508–516.

28. Bikiaris, D., Vassiliou, A., Chrissafis, K., Paraskevopoulos, K. M., Jannakoudakis, J., and Docoslis, A. 2008. Effect of acid treated multi-walled carbon nanotubes on the mechanical, permeability, thermal properties and thermo-oxidative stability of isotactic polypropylene. *Polymer Degradation and Stability* 93:952–967.

29. Zhou, Z., Wang, S., Lu, L., Zhang, Y., and Zhang, Y. 2007. Isothermal crystallization kinetics of polypropylene with silane functionalized multi-walled carbon nanotubes. *Journal of Polymer Science, Part B: Polymer Physics* 45:1616–1624.

30. Miltner, H. E., Grossiord, N., Lu, K., Loos, J., Koning, C. E., and Van Mele, B. 2008. Isotactic polypropylene/carbon nanotube composites prepared by latex technology. Thermal analysis of carbon nanotube-induced nucleation. *Macromolecules* 41:5753–5762.

31. Fereidoon, A., Ahangari, M. G., and Saedodin, S. 2008. Thermal and structural behaviors of polypropylene nanocomposites reinforced with single-walled carbon nanotubes by melt processing method. *Journal of Macromolecular Science, Part B: Physics* 48:196–211.

32. Seo, M. K., Lee, J. R., and Park, S. J. 2005. Crystallization kinetics and interfacial behaviors of polypropylene composites reinforced with multi-walled carbon nanotubes. *Materials Science and Engineering A* 404:79–84.

33. Causin, V., Yang, B. X., Marega, C., Goh, S. H., and Marigo, A. 2009. Nucleation, structure and lamellar morphology of isotactic polypropylene filled with polypropylene-grafted multiwalled carbon nanotubes. *European Polymer Journal* 45:2155–2163.
34. Yin, C. L., Liu, Z. Y., Yang, W., Yang, M. B., and Feng, J. M. 2009. Crystallization and morphology of iPP/MWCNT prepared by compounding iPP melt with MWCNT aqueous suspension. *Colloid and Polymer Science* 287:615–620.
35. Xia, H., Wang, Q., Li, K., and Hu, G. H. 2004. Preparation of polypropylene/carbon nanotube composite powder with a solid-state mechanochemical pulverization process. *Journal of Applied Polymer Science* 93:378–386.
36. Kashiwagi, T., Grulke, E., Hilding, J., Harris, R., Awad, W., and Douglas, J. 2002. Thermal degradation and flammability properties of poly(propylene)/carbon nanotube composites. *Macromolecular Rapid Communications* 23:761–765.
37. Kashiwagi, T., Grulke, E., Hilding, J., Groth, K., Harris, R., Butler, K. et al. 2004. Thermal and flammability properties of polypropylene/carbon nanotube nanocomposites. *Polymer* 45:4227–4239.
38. Song, P., Xu, L., Guo, Z., Zhang, Y., and Fang, Z. 2008. Flame-retardant-wrapped carbon nanotubes for simultaneously improving the flame retardancy and mechanical properties of polypropylene. *Journal of Materials Chemistry* 18:5083–5091.
39. Sarno, M., Gorrasi, G., Sannino, D., Sorrentino, A., Ciambelli, P., and Vittoria, V. 2004. Polymorphism and thermal behaviour of syndiotactic poly(propylene)/carbon nanotube composites. *Macromolecular Rapid Communications* 25:1963–1967.
40. Seo, M. K. and Park, S. 2004. A kinetic study on the thermal degradation of multi-walled carbon nanotubes-reinforced poly(propylene) composites. *Macromolecular Materials and Engineering* 289:368–374.
41. Yang, J., Lin, Y., Wang, J., Lai, M., Li, J. Liu, J. et al. 2005. Morphology, thermal stability, and dynamic mechanical properties of atactic polypropylene/carbon nanotube composites. *Journal of Applied Polymer Science* 98:1087–1091.
42. Jose, M. V., Dean, D., Tyner, J., Price, G., and Nyairo, E. 2007. Polypropylene/carbon nanotube nanocomposite fibers: Process-morphology-property relationships. *Journal of Applied Polymer Science* 103:3844–3850.
43. Gorrasi, G., Romeo, V., Sannino, D., Sarno, M., Ciambelli, P., Vittoria, V. et al. 2007. Carbon nanotube induced structural and physical property transitions of syndiotactic polypropylene. *Nanotechnology* 18:275703.
44. Masuda, J. and Torkelson, J. M. 2008. Dispersion and major property enhancements in polymer/multiwall carbon nanotube nanocomposites via solid-state shear pulverization. *Macromolecules* 41:5974–5977.
45. Tjong, S. C., Liang, G. D., and Bao, S. P. 2007. Electrical behavior of polypropylene/multi-walled carbon nanotube nanocomposites with low percolation threshold. *Scripta Materialia* 57:461–464.
46. Seo, M. K. and Park, S. J. 2004. Electrical resistivity and rheological behaviors of carbon nanotubes-filled polypropylene composites. *Chemical Physics Letters* 395:44–48.
47. Lee, S. H., Cho, E., Jeon, S. H., and Youn, J. R. 2007. Rheological and electrical properties of polypropylene composites containing functionalized multi-walled carbon nanotubes and compatibilizers. *Carbon* 45:2810–2822.
48. Yui, H., Wu, G., Sano, H., Sumita, M., and Kino, K. 2006. Morphology and electrical conductivity of injection-molded polypropylene/carbon black composites with addition of high-density polyethylene. *Polymer* 47:3599–3608.
49. Li, W. H., Chen, X. H., Yang, Z., and Xu, L. S. 2009. Structure and properties of polypropylene-wrapped carbon nanotubes composites. *Journal of Applied Polymer Science* 113:3809–3814.
50. Moore, E. M., Ortiz, D. L., Marla, V. T., Shambaugh, R. L., and Grady, B. P. 2004. Enhancing the strength of polypropylene fibers with carbon nanotubes. *Journal of Applied Polymer Science* 93:2926–2933.

51. Lopez-Manchado, M. A., Valentini, L., Biagiotti, J., and Kenny, J. M. 2005. Thermal and mechanical properties of single-walled carbon nanotubes-polypropylene composites prepared by melt processing. *Carbon* 43:1499–1505.
52. Yang, B. X., Shi, J. H., Pramoda, K. P., and Goh, S. H. 2007. Enhancement of stiffness, strength, ductility and toughness of poly(ethylene oxide) using phenoxy-grafted multiwalled carbon nanotubes. *Nanotechnology* 18:125606.
53. Shi, J. H., Yang, B. X., Pramoda, K. P., and Goh, S. H. 2007. Enhancement of the mechanical performance of poly(vinyl chloride) using poly(n-butyl methacrylate)-grafted multiwalled carbon nanotubes. *Nanotechnology* 18:375704.
54. Yang, B. X., Pramoda, K. P., Xu, G. Q., and Goh, S. H. 2007. Mechanical reinforcement of polyethylene using polyethylene-grafted multiwalled carbon nanotubes. *Advanced Functional Materials* 17:2062–2069.
55. Yang, B. X., Shi, J. H., Li, X., Pramoda, K. P., and Goh, S. H. 2009. Mechanical reinforcement of poly(1-butene) using polypropylene-grafted multiwalled carbon nanotubes. *Journal of Applied Polymer Science* 113:1165–1172.
56. Hwang, G. L., Shieh, Y. T., and Hwang, K. C. 2004. Efficient load transfer to polymer-grafted multiwalled carbon nanotubes. *Advanced Functional Materials* 14:487–491.
57. Blond, D., Barron, V., Reuther, M., Ryan, K. P., Nicolosi, V., Blau, W. J. et al. 2006. Enhancement of modulus, strength and toughness in poly(methyl acrylate)-based composites by the incorporation of poly(methyl methacrylate)-functionalized nanotubes. *Advanced Functional Materials* 16:1608–1614.
58. McIntosh, D., Khabashesku, V. N., and Barrera, E. V. 2006. Nanocomposite fiber systems processed from fluorinated single-walled nanotubes and a polypropylene matrix. *Chemistry of Materials* 18:4561–4569.
59. Prashantha, K., Soulestin, J., Lacrampe, M. F., Krawczak, P., Dupin, G., and Claes, M. 2009. Masterbatch-based multi-walled carbon nanotube filled polypropylene nanocomposites: Assessment of rheological and mechanical properties. *Composites Science and Technology* 69:1756–1763.

10

Polypropylene Nanocomposites with Clay Treated with Thermally Stable Imidazolium Modification*

Vikas Mittal

CONTENTS

10.1 Introduction

Polyolefin nanocomposites are largely synthesized by using melt intercalation method of nanocomposite synthesis where the polymer melt at high temperature is compounded with the organically modified inorganic filler under the action of shear. The compounding temperatures as well as mixing time are two parameters that generally influence the mixing efficiency. To achieve optimal mixing, high compounding temperature as well as long compounding protocols may be required, which, however, can lead to the thermal degradation of the organic modification of the filler. As an example, polypropylene is a material of choice for a wide variety of applications. It is also conventionally incorporated with inorganic fillers to further enhance its properties [1–3] and subsequently its applications. However, the high melt temperatures required along with the high mechanical shear used in the melt intercalation and processing operations poses a concern for the thermal stability of the ammonium modification. Temperatures more than 200°C are generally employed for such processes, which are almost equal or higher than the initial onset of degradation of the alkyl ammonium groups of the surface modifications. It has been observed that the conventionally used alkyl ammonium cations have an onset of

---

* The work was carried out at Institute of Chemical and Bioengineering, Department of Chemistry and Applied Biosciences, ETH Zurich, Zurich, Switzerland.

degradation as low as 180°C by TGA studies [4]. The thermal degradation was observed to follow the Hoffmann degradation path involving the early breakage of weaker C–N bond in the ammonium modification. As the rupture of the C–N bonds would knock off the whole alkyl chains bound to the filer surface, thus, breakage of even a small number of such bonds may be enough to significantly change the thermodynamics of the system. This changes the interfacial interactions between the organic and inorganic phases of the composite because of the changes in the structure of the surfactant. The production of low-molecular-weight species during such degradation reactions can consequently affect the physical and mechanical properties of the polymer in the composite like viscosity, molecular weight, glass transition temperature and flammability, etc. [5–10]. Furthermore, the fundamental theoretical investigations on the thermodynamics and kinetics of polymer melt intercalation and hybrid formation have never undertaken the considerations of the thermal degradation of the organic ammonium ions at higher temperatures in account and have assumed the perfect stability of the organic modification even at high temperatures [11,12].

A number of thermally stable modifications carrying the phosphonium, pyridinium, and imidazolium cations have been reported in the literature to eliminate the problem of low thermal stability of the ammonium-based cations attached to the clay surface [13,14]. Especially, intensive thermal studies have been reported for the imidazolium salts and it has been observed that the montmorillonites modified with imidazolium salts had much better thermal response as compared to the alkyl ammonium cation–modified montmorillonites. The better thermal behavior was also confirmed both in the presence and absence of oxygen as degrading atmosphere [13]. Further studies also reported that the thermal stability of imidazolium salts decreased on increasing length of the alkyl chain attached to the imidazolium group. However, the thermal behavior of the long chain imidazolium salts attached to the surface of the filler was high enough to sustain the higher compounding temperatures [15,16]. The thermal stability was also observed to be directly proportional to the substitution of the imidazolium ions owing to the removal of the ring hydrogen atoms. A few studies using these imidazolium salts for the synthesis of polymer nanocomposites have also been reported, but most of them deal with polystyrene as polymer matrix [17–19].

10.2 Imidazolium Salt and Thermal Stability

The synthesis of required imidazolium salt 1-decyl-2-methyl-3-octadecylimidazolium bromide was achieved by the reaction of 1-decyl-2-methylimidazole with octadecyl bromide in ethyl acetate [32]. The contents were first stirred at room temperature for 2 h under nitrogen followed by the increasing of the temperature to 55°C overnight. A white precipitate of imidazolium salt was obtained on cooling, which was filtered, washed extensively with ethyl acetate, and dried at room temperature under reduced pressure. Figure 10.1 shows the chemical structure of the imidazolium salt.

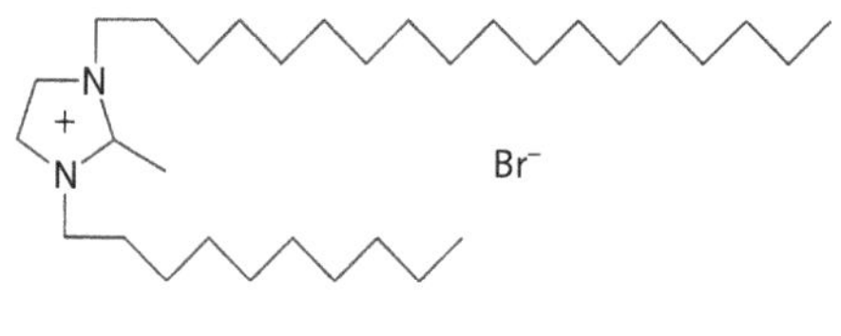

FIGURE 10.1
Structure of the 1-decyl-2-methyl-3-octadecylimidazolium bromide. (Reproduced from Mittal, V., *Eur. Polym. J.*, 43, 3727, 2007. With permission from Elsevier.)

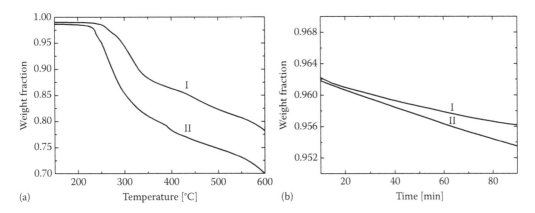

FIGURE 10.2
(a) TGA thermograms and (b) dynamic thermogravimetric analysis; I: Imidazolium treated montmorillonite and II: Ammonium modified montmorillonite. (Reproduced from Mittal, V., *Eur. Polym. J.*, 43, 3727, 2007. With permission from Elsevier.)

The thermal behavior of the organically modified montmorillonite was characterized by using thermogravimetric analysis (TGA), as shown in Figure 10.2a. For comparison, the TGA thermogram of the dimethyldioctadecylammonium-modified montmorillonite was also plotted. The onset of degradation in the ammonium-modified montmorillonite was roughly 50°C earlier than the imidazolium-modified montmorillonite. There was also a presence of sharp degradation, which in the differential TGA plot corresponds to a low-temperature degradation peak, which is a result of some free ammonium cations present in the interlayers, which form pseudo bilayers. It has been reported that this early degradation of the ammonium modification can negatively affect the composite properties [4,20,21]. The treatment with dioctadecyldimethylammonium or higher chain density ammonium ions like trioctadecylmethylammonium generally leads to the formation of the local bilayers with modification molecules physically adsorbed in the interlayer but unattached to the filler surface. It was reported that multiple washing protocols are required in order to completely remove such early degrading materials from the filler surface thus to improve its thermal performance [22]. Not only the onset of degradation temperature was higher for the imidazolium-modified montmorillonite, but the peak degradation temperature was higher than the ammonium-treated montmorillonite by roughly 35°C.

Dynamic thermogravimetric analysis as shown in Figure 10.2b also revealed the better thermal resistance of the imidazolium-modified montmorillonite. To achieve this comparison, the modified fillers were stabilized at 140°C and then heated to 180°C and kept at 180°C under isothermal conditions for 90 min. The rate of degradation was faster in the case of ammonium modification. Though the overall magnitude of degradation is not significantly different in both the fillers as seen from the *y*-axis of the plot, however, even such a small difference assumes significance owing to the thermal degradation taking place by the breakage of the weaker C–N bond. In such scenario, as mentioned above also, breaking of even small number of such bonds can seriously change the structure of the surfactant and its interaction with the polymer and, thus, the building of the interfacial morphology.

10.3 Nanocomposites Morphology

A basal plane spacing of 2.24 nm was observed in the x-ray diffractogram of imidazolium-modified montmorillonite. Nanocomposites prepared with 1, 2, 3, and 4 vol% of modified montmorillonite had filler basal plane spacing of 2.52, 2.28, 2.29, and 2.24 nm in the composites, respectively. It indicates that probably only in 1 vol% composite, there was some extent of polymer intercalation in the silicate interlayers. However, the composites with higher filler fractions had practically no change in the basal plane spacing of the filler indicating no or insignificant polymer intercalation.

However, owing to better interfacial interactions due to the elimination of the thermal degradation of the filler surface modification to a large extent (thus eliminating the unwanted side reactions) and the effect of shearing forces, a decrease in the tactoid thickness and, hence, an increase in the aspect ratio can still be expected. Figure 10.3 demonstrates the microscopic investigation of the polypropylene nanocomposites. Although no intercalation was observed from the x-ray diffraction, but a mixed morphology of the composites was observed in the TEM micrographs. Both single layers as well as stacks of platelets were observed. The filler platelets were also observed to be bent, folded, and misaligned and there was no special orientation at any magnification used.

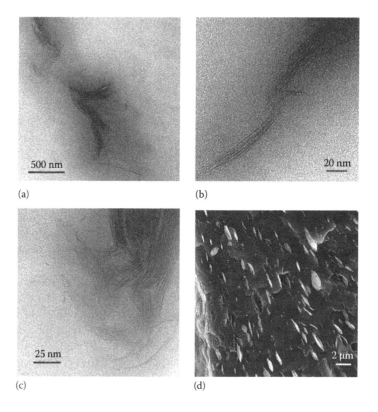

(a) (b) (c) (d)

FIGURE 10.3
(a) through (c) TEM micrographs of polypropylene nanocomposites containing 3 vol% of the imidazolium modified montmorillonite. The dark lines represent the cross-section of alumino-silicate platelets, (d) SEM micrograph of the same composites. (Reproduced from Mittal, V., *Eur. Polym. J.*, 43, 3727, 2007. With permission from Elsevier.)

The SEM micrograph shown in Figure 10.3d also confirmed the uniform distribution of the clay platelets though they were misaligned. Such misalignment is commonly observed in the nanocomposites generated by melt blending; the use of elongation stretching or similar techniques can help to enhance the extent of alignment of platelets, if required. Also important to note here is that the partial exfoliation of the filler could still be achieved even without the use of conventionally added low-molecular-weight compatibilizers for the polyolefin nanocomposites.

10.4 Oxygen Barrier Properties

Significant reduction in the oxygen permeation through the nanocomposite films was observed on the incorporation of imidazolium-treated montmorillonite and the decrease corresponded well with increasing filler volume fraction in the composites [10]. The oxygen permeation of the composites as a function of filler fraction is depicted in Figure 10.4. The oxygen permeation for the pure polypropylene was observed to be $89 \, cm^3 \cdot \mu m/m^2 \cdot day \cdot mmHg$. This value reduced significantly to $48 \, cm^3 \cdot \mu m/m^2 \cdot day \cdot mmHg$ by the incorporation of 4 vol% of organically modified montmorillonite, thus, resulting in a decrease of almost 50% in the oxygen permeation. The oxygen permeation through the composite films with varying fractions of dioctadecyldimethylammonium-modified montmorillonite has also been compared with the imidazolium-based, modified filler composites. The decrease in permeation through the composites with ammonium-treated montmorillonites though corresponded also well with the increase in the filler content, but the overall reduction was significantly lower than the imidazolium-treated filler composites. At a filler volume fraction of 4 vol%, a reduction of only 35% was observed in the oxygen permeation of the composites with ammonium-treated montmorillonite. It should also be noted that the basal plane spacing of 2.24 nm in the imidazolium-modified filler is less than the 2.51 nm interlayer distance in the ammonium-treated montmorillonite, even then the oxygen permeation in the case of the

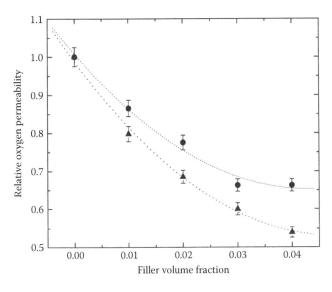

FIGURE 10.4
Relative oxygen permeability of nanocomposites with imidazolium modified montmorillonite as a function of filler volume fraction. The permeation behavior is compared with composites containing dioctadecyldimethylammonium ions [22]; (●): Ammonium and (▲): Imidazolium composites. The dotted lines only serve as guides. (Reproduced from Mittal, V., *Eur. Polym. J.*, 43, 3727, 2007. With permission from Elsevier.)

imidazolium-treated montmorillonite is much better. Thus, if the basal plane spacing in the imidazolium-treated montmorillonite is increased by attaching alkyl chains of longer length or higher chain density imidazolium ions, further enhancements in the oxygen barrier performance of the composites can be achieved. The much better performance of the imidazolium system was attributed to its much better thermal stability, owing to the fact that the permeation properties are very sensitive to the interfacial interactions between the filler and the polymer, which otherwise may be disturbed if the surface modification degrades prematurely.

10.5 Tensile Properties

Tensile properties of the polypropylene nanocomposites with increasing extent of filler content have been reported in Figure 10.5 [10]. As shown in Figure 10.5a, the tensile modulus of the composites was observed to linearly increase with filler volume fraction. A tensile modulus value of 1510 MPa was observed for the pure polymer, which increased to 2007 MPa with 4 vol% content. This indicates an increase of 35% in the modulus as compared to the pure polymer thus confirming the enhanced extent of stress transfer to the inorganic filler platelets. In fact, a good correlation between the decrease of the oxygen permeation and increase in the tensile modulus as a function of increasing filler content existed, which indicated that the optimal design of the surface modification of the inorganic filler leads to enhancement of the composite properties for which the compatibilizer was also not required. However, as the filler was not completely exfoliated and no attractive forces between the filler and the polymer exist, the yield stress was observed to decrease. Strain hardening of the polymer due to the confinement of the polymer chains in the clay tactoids as well as presence of un-exfoliated filler stacks also led to the slight reduction in the yield strain and stress at the break of the nanocomposites.

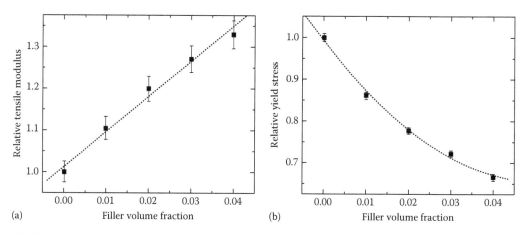

(a) Filler volume fraction   (b) Filler volume fraction

FIGURE 10.5

Relative (a) tensile modulus and (b) yield stress of polypropylene nanocomposites plotted as a function of inorganic volume fraction of filler. The dotted lines serve as guides. (Reproduced from Mittal, V., *Eur. Polym. J.*, 43, 3727, 2007. With permission from Elsevier.)

10.6 Calorimetric and Thermal Properties

It is important to ascertain the calorimetric behavior of the composites in order to relate the effects on the composite properties either to filler amount and its exfoliation in the matrix by the interfacial interactions or to a combination of these with the changes crystalline structure of the polymer. The calorimetric response of the nanocomposites with varying extent of imidazolium-modified montmorillonite confirmed that the filler did not affect the crystallization of polypropylene. There were no observable variations in the onset and peak melting and crystallization temperatures. Also the degree of crystallinity calculated from the enthalpy of melting of polymer in the composite compared with the enthalpy of purely crystalline polymer was found be fairly constant. The thermal behavior of the polypropylene nanocomposites was also characterized by TGA [10], as shown in Figure 10.6a. The thermogram on the extreme left corresponds to pure polypropylene. On the incorporation of even 1 vol% of the organically modified filler, the thermal behavior of the composites was observed to significantly improve. The temperatures of the onset of polymer degradation as well as peak degradation were significantly enhanced thus confirming the role of inorganic filler in improving the thermal resistance of the composites. Addition of organically modified silicates in higher amounts further improved the thermal response of the nanocomposites. Highest thermal stability was exhibited by the composites containing 4 vol% of the organically modified filler. The synergistic improvement in the thermal response of the nanocomposite on the incorporation of the filler is also demonstrated in Figure 10.6b. Curve I corresponds to the thermogram of treated filler whereas curve II corresponds to pure polypropylene. The thermogram of composite with 4 vol% of the organically modified filler was observed to have much better thermal response (onset of degradation and peak degradation) than any of its constituents.

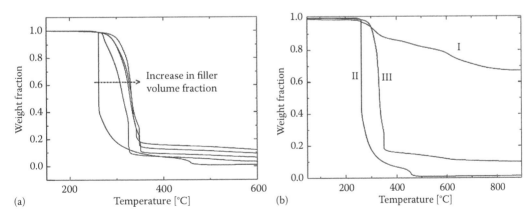

FIGURE 10.6

(a) TGA thermograms of pure polypropylene and polypropylene nanocomposites containing different filler inorganic volume fraction and (b) comparison of the TGA thermograms of pure PP (II), imidazolium clay (I), and the nanocomposite containing 4 vol% of organically modified montmorillonite (III). (Reproduced from Mittal, V., *Eur. Polym. J.*, 43, 3727, 2007. With permission from Elsevier.)

10.7 Mechanical Modeling of Nanocomposites

A number of micro-mechanical models have been developed over the years to predict the mechanical behavior of particulate composites [23–26]. Halpin–Tsai model has received special attention owing to better prediction of the properties for a variety of reinforcement geometries. The relative tensile modulus is expressed as

$$\frac{E}{E_m} = \left(\frac{1 + \zeta \eta \varphi_f}{1 - \eta \varphi_f} \right)$$

where

E and E_m correspond to the elastic moduli of composite and matrix, respectively
ζ represents the shape factor, which is dependent on filler geometry and loading direction
φ_f is the inorganic volume fraction
η is given by the expression

$$\eta = \left(\frac{(E_f/E_m) - 1}{(E_f/E_m) + \zeta} \right)$$

where E_f is the modulus of the filler. The η values need to be correctly defined in order to have better prediction of the properties. For the oriented discontinuous ribbon or lamellae, it is estimated to be twice the aspect ratio. It has been reported to overpredict the stiffness in this case; therefore, its value was reported be 2/3 times the aspect ratio [27]. But still a number of assumptions prevent the theory to correctly predict the stiffness of the layered silicate nanocomposites. Assumptions like firm bonding of filler and matrix, perfect alignment of the platelets in the matrix, uniform shape and size of the filler particles in the matrix make it very difficult to correctly predict the nanocomposites properties. The incomplete exfoliation of the nanocomposites, that is, the presence of a distribution of tactoid thicknesses is another concern. The model has recently been modified in order to accommodate the effect of incomplete exfoliation and misorientation of the filler, but the effect of imperfect adhesion at the surface still needs to be incorporated [28,29].

The tensile properties of polypropylene nanocomposites with varying extents of filler volume fraction were fitted with the Halpin–Tsai model. The solid line in Figure 10.7 shows the fit of the tensile modulus of the composites, here η was considered to be 1. This consideration led to the evaluation of ζ equal to 7.46 [30]. It indicates that possibly in these polyolefin nanocomposite systems, it cannot be simply taken as twice the aspect ratio as generally used [31] as the aspect ratio is not very low as qualitatively observed from the transmission electron micrographs. A new approach where filler particles are replaced by stacks of filler platelets for the property analysis was reported, which simulates the nanocomposite morphology better owing to the presence of tactoids of varying thicknesses, i.e., mixed morphology as seen in the micrographs [28]. Using these considerations to conventional Halpin–Tsai equation, predictions of tensile modulus of the composites as a function of filler volume fraction considering different number of platelets in the stack, could be achieved, as shown in Figure 10.8. The experimental values when compared with the model predictions reveal that relative tensile modulus for 1 and 2 vol% filler containing composites was observed to lie on the curve with 150 platelets in the stack; however,

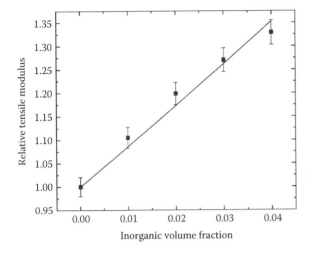

FIGURE 10.7
Relative tensile modulus of OMMT-polypropylene nanocomposites as a function of inorganic volume fraction. The solid line represents the fitting by using unmodified Halpin–Tsai equation. (Reproduced from Mittal, V., *J. Thermoplast. Compos. Mater.*, 22, 453, 2009. With permission from Sage Publishers.)

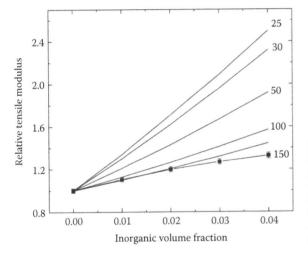

FIGURE 10.8
Relative tensile modulus of OMMT-polypropylene nanocomposites (■, Experimental) compared with the values considering different number of platelets in the stack [27]. (Reproduced from Mittal, V., *J. Thermoplast. Compos. Mater.*, 22, 453, 2009. With permission from Sage Publishers.)

further sagging in the curve was observed for composites with higher volume fractions of the filler owing to the increased extent of unintercalated thick tactoids.

However, the above analysis with the Halpin–Tsai models still does not completely replicate the polyolefin nanocomposites as the models do not incorporate the effect of misoriented platelets on the modulus. Misalignment of platelets leads to the significant change in the properties as compared to the aligned platelets composites. Thus, incorporation of effects in the Halpin–Tsai model along with the incomplete exfoliation have also been studied [27]. Figure 10.9a shows the relative tensile modulus of the polypropylene composites as a function of filler volume fraction plotted against the curves representing different number of platelets in the stack when incomplete exfoliation and platelets misalignment effects were also incorporated [27]. The experimental values, when considered with these predictions, were observed to lie on an average between 100 and 150 platelets in the stack. The tensile modulus of the composites was also predicted by using the models suggested by Brune and Bicerano, which also incorporate the considerations of incomplete exfoliation and misorientation [29]. The property predictions are demonstrated in Figure 10.9b and the experimental data was observed to follow the predictions of 40–50 platelets in the stack in the composites, which may represent the real morphology of the composites as

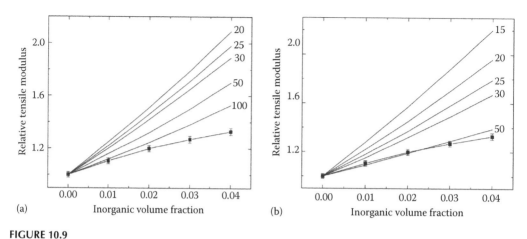

FIGURE 10.9
(a) and (b) Relative tensile modulus of polypropylene nanocomposites (■, Experimental) as a function of inorganic volume fraction compared with the values considering different number of platelets in the stack applying platelet misorientation models [36,40]. (Reproduced from Mittal, V., *J. Thermoplast. Compos. Mater.*, 22, 453, 2009. With permission from Sage Publishers.)

evident from the electron micrographs. However, the limitation of these mechanical models for assuming perfect adhesion at the interface still plagues the analysis of the nanocomposites with these modified models. The polyolefin composites studied in the current work, really lack this adhesion as only weak van der Waals forces can exist in the studied polymer organic monolayer systems. Thus, in such systems, it is of immense need to develop the models, which can consider the various facets of polyolefin nanocomposites so as to correctly predict the nanocomposite properties. It is also important to note that the determination of aspect ratio from the TEM micrographs can also be misleading owing to the bending and misalignment of platelets as shown in the micrographs presented in this study thus requiring a model to estimate overall average aspect ratio.

Mixture designs (using design of experiments) have also been reported in a recent study in order to predict the properties of polymer nanocomposites [31]. Mixtures design can quantify the interactions between the components like amounts of polymer, inorganic filler and the surface modifications. The amount of surface modification ionically bound on the filler can be varied by choosing modifications of different chain densities, i.e., by varying the number of octadecyl chains in the modification. The interactions at the interface significantly affect the morphology and the properties of the composite, thus, varying the amount of components in the composite also leads to variation in the overall properties of the composites. As all the components of the mixture cannot be worked within the ranges of 0%–100% of total weight, a constrained mixture design was, therefore, generated with the constraints set on the polymer from 84% to 100%, inorganic from 0% to 11%, and the corresponding filler modification from 0% to 5% of the total weight of the composite. It was observed in the regression for mixtures design that apart from polymer, organic modification and inorganic filler being statistically important, a two-way interaction between the polymer and the inorganic filler fraction is also significant, thus bringing a nonlinearity in the analysis. Although it has been observed that the surface modification is of importance in bringing about the compatibility between the organic and inorganic phases of the composites, however, it is the resulting compatibilized two-way interaction between the organic polymer phase and the inorganic filler phase, which is of influence on the composite properties. The regression analysis for tensile strength

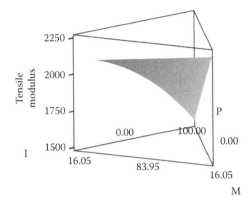

FIGURE 10.10
Mixture surface plot of tensile modulus and of the polypropylene organically modified montmorillonite composites as a function of mixture components. (Reproduced from Mittal, V., *J. Thermoplast. Compos. Mater.*, 21, 9, 2008a. With permission from Sage Publishers.)

led to an R-Sq fit of 98.59% indicating a superior fit of the data. As a result, the residual error was of low magnitude as compared to the main effects. By combining together the estimated coefficients for tensile modulus, the following equation could be generated for the modulus:

$$E = 15.30\ M_P - 109.53\ M_I + 58.96\ M_{OM} + 1.91\ M_P * M_I$$

where
   E is the tensile modulus of the composites in MPa
   M_P is the weight percent of polymer
   M_{OM} is the weight percent of organic modification
   M_I is the weight percent of inorganic filler

The generated equations for tensile modulus can also be pictorially represented as surface plots, as shown in Figure 10.10. The presence of a two-way interaction leads to the curvature in the surface plot for the properties and thus provides the ways to carefully consider the different components of the mixture to achieve an optimum enhancement of the properties. Unlike conventional models, which depend on oversimplified assumptions and are not applicable in reality, these models do not suffer from these limitations and can still predict the composite properties using a set of simple equations.

10.8 Role of Compatibilizer

Low-molecular-weight compatibilizers are commonly added to the polyolefins in order to enhance the interactions between the polar filler surface and the polymer chains by achieving the partial polarization of the matrix through compatibilizer. The compatibilizer being amphiphilic in nature helps to compatibilize the filler and polymer phases and causes enhanced extent of filler delamination. Polypropylene-grafted maleic anhydride (PP-*g*-MA) has been commonly used as a compatibilizer in the polypropylene nanocomposites. Various studies using PP-*g*-MA of different molecular weights with different extents of MA grafting have been reported and it was observed that the filler exfoliation was proportional to the amount of the compatibilizer in the system. To study the effect of compatibilizer in the imidazolium-based montmorillonite filler polypropylene nano-composites, three different kinds of compatibilizers were used. These compatibilizers were selected on the basis of their chemical composition and the location of the polar groups in the polymer chain. PP-*g*-MA1 had a high molecular weight as indicated by its low MFI of 5.5 g (10 min, 230°C, 2.16 kg) with MA content of 0.1 wt%, whereas PP-*g*-MA2 had an M_n of 3900 and had 4 wt% of MA content corresponding to 3.7 MA units per polymer chain, thus, making it high MA compatibilizer (acid number 47 mg KOH/g of PP-*g*-MA). In order to analyze the effect of block copolymers on the performance of nano-composites, PP-*b*-PPG (polypropylene glycol) was synthesized [33]. The resulting copo-lymer is not strictly a diblock copolymer, but still can be represented as one and can be visualized as polypropylene chains containing polar blocks as side chains instead of small MA molecules. Though increased extent of compatibilizer leads to better filler exfoliation, however, its increased amounts in the composites can also lead to deterio-ration of mechanical properties owing to plasticization of the matrix. The amount of compatibilizer was therefore fixed to 2 wt% in correspondence with the similar recently reported study where the deterioration of mechanical properties was observed at higher amounts of compatibilizers and 2 wt% was found to be an optimum amount [33]. A com-patibilizer/organically modified montmorillonite weight ratio of 0.16 was thus used in the composites.

The wide-angle x-ray diffraction patterns of the polypropylene nanocomposites with and without compatibilizer have been shown in Figure 10.11. The diffraction pattern of pure imidazolium–treated montmorillonite has also been plotted for comparison. The presence of basal peaks in all the composites containing different compatibilizers indicated that full exfoliation of the filler did not take place. Also, the increase in the basal plane spacing in the composites as compared to the modified filler or the composite without compatibilizer

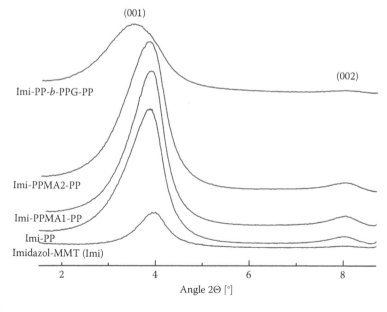

FIGURE 10.11
X-ray diffractograms of imidazolium-treated clay and its 3 vol% composites with and without 2 wt% of com-patibilizer. (Reproduced from Mittal, V., *J. Thermoplast. Compos. Mater.*, 22, 453, 2009. With permission from Sage Publishers.)

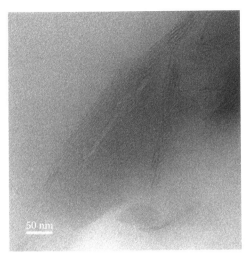

FIGURE 10.12
TEM micrograph of polypropylene clay nanocomposites with 2 wt% of PP-*b*-PPG compatibilizer. (Reproduced from Mittal, V., *J. Thermoplast. Compos. Mater.*, 22, 453, 2009. With permission from Sage Publishers.)

is also minimal indicating that very less amount of compatibilizer intercalated the clay interlayers. Initial lower basal plane spacing of the filler is also one reason of low extent of filler intercalation as compared to the dioctadecyldimethylammonium-modified montmorillonite. Owing to the presence of diffraction peaks in the diffractograms of the composites, it may also be said that the mechanical shear exerted in the compounder may not have been optimum to delaminate all the filler stacks to the nanometer level. Comparing the performance of different compatibilizers, the basal plane spacing of the filler was observed to increase with the increasing polarity of the compatibilizer. PP-*b*-PPG with maximum polarity showed an increase of 0.3 nm as compared to the value of 2.24 for the organically modified clay. Also, as the diffraction peak intensity and width are affected by a number of factors like crystal impurities, sample preparation, etc., and the presence of diffraction peak does not rule out the presence of an exfoliated part of the filler, it is also important to characterize the morphology of the composites by microscopy as well as to relate the composite morphology with the composite properties. Also, even though extensive intercalation of the filler did not take place, delamination of the filler to some extent owing to the shearing forces can surely be expected. The presence of such mixed morphology is not possible to characterize by x-ray diffraction, and other methods of characterization are required to generate the complete picture. Figure 10.12 also shows the TEM micrograph of the polypropylene composites with 2 wt% of PP-*b*-PPG compatibilizer. One can see that the tactoid thickness has been reduced as compared to the TEM micrograph of the composites without compatibilizer. The decrease in tactoid thickness represents increase in aspect ratio of the filler, thus can also be called as increased delamination, though delamination of every platelet to nanometer scale is still not achieved.

All the compatibilizers increased the tensile modulus of the composites as compared to the composite without compatibilizer as well as pure polypropylene owing to increased extent of filler exfoliation on addition of compatibilizers. Values of 2039, 2023, and 1993 were, respectively, measured for PP-*g*-MA1, PP-*g*-MA2, and PP-*b*-PPG containing composites (3 vol% filler) as compared to 1510 for the pure polymer and 1918 MPa for the composite without compatibilizer. Figure 10.13a also shows the correlation between the relative modulus and the basal plane spacing of filler in these composites. The modulus was observed to be roughly unchanged with respect to basal plane spacing indicating only exfoliated layers are responsible for the improvement in modulus. Also the plasticization effects generally associated with the compatibilizers may limit further improvement of the modulus.

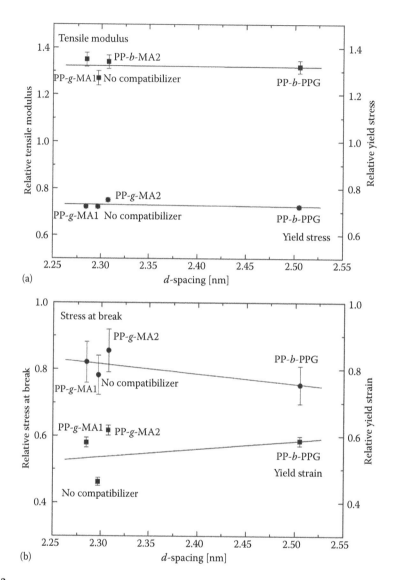

FIGURE 10.13
(a) Relative tensile modulus and yield stress and (b) relative stress at break and yield strain of 3 vol% organically modified montmorillonite—polypropylene nanocomposites with and without 2 wt% compatibilizer. (Reproduced from Mittal, V., *J. Thermoplast. Compos. Mater.*, 22, 453, 2009. With permission from Sage Publishers.)

Increase in the yield strain of the composites with compatibilizer as compared to the composite without compatibilizer also indicated weak plasticization of the polymer. The strain also was observed to marginally increase with increasing basal plane spacing and the polarity of the compatibilizer. Yield stress and the stress at break were also observed to be almost unchanged as the composite without compatibilizer indicating the presence of small extent of brittleness and nonoptimal interfacial interactions. Relative yield stress was also observed to be independent of basal plane spacing, while the break stress was observed to decrease with increase in the basal plane spacing as shown in Figure 10.13. The presence of tactoids, which are not efficient enough for load transfer, can be a reason for this decrease. Only 2 wt% compatibilizer was used in this study, as the properties were observed to deteriorate when the amount of compatibilizer was increased beyond 2 wt% in a similar study on polypropylene nanocomposites [33].

As observed above, the addition of only 2 wt% of the compatibilizers to the systems significantly affected the tensile modulus and other mechanical properties of the composites. Therefore, it is also important to ascertain the effect of the compatibilizer on the calorimetric behavior of the composites similarly as reported above in order to assign the effect on the properties only to the filler exfoliation owing to the compatibilizer or also to the change in crystalline structure of polymers. The calorimetric properties of the composites with various compatibilizers are reported in Table 10.1 [30] along with the pure polymer, mixture of polymer and PP-*g*-MA2 compatibilizer, and the composite without compatibilizer. As is evident from the table, the calorimetric properties are observed to be minimally affected by the addition of the compatibilizer. The onset melting temperatures were in the range of 151°C–153°C and peak melting temperatures were observed to be in the range of 162°C–165°C for all the samples. PP-*g*-MA2 and PP-*b*-PPG were observed to reduce the peak crystallization temperature as compared to pure polypropylene, whereas marginal advancement in the onset of crystallization was observed for PP-*g*-MA1. Degree of crystallization values of the polymer calculated from the enthalpy of melting, however, were fairly constant for all the samples. Therefore, it can be concluded that the observed changes in the composite properties are more related to the organic–inorganic interactions and enhanced exfoliation of the filler owing to the addition of the compatibilizer.

As the oxygen permeation properties are more sensitive to the interfacial interactions between the inorganic and organic phases, it was also of interest to quantify the effect of the compatibilizer addition on the oxygen barrier properties of nanocomposites. The oxygen permeation coefficients through the nanocomposite films with different compatibilizers have been listed in Table 10.2 along with coefficients of filled polypropylene composite without compatibilizer, pure polypropylene, and a blend of polypropylene with 2 wt% PP-*g*-MA2. A value of 57 cm$^3 \cdot \mu$m/m$^2 \cdot$day$\cdot$mmHg was observed for the polypropylene composite with 3 vol% of the filler without compatibilizer as compared to 89 cm$^3 \cdot \mu$m/m$^2 \cdot$day$\cdot$mmHg for the pure polymer. The permeation through the composite films containing different compatibilizers also decreased as compared to the pure polypropylene but

TABLE 10.1

Calorimetric Properties of Pure Matrices and 3 vol% Polypropylene Nanocomposites With and Without 2 wt% Compatibilizer

Compatibilizer (2 wt%)	$T_{m,onset}$[a] (°C)	T_m[b] (°C)	ΔH_mpolymer (J/g)	Crystallinity[c] (X_c)	$T_{c,onset}$[d] (°C)	T_c[e] (°C)
Pure PP	152	162	96	0.58	114	112
PP + PP-*g*-MA2	152	165	94	0.57	112	106
No compatibilizer	153	163	97	0.59	115	111
PP-*g*-MA1	153	162	95	0.58	120	114
PP-*g*-MA2	152	163	93	0.56	114	107
PP-*b*-PPG	151	164	92	0.56	114	108

Source: Reproduced from Mittal, V., *J. Thermoplast. Compos. Mater.*, 22, 453, 2009. With permission from Sage Publishers.

[a] Onset melting temperature.
[b] Peak melt temperature.
[c] Degree of crystallinity calculated using ΔH of 100% crystalline PP = 165 J/g [52,53].
[d] Onset crystallization temperature.
[e] Peak crystallization temperature.

TABLE 10.2

Oxygen Permeation Values of Pure Matrices and 3 vol% Filler—Polypropylene Nanocomposites with and without 2 wt% Compatibilizer

Compatibilizer (2 wt%)	Permeability Coefficient, $cm^3 \cdot \mu m/m^2 \cdot day \cdot mmHg$
pure PP	89
PP + PP-g-MA2	83
No compatibilizer	57
PP-g-MA1	69
PP-g-MA2	61
PP-b-PPG	78

Source: Reproduced from Mittal, V., *J. Thermoplast. Compos. Mater.*, 22, 453, 2009. With permission from Sage Publishers.

the decrease was either lesser or same as the composite without any compatibilizer. It was unexpected as the compatibilizer addition was observed to enhance the exfoliation of the filler platelets; therefore, further decrease in the oxygen permeation is expected. Similar results have also been observed for the composites with dimethyldioctadecylammonium-treated montmorillonite [33]. It was also observed that the blend of polypropylene and the compatibilizer had the permeation coefficient of 83 $cm^3 \cdot \mu m/m^2 \cdot day \cdot mmHg$, which is similar or somewhat better than the pure polymer, thus indicating that the addition of compatibilizer to polymer does not negatively effect the permeation of polypropylene. It can thus be said that the responsible factor in such a scenario is the compatibility of the treated clay with the compatibilizer molecules. As reported earlier, the incompatibility of these components leads to the generation of micro voids or increase in free volume at the interface thus subsequently leading to an increase in the permeation, even though the clay is more exfoliated by the addition of compatibilizer. Thus, the effect of exfoliation leading to decrease in permeation (by decreasing the permeant diffusion owing to the generation of a more tortuous path by the exfoliated platelets) can be counterbalanced with the increase in permeation (enhanced permeant diffusion through the voids at interface) owing to mismatch between the polarity of the composite components leading to no decrease or increase in the permeation as compared to the composites without compatibilizer [33,34]. This is the reason that an improvement in mechanical properties is observed after the addition of even a small amount of compatibilizer; however, a deterioration of interface-sensitive micro properties like gas permeation is observed. It also points that the enhancement of mechanical performance of the composites does not mean that the other composite properties would also be automatically enhanced.

As the compatibilizers added to the composites have low molecular weight, the thermal behavior of composites is also required to be monitored in order to ascertain any degrading effect in the thermal stability of the composites by compatibilizers. Figure 10.14a shows the TGA thermograms of nanocomposites with 3 vol% of the organically modified montmorillonite synthesized with and without the addition of 2 wt% of different compatibilizers. The curves for all the composites were indistinguishable indicating that the thermal behavior of the composites with and without the added compatibilizers was the same. It thus confirms that the thermal behavior of the composites was not affected on the addition of compatibilizers. Figure 10.14b also provides further comparison of the thermal behavior

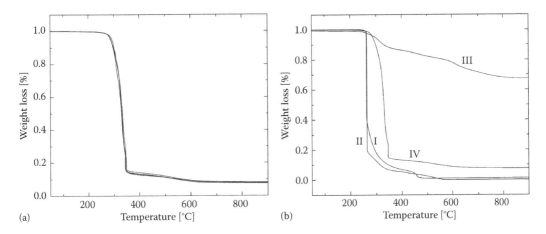

FIGURE 10.14
(a) TGA thermograms of 3 vol% OMMT nanocomposites with and without 2 wt% compatibilizer and (b) TGA thermograms of pure PP (I), PP-*g*-MA2 (II), modified filler (III) and 3 vol% OMMT-polypropylene composite with 2 wt% of PP-*g*-MA2 compatibilizer (IV). (Reproduced from Mittal, V., *J. Thermoplast. Compos. Mater.*, 22, 453, 2009. With permission from Sage Publishers.)

of pure polymer (I), mixture of polymer and 2 wt% of PP-*g*-MA2 (II), imidazolium modified clay (III) and polymer composite with 3 vol% of filler and 2 wt% of PP-*g*-MA2 (IV). The thermal behavior of polymer and its mixture with compatibilizer was very similar indicating that the low molecular weight of compatibilizer did not affect the thermal degradation of polypropylene. It was further confirmed from the TGA thermogram of the composite that synergism in between the organic and inorganic parts of the composites existed as the thermal response of the composite was better than any of the constituents. Similar to the mixture of polymer and the compatibilizer, the compatibilizer addition also did not cause any damaging effect to the thermal response of the composite.

10.9 Conclusions

Thermally stable imidazolium-based surface modification was synthesized and was subsequently exchanged on the surface of montmorillonite platelets. The isothermal and non-isothermal thermogravimetric analysis of the modified montmorillonites confirmed the delayed onset and peak degradation for the imidazolium-modified montmorillonite as compared to the dioctadecyldimethylammonium counterpart commonly used. Polypropylene nanocomposites with varying extents of organically modified montmorillonite were generated by melt compounding and mechanical, thermal, and oxygen barrier properties of the composites were characterized. The oxygen permeation properties were observed to significantly enhance as a decrease of nearly 50% at 4 vol% of filler fraction was achieved, which is much better than dioctadecyldimethylammonium-modified filler composites. The composite morphology was mixed in nature, i.e., clay tactoids of varying thicknesses were present. It was possible to achieve a partial delamination of the filler even without the use of commonly added compatibilizers owing to the better optimization of the interface between organic and inorganic components of the composites. Tensile modulus also linearly increased with filler volume fraction and a good correlation of tensile modulus enhancement with the oxygen permeation reduction was observed.

The addition of organically modified filler did not affect the crystallization behavior of polypropylene under the processing conditions used. The thermal stability of the composites was synergistically improved and the thermal performance was enhanced on increasing the filler volume fraction. The mechanical properties of the nanocomposites could be modeled with Halpin–Tsai models and more recent models incorporating other considerations for the correct prediction of nanocomposite properties like misorientation, incomplete exfoliation, etc. Other possibilities of using mixture designs for modeling the properties also exist where the usual assumptions of conventional mechanical models are not required to be true.

Addition of small amounts of compatibilizer further improved the delamination of the filler in the compatibilizer, which resulted in the tensile modulus of the compatibilized nanocomposites to be improved as compared to the composites without compatibilizer, though a slight plasticization of the matrix was also observed. The oxygen permeation through the compatibilized composites however did not show any improvement as compared to the noncompatibilized composites. Competing phenomena of increase in permeation due to interfacial mismatching and increase in free volume with decrease in permeation due to tortuous permeant pathway generated by the exfoliated platelets were reported to balance each other's effect. Similar to composites without the use of compatibilizer, the addition of compatibilizer also did not affect the crystallization behavior of polypropylene. The thermal behavior of the compatibilized composites was also similarly improved as the noncompatibilized polymer nanocomposites, even though the molecular weight of the compatibilizer was low.

Acknowledgments

The final, definitive versions of this chapter have been published in the *Journal of Thermoplastic Composite Materials*, Vol/Issue, Month/Year by SAGE Publications Ltd./SAGE Publications, Inc., All rights reserved © [as appropriate] http://online.sagepub.com and *European Polymer Journal*, 43(9), 2007, 3727–3736 by Elsevier Limited © 2007.

References

1. Pukanszky, B. 1995. In *Polypropylene: Structure, Blends and Composites*, ed. J. Karger-Kocsis. London, U.K.: Chapman & Hall.
2. Jancar, J. 1999. Mineral fillers in thermoplastics. In *Advances in Polymer Science*, vol. 139. Berlin, Germany: Springer.
3. Radosta, J. A. and Trivedi, N. C. 1987. In *Handbook of Fillers for Plastics*, eds. H. S. Katz and J. V. Milevski. New York: Van Nostrand Reinhold Company.
4. Xie, W., Gao, Z., Pan, W. P., Hunter, D., Singh, A., and Vaia, R. 2001. Thermal degradation chemistry of alkyl quaternary ammonium montmorillonite. *Chemistry of Materials* 13:2979–2990.
5. Fornes, T. D., Yoon, P. J., Keskkula, H., and Paul, D. R. 2001. Nylon 6 nanocomposites: The effect of matrix molecular weight. *Polymer* 42:9929–9940.
6. Gilman, J. W. 1999. Flammability and thermal stability studies of polymer layered-silicate (clay) nanocomposites. *Applied Clay Science* 15:31–49.
7. VanderHart, D. L., Asano, A., and Gilman, J. W. 2001. NMR measurements related to clay-dispersion quality and organic-modifier stability in nylon-6/clay nanocomposites. *Macromolecules* 34:3819–3822.

8. VanderHart, D. L., Asano, A., and Gilman, J. W. 2001. Solid-state NMR investigation of paramagnetic nylon-6/clay nanocomposites. 2. Measurement of clay dispersion, crystal stratification, and stability of organic modifiers. *Chemistry of Materials* 13:3796–3809.

9. Gilman, J. W., Morgan, A., Harris, R. Jr., Jackson, C., and Hunter, D. 2000. Phenolic cyanate ester clay nanocomposites: Effect of ammonium ion structure flammability and nano-dispersion. *Polymer Materials Science and Engineering Preprints* 82:276–277.

10. Mittal, V. 2007. Gas permeation and mechanical properties of polypropylene nanocomposites with thermally-stable imidazolium clay. *European Polymer Journal* 43:3727–3736.

11. Vaia, R. A. and Giannelis, E. P. 1997. Lattice model of polymer melt intercalation in organically-modified layered silicates. *Macromolecules* 30:7990–7999.

12. Lee, J. Y., Baljon, R. C., and Loring, R. F. 1999. Spontaneous swelling of layered nanostructures by a polymer melt. *Journal of Chemical Physics* 111:9754–9760.

13. Awad, W. H., Gilman, J. W., Nyden, M., Harris, R. H. Jr., Sutto, T. E., Callahan, J., Trulove, P. C., DeLong, H. C., and Fox, D. M. 2004. Thermal degradation studies of alkyl-imidazolium salts and their application in nanocomposites. *Thermochimica Acta* 409:3–11.

14. Zhu, J., Morgan, A. B., Lamelas, F. J., and Wilkie, C. A. 2001. Fire properties of polystyrene-clay nanocomposites. *Chemistry of Materials* 13:3774–3780.

15. Ngo, H. L., LeCompte, K., Hargens, L., and McEwan, A. B. 2000. Thermal properties of imidazolium ionic liquids. *Thermochimica Acta* 357–358:97–102.

16. Wilkes, J. S., Levisky, J. A., Wilson, R. A., and Hussey, C. L. 1982. Dialkylimidazolium chloroaluminate melts: A new class of room-temperature ionic liquids for electrochemistry, spectroscopy and synthesis. *Inorganic Chemistry* 21:1263–1264.

17. Bottino, F. A., Fabbri, E., Fragala, I. L., Malandrino, G., Orestano, A., Pilati, F., and Pollicino, A. 2003. Polystyrene-clay nanocomposites prepared with polymerizable imidazolium surfactants. *Macromolecular Rapid Communications* 24:1079–1084.

18. Gilman, J. W., Awad, W. H., Davis, R. D., Shields, J., Harris, R. H. Jr., Davis, C., Morgan, A. B., Sutto, T. E., Callahan, J., Trulove, P. C., and DeLong, H. C. 2002. Polymer/layered silicate nanocomposites from thermally stable trialkylimidazolium-treated montmorillonite. *Chemistry of Materials* 14:3776–3785.

19. He, A., Hu, H., Huang, Y., Dong, J. Y., and Han, C. C. 2004. Isotactic poly(propylene)/monoalkylimidazolium-modified montmorillonite nanocomposites: Preparation by intercalative polymerization and thermal stability study. *Macromolecular Rapid Communications* 25:2008–2013.

20. Morgan, A. B. and Harris, J. D. 2003. Effects of organoclay Soxhlet extraction on mechanical properties, flammability properties and organoclay dispersion of polypropylene nanocomposites. *Polymer* 44:2313–2320.

21. Shah, R. and Paul, D. R. 2006. Organoclay degradation in melt processed polyethylene nanocomposites. *Polymer* 47:4075–4084.

22. Osman, M. A., Mittal, V., and Suter, U. W. 2007. Poly(propylene)-layered silicate nanocomposites: Gas permeation properties and clay exfoliation. *Macromolecular Chemistry and Physics* 208:68–75.

23. Kerner, E. H. 1956. The elastic and thermo-elastic properties of composite media. *Proceedings of the Physical Society B* 69:808–813.

24. Halpin, J. C. 1992. *Primer on Composite Materials Analysis*. Lancaster, U.K.: Technomic.

25. Hashin, Z. and Shtrikman, S. 1963. A variational approach to the theory of the elastic behavior of multiphase materials. *Journal of the Mechanics and Physics of Solids* 11:127–140.

26. Halpin, J. C. 1969. Stiffness and expansion estimates for oriented short fiber composites. *Journal of Composite Materials* 3:732–734.

27. van Es, M., Xiqiao, F., van Turnhout, J., and van der Giessen, E. 2001. In *Specialty Polymer Additives: Principles and Application*, eds. S. Al-Malaika, A. W. Golovoy, and C. A. Wilkie. Malden, MA: Blackwell Science.

28. Fornes, T. D. and Paul, D. R. 2003. Modeling properties of nylon 6/clay nanocomposites using composite theories. *Polymer* 44:4993–5013.

29. Brune, D. A. and Bicerano, J. 2002. Micromechanics of nanocomposites: Comparison of tensile and compressive elastic moduli, and prediction of effects of incomplete exfoliation and imperfect alignment on modulus. *Polymer* 43:369–387.
30. Mittal, V. 2009. Polypropylene nanocomposites with thermally-stable imidazolium modified clay: Mechanical modeling and effect of compatibilizer. *Journal of Thermoplastic Composite Materials* 22:453–474.
31. Osman, M. A., Rupp, J. E. P., and Suter, U. W. 2005. Tensile properties of polyethylene-layered silicate nanocomposites. *Polymer* 46:1653–1660.
32. Mittal, V. 2008. Modeling the behavior of polymer-layered silicate nanocomposites using factorial and mixture designs. *Journal of Thermoplastic Composite Materials* 21:9–26.
33. Mittal, V. 2008. Mechanical and gas permeation properties of compatibilized polypropylene-layered silicate nanocomposites. *Journal of Applied Polymer Science* 107:1350–1361.
34. Osman, M. A., Mittal, V., Morbidelli, M., and Suter, U. W. 2003. Polyurethane adhesive nanocomposites as gas permeation barrier. *Macromolecules* 36:9851–9858.

Functional Polyolefins for Polyolefin/Clay Nanocomposites

Elisa Passaglia and Serena Coiai

CONTENTS

11.1 Introduction

The inertness, chemical stability, and solvent resistance are some of the most important features of polyolefins (POs), which are used for a wide range of applications even if they have drawbacks when wettability, adhesion, or surface/interfacial interactions with polar materials are requested, in particular in the field of blending with polar polymers and in the preparation of PO/inorganic composites.

Owing to the hydrophobic nature of POs, to overcome the lack of possible interactions with a polar substrate, suitable functionalities are inserted during the polymer preparation and/or by post-modification processes, which are generally performed in the melt by radical reactions. Acid, ester, anhydride, hydroxy, amino, oxazoline, and other functionalities have been grafted/anchored onto—or embodied into—the backbone of POs (Boffa and Novak 2000, Passaglia et al. 2009) and these functional POs have been successfully used for providing effective interactions with polar compounds. In particular, depending on the nature of the polar substrates (technopolymers or inorganic compounds), different types of interaction from the van der Waals and dispersion forces to the acid–base or ionic interactions as well as covalent bonds can be established according to the chemistry at the interface between the different materials.

In the last decades, nanostructured inorganic substrates, and in particular nanostructured layered silicates (clays), have found great importance owing to their capability of being dispersed at nanoscale level in a polymer matrix. The resulting polymer nanocomposites possess unique properties, not shared by their conventional microcomposites, which are attributed to the nanometer-size features joined to the extraordinarily high surface area of the dispersed clay (Alexandre and Dubois 2000, Vaia 2000, Sinha Ray and Okamoto 2003, Pavlidou and Papaspyrides 2008, Mittal 2009, Coltelli 2010). On this matter, PO-layered silicate nanocomposites have gained great attention owing to the nanostructured morphology which improves or adds new physical properties, such as the tensile strength, modulus, heat distortion temperature increase, and the gas permeability decrease (barrier properties).

As all the other nanostructured polymer materials, PO–clay nanocomposites are prepared by incorporating finely dispersed layered silicates in a PO matrix (Ciardelli et al. 2008). However, the nanolayers are not easily dispersed due to their preferred face-to-face stacking in agglomerated tactoids, whose dispersion into monolayers is hindered in the case of POs by considering the intrinsic incompatibility between the hydrophilic layered silicates and hydrophobic macromolecules: the pure/starting components are not able to interact and no interactions favoring the dispersion at the interface occur. Therefore, layered silicates first need to be organically modified to produce polymer-compatible clay (organoclay or organo-layered silicate, OLS): generally the replacement of the inorganic exchangeable cations in the cavities or galleries of the native clay silicate structure by alkylammonium surfactants can compatibilize the surface chemistry of the clay with most of plastics, and it is an essential requirement in the case of POs (Figure 11.1).

For long-chain onium-exchanged organoclays, the galleries swollen by these surfactants show a d-spacing increasing with the length of alkyl chains (Vaia et al. 1994, Wang et al. 2001). In addition, this modification is able to render more hydrophobic the surface of the layers that is a basic necessity, particularly for the POs melt intercalation process. Depending on the charge density of the layered silicate, on the chain length and nature of cationic surfactant and (really important) on the volume ratio between PO and OSL, intercalated or in a few cases exfoliated morphologies have been obtained (Balazs et al. 1998, 1999, Koo et al. 2002, Passaglia et al. 2005, Ciardelli et al. 2008).

From a thermodynamic point of view, the intercalation of a polymer into the galleries of OLS can be approached by considering entropic and enthalpic effects: the entropy loss due to the confinement (the macromolecules pass from 3D coils to 2D structures) has to be compensated by the entropy gain of the surfactant chain and by the balance of different possible interactions: polymer–surfactant, polymer–silicate surface, and surfactant–silicate surface interactions (Ciardelli et al. 2008) (Figure 11.2).

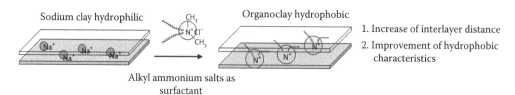

FIGURE 11.1
Clay modification reaction.

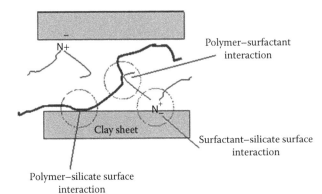

FIGURE 11.2
Possible different interactions in a polymer-intercalated nanocomposite system.

OLSs are reported to be adequate in offering a surplus enthalpy for promoting their dispersion into an apolar matrix, but it is necessary to maximize the magnitude and number of favorable polymer–silicate surface interactions with respect to the apolar alkyl chain (surfactant)–polymer interactions (Vaia and Giannelis 1997a,b, Balazs et al. 1998, Ciardelli et al. 2008). In other words, it is necessary that the surface polarity of both macromolecules and clay be matched in order for the PO to fully wet and intercalate clay tactoids providing specific polymer chain–silicate surface interactions. Therefore, POs with grafted functionalities suitable for interacting with silicate surface can be used as matrix and/or as compatibilizers or coupling agents.

The specific interactions between organoclays and polymer melts can be estimated from the effective Hamaker constants (Hamaker 1937) generally calculated by using the group contribution method and surface tension data (Bhattacharya et al. 2008). In agreement with the thermodynamic theory, Bhattacharya et al. highlighted a negative effective Hamaker constant value for polyethylene, defining consequently, the OLS nanodispersion in this matrix a serious challenge. In fact, this value negatively affects the work of adhesion, a quantitative indication of interfacial bond strength between the silicate surface and polymer matrix, only depending on the dispersion forces owing the lack of other possible interactions (like as the hydrogen bonds) (Kamal et al. 2009).

No similar results/data about functional POs and OLS have been reported, but the role of the chemistry at the interface and of specific interfacial interactions, provided by inserted functionalities, has been experimentally proved by indirect measurements: the effects of the presence of polar groups and their quantity onto the final morphology development and ultimate properties of related nanocomposites have been really assessed (Ciardelli et al. 2008). Two main important effects of the presence of polar groups onto POs during the melt intercalation and after the preparation of the nanocomposites can be highlighted from results reported and discussed in the literature:

1. The polar groups owing to their specific/intrinsic capability to provide interactions with functionalities of OLS layers surface improve compatibility and then the final morphology by helping/favoring the OLS dispersion.

2. The established interactions at the interface are able to thermodynamically stabilize the system and to maintain unchanged the assessed/reached morphologies also during reprocessing of the nanocomposites.

Regarding the first aspect, generally, the possible interactions investigated for a functional PO and an OLS involve the formation of effective hydrogen bonds between the OH groups present on the layer surface (preferentially at the edge) and the carbonyl groups conventionally characteristic of these polymer materials. When the matrix is a fully functional PO, the establishment of such an active interphase leads to an exfoliated morphology. The importance of the presence of carbonyl/carboxylic groups is brought out by the fact that the same unmodified polymer provides the formation of only intercalated nanostructures (Passaglia et al. 2005).

At the same time, the extensive literature about the preparation of polypropylene (PP)-based nanocomposites by using a compatibilizer highlighted that the first step of the process is exactly the intercalation of the compatibilizer/coupling agent (Kato et al. 1997, Kawasumi et al. 1997, Hasegawa et al. 1998, Wang et al. 2004, Dubnikova et al. 2007): this is generally a PP oligomer functionalized with maleic anhydride (PP-*g*-MAH) and the results about the intercalation process and morphology development suggest that the compatibilizer is the only polymer phase able to really interact with the silicate surface of OLS; the following step is the dispersion of intercalated/exfoliated tactoids into the unfunctionalized PP (Figure 11.3).

With reference to the second important role exerted by the polar groups of functional POs, the morphology stability of nanocomposites reached through the help of strong shear forces during the preparation, without the presence of the functional PO, has been only in few cases investigated; these systems, however, remain thermodynamically unstable. Isotactic polypropylene (iPP) or linear low-density polyethylene (LLDPE) nanocomposite materials can be prepared by melt mixing of the polymers with OLS using high shear forces. If such a mixture is heated (e.g., during processing) to a temperature above the melting temperature, immediately a (partial) re-agglomeration of the particles takes place (Fisher 2003, Hotta and Paul 2004). In the case of a iPP-based nanocomposite prepared as above, the starting XRD analysis showed the absence of any small-angle reflection signal indicating an intrinsic homogeneous and exfoliated material; after melting of the iPP crystals, the material morphology changed almost immediately and the appearance of the small-angle reflection due to stacked layer packing occurred (Figure 11.4); as suggested by the author, this effect is due to "thermodynamic incompatibility of the surface-modified sheets and the polymer matrix."

Other evidences of the poor morphological stability of unfunctionalized PO nanocomposites prepared without compatibilizers can be brought out by taking into account papers reporting PE-based nanocomposites prepared by using the polymerization filling technique (PFT) (Alexandre et al. 2000, 2002). Pristine and unmodified clays first were treated with MAO (methylallumoxane) before being contacted by a titanium-based constrained geometry catalyst. This methodology consists of attaching the polymerization catalyst to the surface and into the interlayer space of the silicate and polymerizing ethylene *in situ*.

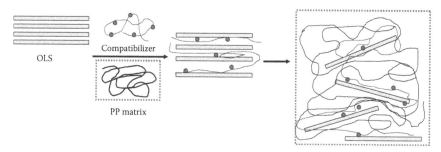

FIGURE 11.3
Schematic representation of PP intercalation process in the presence of a compatibilizer.

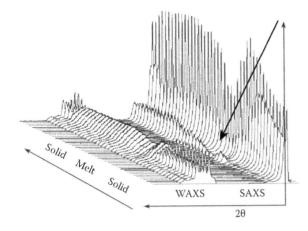

FIGURE 11.4
XRD during melting/crystallization of PP matrix of iPP-clay nanocomposite. (Reprinted from Fisher, H., *Mater. Sci. Eng.*, C23, 763, 2003. With permission.)

The growing polymer chains near the surface move the stacked layers by overcoming the interlayer forces between the platelets and then generating morphologies with different, but generally good, dispersion levels. The exfoliated morphology is, however, thermodynamically unstable: due to the lack of specific surface–polymer chain interactions, for sufficiently fluid PE matrices, the exfoliated structure collapses in the melt (during processing) into a non-regular structure, showing new platelets aggregation.

This introductory discussion highlights how the functionalities present on a backbone of a PO can be effective in granting nanocomposites formation and stability, even if in most of the cases their concentration is really low, less than 1% by mol, which means one polar group per hundred monomeric units. By assuming these important functions as clarified, the main subject of this chapter is to review recent scientific and technological advances in the literature for contributing to the understanding the chemistry at interface, that is, the nature of interactions of silicate surface-functional groups according to their molecular structure, concentration, and distribution along the macromolecules of the polymer matrix; finally the role of these interactions at the interface in changing the bulk properties of the PO matrix (that can be also ascribed to confinement evidences) is described together with their effects onto the ultimate characteristics of such nanostructured materials.

11.2 The Role of the Functional Groups on Preparation of Polyolefin Nanocomposites: Effect on Interfacial Interactions and Clay Dispersion

As discussed in Section 11.1, one of the most critical conditions for the preparation of intercalated and/or exfoliated nanocomposites is the presence of polar-type interactions other than van der Waals forces. Hence, polar polymers (for example, polyamides, poly(ethylene oxide), and poly(vinyl alcohol)) containing groups capable of associative type interactions, such as Lewis acid–base interactions or hydrogen bonds, successfully promote the intercalation of macromolecular chains into the silicate galleries, also by conventional melt processing techniques (Vaia et al. 1995, Vaia and Giannelis 1997b, Strawhecker and Manias 2000, Manias et al. 2001). On the contrary, in the case of non-polar polymers, like POs, the preparation of well-exfoliated nanocomposites typically requires the use of compatibilizers, even if the clay has been organically modified, as the polymer matrix is hydrophobic and lacks interactions with the clay surface (Manias et al. 2001, Wang et al. 2001, Hotta and Paul 2004, Preston et al. 2004, Zanetti and Costa 2004, Zhai et al. 2004).

Favorable enthalpic contributions are necessary to overcome the entropic penalties due to the confinement of the polymer between the silicate sheets (Vaia and Giannelis 1997a,b, Balazs et al. 1998). However, this condition is respected only when polymer–clay interactions are more favorable than surfactant–clay interactions. This represents a challenge for PO/OLS nanocomposites because PO–clay interactions are as poor as the surfactant alkyl–clay interactions; for this reason, the system is under "theta conditions" of mixing and the entropic barrier prevents any dispersion of the inorganic fillers.

The importance of polar interactions for the dispersion of OLSs into POs can be appreciated considering the preparation of exfoliated ethylene vinyl acetate copolymer (EVA)-OLS-based nanocomposites via melt intercalation. Indeed, the polar vinyl acetate moieties, distributed along the polymer chain, allow the macromolecules to intercalate into the inter-galleries of OLS. In fact, it was demonstrated that by mixing an OLS with EVA copolymers, having a different amount of vinyl acetate (i.e., 9, 18, and 28 wt%), extensive exfoliation occurred when more polar EVA18 and EVA28 were used, while only intercalated structures were obtained using the less polar EVA9 (Chaudhary et al. 2005). This suggests that higher amounts of polar groups reduce the thermodynamic energy barrier for clay–polymer interactions, allow a relatively higher number of polymer chains to migrate and stabilize within the clay platelets, and finally form partially exfoliated and/or disordered intercalated states.

Therefore, the solution for creating an excess of favorable interactions in PO/OLS systems is to introduce polar groups into the matrix or to add an amount of a compatibilizer bearing polar groups, which are attracted by clay surfaces to a greater degree than methylene and methine groups of POs. In particular, an optimal compatibilizer is that bearing a fragment highly attracted by the clay surface, which allows favorable polymer/clay interactions, and a longer fragment not attracted by the sheets, but rather attempting to gain entropy by pushing the layers apart and, after the exfoliation, sterically hindering the layer surfaces from coming into close contact, thus stabilizing the final morphology.

Different types of compatibilizers promoting adhesion between PO and clay such as functional oligomers, block copolymers containing polar units, and functionalized polymers with telechelic or randomly grafted functional groups were used (Pinnavaia and Beall 2000, Manias et al. 2001, Sinha Ray and Okamoto 2003, Ciardelli et al. 2008, Harrats and Groeninckx 2008, Pavlidou and Papaspyrides 2008). Among these, the most popular compatibilizers for PO/OLS nanocomposites are undoubtedly POs functionalized by free radical–initiated post-modification reactions. The functional POs are commonly used as coupling agents due to their relatively low cost, adaptability for melt processes, and the large grades' availability. The free-radical functionalization is a widespread methodology, generally carried out in the melt rather than in solution, and currently used on an industrial scale to introduce functionalities onto a PO. It involves the treatment of the PO with a peroxide as radical initiator and a monomer able to graft (generally an unsaturated monomer like as maleic anhydride or its derivatives) bearing suitable functionalities (Ciardelli et al. 1997, 1998, 2004, Moad 1999, Passaglia et al. 2003a,b). The functional monomer (M) is grafted onto the PO backbone by generation of a macroradical, which is successively converted into a non-radical inserted polar group through combination of grafting and transfer reactions (Figure 11.5).

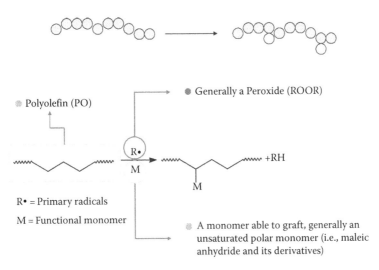

FIGURE 11.5
Ideal mechanism of PO functionalization by radical-mediated post-modification reaction.

Despite difficulties in controlling side reactions (i.e., degradation and chain extension) (Passaglia et al. 2009), this approach is rather popular and it is applied for producing the most part of commercial functional POs, allowing excellent control of the functionalization degree (the functionalization degree FD is defined as the number of grafted functional groups per 100 mol of monomeric units), which can reach values ranging between 0.1 and 1 mol%.

11.2.1 Insight into the Nature of the Functional Group

Initial attempts to improve the interactions between nonpolar POs and layered silicates were carried out using end-functionalized oligomers as mediators of the matrix polarity (Kurokawa et al. 1996, 1997, Kato et al. 1997, Kawasumi et al. 1997, Usuki et al. 1997, Hasegawa et al. 1998, Oya et al. 2000, Wang et al. 2003, Osman et al. 2005).

Usuki et al. first reported the successful preparation of polypropylene (PP)/OLS nanocomposites assisted by a functional oligomer as compatibilizer (Usuki et al. 1997). In particular, a PP oligomer with telechelic OH groups (PP-*t*-OH) was intercalated between the sheets of an alkylammonium exchanged montmorillonite (MMT) via solution blending. The favorable interactions between the silicate layers and the OH groups promoted the oligomer penetration between the clay sheets expanding the basal spacing as the mixing ratio PP-*t*-OH/OLS increased (from 1 to 10), till the disappearance of the basal reflection peak (Figure 11.6). Thereafter, the PP-*t*-OH/OLS masterbatch was melt mixed with PP to obtain quasi-exfoliated PP/OLS hybrids. Indeed, the PP polymer and PP-*t*-OH/OLS masterbatch were under theta conditions and the extrusion promoted mixing through mechanical shears (Manias et al. 2001).

Similar results were achieved by Toyota researchers (Kato et al. 1997) using an octadecyl ammonium–modified layered silicate and three different kinds of PP oligomers: two types of maleic anhydride (MAH)–functionalized PP oligomers containing different amounts of MAH groups (acid value: 52 and 7 mg KOH g^{-1}, respectively) and a hydroxyl-modified PP oligomer (OH value: 54 mg KOH g^{-1}). While for the unmodified PP/OLS composite the wide-angle x-ray diffraction (WAXD) pattern showed that the basal (001) peak did not shift at all, in the case of modified oligomer/OLS samples, the (001) plane peak moved to

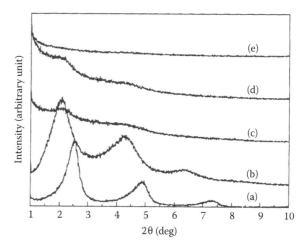

FIGURE 11.6
X-ray diffraction patterns of mixtures of OLS and polypropylene diol oligomer (PP-OH). (a) OLS; (b) PP-OH/OLS = 1; (c) PP-OH/OLS = 3; (d) PP-OH/OLS = 5; and (e) PP-OH/OLS = 10. (Reprinted from Usuki, A. et al., *J. Appl. Polym. Sci.*, 63, 137, 1997. With permission.)

lower angles indicating the increase in interlayer spacing and evidencing the intercalation capability of the functional oligomers. The WAXD analysis highlighted also that the intercalation of the functional oligomers between the sheets was independent of the nature of the functional group (i.e., MAH or OH groups), but at least one carboxyl or hydroxyl group per 25 monomeric units was necessary to promote intercalation. Interestingly, the driving force of the intercalation was considered to have originated from the hydrogen bonds between the oligomer polar groups and polar groups of the silicate (Figure 11.7).

Similarly, hydroxyl-functional PP compatibilizers, prepared by copolymerization of propylene and 10-undecen-1-ol and having a content of OH groups between 1.5 and 2.6 wt%, were compared with a commercial MAH-functionalized polypropylene (high-molecular-weight functional POs are more commonly used than oligomers to prepare nanocomposites by conventional melt-processing techniques) (Ristolainen et al. 2005). Accordingly, PP/organoclay nanocomposites were prepared by melt blending, and two PP/compatibilizer/organoclay ratios, 90/5/5 and 70/20/10, were investigated. It appeared that composites with 70/20/10 weight ratio were more extensively exfoliated than 90/5/5. Moreover, PP-*co*-OH was an effective compatibilizer in the PP/compatibilizer/organoclay (70/20/10) composites, enabling the formation of the desired nanostructure, even if the PP-*g*-MA compatibilizer promoted a more extensive exfoliation of the organoclay. Indeed, the low molecular weight of the PP-*co*-OH compatibilizers did not facilitate the exfoliation more than the PP-*g*-MA compatibilizer.

In addition, MAH-functionalized polyethylene (PE), polypropylene (PP), and ethylene-propylene copolymers (EPM) (Hasegawa et al. 1998, 2000) were successfully intercalated into the clay galleries, like the functional oligomers. Nonetheless, the degree of intercalation/exfoliation appeared dependent on the number of functional groups as well as on molecular weight and polymer architecture, as will be discussed later. Similarly, effective interactions between functional POs and OLS lamellae were also evidenced by blending a diethyl maleate (DEM)–functionalized EPM (EPM-*g*-DEM, with 1 mol% grafted functional groups) with increased amounts (up to 20 wt%) of OLS. In the case of the 5 wt% OLS nanocomposite prepared using EPM-*g*-DEM as matrix, the clay was almost completely exfoliated (Passaglia et al. 2005). It was hypothesized that the hydrogen bonds between the OH groups present on the OLS surface and the carbonyl groups of the functional EPM allowed both optimization and stabilization of clay dispersion (Figure 11.8). Most likely, the compatibilizer was placed between the polar silicate layers and contemporarily mixed

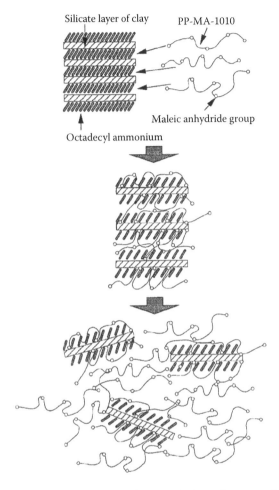

FIGURE 11.7
Schematic representation of the intercalation process of PP-*g*-MAH oligomer into the organized OLS. (Reprinted from Kato, M. et al., *J. Appl. Polym. Sci.*, 66, 1781, 1997. With permission.)

with the nonpolar polymer chains, thus creating a concentration gradient suitable for providing an exfoliated morphology and avoiding phase separation.

A direct comparison between DEM-functionalized PP and MAH-functionalized PP was reported by García-López (García-López et al. 2003): MAH-grafted groups, more polar than DEM grafted groups and possessing a permanent dipole moment, due to the rigid five member ring, seemed to interact more strongly with OLS layers, promoting and stabilizing the clay dispersion. Moreover, this functional PP reacted with the alkyl amine cations of the OLS providing the imide bond formation. Indeed, the modified clay surfactant is able to react as a nucleophile with carbonyl groups and owing to the ring strain the reactivity of MAH carbonyl groups seems higher than that of DEM.

Manias confirmed the efficient interaction of both MAH and OH groups with clay layers by comparing the ability of random PP copolymers having a low amount of *p*-methylstyrene, maleic anhydride styrene, and hydroxyl-containing styrene groups, respectively, to disperse clay at the nanometric level. They were used as matrix of the corresponding nanocomposite (Manias et al. 2001). Even in the absence of mechanical shears, under static melt intercalation conditions, all the copolymers provided an increase in the OLS interlayer spacing, thus demonstrating that the interaction of functional groups with the clay is thermodynamically favorable, and allowing the formation of composites with intercalated OLS tactoids in coexistence with exfoliated/disordered stacks.

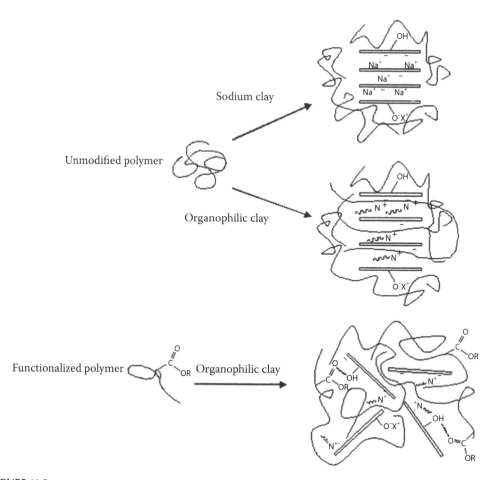

FIGURE 11.8
Effect of polar interaction between compatibilizer and clay layers on the morphology of the composites. (Reprinted from Passaglia, E. et al., *Polym. Int.*, 54, 1549, 2005. With permission.)

In particular, MAH groups, even if in a small amount (0.5 mol%), were more efficient than the other groups in creating strong interactions with the clay layers, preventing the polymer detachment from the inorganic layers during high-temperature processing, similarly to the case of polar polymers, such as poly(vinyl alcohol) and nylon-6 (Strawhecker and Manias 2000).

In addition to MAH and MAH derivatives, grafted PO and oligomers, ammonium-terminated PP samples (PP-t-NH$_3^+$) were also used as compatibilizers preparing well-exfoliated PP/PP-t-NH$_3^+$/OLS 5 wt% nanocomposites via a long-time static melt intercalation step (Wang et al. 2003). The cation exchange between the terminal hydrophilic -NH$_3^+$ groups and sodium cations at the clay surface seems to favor the nanocomposite formation. However, Cui and Paul (2007) did not evidence advantages in the level of clay dispersion by comparing PP-g-MAH over an amine-functionalized PP (PP-g-NH$_2$), prepared by reacting PP-g-MAH with a diamine, as well as for the corresponding protonated sample (PP-g-NH$_3^+$). Nonetheless, they supposed that the high molecular weight of the NH$_2$ and NH$_3^+$ functional PPs, owing to cross-linking side reactions occurring during the PP functionalization, limited the intercalation within clay platelets.

Polyethylene chains functionalized with dimethyl ammonium chloride either as single end group or as multiple functional groups along the chain were also investigated and compared to a diblock copolymer of poly(ethylene-*block*-methacrylic acid) and a MAH-grafted polyethylene with a low functionalization degree (Chrissopoulou et al. 2005, 2008). Results showed that one functional quaternary ammonium end group is not enough to alter the clay structure of both OLS and Na-MMT, and also additives with larger amount of these functional ammonium groups along the chain led to immiscible systems. Instead, carboxyl acid groups of the polyethylene-*block*-poly(methacrylic acid) are able to promote the intercalation of copolymer chains between the OLS layers. However, again PE-*g*-MAH compatibilizer induces the best dispersion of the OLS.

Osman et al. (2005) investigated two different amphiphilic random PE-copolymers, poly(ethylene-*co*-vinyl alcohol) (PE-*r*-VOH) and poly(ethylene-*co*-methacrylic acid) (PE-*r*-MAA), having several polar groups randomly distributed along the polymer chain, as compatibilizers of high-density polyethylene (HDPE)/OLS nanocomposites. These two copolymers were chosen in virtue of their ability to form hydrogen bonds with aluminosilicates and also for investigating the influence of the basic/acid nature of the hydroxyl groups. Indeed, the hydroxyl groups of PE-*r*-VOH are of basic nature, while those of PE-*r*-MAA have an acidic character. These compatibilizers were added in a very low amount to the HDPE/OLS system so that the crystallization and mechanical properties of the matrix were not impaired. Even if a complete exfoliation was not achieved, neither with the PE-*r*-VOH nor with PE-*r*-MAA, both the compatibilizers increased the d-spacing of the OLS. In particular, the PE-*r*-VOH copolymer with a high number of polar units was more intercalated, but without providing exfoliation of the system. Most likely, a simple PE-*r*-VOH chain was able to form multiple hydrogen bonds with different silicate layers, so that clay platelets were bridged hindering the exfoliation. Instead, no bridging between the silicate layers took place with PE-*r*-MAA, probably due to the low molar fraction of polar units or to the acidic nature of OH groups. The results about the barrier properties (oxygen permeability) highlighted that only the exfoliated layers contribute to properties enhancement in nanocomposites and not the intercalated tactoids.

Finally, specially designed block copolymers containing hydrophilic block strongly interacting with the clay were also investigated. In particular, poly(ethylene-*block*-ethylene oxide) (PE-*b*-PEO) nonionic surfactants were used as compatibilizers of PP/OLS nanocomposites, the clayophilic PEO block being able to intercalate unmodified clays (Vaia et al. 1997, Strawhecker and Manias 2003, Moad et al. 2006) whereas the PE block provides compatibility with the PO matrix. PE-*b*-PEO copolymers with linear PE chains and branched structures were mixed with both modified and unmodified MMT and PP. PEO-based surfactants broke down clay agglomerates, provided a small expansion of the Na-MMT structure, and yielded a partially exfoliated structure mainly composed of clay tactoids. Moreover, PEO-based surfactants with longer PEO blocks promoted greater intercalation of the silicate layers rather than exfoliation, as a result of the stabilization of the layered silicate structure as tactoids. A PE-*b*-PEO copolymer with the same molecular weight and composition of end-functionalized PE oligomer with a small polar head group was also used as compatibilizer of HDPE/OLS nanocomposites (Osman et al. 2005). The oligomer easily penetrated between OLS sheets increasing the *d*-spacing by growing its concentration. However, even for the higher concentrations, a complete exfoliation was not achieved, since XRD patterns still evidenced the intercalated clay (001) reflection. The strong interaction between the oligomer and clay layers was confirmed by a significant increase of the composite strength. Indeed, by increasing the amount of oligomer per fixed quantity of OLS, the elastic modulus increased linearly as well as yield stress and stress at break, after an initial small drop resulting from the copolymer plasticizing effect. Obviously, more the copolymer interacted with the clay, more the plasticizing effect can be reduced.

11.2.2 Number of Functionalities and Distribution of Functional Groups

In addition to the nature of the functional group, an important aspect for evaluating the efficiency of a functional polyolefin in establishing stable interactions with the clay and favoring its dispersion is the number of functional groups and their distribution on the polymer backbone. In particular, in the case of the most common PO-*g*-MAH samples, the hydrophilicity increases with the amount of MAH-grafted groups thus improving the compatibilizer–clay interaction. Typically, the level of grafting ranges between 0.1 and 2 wt% (Biswas and Sinha Ray 2001, Marchant and Jayaraman 2002, Yang and Ozisik 2006) and 0.1 wt% seems to be a limit value for achieving a good improvement in morphology (Wang et al. 2001). In the case of both PE and PP/OLS nanocomposites, if the grafting level is higher than this limit value, a shift or even a disappearance of the basal reflection peak of the OLS on the XRD diffraction pattern can be observed (Kawasumi et al. 1997, Hasegawa et al. 1998, Wang 2001, Marchant and Jayaraman 2002).

Since the first studies about the use of MAH functional PP oligomers as compatibilizers of PP/OLS nanocomposites (Kato et al. 1997, Kawasumi et al. 1997, Hasegawa et al. 1998, Manias 2001), it was observed that a very low MAH content does not promote the nanocomposite formation and in contrast a very high MAH content makes the functional oligomer/OLS masterbatch so robust that clay does not mix further with neat PP. Similar results were also achieved in the case of PE/OLS composites prepared in a discontinuous mixer by using PE functionalized with MAH as compatibilizer (Wang et al. 2001). It was observed that only above a critical grafting level of MAH (0.1 wt%) and by using OLS modified with alkyl ammonium chains with a number of ethylene groups higher than 16, a good clay dispersion with partial exfoliation can be achieved. However, the relative amount of grafted MAH groups cannot exceed a given value, in order to retain some miscibility between the functional PO and PP matrix. This aspect is well evident for functional PP oligomers. Indeed, it was demonstrated that only if the miscibility with PP is good enough to disperse at the molecular level, the exfoliation of the intercalated oligomer/clay can take place efficiently. In particular, the macrophase separation texture was avoided only by mixing PP with the oligomer having the lower number of functional groups per chain, thus achieving best nanocomposite morphologies and reinforcement effects (Kawasumi et al. 1997). Nonetheless, in these conditions, the amount of oligomer necessary for achieving an appreciable increase of the PP matrix properties is elevated (20 wt%) thus implying phase separation even if the functionalization degree is low. This is one of the reasons for the more popular use of functional POs as compatibilizers instead of oligomers, despite their easy penetration between clay lamellae. However, even using high-molecular-weight functional polyolefins like in the case of PP/PP-*g*-MAH/OLS composites, if the number of grafted functional groups spread along the polymer chain is too high, it is not possible to obtain an increase in the interlayer spacing, leading rather to the dispersion of PP-*g*-MAH intercalated clay in the PP matrix (Alexandre and Dubois 2000, Pavlidou and Papaspyrides 2008).

The distribution of the functional groups onto the polymer chain also plays a fundamental role. Except functional oligomers and copolymers, more common compatibilizers

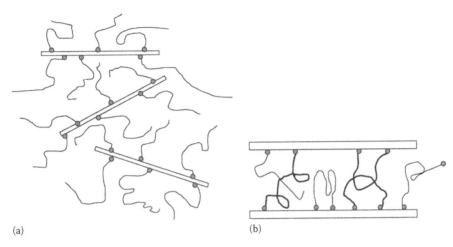

(a)   (b)

FIGURE 11.9
Illustration of the molecular structures of (a) chain-end functionalized polyolefin and (b) side-chain function-alized polyolefin located between clay interlayers (Reprinted from Xu, L. et al., *Macromolecules*, 42, 3795, 2009. With permission.)

are polymers in which the functional groups are distributed randomly along the back-bone or functionalized at the chain end. Interestingly, both simulation studies and experimental results (Ginzburg and Balazs 2000, Sinsawat et al. 2003, Wang et al. 2003) provided evidences that end-functionalized polymers—unlike more functional polymer systems—assume a special configuration on the clay surface which allow complete exfoliation of the clay. In particular, in the case of PP/OLS nanocomposites, it was shown that by using an ammonium group–terminated PP (PP-t-NH$_3^+$) as compatibilizer, the terminal hydrophilic NH$_3^+$ groups interact with the OLS through ionic interaction, resulting in end-tethering of the polymers and exfoliation of the structure (Figure 11.9a) (Wang et al. 2003, Xu et al. 2009).

However, this single interaction site of the polymer with the silicate surface is insufficient for the formation/stabilization of a long-lived silicate network, mediated by polymer chains that bridge across nanoparticles. Therefore, the linear viscoelastic melt rheological data of PP nanocomposites prepared by using PP-t-NH$_3^+$ as compatibilizer (28 wt%), OLS (5 wt%) and three different types of PP having low, medium, and high molecular weight (MW), respectively, displayed a liquid-like response in the low-frequency region for all the three nanocomposites (Figure 11.10).

Evidently, the absence of a long-lived 3D silicate network and the inherent lack of attractive interactions between PP and the layered silicate resulted in the dissipation of the stress via relaxations of the polymer chains, which is manifested macroscopically in a liquid-like behavior. In contrast, side chain–functionalized POs as well as PO block copolymers or random functionalized POs can form multiple contacts with each of the clay surfaces; this results not only in the polymer chains being aligned parallel to the clay surfaces, but also consecutive clay platelets being bridged, thus promoting the production of intercalated structures (Figure 11.9b). This bridging effect was established, for example, by using side chain–functionalized POs characterized by large (>2 wt%) amounts of grafted polar groups (Ginzburg and Balazs 2000, Sinsawat et al. 2003, Wang et al. 2003). In particular, Xu et al. reported that PP-g-MAH compatibilizers were able to form at least two contacts per chain with the silicate layers and further, owing to physical entanglements of such doubly

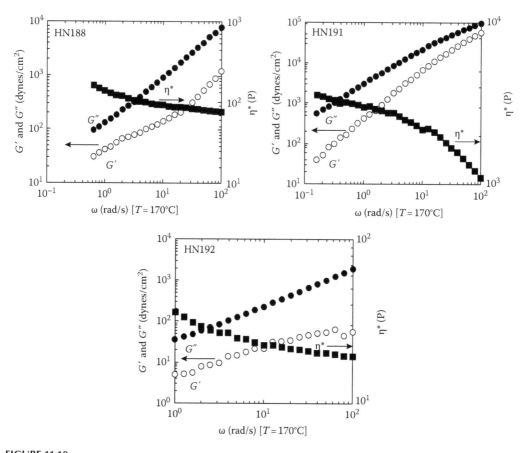

FIGURE 11.10

Melt-state dynamic linear viscoelasticity of the ammonium-functionalized nanocomposites at 170°C. (Reprinted from Xu, L. et al., *Macromolecules*, 42, 3795, 2009. With permission.)

tethered polymers, lead to effective bridging interactions between the nanoparticles and to the development of an extended and possibly hierarchical superstructure, macroscopically manifested in a rheological solid-like behavior. Indeed, the melt-state viscoelastic data for three nanocomposites prepared by diluting a PP-*g*-MAH/OLS masterbatch with neat PP polymers of varied MW evidenced a rheological behavior between that of a liquid and that of a solid at low oscillatory shear frequencies (Figure 11.11). In particular, at all frequencies the values of the storage modulus for the nanocomposite prepared with the high-MW PP were significantly higher than those of medium and low-MW PPs suggesting that there was a significant additional reinforcement in the case of high-MW PP, which was consistent with better dispersion of the organoclay in high-MW PP compared with medium PP and low-MW PP.

A comparable solid-like response, presumably owing to a similar filler network and same qualitative rheological trend, was observed in EVA/MMT-filled nanocomposites, but only when the EVA copolymer contained a high polar comonomer fraction (>28%). In addition, the polymer/nanoparticle interactions of PE nanocomposites prepared by diluting a EVA/MMT masterbatch in PE were substantially weaker than the ones of the respective

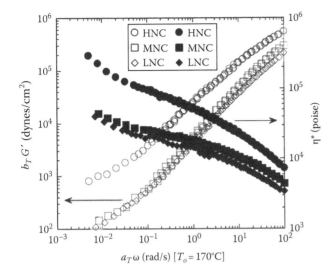

FIGURE 11.11
Melt-state dynamic linear viscoelasticity of the MAH-functionalized nanocomposites diluted from masterbatch superposed at 170°C. (Reprinted from Xu, L. et al., *Macromolecules*, 42, 3795, 2009. With permission.)

PE-*g*-MAH/MMT masterbatch-based nanocomposites. These trends for the EVA-based nanocomposites further attest to the effectiveness of the PO-*g*-MAH in stabilizing OLS filler networks.

11.2.3 The Effect of Functional Polyolefin Molecular Weight

The molecular weight and structure of the compatibilizer as well as melt viscosity and rheological properties are responsible for composites' final morphology (Yang and Ozisik 2006). The best results are generally obtained by using a compatibilizer with similar rheological properties of the matrix; otherwise, the miscibility between the two polymers might be compromised and the intercalation/exfoliation process inhibited (Coiai et al. 2009).

Koo et al. first examined the effect of the MW of functional POs on the intercalation kinetics and on the final morphology of nanocomposites (Koo et al. 2003). Two PP-*g*-MAH samples at low and high MW (59,000 and 18,500 mol/g, respectively) with the same grafting content of MAH (1.8%–1.9%) were used for preparing PP-*g*-MAH/OLS composites containing 1–10 wt% OLS. WAXD and TEM results evidenced that the high-MW PP-*g*-MAH intercalated slowly between the silicate layers, while the low-MW PP-*g*-MAH exfoliated rapidly into the OLS. The slow intercalation of the high-MW compatibilizer was explained as due to low mobility of polymer chains. On the contrary, Hasegawa and Usuki found that PP-*g*-MAH samples having different MAH grafting level and MW were all intercalated into the OLS galleries, and in some cases, owing to the induced shears of the extrusion process, exfoliation also occurred (Hasegawa and Usuki 2004). Recently, it was reported that a low-MW PP-*g*-MAH sample with high functionalization degree successfully intercalated into the galleries of an OLS, but it provided PP nanocomposites with decreased thermomechanical stability and low stiffness at high temperature because of its limited miscibility with the nonpolar matrix (Dubnikova et al. 2007).

Further studies evidenced that depending on the characteristics of the PP-*g*-MAH (MW and grafting content), a different extent of interaction with the clay and miscibility with the matrix can be reached, thus influencing the degree of clay dispersion and

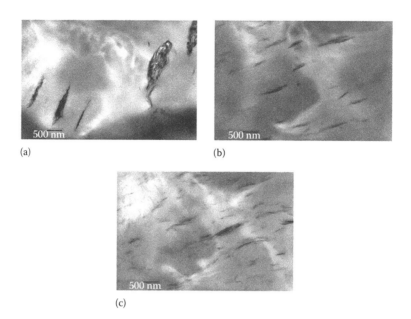

(a)   (b)

(c)

FIGURE 11.12
TEM micrographs of (a) PP/OLS, (b) PP/PP-*g*-MAH (9000 g/mol, 3.5% MAH-grafted groups)/OLS, and (c) PP/PP-*g*-MAH (330,000 g/mol, 0.5% MAH-grafted groups)/OLS nanocomposites at magnification of 15,000×. (Re-elaborated and reprinted from Perrin-Sarazin, F. et al., *Polymer*, 46, 11624, 2005. With permission.)

final properties of nanocomposites (Perrin-Sarazin et al. 2005). In particular, the use of PP-*g*-MAH with a low MW (9000 g/mol) and high grafting content (3.8%) resulted in a relatively good and uniform intercalation, without exfoliation, as observed by TEM images (Figure 11.12).

Such PP-*g*-MAH could interact largely with clay particles and intercalate easily compared with the coarse dispersion obtained by blending OLS with neat PP (Figure 11.13). Nonetheless, the lack of miscibility with the PP matrix can explain the morphological results.

On the contrary, by using a PP-*g*-MAH with high MW (330,000 g/mol) and low grafting content (0.5%), a more heterogeneous intercalation with some exfoliation evidences were achieved as evidenced by TEM analysis (Figure 11.12). In this case, the PP-*g*-MAH interacted to a lesser extent with the clay due to its lower grafting content leading to some limited intercalation (Figure 11.13). However, its higher MW and better miscibility with PP allowed to achieve some larger level of intercalation and a partial exfoliation.

From the results discussed above, a good balance between the functionalization degree (FD) and MW is necessary to reach a good level of clay dispersion till exfoliation. In the case of PP, a high amount of compatibilizer is required to obtain an increase in the OLS basal spacing, but this may cause a deterioration in the properties of the nanocomposite due to the low MW of the commonly used PP-*g*-MAH samples. Unfortunately, functional PPs with high molecular weights are yet not available commercially. The conventional radical functionalization process with peroxide and MAH in the case of PP causes a dramatic decrease in the MW of the polymer, leading to severe damage of its rheological and mechanical properties. Notably, PP is very sensitive to degradation reactions when treated with peroxides above its melting temperature, even

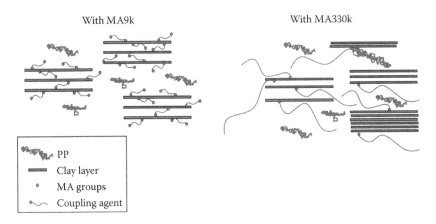

FIGURE 11.13
Scheme of the clay intercalation process with two PP-*g*-MAH compatibilizers at low and high molecular weight as well as high and low MAH grafting degree. (Reprinted from Perrin-Sarazin, F. et al., *Polymer*, 46, 11624, 2005. With permission.)

in the presence of commonly used maleate functionalizing agents (Ruggeri et al. 1983, Moad 1999, Passaglia et al. 2003b, Ciardelli et al. 2004). Consequently, conventional PP-*g*-MAH compatibilizers have in general a low FD and MW. The problem of obtaining appropriately functional PP can be overcome by a new radical functionalization approach involving the use of a furan derivative (butyl 3-(2-furyl)propenoate, BFA). This is added as coagent during the radical functionalization of PP with MAH, and it can yield a wide range of PP-*g*-MAH samples with different FDs and MWs, by controlling the macroradical formation and content (Coiai et al. 2004, Augier et al. 2006, Ciardelli et al. 2006). These new functional PP samples were recently tested in the preparation of PP/OLS nanocomposites, both as matrix or as compatibilizer, with the aim to study the effects of both polarity and chain structure/architecture on clay dispersion and ultimate properties of the corresponding nanocomposites. PP-*g*-MAH with low MW and a high FD (>2 mol%) showed an excellent capability to disperse clay at the nanometric level, especially when used as the matrix of the corresponding nanocomposite. Samples characterized by a high FD value (>2 mol%) and a branched structure/architecture with high MW produced nanocomposites with a lower degree of exfoliation. However, those nanocomposites with a composition of 90/5/5 PP/functional PP/OLS, where the compatibilizer was characterized by a high FD and it was prepared by using the grafting procedure to avoid PP degradation reactions, provided the best performance in terms of morphological and thermomechanical properties. These results not only confirm the important role of the FD, but also highlight the fact that the control of molecular weight and structure/architecture during functionalization ensures a good compatibility of the compatibilizer with the PP matrix, which in turn has a positive effect on the ultimate properties of the PP/OLS. In particular, elongation at break point (which usually is poor for similar systems) reached values in excess of 500%, with an excellent reproducibility (Augier et al. 2009).

By summarizing the main important evidences, we report in Table 11.1 a few examples of functional olefin material–based oligomers and polymers, used as matrix and/or compatibilizer in PO/clay nanocomposites: they are reported on the basis of the type of functionality and its distribution along the backbone, molecular weight, and kind of interactions (as suggested by the authors) with layered silicate platelets. This is not exhaustive

TABLE 11.1

Some Examples of Functionalities in Functional Polyolefins Used as Matrix and/or Compatibilizer of PO/Clay Nanocomposites

Types of Functionality	Distribution of Functionalities	Kind of Oligomer/Polymer	Functionalization Level	Molecular Weight (Da)	Kind of Supposed Interactions of Functionality	Reference(s)
MAH (n < 1000)	Randomly grafted onto olefin oligomers	PP-g-MAH	Acid value 26 mg KOH/g	$M_w = 40{,}000$	Hydrogen bonds with oxygen groups of silicates	Kawasumi et al. (1997)
		PP-g-MAH	Acid value 52 mg KOH/g	$M_w = 30{,}000$	Hydrogen bonds with oxygen groups of silicates	Hasegawa et al. (1998), Kato et al. (1997), Kawasumi et al. (1997)
		PP-g-MAH	Acid value 7 mg KOH/g	$M_w = 12{,}000$	Hydrogen bonds with oxygen groups of silicates	Kato et al. (1997)
MAH (n > 1000)	Randomly grafted onto POs	PP-g-MAH	1.2 wt%	Not available	Generally polar interactions	García-López et al. (2003)
		ULDPE-g-MAH	0.5–1 wt%	$M_w = 129{,}000$	Not evidenced	Morawiec et al. (2005)
		PP-g-MAH	0.6 wt%	Not available	Generally polar interactions	Othman et al. (2006), Lertwimolnun and Vergnes (2005)
		PP-g-MAH	0.55–4.8 wt%	$M_w = 9{,}100$–330,000	Generally polar interactions	Wang et al. (2003, 2004)
		PP-g-MAH	0.6 wt%	Not available	Hydrogen bonds	Chrissopoulou et al. (2005, 2008)
		PP-g-MAH	0.5–3.5 mol%	$M_w = 330{,}000$	Not evidenced	Perrin-Sarazin et al. (2005)
		PP-g-MAH	1.2 wt%	Not available	Hydrogen bonds and MAH reactions with the alkyl amine cations (imide formation)	García-López et al. (2003)

Structure	Distribution	Polymer	Content	Molecular weight	Interactions	Reference
MAH		PP-g-MAH	3.5 wt%	Not available	MAH reactions with the alkyl amine cations (imide formation)	Száazdi et al. (2005)
		PE-g-MAH	<1 mol%	Not available	Interactions with clay surface especially Si–OH and Al–OH groups at the borders of layers	Mainil et al. (2006)
		PE-g-MAH	0.1–0.85 wt%	Not available	Generally polar interactions	Wang et al. (2001)
		PE-g-MAH	0.85 wt%	Not available	Generally polar interactions	Kim (2006)
	Randomly distributed in olefins copolymer	PP-co-MAH from copolymerization with α-methylstyrene followed by modification	0.5 mol%	$M_w = 200{,}000$	Generally polar interactions	Manias et al. (2001)
OH	Randomly grafted onto olefins oligomer	PP-g-OH	Acid value 54 mgKOH/g	$M_w = 20{,}000$	Hydrogen bonds with oxygen groups of silicates	Kato et al. (1997)
OH	End-functionalized olefin oligomers	PP-t-OH	Not available	Not available	Not evidenced	Usuki et al. (1997), Manias et al. (2001)
OH	Randomly distributed in olefin copolymers	PP-co-OH from copolymerization with 10-undecen-1-ol	1.5–2.6 wt%	$M_w = 40{,}000$–$107{,}000$	Not evidenced	Ristolainen et al. (2005)

= —$(CH_2)_n$—(n = 0 or 10)
–Ph

(continued)

TABLE 11.1 (continued)

Some Examples of Functionalities in Functional Polyolefins Used as Matrix and/or Compatibilizer of PO/Clay Nanocomposites

Types of Functionality	Distribution of Functionalities	Kind of Oligomer/Polymer	Functionalization Level	Molecular Weight (Da)	Kind of Supposed Interactions of Functionality	Reference(s)
		PP-*co*-OH from copolymerization with α-methyl styrene followed by modification	0.5 mol%	$M_w = 200,000$	Generally polar interactions	Manias et al. (2001)
		PE-*co*-OH from copolymerization with vinyl alcohol	0.69 mol%; 0.5 mol%	Not available	Hydrogen bonds with aluminosilicates by hydroxyl groups with basic nature	Osman et al. (2005), Wang et al. (2003)
DEM	Randomly grafted onto polyolefins	PP-*g*-DEM	0.9 wt%	Not available	Generally polar interaction	García-López et al. (2003)
		EPM-*g*-DEM	1–1.6 mol%	$M_w = 300,000$	Hydrogen bonds with OH groups of silicates	Passaglia et al. (2005, 2008b), Coltelli et al. (2010)
Other ester groups	Randomly grafted onto polyolefins	PP-*g*-MAH-*g*-BFA from grafting reaction with MAH and butyl furanyl propenoate (BFA)	0.24–0.93 mol%	$M_w =$ 230,000–260,000	Not discussed	Augier et al. (2009)

Structure	Distribution	Polymer	Content	Molecular weight	Interaction	Reference
$-NH_3^+$ or $-N(R)_2H^+$; NH_3^+ $(NH(R_2)^+)$ and succinimide (N-substituted) structure	Randomly grafted onto POs	PP-g-NH₃⁺ from reaction with 1,12-diaminododecane and PP-g-MAH	1 wt% starting content of MAH	$M_w = 90{,}000$ of starting PP-g-MAH	Intercalating/silicate modifying polymer with exchangeable cations	Cui and Paul (2007)
$-NH_3^+$ or $-N(R)_2H^+$; NH_3^+ ... n	End-functionalized POs	PP-t-NH₃⁺ from catalysis	0.02–0.19 mol%	$M_n = 58{,}900$–$24{,}200$	Intercalating/silicate modifying polymer with exchangeable cations	Wang et al. (2003)
		PP-t-NH₃⁺ from catalysis	Not reported	$M_w = 135{,}500$	Intercalating/silicate modifying polymer with exchangeable cations	Xu et al. (2009)
$-COOH;$ $-COOR;$ $(CH_2-CH_2O)_n-OH$	Blocky distributed in block copolymers/oligomers	PP-b-PMMA	1.0–5.0 mol%	$M_w = 200{,}000/220{,}000$	Generally polar interactions	Manias et al. (2001)
		PE-b-PMAA	Not available	$M_w = 13{,}800$	Generally polar interactions	Chrissopoulou et al. (2005)
$(CH_2-CH_2-O)_y-OH$ with $COOH(R)$		PE-b-PEO	Not available	$M_n = 500/1600$	Generally polar interactions due to clayophilic block	Moad et al. (2006)
		PE-b-PEO	0.20 mol%	$M_n = 575$	Generally polar interactions due to clayophilic block which is end-tethered.	Osman et al. (2005)
$-COOH;$ $-COOR;$ $COOH(R)$	Randomly distributed in copolymers/oligomers	PE-co-PMAA from copolymerization with methacrylic acid	0.11 mol%	$M_n = 5{,}700$–$68{,}000$	Hydrogen bonds with aluminosilicates by hydroxyl groups with acid nature	Osman et al. (2005)

with regards to literature references, and we underline that the examples are arbitrarily chosen as considered suitable to provide the reader with illuminating information about the main topic.

11.2.4 From the Two-Step to the One-Step Process: Advantages and Drawbacks

Up to now, we have examined the role, properties, and finally correct selection of functional POs for the preparation of nanocomposites by a conventional approach, which is the use of those polymers as matrices or as compatibilizers. In particular, in most reports, it was observed that the best stack delamination is achieved when a masterbatch of the functional PO (compatibilizer) with a large proportion of clay (10–50 wt%) is prepared, followed by dilution in the nonpolar PO matrix (masterbatch process). The advantage of this method, compared to a direct melt blending of all ingredients together, is that the intercalation of the polar polymer chains between the clay stacks is facilitated and a pre-intercalated material is obtained. Obviously, it is necessary to grant compatibility/miscibility of polymer in the masterbatch with polymer constituting the matrix.

The possibility of obtaining, in a single step, both functionalization of PO and nanocomposite formation is rarely discussed. Such a method would be especially interesting: its successful application would not only avoid selecting the compatibilizer but also save a preparative step. In fact, as noted above, selection of the best compatibilizer may require special care, as this must be sufficiently polar to allow interaction with the clay, but not so hydrophilic as to favor phase separations. Not least, it should have similar rheological properties as the matrix, in order to avoid any detriment of the composite ultimate mechanical properties.

Tjong et al. (2002) were the first to describe an example of one step approach for preparing PP/vermiculite nanocomposites. Accordingly, MAH was introduced between the silicate galleries via interaction between the carboxylic groups of the diacid MAH form and clay platelets; afterward the modified silicate was melt blended with PP and peroxide. In this way, the radical grafting of MAH to the PP backbone occurred *in situ* into the galleries and the functional PP chains were promptly available for interacting with the layers and PP-vermiculite nanocomposites with an intercalated or exfoliated structure were obtained. Both the ternary molecular structure of vermiculite/MAH/PP, where grafted MAH acted as a bridge to bond the vermiculite and PP (Figure 11.14), and the shear forces within the extruder were proposed as the driving force for clay dispersion.

The concept of nanoreactor by preconfining the peroxide between the OLS lamellae to functionalize the PP was introduced by Shi et al. Indeed, in this way, the selectivity of MAH free radical grafting to PP chains was improved as the release of primary radicals

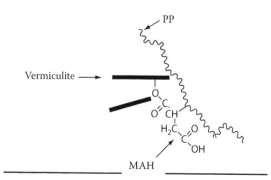

FIGURE 11.14
Ternary vermiculite/MAH/PP molecular structure. (Reprinted from Tjong, S.C. et al., *Chem. Mater.*, 14, 44, 2002. With permission.)

is slowed down and consequently there is a control of PP chain scission (Shi et al. 2006). However, no data about the clay dispersion were discussed.

Recently, the two-step process was directly compared with the one step for the preparation of EPM and PP OLS nanocomposites (Passaglia et al. 2008c). The results obtained by the two-step process indicated that, using the functional PO as matrix, only a fraction of the polar groups on the polyolefin macromolecules would be effective in establishing a bonding with the silicate surface. Instead, if the functionalization step is performed during mixing of the PO with the clay, the small polar molecules can more easily penetrate the clay channel, with grafting to the organic macromolecules occurring successively. In particular, comparative experiments were carried out according to the one-step procedure, following two different routes. In one case, the functionalizing reagents (initiator, DEM and, occasionally, MAH) and OLS were premixed and added contemporarily to the molten polymer in a discontinuous mixer, such that functionalization and intercalation/exfoliation occurred simultaneously. In the second case, the functionalizing reagents were added to the molten polymer, with the OLS being added after a few minutes. Here, the procedure was essentially similar to the two-step method, although for kinetic reasons there was a degree of overlapping within the timescale between the grafting of the functional groups to PO and intercalation. Even if the results obtained evidenced that the one-step process is very complex, the simultaneous addition of the functionalizing reagent, peroxide, and OLS to the molten EPM proved to be effective. Nonetheless, an optimal ratio between the FD and clay basal spacing enlargement appeared to be established.

11.3 Evidences of Polymer/Clay Interaction and Confinement in Polyolefin/Clay Nanocomposites

11.3.1 Structural Characterization by Spectroscopic Analysis

It has been generally assessed that the mechanism of interaction among the components of PO/clay nanocomposites involves the functionalities of the functional polymer used as matrix or compatibilizer. The most frequent explanation is related to the hydrophilic/hydrophobic balance of components involving some kind of undefined polar interaction between the silicate layers and the functional polymer (Alexandre and Dubois 2000, Sinha Ray and Okamoto 2003, Pavlidou and Papaspyrides 2008). The direct intercalation/interaction of a wide numbers of MAH-grafted low-molecular-weight compounds with layered silicates has been studied suggesting that the anhydride can promote the intercalation even if this is not a modeling of polymer intercalation (Sibold et al. 2007).

The beneficial effects on nanocomposite final properties as well as on dispersion level of anhydride, acid, ester, and/or hydroxy groups grafted or embodied onto the backbone of POs are well documented, even if the nature of the interfacial coupling reactions requires further clarification and in-depth examination. Hydrogen bridges as well as covalent, hydrophobic, and ionic bond formations at the interface between a filler and a PO are generally invoked and are required to transfer stresses during processing in order to achieve improved exfoliation and stabilization of uniformly dispersed anisotropic nanoparticles (Reichert et al. 2000).

An often cited rationalization, presented for the first time by Kato et al. (1997), but also reported in other reports (Usuki et al. 1997, Mishra et al. 2005), explains the exfoliation and improved properties by formation of hydrogen bonds between the oxygen groups of

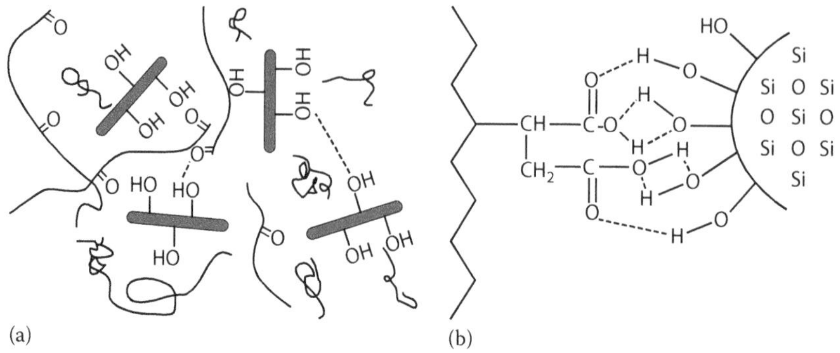

(a)   (b)

FIGURE 11.15
(a) Schematic representation of percolating network of tactoids in the presence of carbonyl-functionalized poly-mer highlighting a competition for association with the clay surface. (Reprinted from Thompson, M.R. and Yeung, K.K., *Polym. Degrad. Stab.*, 91, 2396, 2006. With Permission.); (b) Interaction between the maleic anhydride groups of the functionalized PP and the surface hydroxyl functionalities of the silica nanoparticles. (Reprinted from Bikiaris, D.N., *Eur. Polym. J.*, 41, 1965, 2005. With permission.)

the silicate layers and the polar groups of the functional polymer. In the wide literature scenario, there are schemes reporting the hydrogen bond formation between the carbonyl groups of grafted functionalities and OH groups present on clay layers (Figure 11.15a).

This kind of interactions can be really effective in the case of silica nanoparticles (Figure 11.15b) where the OH groups are located on the surface of the layers (Bikiaris et al. 2005). Instead, in the case of layered silicates, the hydroxyl functionalities are located in the octa-hedral sheet of aluminum linked by shared oxygen atoms to tetrahedral sheets of silica that constitute the surface of the layers. By taking into account their structure, the layered silicates contain active OH groups only at the edges and their interaction with polymer functionalities, though useful for the intercalation, cannot completely explain results, spe-cially the exfoliation. Such structures, as reported in Figure 11.15a, are generally inferred by indirect measurements like the rheological behavior of nanocomposites showing a solid-like behavior attributed to the development of an extended and possibly hierarchical superstructure, where platelets and polymer chains are strongly interacting and intercon-nected through interaction between functionalities (Lim and Park 2001, Thompson and Yeung 2006, Wang et al. 2006, Xu et al. 2009). In fact, few techniques are really focused to investigate which kind of effective interactions occur and are established during the inter-calation of functional POs between the layers of an OLS. The IR spectroscopy is one of the most common due to ease of use and application to solid materials, and it was applied for characterizing the state of intercalation and exfoliation in polymer nanocomposites, by the analysis of Si–O stretching modes, which give rise to strongly absorbing bands in the 1100–1000 cm^{-1} region (Cole 2008). Some of these modes involve the basal oxygens of the silicon oxygen tetrahedra (these correspond to Si–O–Si linkages at the surface of the clay layers) and have a transition moment lying in the plane of the layer; for this reason, they were designated "in-plane" and generally provide three different peaks, for which the position of the maximum depends on the structure of the layered silicate. Others involve the apical oxygens (corresponding to the Si–O bonds directed toward the octahedrally coordinated aluminum ions) at the center of the layer; the transition moment of these vibrations is perpendicular to the layer and for this reason they were designated "out of plane." The intrinsically high absorptivity of these bands resulted in significant variations when the clay is dispersed in a "swelling" media (water or solvent or a polymer matrix). As an example, IR spectra of pure organoclay (Cloisite15A, intercalated with 43 wt% dimethyl

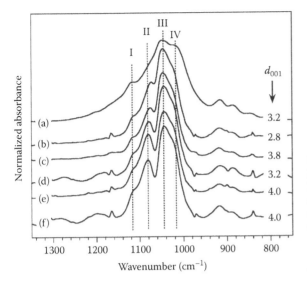

FIGURE 11.16
FTIR spectra of 2% Cloisite15A: (a) dispersed by hand in Nujol; (b–f) in PP nanocomposite films containing different amount of compatibilizing agents (d–f) differing in increased level of maleic anhydride grafted functionalities (c and d); (b) is the composite obtained without compatibilizer. (Reprinted from Cole, K.C., *Macromolecules*, 41, 834, 2008. With permission.)

dihydrogenated tallow) and of PP nanocomposites prepared by dispersing the organoclay and PP-g-MAH functional polyolefins with a different amount of grafted polar groups are reported in Figure 11.16 with the corresponding interlayer distances (d_{001} on the right).

Compared to the unprocessed clay (Figure 11.16a), the clay that was processed with PP in the extruder (Figure 11.16b through f) gave a narrower band envelope and showed a visible out-of-plane vibrational mode (peak II). Even without the compatibilizing agent, this peak can be seen at 1075 cm^{-1} (Figure 11.16b). When 4% of compatibilizer is present in the composition, it became more pronounced and the maximum shifted to 1077 cm^{-1} (Figure 11.16c). By increasing the content of grafted functionalities (MAH), it became still more evident and shifted to 1079 cm^{-1} (Figure 11.16d). The trend continued if the concentration of the compatibilizer was increased, with peak II continuing to shift slightly to 1080 cm^{-1} (Figure 11.16e) and 1081 cm^{-1} (Figure 11.16f). This apparent growth and/or shift of peak II was very similar to that observed for swelling with water. The author associated this trend with the increasing efficiency of intercalation/exfoliation, and the XRD data included in Figure 11.16 tended to support this explanation. It is also important to underline that the change in position and width of the peak arising from the out-of-plane Si–O vibrational mode was dependent on the amount of compatibilizing agent present in the material. By taking into account this experimental result, the author supposed that the chemistry of the compatibilizing agent also plays a role and the presence of a higher amount of polar MAH groups may result in enhanced interaction with the Si–O dipoles.

Very similar results were recently obtained for LLDPE nanocomposites produced again from Cloisite15A and PE-g-MA as compatibilizer (Dintcheva et al. 2009). ATR-FTIR has been used to determine the extent of organoclay delamination by monitoring the in-plane and out-of-plane absorptions of the Si–O bond at around 1075–1045 cm^{-1}. The fact that Si–O absorption bands, which were overlapped in the neat clay, became much more resolved in the compatibilized analogue sample (particularly for the out-of-plane vibrational mode) confirmed the effect of the functional PO onto the dispersion level, but once again suggested the occurrence of a specific interaction of grafted functionalities (the carbonyl of the anhydride) with the oxygen of Si–O dipoles of the layers.

Some spectroscopic evidences of this interaction can be inferred from studies regarding poly(methylmethacrylate) (PMMA) nanocomposites prepared by employing layered

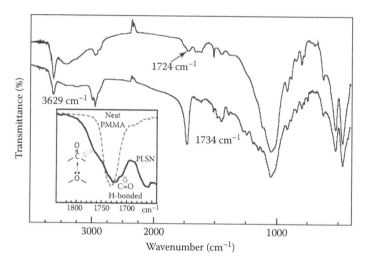

FIGURE 11.17
FTIR spectra of PMMA nanocomposites after (top spectrum) and before (bottom spectrum) extraction by THF. The inset compares the carbonyl stretching band of neat PMMA (dash line) and the extracted sample (solid line). (Reprinted from Chang, C.C. and Hou, S.S., *Eur. Polym. J.*, 44, 1337, 2008. With permission.)

silicate modified with tyramine hydrochloride (Chang and Hou 2008). The FTIR analysis of polymer and collected nanocomposites showed the C=O stretching of intercalated PMMA molecules broader than that of neat PMMA and two clear shoulders toward lower and higher wavenumbers can be observed in the spectra: the first is due to H-bond C=O stretching with OH of tyramine, but the latter was attributed to an interaction of C=O with the surface oxygens of silicate layers (Figure 11.17). This kind of interaction is feasible where the lone pair of electrons associated with the surface oxygen is donated toward the electron-poor carbonyl atom of the functional group.

Similar interactions can be thus invoked for explaining the results about the improved exfoliation levels obtained for functional POs (with respect to the unfunctionalized polymers) containing carbonyl groups (ester, acid, and anhydride functionalities) and even if not exhaustive, they can partially clarify the intercalation mechanism. EPM functionalized with diethylmaleate (EPM-*g*-DEM) was intercalated into OLS (montmorillonite MMT modified by dimethyl-dihydrogenated tallow-quaternary ammonium chloride) providing nanocomposites with different dispersion level and morphology dependent on the content of inorganic substrate. After extraction with toluene to remove polymer chains not really interacting with silicate surface, the ATR analysis of carbonyl ester group of the EPM-*g*-DEM fraction contained in the residue evidenced a shift of the stretching band and an enlargement with respect to neat EPM-*g*-DEM (Figure 11.18) (Passaglia et al. 2008a). A shoulder at about 1715 cm⁻¹ due to a new absorption at lower energy suggested the C=O participation to a hydrogen bound with the silicate surface, and a weak shoulder at higher wavenumbers was highlighted owing to the donary bond of surface oxygen of silicate.

Specific interactions between MAH groups grafted onto EPDM chains and the surface of halloysite nanotubes (HNT, aluminosilicate with nanotubular structure) have been evidenced, involving both the oxygens of Si–O and Al–OH (Pasbakhsh et al. 2009). The Al–OH group is located inside the tubes while the outer surface of HNTs is covered by the Si–O. HNT is a dioctahedral 1:1 layered aluminosilicate consisting of two different interlayer surfaces. Aluminum atoms make an octahedral structure with oxygen atoms, while OH groups are situated on one side of the lamella, and silicon–oxygen atoms are located on the other side of the lamella (Figure 11.19).

According to the ATR-FTIR spectra of HNTs, the absorption peaks around 912 and 1032 cm⁻¹ are related to the Al–OH vibrations and Si–O stretching bands, respectively.

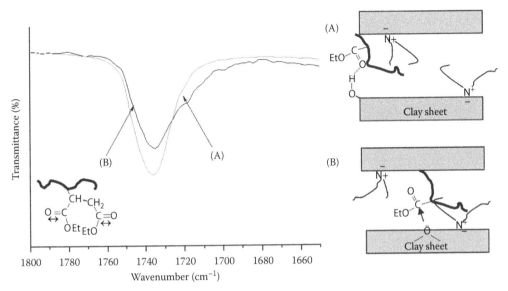

FIGURE 11.18
ATR spectra of EPM-*g*-DEM fraction in the residue to toluene extraction of the nanocomposite (B) and of pristine functional polyolefin (A), enlargement of C=O stretching, and schematic representation of possible occurring interactions.

For the FTIR spectra of compatibilized EPDM/HNT nanocomposite with 5 phr of HNT loading, the absorption bands of Si–O stretching and Al–OH are shifted to 1027 and 932 cm^{-1}, respectively, both related to the formation of hydrogen bonds between outer and inner surfaces of the HNTs with EPDM-g-MAH, then suggesting the possibility of interaction of grafted polar functionalities with polar groups of the inner octahedral sheet.

The direct favorable interactions between the functionalities of PO and the polar groups of the inorganic layer surface is not completely clarified, and the sole explanation for the improved exfoliation results is, in some cases, not comprehensive, because it does not take into consideration the presence of the aliphatic chains, which cover the majority of the silicate surface. Generally, the surfactants are quaternary ammonium salts bearing one or two long alkyl chains and methyl substituents, providing weak "unfavorable" interactions with the backbone of the molten PO (Vaia and Giannelis 1997a). In some cases, the surfactant is differently structured and specific interactions with functional POs can be provided.

Model reactions carried out with components frequently used for the preparation of intercalated or exfoliated PP nanocomposites prove that maleinated polypropylene (PP-*g*-MAH) can react chemically with the surfactant used for the organophilization of the filler if this latter contains active hydrogen groups (Százdi et al. 2005). The reaction of hexadecylamine and PP-*g*-MAH was highlighted by FTIR spectroscopy (and also by MALDI-TOF experiments): the anhydride groups were consumed and amide groups were formed in the reaction of PP-*g*-MAH with the surfactant adsorbed on the surface of the silicate in ionic form. As a result of this reaction at the interface, the surfactant was removed from the surface and hydrogenated silicate sites were left behind. In these conditions, the surface may interact either with the anhydride or the amide groups by dipole–dipole interactions, but in any case the compatibilizer/surfactant reaction at the interface seems to play a crucial role in structure formation of PP nanocomposites containing functional polymer.

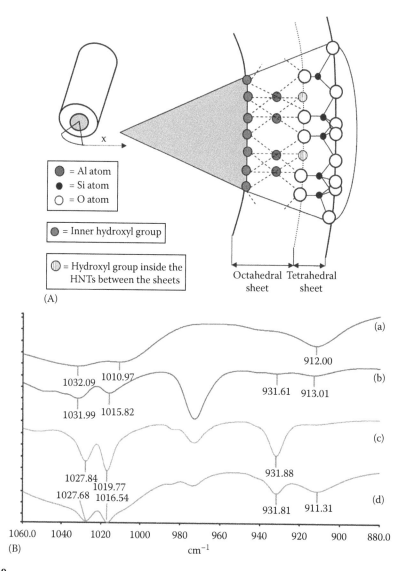

FIGURE 11.19
Structure of HNT (A) and FTIR spectra (B) of (a) HNT, (b) uncompatibilized composite, (c–d) compatibilized nanocomposites with different concentration of HNT. (Re-Elaborated and reprinted from Pasbakhsh, P. et al., *Polym. Test.*, 28, 548, 2009. With permission.)

Similar results were also obtained and discussed by using HDPE functionalized with vinyl triethoxysilane as matrix of nanocomposites obtained by dispersing bentonite modified with octadecyl trimethyl ammonium (BT). Even if the surfactant was not containing active hydrogen, the functional HDPE chain contained hydroxyl groups, which were considered to be reactive with the surfactant (Fang et al. 2006). This was confirmed by FT-IR spectra (Figure 11.20) where Figure 11.20a is the curve of pure BT, 1042.1 cm^{-1} is the characteristic peak of octadyl trimethyl ammonium; Figure 11.20b and c are the curves of HDPE/BT and functional-HDPE/BT after extraction with xylene and chloroform, respectively.

Comparing Figure 11.20a with b, we can observe that the two curves are almost the same, indicating that the extracted residue of HDPE/BT is only BT, and the polyethylene

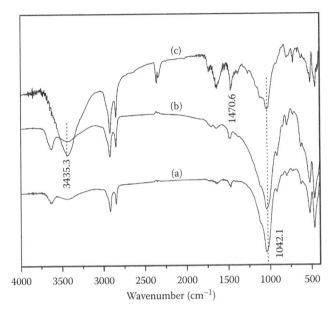

FIGURE 11.20
FTIR spectra of (a) BT, (b) HDPE/BT, and (c) functional HDPE/BT after solvent extraction procedure. (Reprinted from Fang, Z. et al., *J. Mater. Sci.*, 41, 5433, 2006. With permission.)

was totally extracted. The result infers that the chain of polyethylene did not bond with BT. On the contrary, the change in curve (c) is evident: first, the intensity of the amine peak reduced, revealing that part of the amine reacted with hydroxyls of silane; second, there is a characteristic peak of polyethylene at $1470.6\,cm^{-1}$; third, there is a strong peak of hydroxyl at $3435.3\,cm^{-1}$. These results indicate that the functional HDPE was chemically bonded with BT.

In other cases, the surfactant is specifically bearing functionalities to generate chemical reactions with the functional POs with the aim to establish covalent bonds favoring the intercalation process and the stabilization of final morphology. For instance, the IR spectra of nanocomposites obtained from PP-g-MAH and montmorillonite modified by different α,ω hydroxy-amines show the absorption bands characteristic of the anhydride group (1865 and $1790\,cm^{-1}$) and a new absorption peak at $1730\,cm^{-1}$ owing to the esterification reaction taking place between the anhydride and the hydroxy groups of OLS (Jang et al. 2005) and providing nanocomposites with good dispersion level and enhanced mechanical properties (Figure 11.21).

The use of a sylane derivative was also approached to directly modify the surface of inorganic layer by "anchoring" amino groups reacting with PP-g-MAH (Du et al. 2006); this reaction allowed to chemically graft PP chains onto the surface of halloysite and then the g-halloysite was compounded with PP to form composites. The occurrence of reaction and amide formation was proved by model reactions and FT-IR spectroscopy: the g-halloysite was dispersed in PP matrix in the form of microscale clusters, which showed strong interfacial adhesion with PP matrix providing new improved mechanical performance.

11.3.2 Solubility and Morphological Stability

The variation of polymer chains' solubility owing to specific interactions with inorganic substrates was clearly evidenced, particularly in the field of filled elastomers: The interaction between filler and elastomer materials is revealed by formation of a polymer fraction not extractable (insoluble) called "bound rubber" that seems to play a significant role in

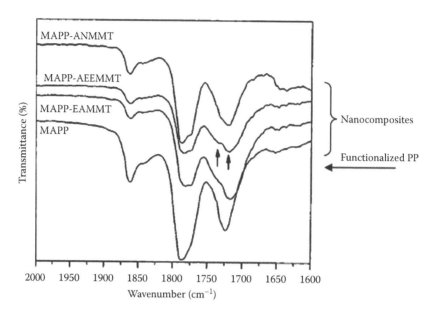

FIGURE 11.21
FTIR spectra of PP-*g*-MAH and PP-*g*-MAH/OLS nanocomposites (Re-elaborated and reprinted from Jang, L.W. et al., *J. Appl. Polym. Sci.*, 98, 1229, 2005. With permission.)

the reinforcement property of provided composites, particularly with carbon black and amorphous silica microparticles (Kraus 1971, Dutta et al. 1994). In this latter case, when the polymer matrix is a functional PO, the amount of the bound rubber is increased due to the hydrogen bond of carbonyl groups with the OH groups on the surface of the silica particles whose interaction is represented in Figure 11.15b. The amount of bound rubber is reported to increase linearly with the filler loading (Passaglia et al. 2001).

The solvent extraction methodology applied to microcomposites for the bound rubber evaluation can be considered an useful and suitable tool to selectively isolate the polymer really interacting and/or trapped/confined between the platelets of a clay and therefore to highlight and in some way to quantify the extent of interactions and their effect onto bulk properties of this particular polymer fraction (Figure 11.22).

Few examples of this methodology application are discussed in the literature, particularly for PO/clay materials, most of them reporting results about PO–clay hybrids prepared

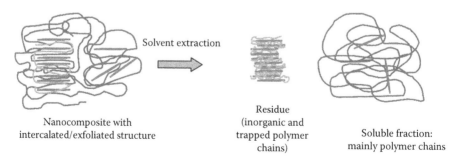

FIGURE 11.22
Schematic representation of isolation of polymer chain trapped/confined between clay platelets by solvent extraction procedure.

by in situ polymerization (Du et al. 2002, Shin et al. 2003, Cui and Woo 2008). Independent of the type of clay (montmorillonite, palygorskite, etc.), kind of catalyst (Ni, Ti, or V-based), and experimental conditions, a fraction of polyethylene seemed to be trapped inside the layers and its amount increased by increasing the content of the inorganic clay in the composite. Even if any specific interaction between polyethylene and layer surfaces was evidenced, the formation of two different fractions of polyethylene phase is invoked to explain these results: the PE chains growing directly from the surface due to the presence of several "activated" points on each platelet; this phase is called "macromolecular comb" and it is the fraction of insoluble polymer that interacts with the second fraction of PE chains that is generated by the catalyst not "anchored" on the surface and then extractable.

For PO/clay nanocomposites obtained by melt mixing, a fraction of insoluble polymer chains is collected only in the presence of suitable functionalities grafted onto the macro-molecules (Fang et al. 2006, Passaglia et al. 2008b) (Figure 11.19).

In the case of EPM-*g*-DEM composites, the amount of bound PO, obtained with montmorillonite was much larger (Figure 11.23) than with silica microparticles (Passaglia et al. 2001). This difference was kept constant also by normalizing the content of bound PO with respect to the surface area of the two inorganic substrates.

A very similar solvent extraction behavior was discussed for intercalated PMMA/OLS nanocomposites (Li et al. 2003). In the case of EPM-*g*-DEM, the authors underlined a very general high content of insoluble PO suggesting that even if the EPM-*g*-DEM contains a substantially lower amount of ester groups (1.8 for 100 mol of monomeric units) than PMMA (100 for 100 mol of monomeric units), the interactions occurring between the polymer and the silicate surface are able to grant a part of PO chains to stay immobilized inside and/or on the layered silicate. At the same time, an evident effect of the inorganic content can be observed: by increasing the loading, the amount of trapped polymer fraction seemed to be enlarged (until a limit value) and the

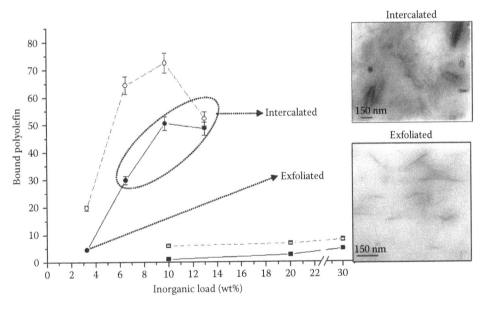

FIGURE 11.23
Bound polyolefin (filled symbols) vs. inorganic load for OMMT and amorphous silica composites and data (open symbols) normalized with respect to the inorganic surface area. TEM analysis of exfoliated and intercalated samples. (Re-elaborated and reprinted from Passaglia, E. et al., *Eur. Polym. J.*, 44, 1296, 2008b.)

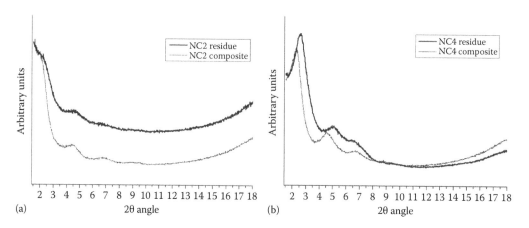

FIGURE 11.24
XRD analysis of (a) NC2 (10 wt% of OLS) and its residue to toluene extraction; (b) NC4 (20 wt% of OLS) and its residue to toluene extraction. (Reprinted from Passaglia, E. et al., *Eur. Polym. J.*, 44, 1296, 2008b.)

morphology varied. "Bridging behavior" of polymer chains interacting with stacked layers in an intercalated structure (see also Figure 11.9b) could be characterized by a decreased solubility with respect to polymer chains interacting with layers in an exfoliated structure (Figure 11.9a).

The XRD analysis of extraction residues (Figure 11.24) showed a substantially unchanged interlayer distance/morphology for sample containing 10 wt% of OLS confirming a homogeneous nanostructure/morphology for the whole composite, whereas sample with 20 wt% of OLS showed an interlayer distance decreasing (from 4.07 nm for the composite to 3.83 nm for the residue) after extraction.

Two polymer phases seemed to form the intercalated structure in this case: the chains really strongly interacting with the silicate platelets within the layers and intercalated (for entanglement with macromolecules strongly interacting with platelets) but extractable chains, suggesting a gradient of the interaction degree at high OLS percentage.

Accordingly, sample with 20 wt% of OLS showed a bound PO/surface area content similar to that of sample with 10 wt% of OLS due presumably to the poor degree of dispersion as evidenced particularly by XRD analysis before and after toluene extraction.

Very similar results were obtained also in the case of PP nanocomposites using PP-*g*-MAH as coupling agent (Muksing et al. 2010); for intercalated nanocomposites, the extraction with polymer solvent generates a collapse of the intercalated structures with high content of inorganic filler owing to the removal of entangled chains fraction.

11.3.3 Thermal Properties: Crystallization and Glass-Transition Behavior of Confined Polymer Chains

It is well known that in polymer/clay nanocomposites, the polymer crystallization behavior depends not only on polymer molecular structure, but also on type and content of fillers, morphology, and polymer/clay interactions. In particular, the crystallization behavior of the polymer matrix can be affected by two competing factors: first, the clay can provide heterogeneous nuclei during nucleation step; then, interfacial interactions between the clay surface and polar groups grafted on the polymer may restrict the conformational transitions of molecules during growth step.

Generally, the addition of a rather small quantity of filler (<1 wt%) into crystallizable polymers, such as PP, PET, and PA, speeds up the crystallization process, reduces spherulite size, and changes the final morphology, but in the case of PE the flexible nonpolar chains crystallize so rapidly that a small quantity of the filler does not act as nucleating agent (Kim 2006). Nevertheless, if a large amount of filler (>1 wt%) is introduced into a PE matrix, the morphology and crystallization behavior of PE may be altered as well if a functional PE is used as matrix or as compatibilizer. Accordingly, the crystallization behavior of exfoliated and intercalated PE-*g*-MAH/OLS nanocomposites prepared by solution blending was compared (Xie et al. 2006). WAXD and TEM analysis evidenced that the morphology of the samples changed from exfoliated to intercalated by gradually increasing the amount of OLS. Indeed, the morphology is decided by the equilibrium between PE-*g*-MAH/clay interfacial interactions and van der Waals attraction of clay–clay; hence, it depends on the ratio between the functional PO and OLS. Although the authors observed that the clay reduced the size of spherulites, it had no significant effect on the total non-isothermal crystallization rate. Otherwise, study on the dynamic mechanical properties showed that the storage modulus of the hybrids was around 30% higher than that of the polymer matrix. Moreover, the motions were confined by the strong interactions between the polymer and the clay.

Indeed, it is reported that in intercalated polymer/clay nanocomposites, the polymer chains confined in a 2D space may show a different crystallization kinetics compared with that of free polymer chains (Giannelis et al. 1999, Manias et al. 2000). In the specific case of intercalated poly(ethylene oxide) (PEO)/clay nanocomposites, the crystallization of polymer chains confined between lamellae was even suppressed (Kuppa et al. 2003). In the case of intercalated and exfoliated PE/MMT nanocomposites, prepared by in situ polymerization using MMT-supported metallocene as catalyst, the DSC analysis evidenced that the exfoliated sample exhibited higher crystallization temperature than the neat PE, evidencing a nucleation effect of the clay (Xu et al. 2005). On the contrary, the intercalated sample had lower crystallization temperature than the neat PE due to the confinement that suppresses the crystallization. Moreover, it was observed that the intercalated sample had a longer induction period and a faster crystallization rate, indicating coexistence of suppression and nucleation effects, in agreement with a smaller crystallization activation energy determined with the Kissinger method.

Finally, the Avrami equation modified by Jeziorny (11.1) was applied for studying the non-isothermal crystallization kinetics (Xu et al. 2005):

$$1 - X(t) = \exp(-Z_t t^n) \tag{11.1}$$

By the Avrami plots, collected at different cooling rates, the Avrami exponent n, which depends on the type of nucleation and growth process parameters, for both exfoliated and intercalated PE/MMT nanocomposites was determined. It was found that the Avrami exponents for the exfoliated sample were around 3.0, indicating a 3D growth of the PE crystals. In contrast, in the case of the intercalated nanocomposite, the Avrami exponents were close to 2.0, suggesting a 2D growth of the PE crystals. This last result showed that in the intercalated sample, the PE crystals were confined between the clay layers, and crystal stems were constrained to grow parallel to the MMT layers and not perpendicular.

Similar results have been also collected by studying the crystallization behavior of PE-*g*-DEM/OLS nanocomposites prepared with a different amount of montmorillonite (MMT) (Passaglia et al. 2008a). The diffraction of d_{110} and d_{200} lattice planes of orthorhombic polyethylene was investigated. The relative intensity of the d_{110} peak decreased with increasing MMT amount in PE-*g*-DEM composites (Figure 11.25), as a result of molecular

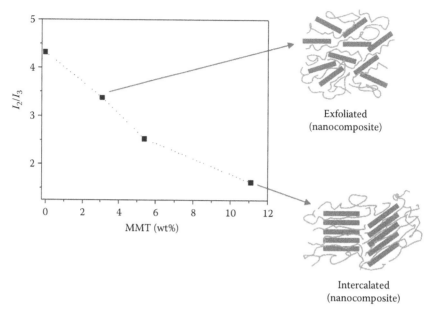

FIGURE 11.25
Ratio between XRD d_{200} and d_{110} intensity peaks of orthorhombic polyethylene polymorph in PE-*g*-DEM-based composites at different content of MMT and morphology.

axis orientation along a preferential direction in agreement with the morphology and the occurrence of a 2D growth of LDPE crystals inside the MMT gallery with stems parallel to the MMT layers.

The effect of PE-*g*-MAH as compatibilizer of PE/OLS nanocomposites was investigated by varying its concentration. PE/PE-*g*-MAH/OLS samples showed an increase in both the exothermic peak temperature and activation energy compared with PE/OLS composites prepared with the same clay concentration, but without the compatibilizer. Hence, the clay was effective as nucleating agent and the composite system with PE-*g*-MAH was more active in nucleation process (Kim 2006). These effects were correlated with the increase of the melt viscosity in the case of PE/PE-*g*-MAH/OLS samples due to the confinement effect on the motion of the polymer chains and stronger interactions between polymer and clay.

The glass transition temperature (T_g) of polymer material in nanocomposites is strongly affected by the addition of nanoparticles and, particularly when there is a good filler–particle interaction, T_g of the amorphous polymer tends to increase by decreasing the size of the particles or by increasing the filler content; this behavior is generally associated to the confinement effect generating a reduction in chain mobility until suppression of cooperative segmental motions of the confined macromolecules. These effects, well known for polar polymers (Giannelis et al. 1999), were also evidenced for POs particularly for amorphous matrices. The T_g of natural rubber increases in the presence of organoclay due to the confinement of the elastomer segment into the organoclay nanolayers (Wang et al. 2005).

Accordingly, EPM-*g*-DEM organoclay (Passaglia et al. 2008b) and PP/PP-*g*-MAH/organoclay nanocomposites (Muksing et al. 2010) prepared by intercalation/exfoliation processes in the melt show larger T_g as detected by DSC measurements with increased silicate content (Figure 11.26). The isolation of polymer trapped inside the layers by solvent extraction procedures allowed to deepen inside the thermal characteristics

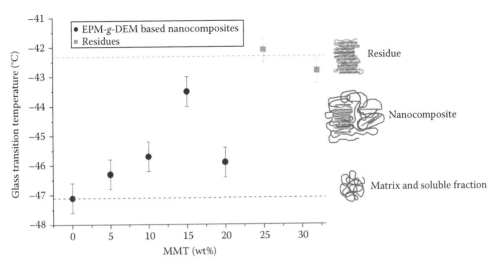

FIGURE 11.26
T_g values of EPM-g-DEM/OLS nanocomposites and their residues to solvent extraction.

of this fraction. Different T_g values of bulk and confined polymer chains, were indeed collected: higher T_g values were detected for the polymer in the residue providing further evidences of polymer–layer interactions and confirming the reduction of mobility of polymer chains trapped into the platelets.

Owing to the presence of functional groups characterized by a dipole (the ester grafted groups), dielectric measurements on nanocomposites and their residues were performed evidencing that the α-relaxation peak disappears in such last samples. The cooperative dynamics of the dipolar groups of the macromolecular chains interacting with and confined into the layers is strongly reduced.

11.3.4 Effects of Confined Polymer onto Prediction of Barrier Properties

The effects of interactions between functional PO chains and clay platelets, generating "confined" polymer fractions, onto the final performances of nanocomposites with particular reference to mechanical properties and thermal stability have been discussed in literature with wide reviews on the topic (Alexandre et al. 2002, Gopakumar et al. 2002, Sinha Ray and Okamoto 2003, Preston et al. 2004, Százdi et al. 2006, Tjong 2006, Ciardelli et al. 2008). On the contrary, in absence of functionalities onto the backbone of POs, even if a certain degree of exfoliation can be reached by shearing forces, the lack of specific polar interactions negatively affects the general mechanical properties (Mittal 2007).

The barrier properties of polymer/layered silicate nanocomposites are generally justified and theoretically treated by considering models geometrically based like the "tortuous path," that is, when impermeable nanoparticles are incorporated into a polymer, the permeating molecules are forced to wiggle around them in a random walk, and hence diffuse by a tortuous pathway.

Models able to describe permeability in nanofilled polymers are developed by taking into account the general tortuosity arguments elaborated by considering correlation between the sheet length, concentration, relative orientation, and state of aggregation, all parameters proving guidance for preparing better barrier materials using the nanocomposites approach (Bharadwaj 2001, Gusev and Lusti 2001, Lu and Mai 2005, Sorrentino et al. 2006).

The reduction of permeability arises from the longer diffusive path that the penetrants must travel in the presence of the filler (layered silicate in the present case). A sheet-like morphology is particularly efficient as it maximizes the path length. The tortuosity factor (f or τ depending on the symbology) is defined as the ratio of the actual distance (d^l) that a penetrant must travel to the shortest distance (d) that it would have traveled in the absence of the layered silicate and is expressed in terms of the length (L), width (W), and volume fraction of the sheets (ϕ).

$$\tau = \frac{d^l}{d} = 1 + \frac{L}{2W}\phi \tag{11.2}$$

The effect of tortuosity on the permeability is expressed as

$$\frac{P_s}{P_p} = \frac{(1-\phi)}{\tau} \tag{11.3}$$

where P_s and P_p represent the permeabilities of the permeant in polymer-filled composite and of the permeant in the pure matrix.

Accordingly, for example, the oxygen permeability of EPM-*g*-DEM/OLS nanocomposites decreased with increasing MMT content as in the corresponding composites prepared with PE-*g*-DEM (Passaglia et al. 2008a), but the permeation dependence on the MMT amount seemed to be not predictable on the basis of the sole aspect ratio reduction. The inability of models based on tortuous paths to predict the permeation reduction for EPM-*g*-DEM/OLS nanocomposites evidenced limitations due to a too much simplified approach. The different interactions gradient (polymer chains/silicate surface like and/or polymer chains/polymer chains as physical absorption and entanglements) can change the density and the

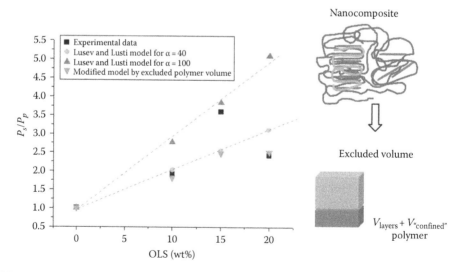

FIGURE 11.27
Comparison between oxygen permeation reduction observed in EPM-*g*-DEM composites (dark square), as predicted by the Gusev and Lusti model for disk-like particles with aspect ratio 60 and 100 (square) and as predicted by assuming that the polymer phase adsorbed on particle surface has a negligible permeability to oxygen; a schematic representation of excluded volume is reported.

mobility of polymer chains (particularly of polymer trapped) and then their permeability characteristics. In fact, EPM-g-DEM nanocomposites displayed an increase of T_g and a reduction of the solubility with MMT loading suggesting the existence of a large fraction of polymeric phase confined on the surface and/or within the galleries of MMT.

This phase has presumably mobility and therefore gas permeability properties differing from the bulk. Based on this consideration, the oxygen permeation through the confined polymer phase was tentatively assumed to be negligible with respect to the bulk one, and a modified version of the tortuous path model of Gusev and Lusti taking into account the total non permeable volume (silica plus volume of confined polymer chains) was derived (Figure 11.27).

Even if data calculated by using this modified model are not completely in agreement with the experimental data, the concept of confined polymer chains as characterized by different permeability from the matrix, has to be assumed in order to correctly predict the barrier properties of nanocomposites.

11.4 Conclusions

In this chapter, we have revised and discussed the main aspects concerning the use of functional POs as matrix or compatibilizer of PO/clay nanocomposites.

It is somewhat evident that the PO hydrophobicity makes almost difficult to obtain intercalated and/or exfoliated nanocomposites without chemical modification of the clay or polymer, or both. Indeed, even if a good dispersion of the clay into a PO can be achieved simply by applying strong shear forces during the process, the structure will be unstable and re-agglomeration of the particles may occur quite rapidly if an effective interaction between the polymer and clay surface is not ensured. The establishment of thermodynamically favorable surface interactions between the nanodispersed and continuous polymer phase is determinant to reach the desired morphology, and also plays an important role in maintaining the nanostructure of the final material during its application. For this reason, the hydrophilic clay is often changed to hydrophobic and typically a compatibilizer (the functional polyolefin) capable of locating itself at the interface is added.

Functional POs bearing acid, ester, anhydride, hydroxy, amino, and other functionalities side-grafted, end-tethered, or copolymers with random or block polar units were successfully intercalated between clay layers. Among all the functional groups, MAH-grafted groups, even if in a small amount were generally more efficient than the others in creating strong interactions with clay layers, thus promoting intercalation and exfoliation. Even if the nature of interactions requires further clarification, hydrogen bonds with OH groups (present on the edges of platelets), donary bonds with tetrahedra oxygen of silicate surface, and probably the reactions with amino cation surfactants are all reasonable explanations for the exfoliation mechanism. However, the relative amount of grafted MAH groups could not exceed a given value in order to retain some miscibility between the functional PO and the matrix, even if for a higher amount of functional groups the hydrophilicity of the functional polyolefin increases and polymer/clay interaction is improved.

By considering the distribution along the backbone of functionalities, it was evidenced that end-functionalized polymers assume a special configuration on the clay surface which allows complete exfoliation of the clay. However, this single interaction site is insufficient for the formation/stabilization of a long-lived silicate network, mediated by polymer chains that bridge across nanoparticles, as manifested macroscopically by the liquid-like behavior of the resulting PO nanocomposites. In contrast, side chain–functionalized POs as well as PO block copolymers or random-functionalized POs were seen to form multiple contacts with each clay surface, thus resulting in effective bridging interactions between the nanoparticles and in the development of an extended and possibly hierarchical super-structure, as shown by the rheological solid-like behavior.

Moreover, the molecular weight and structure of the functional PO as well as melt vis-cosity and rheological properties were found responsible for the final morphology of com-posites. Best results are generally achieved by employing polymer with molecular weight similar to that of the matrix; otherwise the miscibility between the two POs might be compromised and the intercalation/exfoliation process inhibited.

Independently of their nature, the established interactions at interface change the bulk properties of PO (and even of functional PO): an amount of unextractable polymer in both MAH and DEM functionalized PO/OLS nanocomposites was collected confirming the functional PO/clay interaction and confinement and allowing to discriminate between functional PO chains interacting with the layers and those only intercalated as these last were easily removed with a reduction of the interlayer distance.

Thermal properties of functional PO/OLS nanocomposites can be also nicely correlated with the level of interaction between functional chains and clay or confinement. Both crys-tallization and glass transition are generally modified. Indeed, one of the more common effect is the nucleating activity of the clay which speeds up the crystallization process, reduces spherulite size, and changes the final morphology. However, in the case of inter-calated PE/clay nanocomposites, the confinement of chains between the layers suppresses the crystallization showing a lower crystallization temperature than neat PE and exfoli-ated nanocomposites. The T_g is also strongly affected by good filler–particle interactions: in particular, T_g tends to increase with the interaction between the functional PO and clay layers. This was specifically evidenced for functional PO chains unextractable and trapped between the layers confirming the reduction of mobility of polymer chains trapped into the platelets. Finally, the reduced mobility of confined polymer chains can be also invoked to explain the results about oxygen permeability, not perfectly rationalized by the theoreti-cal data predicted from mathematical models.

Acknowledgment

The authors wish to thank Dr. Monica Bertoldo for the helpful discussion.

References

Alexandre, M. and Dubois, P. 2000. Polymer-layered silicate nanocomposites: Preparation, properties and uses of a new class of materials. *Materials Science and Engineering* R28:1–63.

Alexandre, M., Martin, E., Dubois, P., Garcia-Marti, M., and Jerome, R. 2000. Use of metallocenes in the polymerization-filling technique with production of polyolefin-based composites. *Macromolecular Rapid Communications* 21:931–936.

Alexandre, M., Dubois, P., Sun, T., Garces, J. M., and Jerome, R. 2002. Polyethylene-layered silicate nanocomposites prepared by the polymerization-filling technique: Synthesis and mechanical properties. *Polymer* 43:2123–2132.

Augier, S., Coiai, S., Gragnoli, T., Passaglia, E., Pradel, J. L., and Flat, J. J. 2006. Coagent assisted polypropylene radical functionalization: Monomer grafting modulation and molecular weight conservation. *Polymer* 47:5243–5252.

Augier, S., Coiai, S., Pratelli, D., Conzatti, L., and Passaglia, E. 2009. New functionalized polypropylenes as controlled architecture compatibilizers for polypropylene layered silicates nanocomposites. *Journal of Nanoscience and Nanotechnology* 9:4858–4869.

Balazs, A. C., Singh, C., and Zhulina, E. 1998. Modeling the interactions between polymers and clay surfaces through self-consistent field theory. *Macromolecules* 31:8370–8381.

Balazs, A. C., Singh, C., Zhulina, E., and Lyatskaya, Y. 1999. Modelling the phase behaviour of polymer/clay nanocomposites. *Accounts of Chemical Research* 32:651–657.

Bharadwaj, R. K. 2001. Modeling the barrier properties of polymer-layered silicate nanocomposites. *Macromolecules* 34:9189–9192.

Bhattacharya, S. N., Kamal, M. R., and Gupta, R. K. 2008. *Polymer Nanocomposites: Theory and Practice.* Munich, Germany: Hanser.

Bikiaris, D. N., Vassiliou, A., Pavlidou, E., and Karayannidis, G. P. 2005. Compatibilization effect of PP-g-MA copolymer on iPP/SiO$_2$ nanocomposites prepared by melt mixing. *European Polymer Journal* 41:1965–1978.

Biswas, M. and Sinha Ray, S. 2001. Recent progress in synthesis and evaluation of polymer-montmorillonite nanocomposites. *Advances in Polymer Science* 155:167–221.

Boffa, L. S. and Novak, B. M. 2000. Copolymerization of polar monomers with olefins using transition-metal complexes. *Chemical Reviews* 100:1479–1493.

Chang, C. C. and Hou, S. S. 2008. Intercalation of poly(methyl methacrylate) into tyramine-modified layered silicates through hydrogen-bonding interaction. *European Polymer Journal* 44:1337–1345.

Chaudhary, D. S., Prasad, R., Gupta, R. K., and Bhattacharya, S. N. 2005. Clay intercalation and influence on crystallinity of EVA-based clay nanocomposites. *Thermochimica Acta* 433:187–195.

Chrissopoulou, K., Altintzi, I., Anastasiadis, S. H., Giannelis, E. P., Pitsikalis, M., Hadjichristidis, N., and Theophilou, N. 2005. Controlling the miscibility of polyethylene/layered silicate nanocomposites by altering the polymer/surface interactions. *Polymer* 46:12440–12451.

Chrissopoulou, K., Altintzi, I., Andrianaki, I., Shemesh, R., Retsos, H., Giannelis, E. P., and Anastasiadis, S. H. 2008. Understanding and controlling the structure of polypropylene/layered silicate nanocomposites. *Journal of Applied Polymer Science* 46:2683–2695.

Ciardelli, F., Aglietto, M., Ruggeri, G., Passaglia E., and Castelvetro, V. 1997. Functionalization of polyolefins in the melt through reaction with molecules and macromolecules. *Macromolecular Symposia* 118:311–316.

Ciardelli, F., Aglietto, M., Passaglia, E., and Ruggeri, G. 1998. Molecular and mechanistic aspects of the functionalization of polyolefins with ester groups. *Macromolecular Symposia* 129:79–88.

Ciardelli, F., Aglietto, M., Coltelli, M.-B., Passaglia, E., Ruggeri G., and Coiai, S. 2004. Functionalization of polyolefins in the melt. In *Modification and Blending of Synthetic and Natural Macromolecules*, F. Ciardelli and S. Penczeck (eds.), pp. 47–71. Dordrecht, the Netherlands: Kluwer.

Ciardelli, F., Passaglia, E., and Coiai, S. 2006. A process of controlled radical grafting of a polyolefin. University of Pisa, Pisa, Italy, U.S. 2006 148,993 A1.

Ciardelli, F., Coiai, S., Passaglia, E., Pucci, A., and Ruggeri, G. 2008. Reactive blending of polyolefins to nanostructured functional thermoplastic materials. *Polymer International* R57:805–836.

Coiai, S., Passaglia, E., Aglietto, M., and Ciardelli, F. 2004. Control of degradation reactions during radical functionalization of polypropylene in the melt. *Macromolecules* 37:8414–8423.

Coiai, S., Scatto, M., Bertoldo, M., Conzatti, L., Andreotti, L., Sterner, M., Passaglia, E., Costa, G., and Ciardelli, F. 2009. Study of the compounding process parameters for morphology control of LDPE/layered silicate nano-composites. *e-Polymers*, no. 050:1–18. http://www.e-polymers.org/journal/papers/scoiai_280509.pdf

Cole K. C. 2008. Use of infrared spectroscopy to characterize clay intercalation and exfoliation in polymer nanocomposites. *Macromolecules* 41:834–843.

Coltelli, M.-B., Coiai, S., Bronco, S., and Passaglia, E. 2010. Nanocomposites based on phyllosilicates: From petrochemicals to renewable thermoplastic matrices. In *Advanced Nanomaterials*, K. E. Geckeler and H. Nishide (eds.), pp. 403–458. Weinheim, Germany: Wiley-VCH.

Cui, L. and Paul, D. R. 2007. Evaluation of amine functionalized polypropylene as compatibilizers for polypropylene nanocomposites. *Polymer* 48:1632–1640.

Cui, L. and Woo, S. I. 2008. Preparation and characterization of polyethylene (PE)/clay nanocomposites by in situ polymerization with vanadium-based intercalation catalyst. *Polymer Bulletin* 61:453–460.

Dintcheva, N. Tz., Al-Malaika, S., and La Mantia, F. P. 2009. Effect of extrusion and photo-oxidation on polyethylene/clay nanocomposites. *Polymer Degradation and Stability* 94:1571–1588.

Du, Z., Zhang, W., Zhang, C., Jing, Z., and Li, H. 2002. A novel polyethylene/palygorskite nanocomposite prepared via in situ coordinated polymerization. *Polymer Bulletin* 49:151–158.

Du, M., Guo, B., Liu, M., and Jia, D. 2006. Preparation and characterization of polypropylene grafted halloysite and their compatibility effect to polypropylene/halloysite composite. *Polymer Journal* 38:1198–1204.

Dubnikova, I. L., Berezina, S. M., Korolev, Y. M., Kim, G. M., and Lomakin, S. M. 2007. Morphology, deformation behavior and thermomechanical properties of polypropylene/maleic anhydride grafted polypropylene/layered silicate nanocomposites. *Journal of Applied Polymer Science* 105:3836–3850.

Dutta, N. K., Choudhury, N. R., Haidar, B., Vidal, A., Donnet, J. B., Delmotte, L., and Chezeau, J. M. 1994. High resolution solid-state NMR investigation of the filler-rubber interaction: 1. High speed ^{1}H magic-angle spinning NMR spectroscopy in carbon black filled styrene-butadiene rubber. *Polymer* 35:4293–4299.

Fang, Z., Xu, Y., and Tong, L. 2006. On promoting dispersion and intercalation of bentonite in high density polyethylene by grafting vinyl triethoxysilane. *Journal of Materials Science* 41:5433–5440.

Fisher, H. 2003. Polymer nanocomposites: From fundamental research to specific applications. *Materials Science and Engineering* C23:763–772.

García-López, D., Picazo, O., Merino, J. C., and Pastor, J. M. 2003. Polypropylene–clay nanocomposites: Effect of compatibilizing agents on clay dispersion. *European Polymer Journal* 39:945–950.

Giannelis, E. P., Krishnamoorti, R., and Manias, E. 1999. Polymer-silicate nanocomposites: Model systems for confined polymers and polymer brushes. *Advances in Polymer Science* 138:107–147.

Ginzburg, V. V. and Balazs, A. C. 2000. Calculating phase diagrams for nanocomposites: The effect of adding end-functionalized chains to polymer/clay mixtures. *Advanced Materials* 23:1805–1809.

Gopakumar, T. G., Lee, J. A., Kontopoulou, M., and Parent, J. S. 2002. Influence of clay exfoliation on the physical properties of montmorillonite/polyethylene composites. *Polymer* 43:5483–5491.

Gusev, A. A. and Lusti, H. R. 2001. Rational design of nanocomposites or barrier applications. *Advanced Materials* 13:1641–1643.

Hamaker, H. C. 1937. The London-van der Waals attraction between spherical particles. *Physica* 4:1058–1075.

Harrats, C. and Groeninckx, G. 2008. Features, questions and future challenges in layered silicates clay nanocomposites with semicrystalline polymer matrices. *Macromolecular Rapid Communications* 29:14–26.

Hasegawa, N. and Usuki, A. 2004. Silicate layer exfoliation in polyolefin/clay nanocomposites based on maleic anhydride modified polyolefins and organophilic clay. *Journal of Applied Polymer Science* 93:464–470.

Hasegawa, N., Kawasumi, M., Kato, M., Usuki, A., and Okada, A. 1998. Preparation and mechanical properties of polypropylene-clay hybrids using a maleic anhydride-modified polypropylene oligomer. *Journal of Applied Polymer Science* 67:87–92.

Hasegawa, N., Okamoto, H., Kawasumi, M., Kato, M., Tsukigas, A., and Usuki, A. 2000. Polyolefin–clay hybrids based on modified polyolefins and organoclay. *Macromolecular Materials and Engineering* 280/281:76–79.

Hotta, S. and Paul, D. R. 2004. Nanocomposites formed from linear low density polyethylene and organoclays. *Polymer* 45:7639–7654.

Jang, L. W., Kim, E. S., Kim, H. S., and Yoon, J.-S. 2005. Preparation and characterization of polypropylene/clay nanocomposites with polypropylene-graft-maleic anhydride. *Journal of Applied Polymer Science* 98:1229–1234.

Kamal, M. R., Calderon, J. U., and Lennox, R. B. 2009. Surface energy of modified nanoclays and its effect on polymer/clay nanocomposites. *Journal of Adhesion Science and Technology* 23:663–688.

Kato, M., Usuki, A., and Okada, A. 1997. Synthesis of polypropylene oligomer-clay intercalation compounds. *Journal of Applied Polymer Science* 66:1781–1785.

Kawasumi, M., Hasegawa, N., Kato, M., Usuki, A., and Okada, A. 1997. Preparation and mechanical properties of polypropylene-clay hybrids. *Macromolecules* 30:6333–6338.

Kim, Y. C. 2006. Effect of maleated polyethylene on the crystallization behavior of LLDPE/clay nanocomposites. *Polymer Journal* 38:250–257.

Koo, C. M., Ham, H. T., Kim, S. O., Wang, K. H., and Chung, I. J. 2002. Morphology evolution and anisotropic phase formation of the maleated polyethylene layered silicate nanocomposites. *Macromolecules* 35:5116–5122.

Koo, C. M., Kim, M. J., Choi, M. H., Kim, S. O., and Chung I. J. 2003. Mechanical and rheological properties of the maleated polypropylene-layered silicate nanocomposites with different morphology. *Journal of Applied Polymer Science* 88:1526–1535.

Kraus, G. 1971. Reinforcement of elastomers by carbon black. *Advances in Polymer Science* 8:155–237.

Kuppa, V., Menakanit, S., Krishnamoorti, R., and Manias, E. 2003. Simulation insights on the structure of nanoscopically confined poly(ethylene oxide). *Journal of Polymer Science, Part B: Polymer Physics* 41:3285–3298.

Kurokawa, Y., Yasuda, H., and Oyo, A. 1996. Preparation of a nanocomposite of polypropylene and smectite. *Journal of Materials Science Letters* 15:1481–1483.

Kurokawa, Y., Yasuda, H., Kashiwagi, M., and Oyo, A. 1997. Structure and properties of a montmorillonite/polypropylene nancomposite. *Journal of Materials Science Letters* 16:1670–1672.

Lertwimolnun, W. and Vergnes, B. 2005. Influence of compatibilizer and processing conditions on the dispersion of nanoclay and a polypropylene matrix. *Polymer* 46:3462–3471.

Li, Y., Zhao, B., Xie, S., and Zhang, S. 2003. Synthesis and properties of poly(methyl methacrylate)/montmorillonite (PMMA/MMT) nanocomposites. *Polymer International* 52:892–898.

Lim, Y. T. and Park, O. O. 2001. Phase morphology and rheological behavior of polymer/layered silicate nanocomposites. *Rheologica Acta* 40:220–229.

Lu, C. and Mai, Y. W. 2005. Influence of aspect ratio on barrier properties of polymer-clay nanocomposites. *Physical Review Letters* 95:088303:1–4.

Mainil, M., Alexandre, M., Monteverde, F., and Dubois, P. 2006. Polyethylene organo-clay nanocomposites: The role of the interface chemistry on the extent of clay intercalation/exfoliation. *Journal of Nanoscience and Nanotechnology* 6:337–344.

Manias, E., Chen, H., Krishnamoorti, R., Genzer, J., Kramer, E. J., and Giannelis, E. P. 2000. Intercalation kinetics of long polymers in 2 nm confinements. *Macromolecules* 33:7955–7966.

Manias, E., Touny, A., Wu, L., Strawhecker, B. L., and Chung, T. C. 2001. Polypropylene/montmorillonite nanocomposites. Review of the synthetic routes and materials properties. *Chemistry of Materials* 13:3516–3523.

Marchant, D. and Jayaraman, K. 2002. Strategies for optimizing polypropyleneclay nanocomposite structure. *Industrial and Engineering Chemistry Research* 41:6402–6408.

Mishra, J. K., Hwang, K.-J., and Ha, C.-S. 2005. Preparation, mechanical and rheological properties of a thermoplastic polyolefin (TPO)/organoclay nanocomposite with reference to the effect of maleic anhydride modified polypropylene as a compatibilizer. *Polymer* 46:1995–2002.

Mittal, V. 2007. Polypropylene-layered silicate nanocomposites: Filler matrix interactions and mechanical properties. *Journal of Thermoplastic Composite Materials* 20:575–599.

Mittal V. 2009. Polymer layered silicate nanocomposites. *Materials* 2:992–1057.

Moad, G. 1999. The synthesis of polyolefin graft copolymers by reactive extrusion. *Progress in Polymer Science* 24:81–142.

Moad, G., Dean, K., Edmond, L., Kukaleva, N., Li, G., Mayadunne, R. T. A., Pfaendner, R., Schneider, A., Simon, G. P., and Wermter, H. 2006. Non-ionic, poly(ethylene oxide)-based surfactants as intercalants/dispersants/exfoliants for poly(propylene)-clay nanocomposites. *Macromolecular Materials and Engineering* 291:37–52.

Morawiec, J., Pawlak, A., Slouf, M., Galeski, A., Piorkowska, E., and Krasnikowa, N. 2005. Preparation and properties of compatibilized LDPE/organo-modified montmorillonite nanocomposites. *European Polymer Journal* 41:1115–1122.

Muksing, N., Coiai, S., Conzatti, L., Passaglia, E., Magaraphan, R., and Ciardelli, F. 2010. Morphology development and stability of polypropylene/organoclay nanocomposites. *Journal of Nanoscience and Nanotechnology* 10:5814–5825.

Osman, M. A., Rupp, J. E. P., and Suter, U. W. 2005. Effect of non-ionic surfactants on the exfoliation and properties of polyethylene-layered silicate nanocomposites. *Polymer* 46:8202–8209.

Othman, N., Ismail, H., and Mariatti, M. 2006. Effect of compatibilisers on mechanical and thermal properties of bentonite filled polypropylene composites. *Polymer Degradation and Stability* 91:1761–1774.

Oya, A., Kurokawa, Y., and Yasuda, H. 2000. Factors controlling mechanical properties of clay mineral/polypropylene nanocomposites. *Journal of Materials Science* 35:1045–1050.

Pasbakhsh, P., Ismail, H., Ahmad Fauzi, M. N., and Abu Bakar, A. 2009. Influence of maleic anhydride grafted ethylene propylene diene monomer (MAH-g-EPDM) on the properties of EPDM nanocomposites reinforced by halloysite nanotubes. *Polymer Testing* 28:548–559.

Passaglia, E., Bertuccelli, W., and Ciardelli, F. 2001. Composites from functionalized polyolefins and silica. *Macromolecular Symposia* 176:299–315.

Passaglia, E., Corsi, L., Aglietto, M., Ciardelli, F., Michelotti, M., and Suffredini, G. 2003a. One-step functionalization of an ethylene/propylene random copolymer with two different reactive groups. *Journal of Applied Polymer Science* 87:14–23.

Passaglia, E., Coiai, S., Aglietto, M., Ruggeri, G., Rubertà, M., and Ciardelli, F. 2003b. Functionalization of polyolefins by reactive processing: Influence of starting reagents on content and type of grafted groups. *Macromolecular Symposia* 198:147–159.

Passaglia, E., Sulcis, R., Ciardelli, F., Malvaldi, M., and Narducci, P. 2005. Effect of functional groups of modified polyolefins on the structure and properties of their composites with lamellar silicates. *Polymer International* 54:1549–1556.

Passaglia, E., Bertoldo, M., Ceriegi, S., Sulcis, R., Narducci, P., and Conzatti, L. 2008a. Oxygen and water vapor barrier properties of MMT nanocomposites from low density polyethylene or EPM with grafted succinic groups. *Journal of Nanoscience and Nanotechnology* 8:1690–1699.

Passaglia, E., Bertoldo, M., Ciardelli, F., Prevosto, D., and Lucchesi, M. 2008b. Evidences of macromolecular chains confinement of ethylene–propylene copolymer in organophilic montmorillonite nanocomposites. *European Polymer Journal* 44:1296–1308.

Passaglia, E., Bertoldo, M., Coiai, S., Augier, S., Savi, S., and Ciardelli, F. 2008c. Nanostructured polyolefins/clay composites: Role of the molecular interaction at the interface. *Polymers for Advanced Technologies* 19:560–568.

Passaglia, E., Coiai, S., and Augier, S. 2009. Control of macromolecular architecture during the reactive functionalization in the melt of olefin polymers. *Progress in Polymer Science* 34:911–947.

Pavlidou, S. and Papaspyrides, C. D. 2008. A review on polymer-layered silicate nanocomposites. *Progress in Polymer Science* 33:1119–1198.

Perrin-Sarazin, F., Ton-That, M. T., Bureau, M. N., and Denault, J. 2005. Micro- and nano-structure in polypropylene/clay nanocomposites. *Polymer* 46:11624–11634.

Pinnavaia, T. J. and Beall, G. W. 2000. *Polymer-Clay Nanocomposites*, New York: John Wiley & Sons, Inc.

Preston, C. M. L., Amarasinghe, G., Hopewell, J. L., Shanks, R. A., and Mathys, Z. 2004. Evaluation of polar ethylene copolymers as fire retardant nanocomposite matrices. *Polymer Degradation and Stability* 84:533–544.

Reichert, P., Nitz, H., Klinke, S., Brandsch, R., Thomann, R., and Mülhaupt, R. 2000. Poly(propylene)/organoclay nanocomposite formation: Influence of compatibilizer functionality and organoclay modification. *Macromolecular Materials and Engineering* 275:8–17.

Ristolainen, N., Vainio, U., Paavola, S., Torkkell, M., Serimaa, R., and Seppälä, J. 2005. Polypropylene/organoclay nanocomposites compatibilized with hydroxyl-functional polypropylenes. *Journal of Polymer Science, Part B: Polymer Physics* 43:1892–1903.

Ruggeri, G., Aglietto, M., Petragnani, A., and Ciardelli, F. 1983. Some aspects of polypropylene functionalization by free radical reactions. *European Polymer Journal* 19:863–866.

Shi, D., Hu, G. H., and Li, R. K. Y. 2006. Concept of nano-reactor for the control of the selectivity of the free radical grafting of maleic anhydride onto polypropylene in the melt. *Chemical Engineering and Science* 61:3780–3784.

Shin, S. Y. A., Simon, L. C., Soares, J. B. P., and Scholz, G. 2003. Polyethylene-clay hybrid nanocomposites: In situ polymerization using bifunctional organic modifiers. *Polymer* 44:5317–5321.

Sibold, N., Dufour, C., Gourbilleau, F., Metzner, M. N., Lagrève, C., Le Puart, L., Madec, P. J., and Pham, T. N. 2007. Montmorillonite for clay-polymer nanocomposites: Intercalation of tailored compounds based on succinic anhydride, acid and acid salt derivatives: A review. *Applied Clay Science* 38:130–138.

Sinha Ray, S. and Okamoto, M. 2003. Polymer/layered silicate nanocomposites: A review from preparation to processing. *Progress in Polymer Science* 28:1539–1641.

Sinsawat, A., Anderson, K. L., Vaia, R. A., and Farmer, B. L. 2003. Influence of polymer matrix composition and architecture on polymer nanocomposite formation: Coarse-grained molecular dynamics simulation. *Journal of Polymer Science, Part B: Polymer Physics* 41:3272–3284.

Sorrentino, A., Tortora, M., and Vittoria, V. 2006. Diffusion behavior in polymer–clay nanocomposites. *Journal of Polymer Science, Part B: Polymer Physics* 44:265–274.

Strawhecker, K. and Manias, E. 2000. Structure and properties of poly(vinyl alcohol)/Na$^+$ montmorillonite nanocomposites. *Chemistry of Materials* 12:2943–2949.

Strawhecker, K. E. and Manias, E. 2003. Crystallization behavior of poly(ethylene oxide) in the presence of Na$^+$ montmorillonite fillers. *Chemistry of Materials* 15:844–849.

Százdi, L., Pukánszky, B. Jr., Földes, E., and Pukánszky, B. 2005. Possible mechanism of interaction among the components in MAPP modified layered silicate PP nanocomposites. *Polymer* 46:8001–8010.

Százdi, L., Pukánszky, B. Jr., Vancso, G. J., and Pukánszky, B. 2006. Quantitative estimation of the reinforcing effect of layered silicates in PP nanocomposites. *Polymer* 47:4638–4648.

Thompson, M. R. and Yeung, K. K. 2006. Recyclability of a layered silicate-thermoplastic olefin elastomer nanocomposite. *Polymer Degradation and Stability* 91:2396–2407.

Tjong, S. C. 2006. Structural and mechanical properties of polymer nanocomposites. *Materials Science and Engineering R* 53:73–197.

Tjong, S. C., Meng, Y. Z., and Hay, A. S. 2002. Novel preparation and properties of polypropylene-vermiculite nanocomposites. *Chemistry of Materials* 14:44–51.

Usuki, A., Kato, M., Okada, A., and Kurauchi, T. 1997. Synthesis of polypropylene-clay hybrid. *Journal of Applied Polymer Science* 63:137–139.

Vaia, R. A. 2000. Structural characterization of polymer-layered silicate nanocomposites. In *Polymer-Clay Nanocomposites*, T. J. Pinnavaia and G. W. Beall (eds.), pp. 229–263. Chichester, U.K.: John Wiley & Sons.

Vaia, R. A. and Giannelis, E. P. 1997a. Lattice model of polymer melt intercalation in organically modified layered silicates. *Macromolecules* 30:7990–7999.

Vaia, R. A. and Giannelis, E. P. 1997b. Polymer melt intercalation in organically-modified layered silicates: Model predictions and experiment. *Macromolecules* 30:8000–8009.

Vaia, R. A., Teukolsky, R. K., and Giannelis, E. P. 1994. Interlayer structure and molecular environment of alkylammonium layered silicates. *Chemistry of Materials* 6:1017–1022.

Vaia, R. A., Vasudevan, S., Krawiec, W., Scanlon, L. G., and Giannelis, E. P. 1995. New polymer electrolyte nanocomposites. Melt intercalation of poly(ethylene oxide) in mica-type silicates. *Advanced Materials* 7:154–156.

Vaia, R. A., Sauer, B. B., Tse, O. K., and Giannelis, E. P. 1997. Relaxations of confined chains in polymer nanocomposites: Glass transition properties of poly (ethylene oxide) intercalated in montmorillonite. *Journal of Polymer Science, Part B: Polymer Physics* 35:59–67.

Wang, K. H., Choi, M. H., Koo, C. M., Choi, Y. S., and Chung, I. J. 2001. Synthesis and characterization of maleated polyethylene/clay nanocomposites. *Polymer* 42:9819–9826.

Wang, Z. M., Nakajima, H., Manias, E., and Chung, T. C. 2003. Exfoliated PP/clay nanocomposites using ammonium-terminated PP as the organic modification for montmorillonite. *Macromolecules* 36:8919–8922.

Wang, Y., Chen, F.-B., Li, Y.-C., and Wu, K.-C. 2004. Melt processing of polypropylene/clay nanocomposites modified with maleated polypropylene compatibilizers. *Composites: Part B* 35:111–124.

Wang, Y., Zhang, H., Wu, Y., Yang, J., and Zhang, L. 2005. Preparation and properties of natural rubber/rectorite nanocomposites. *European Polymer Journal* 41:2776–2783.

Wang, K., Liang, S., Deng, J., Yang, H., Zhang, Q., Fu, Q. et al. 2006. The role of clay network on macromolecular chain mobility and relaxation in isotactic polypropylene/organoclay nanocomposites. *Polymer* 47:7131–7144.

Xie, Y., Yu, D., Kong, J., Fan, X., and Qiao, W. 2006. Study on morphology, crystallization behaviors of highly filled maleated polyethylene-layered silicate nanocomposites. *Journal of Applied Polymer Science* 100:4004–4011.

Xu, J.-T., Wang, Q., and Fan, Z.-Q. 2005. Non-isothermal crystallization kinetics of exfoliated and intercalated polyethylene/montmorillonite nanocomposites prepared by in situ polymerization. *European Polymer Journal* 41:3011–3017.

Xu, L., Nakajima, H., Manias, E., and Krishnamoorti, R. 2009. Tailored nanocomposites of polypropylene with layered silicates. *Macromolecules* 42:3795–3803.

Yang, K. and Ozisik, R. 2006. Effects of processing parameters on the preparation of nylon 6 nanocomposites. *Polymer* 47:2849–2855.

Zanetti, M. and Costa, L. 2004. Preparation and combustion behaviour of polymer/layered silicate nanocomposites based upon PE and EVA. *Polymer* 45:4367–4373.

Zhai, H., Xu, W., Guo, H., Zhou, Z., Shen, S., and Song, Q. 2004. Preparation and characterization of PE and PE-g-MAH/montmorillonite nanocomposites. *European Polymer Journal* 40:2539–2545.

12

Preparation and Properties of Polyolefin/ Needle-Like Clay Nanocomposites

Emiliano Bilotti, Jia Ma, and Ton Peijs

CONTENTS

12.1 Introduction

Most of the research on polymer/clay nanocomposites reported in the literature has been focused on platelet-like clays, usually smectite clays such as montmorillonite (MMT) [1–4]. Considerably less work has been carried out on other forms of clays (Figure 12.1).

This chapter reviews the use of the sepiolite/palygorskite group of clays as a nanofiller for polymer nanocomposites. Sepiolite and palygorskite are characterized by a needle-like or fiber-like shape. This peculiar shape offers unique advantages in terms of mechanical reinforcement while, at the same time, it allows to study the effect of the nanofiller's shape on the final composite properties. The importance of the nanofiller shape for the composite properties is analyzed in Section 12.2, introducing the rationale of the whole chapter. After a general description of needle-like nanoclays in Section 12.3, the chapter develops into a main part (Section 12.4), reviewing the preparation methods and physical properties of polyolefin/needle-like clay nanocomposites.

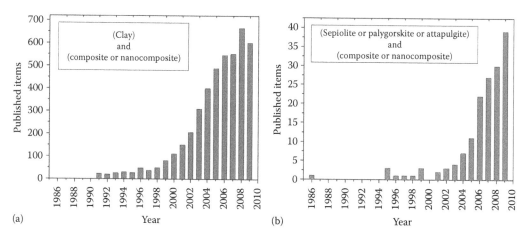

FIGURE 12.1

Published items in the field of (a) nanoclay composite and (b) needle-like clay composites, according to Web of Science, updated to end of 2009. The search criteria are indicated in each graph.

12.2 The Importance of Nanofiller Shape

The ability to control the dispersion of nanoclays in a polymeric phase is one of the key issues that affect the performance of the final material and the very possibility to obtain a nanocomposite. However, the exfoliation of layered silicates and the preparation of homogeneous nanoclay composites is, in general, seriously limited by the strong tendency of platelet-like clays to agglomerate due to their extended contact surface. Although particle aggregation depends on many factors such as size, chemical composition, and processing condition, the shape is also important. It can be demonstrated that the specific surface area of fiber-like fillers is lower than platelet-like fillers of the same aspect ratio. The relatively small contact surface of fibrous fillers and hence their reduced tendency to agglomerate can lead to better mechanical properties [5].

This can be observed in Figure 12.2, which shows the surface area-to-volume ratio (A/V) as a function of the aspect ratio ($a = l/d$) of a cylindrical particle (disk/platelet-like or cylinder/fiber-like), for a given particle volume. Values of $l/d < 1$ correspond to platelet-like particles, while $l/d > 1$ correspond to rod-like particles. It can be seen that A/V increases faster for platelet-like particles than for rod-like particles with respect to their aspect ratio [6]. Hence, for an equivalent volume of particles and for the same aspect ratio, platelet-like particles have higher contact surfaces, which makes them more difficult to be dispersed.

Following from the same considerations, it can be demonstrated that a fiber-like shape, instead of a platelet-like, can minimize the sharp increase of viscosity due to the addition of nanoclays to polymer melts, and therefore *improve the processability* of the nanocomposite. Different factors contribute to the viscosity of a nanocomposite: (a) the polymer–polymer network, (b) the clay–clay network, and (c) the polymer–clay interaction. Supposing that the chemistry of the system is fixed, the first contribution is invariant and the second depends only on the particle aspect ratio, according to percolation theory [7]: the higher the aspect ratio, the higher the viscosity of the composite. Comparing the effect of two nanoclay particles with the same aspect ratio, one with a rod-like shape and the other with

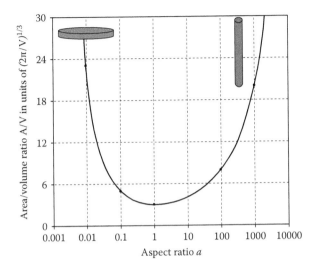

FIGURE 12.2
Surface area-to-volume ratio (A/V) as a function of the aspect ratio ($a = l/d$) for disk-like and rod-like cylindrical particles. (Reproduced from Fischer, H., *Mat. Sci. Eng. C*, 23, 763, 2003. With permission of Elsevier.)

a platelet-like shape, it can be seen that the only difference on the composite viscosity is in the last contribution: the polymer–clay interaction. Since rod-like clays have lower specific surface areas, the interaction with the polymer phase is relatively smaller and the viscosity of the composite lower.

An additional advantage of using a fiber-like reinforcement over a platelet-like reinforcement is its *higher reinforcing efficiency* in case of unidirectionally aligned systems [8], as can be demonstrated by micromechanical models like Halpin-Tsai [9]. A particle is said to reinforce efficiently a polymeric matrix if the increase in Young's modulus is close to the theoretical limit given by the rule of mixtures [10].

Figure 12.3a shows the Halpin-Tsai micromechanical predictions for unidirectional PP composites filled with 5 vol% of aligned 1D needle-like and 2D cylindrical platelet-like nanoclays of different aspect ratios, assuming a Young's modulus of 200 GPa for both

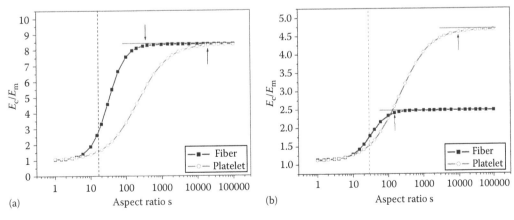

FIGURE 12.3
Reinforcement of PP by 5 vol% of fiber-like and platelet-like filler in the case of (a) unidirectional alignment and (b) random orientation of the filler. The dashed vertical line shows the average sepiolite aspect ratio. Arrows show the smallest aspect ratio necessary to reach theoretical reinforcement (rule of mixtures—horizontal solid line) [11].

nanoclays. It clearly shows that a lower aspect ratio is required in the case of 1D fillers compared to 2D fillers to achieve a maximum composite Young's modulus.

Both fillers, for aspect ratios sufficiently high, have similar levels of ultimate reinforcement, reaching the same plateau, which corresponds to the rule of mixtures (horizontal solid lines). However fiber-like fillers approach maximum reinforcement already for aspect ratios of 100, while platelet-like fillers need aspect ratios higher than 2000. It can therefore be concluded that, in the unidirectional case, fibers are more effective than platelets. This is, however, different for the situation of randomly distributed fillers. Figure 12.3b shows the reinforcement of 3D randomly oriented fiber-like and platelet-like fillers, of different aspect ratios, in a polypropylene (PP) matrix. The observation that fiber fillers reach the maximum reinforcement for aspect ratios much smaller than those necessary in the case of platelet filler still holds. This effect is also more prominent since randomly distributed platelets need an aspect ratio of 10,000. Nevertheless, the plateau relative to platelet fillers is twice as high as for fibers filler. It can therefore be concluded that, in the case of randomly oriented filler, platelets are more effective than fibers but only for aspect ratios higher than 100.

12.3 Needle-Like Nanoclays: Crystal Structure, Morphology, and Properties

In geology, the term clay includes particles less than $2\,\mu m$ in size, which are commonly the result of weathering and secondary sedimentary processes [12,13]. The shape of clay minerals can be very diverse and is a distinctive characteristic of any particular clay. For instance, kaolinite (Figure 12.4a) usually shows hexagonal flake-shaped unites with a ratio of areal diameter to thickness (aspect ratio) of 2–25:1, while most of smectite mineral particles (Figure 12.4b) have an irregular flake shape and a much higher aspect ratio (100–300). Halloysite minerals (Figure 12.4c) show a tubular shape, while the family of sepiolite and palygorskite (also known as attapulgite) (Figure 12.4d) are characterized by an elongated fibrous shape and aspect ratios of 10–80.

Sepiolite and palygorskite (attapulgite) are hydrated magnesium and aluminum silicates, and their typical formula is, respectively, $Mg_4Si_6O_{15}\cdot(OH)_2\cdot6H_2O$ and $(MgAl)_5$ $(SiAl)_8O_{20}\cdot(OH)_2\cdot8H_2O$. Sepiolite or palygorskite clays are included in the phyllosilicate group because they contain a continuous 2D tetrahedral layer of composition Si_2O_5 [12,13]. They differ, however, from the other layered silicates because of the lack of a continuous octahedral layer (Figure 12.5b and c). They can be imagined as formed of blocks structurally similar to layered clay minerals (i.e., MMT), composed of two tetrahedral silica sheets and a central octahedral sheet containing Mg (or Al), but continuous only in one direction (c-axis). More blocks are linked together along their longitudinal edges by Si–O–Si bonds and this creates channels along the c-axis (Figure 12.5a and c). Due to the discontinuity of the external silica sheets, silanol groups (SiOH) are situated at the edges of these minerals.

The dimensions of a single nanoclay fiber can vary between 0.2 and $4\,\mu m$ in length, 10 and $40\,\text{Å}$ in width, 5 and $20\,nm$ in thickness, with open channels of dimensions $3.6\,\text{Å} \times 10.6\,\text{Å}$ running along the axis of the particle (Figure 12.5a). Due to this particular system of external as well as internal surfaces, needle-like clay possess very high specific surface areas (about $300\,m^2/g$) and a high sorption capacity. There are three

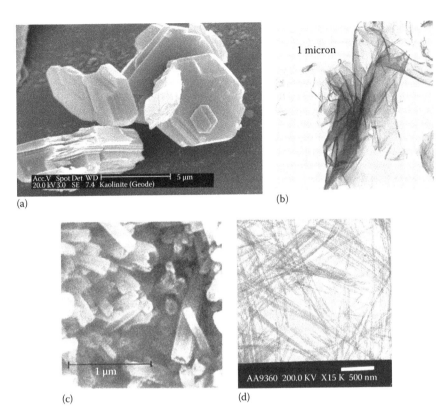

FIGURE 12.4
Electron microscope micrographs of different clays: (a) kaolinite [14], (b) montmorillonite [15], (c) halloysite [14], and (d) sepiolite [11].

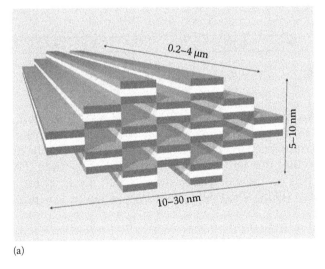

(a)

FIGURE 12.5
Structure of needle-like nanoclays: (a) schematic representation of a single needle-like clay. (Reproduced from Bilotti, E. et al., *J. Appl. Polym. Sci.*, 107, 1116, 2008. With permission from Wiley.) Suggested mineralogical structure of (b) sepiolite and (c) attapulgite. (From Grim, R.E., *Clay Mineralogy*, McGraw–Hill, New York, 1968.)

(continued)

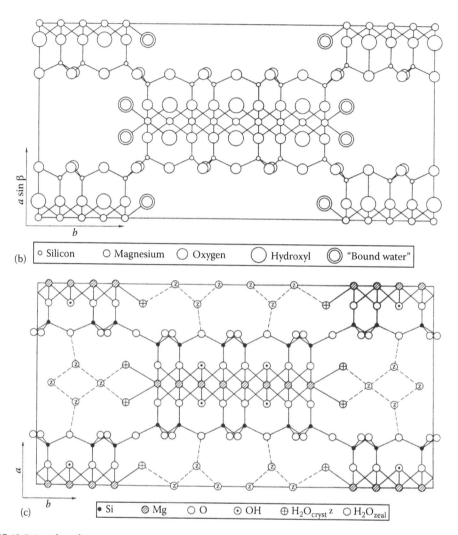

FIGURE 12.5 (continued)

sorption sites: (1) oxygen ions on the tetrahedral sheet, (2) a small amount of cation-exchange sites (0.1–0.6 mequiv/100 g), and (c) the already mentioned SiOH groups. Adsorption is also influenced by the size, shape, and polarity of the molecules involved. Neither large molecules nor those of low polarity can penetrate the channels, though they can be adsorbed on the external surface, which accounts for about 50%–60% of the total surface area [17,18]. The SiOH groups act as neutral sorption sites suitable for organic species.

Water molecules of different nature are normally present on needle-like clays. These can be well distinguished via thermo-gravimetric (TGA) test (Figure 12.6). By increasing the temperature up to 1000°C, four main weight losses are observed which can be attributed respectively—for ascending temperature—to the release of adsorbed and zeolitic water, the release of the first structural water, the release of the second structural water, and the dehydroxylation of the Mg–OH groups [19].

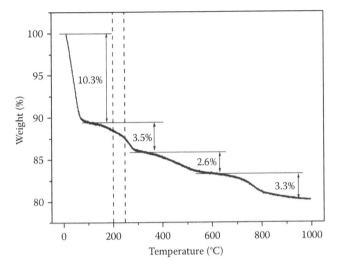

FIGURE 12.6
Dehydration of sepiolite clays under temperature scan of 20 K/min in inert atmosphere. The dashed lines represent the temperature window at which clays are typically subjected to during polyolefin composites preparation (extrusion and compression/injection moulding) [11].

The water adsorbed on external surfaces and the zeolitic water from the nanoporous tunnels is removed at relatively low temperatures. The elimination of coordinated, structural water starts when the zeolitic water is lost and is completed when dehydroxylation begins. Folding of the sepiolite crystals occurs when some structural water has been removed. This process, reversible for temperatures below 350°C, becomes irreversible once all the structural water molecules are removed and partial dehydroxylation has occurred, forming an anhydride form. Finally, the remaining Mg–OH hydroxyl groups are released at ~850°C.

Apart from the outstanding sorptive/desorptive capacity, needle-like clays are also known for their colloidal properties. When dispersed in a liquid, they form a structure of randomly intermeshed elongated particles, which is governed by secondary bonds. This structure is stable even in conditions of high salt concentrations, which usually produce flocculation of other clay suspensions, such as bentonite. Needle-like nanoclay provides a pseudoplastic and thixotropic behavior, which makes it a valuable material in different applications in order to improve processability, application, or handling of specific products. A brief summary of the main production sites and quantities of sepiolite and palygorskite clays, along with indicative prices, are presented in Table 12.1. Table 12.2 presents some general industrial applications of the same clays.

12.3.1 Health and Safety

The use of nanoparticles has recently raised several health issues. It is therefore fundamental to understand the risks associated with needle-like nanoclays and their use as nanofiller for polymeric matrices. The health and safety assessments of sepiolite from the River Tajo basin, Madrid, Spain (commercialized by Tolsa (Spain)), for instance, does not show any health hazards after epidemiological, in vitro, and in vivo studies [22,23]. Sepiolite is even registered by the EU as an additive for animal feed [24,25]. However, if there were any risks, they would be expected to be associated with inhalation through the respiratory system. In this respect, an aspect of concern can be the similarity of needle-like nanoclays morphology with asbestos, a notorious carcinogen. Sepiolite particles have an average length of 1–2 μm, while asbestos fibers have a much

TABLE 12.1

Production and Prices of Sepiolite and Palygorskite Clays Worldwide

Clay Mineral	Production (tpa)	Country	Producer/Comments	Price ($/tonne)[a]
Palygorskite (or Attapulgite)	600,000	United States, Senegal, Spain, Australia, South Africa	U.S. main producer with 250,000 tonnes and Senegal at 180,000 tonnes	$120–700
	80,000	China	Mingmei (Ming Tech Inc.)	$100–1200
	50,000	Greece	Geohellas	—
Sepiolite	600,000	Spain	Tolsa SA	$32–900
		Namibia	Afhold Ltd. New Entry	—

Source: Wilson, I., *Special Clays*, Clays, March 2008.

[a] The price is only indicative since it significantly depends on the degree of processing and treatment as well as final customer/application area. The target price for needle-like clays as nanofillers could be in the range of 3–4 €/kg, compared with the 5–18 €/kg for organically modified MMT [21,22].

TABLE 12.2

Traditional Industrial Applications of Sepiolite and Palygorskite and Their Physicochemical Properties

Application	Characteristic
Cat and pet litters	Light weight, high liquid absorption, odor control
Industrial absorbents	High liquid absorption, mechanical strength in wet conditions, nonflammability, chemical inertness
Carrier for chemicals	Absorption of active chemicals and easiness and effectiveness in delivering them
Bitumens	Control of rheological properties in heat application systems, improved fire resistance
Rheological additives	Stability, pseudo-plasticity, and thixotropy in paints, adhesives, mastics, and sealants

longer particle length, even of millimeters. Only fibers with a length longer than 5 μm are considered a possible health hazard. It is noted that some palygorskite clays can be longer than the 1–2 μm of the sepiolite. The two minerals (asbestos and sepiolite/palygorskite) are also quite easy to distinguish (i.e., using XRD) and they are not usually contaminated by each other. In fact, they have a completely different geological origin. Most of the sepiolite clays have a sedimentary geological origin. They have been formed, around 15 million years ago, by chemical precipitation in shallow lakes in periods of arid climate when the concentration of elements (Si, Mg, Al mainly) were suitable. These conditions are quite rare and this is one of the reasons why there are so few commercial sepiolite deposits in the world. On the contrary, asbestos originated in conditions of higher pressure and temperature that produce well crystallized and very long particles. In fact, the conditions for the formation of sedimentary sepiolite are not compatible with the formation of asbestos and therefore sepiolite cannot occur alongside asbestos. There is another sepiolite type, very rare, that is formed in hydrothermal conditions and whose particles have a longer length. This sepiolite type could be contaminated with asbestos since the conditions for the formation of this particular sepiolite are compatible with the formation of asbestos.

12.4 Nanocomposites Based on Needle-Like Nanoclays

12.4.1 Preparation Methods

The preparation of needle-like clay nanocomposites has followed strategies similar to the most common platelet-like clay nanocomposites. The preparation methods can be analogously classified in: 1) solution processing, 2) in situ polymerization, and 3) melt compounding. *Solution processing* is based on a solvent system in which the polymer is soluble and, at the same time, the nanoclay is able to disperse in. In general, clays are first dispersed in a solvent (or mixture of solvents) to form a homogeneous suspension in which the polymer is successively added. The process ends with the evaporation of the solvent or the precipitation of the mixture, which trap the polymer chains between the clays. *In situ polymerization* involves the dispersion of clays in a monomer, or monomer solution, which is successively polymerized directly in the presence of clays. During *melt processing*, a mixture of polymer and nanoclay is annealed above the softening temperature of the polymer, usually under vigorous shearing induced by traditional polymer processing techniques like twin-screws extrusion.

The good solvent resistance of polyolefins has effectively prevented the use of solution processing as a viable and industrially acceptable preparation method for nanoclay composites. Polyolefin/nanoclay composites have been almost exclusively prepared via melt compounding, with few cases of in situ polymerization. In the following sections, these preparation methods are analyzed in more detail.

12.4.1.1 In Situ Polymerization

The in situ polymerization of PA6 in the presence of nanoclays was the first successful example of nanocomposite preparation, which opened the way to the new research field of polymer hybrids or polymer/clay nanocomposites, which is still very active even after more than two decades [26]. The winning strategy was the choice of the monomer, ε-caprolactam, which is a polar compound and can be easily intercalated between individual clay particles. The in situ polymerization of polyolefins, instead, is complicated by the nonpolar nature of the starting monomers and the difficulties in obtaining good nanoclay dispersion. In fact, although the in situ growth of polymer chains from the clay surface and/or between clays can be a dispersing mechanism, having a good initial dispersion is still essential and ultrasounds can be used to improve that [27]. Before a polyolefin polymerization can be carried out, it is also necessary to add a suitable catalyst. Rong et al. [28] supported a Ziegler–Natta catalyst directly onto the palygorskite nanoclay surface (Figure 12.7a), from which the polyolefin polymerization was carried out, and successfully prepared a polyethylene/palygorskite nanocomposite.

Activation of the clay surface with the catalyst ($TiCl_4$) is the most critical step. Activation is facilitated by the high specific surface area and strong absorptive capacity of palygorskite but the reaction conditions needs to be carefully controlled. In fact, $TiCl_4$ can react in a number of ways with the clay [28]:

i. $TiCl_4$ can react with the water present on/in the palygorskite and forms titanium oxides or their derivatives:

$$TiCL_4 + H_2O \rightarrow TiO_2 + HCl$$

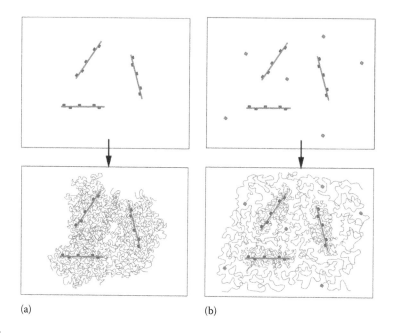

(a)   (b)

FIGURE 12.7
Schematic representation of in situ polymerization of PE: (a) directly and exclusively from the Ziegler–Natta catalyst supported on the palygorskite surface ("macromolecular comb" structure) or (b) also from a second dispersed catalyst.

ii. $TiCl_4$ can react with the silanol groups present on the clay surface:

$$TiCl_4 + Si - OH \rightarrow Si - OTiCl_3 + HCl$$

iii. $TiCl_4$ can coordinate with the magnesium vacancies in the clay structure:

$$Mg\overset{\square}{\underset{\square}{<}} \; + \; TiCl_4 \; \longrightarrow \; Mg\overset{Cl}{\underset{Cl}{<}}Ti\overset{Cl}{\underset{Cl}{<}}$$

If water is present, $TiCl_4$ will preferentially react following reaction (i) and non-catalytic titanium species will be formed. In order to promote catalytic titanium species (reaction iii), it is essential to remove water before clay activation. The dehydration process was already described in Figure 12.6. With increasing temperatures, water molecules of different nature (from adsorbed water to structural water) can be gradually removed. Rong et al. [28] demonstrated that an optimum temperature exists (between 500°C and 600°C), in correspondence of which water removal is maximized but collapse of the clay crystal structure is mostly prevented. Table 12.3 shows the effect of the dehydration temperature on the activity of clay during polymerization. With increasing temperatures, less titanium reacted but of a more catalytic species, which resulted in a higher polymerization activity. Temperatures exceeding 600°C, instead, are detrimental because of the reduction of the clay surface area and sorption capacity caused by the modification of the clay structure.

After the palygorskite clay were conditioned and activated, the in situ polymerization of polyethylene was carried out. A solvent (heptane), a cocatalyst (AlR_3), and the monomer (ethylene, under a pressure of 1 atm), together with the activated clay, were added to the

TABLE 12.3

Effect of Palygorskite Dehydration Temperature on the Amount of Ti Reacted during the Activation Process and on the Polymerization Activity

	Dehydration Temperature [°C]				
Measurement	100	200	300	500	800
Titanium content [wt%]	8.97	7.14	2.03	1.06	0.54
Activity [gPE (gTi h)$^{-1}$]	32	75	256	966	904

Source: Reproduced from Rong, J., *J. Appl. Polym. Sci.*, 82, 1829, 2001. With permission from Wiley.

Preparation condition: excess of TiCl$_4$ is added into the reaction system during activation.

Polymerization conditions: 40°C, Al/Ti = 15.

TABLE 12.4

Effect of Polymerization Temperature on the Ethylene Concentration, Polymerization Activity, and Molecular Weight of the Resulting PE

	Polymerization Temperature [°C]			
Measurement	30	40	50	60
Ethylene concentration [mol L^{-1}]	0.1405	0.1295	0.1190	0.1080
Molecular weight, Mw [10^6]	5.7	5.0	5.3	4.5
Activity [gPE (gTi h)$^{-1}$]	1311	1826	2090	1334

Source: Reproduced from Rong, J., *J. Appl. Polym. Sci.*, 82, 1829, 2001. With permission from Wiley.

Note: Activation conditions: 40°C 4 h, Ti content = 0.3 wt%.

reactor and the polymerization was conducted at different temperatures. The optimum amount of cocatalyst for the polymerization activity gives a ratio of Al/Ti of 10–30. Also the nature of the cocatalyst is important. The order of activity was triisobutylaluminium > triethylaluminium > trimethylaluminium. Temperature has a great influence on the polymerization. A higher temperature results in a higher activity but a lower ethylene concentration (Table 12.4).

It can be observed that the Mw of all samples—measured via viscometry—is very high. The accuracy of the data is dubious since the viscosity of the nanocomposite/decalin solution is strongly influenced by the presence of the clay, apart from the Mw of the synthesized polymer. Nevertheless, it is indicative of the poor processability of the PE nanocomposite.

To overcome this problem, a different in situ polymerization method was designed in order to have simultaneously two separate groups of PE molecular chains: on one hand PE chains grafted onto the clay surface, to form a "macromolecular comb," and improving the compatibility of clay to PE; on the other hand, PE chains are not bounded to the clay, which provided processability (Figure 12.7b). This was obtained by Du et al. [29] via a mixed catalyst system TiCl$_4$/MgCl$_2$/AlR$_3$. During the polymerization of PE macromolecules catalyzed by TiCl$_4$ present on the palygorskite surface, a second catalyst (MgCl$_2$) was introduced into the reaction in order to produce a group of PE macromolecules unbounded to the clay.

12.4.1.2 Melt Compounding

The in situ synthesis of polymers in the presence of nanoclays can produce good nanofiller dispersions but is often difficult to control and its production times are relatively long. Melt processing is certainly the most interesting preparation method from an application and economical point of view since it can be readily implemented into traditional polymer processing routes and provide reliable and reproducible products.

However, the simple melt mixing of polyolefins with natural clays, does not guarantee a sufficient level of dispersion of the nanoparticles, which are often present in the form of micron-size agglomerates. In order to overcome this problem, two main strategies have been followed: surface functionalization of needle-like clays (usually by alkylsilanes) or addition of a third polymeric phase (usually maleic anhydrite modified PP: PP-g-MA), which acts as a compatibilizer between the matrix and nanofiller. Both methods tend to modify the surface energies of the nanocomposite system, in order to reduce the interparticle interaction and improve the dispersion. In the case of a reactive surface treatment only, the polymer–clay interaction is expected to be enhanced, along with better nanoclay dispersion, which is very important for the final mechanical properties.

Among the first examples in the scientific literature of needle-like clay nanocomposites prepared by melt compounding were a series of publications by Acosta et al. [30–32]. Different concentrations of pristine or surface-treated sepiolite (10, 25, and 40 wt%) in PP were obtained via an internal mixer [30]. Normal processing conditions were temperature of 200°C, rotor speed of 60 rpm, and mixing time of 15 min. The sepiolite functionalization consisted of different aliphatic organic acids grafted to the clay surface by means of a condensation reaction (esterification) of the superficial hydroxyl groups (silanol groups) with the acid groups of the reagents. Only limited information were provided to demonstrate the success and extent of this reaction. Unfortunately, the dispersion of the pristine and functionalized sepiolite in PP was not evaluated at all, but it is rather unlikely that major agglomeration could be avoided at such high nanofiller content (10–40 wt%). The same authors presented the production of a hybrid composite in which different amounts of sepiolite were added to a PP/glass fiber composite [33]. The approach was intended to examine any synergistic effects between the two fillers and also to partially replace the glass fibers with a cheaper material, with obvious economic benefits. The binary systems PP/glass fiber (glass fiber content: 10, 20, and 30 wt%) and PP/sepiolite (sepiolite content: 10, 25, and 49 wt%) were melt compounded in a roll mixer at 190°C, in the required amounts in order to obtain weight ratios of 10/20 and 20/10 of glass fiber/sepiolite in the final hybrid composites. Although a poor dispersion of sepiolite (clay agglomerates: 15–20 μm) was shown in SEM micrographs, interesting mechanical properties were found in the ternary systems PP/glass fiber/sepiolite, in particular for what concerns the flexural strength.

Wang and Sheng [34,35] reported on the preparation of PP/attapulgite nanocomposites. The needle-like clays were organo-modified by first reacting with γ-methacryloxypropyl trimethoxysilane followed by a grafting of butyl acrylate. Different amounts of org-attapulgite clays were then melt blended with PP in a mixer apparatus at a temperature of 185°C and a residence time of 10 min at 32 rpm, to obtain nanocomposites with filler concentrations from 1 to 7 wt%. Figure 12.8 refers to a 5 wt% PP/attapulgite. The TEM picture shows a mixed dispersion state, with areas with individual attapulgite needles or small bundles but also areas in which attapulgite clays are agglomerated.

Recently, Bilotti et al. [16] reported on the effect of different surface-active polymers added as compatibilizers to melt-compounded PP/sepiolite nanocomposites.

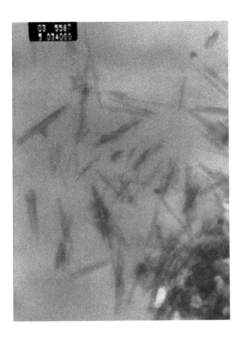

FIGURE 12.8
TEM picture of PP/attapulgite nanocomposites with 5 wt%
filler content [35]. Although no scale bar appears in the
micrograph, the diameter of the attapulgite is estimated at
40 nm. (Reproduced from Wang, L.H. and Sheng, J., *Polymer*,
46, 6243, 2005. With permission from Elsevier.)

A functionalized PP-acid and a PP-PEO di-block copolymer were used and compared
to the more common PP-*g*-MA. The three polymers were characterized by the presence
of different functional groups (but also different molecular weights) and hence different affinities with the nanoclay surface, which affected the resulting nanocomposites
morphologies (Figure 12.9). Figure 12.9a presents an SEM micrograph of a PP/sepiolite
nanocomposite with 2.5 wt% filler content. Micrometric agglomerates of sepiolite in the
PP matrix are evident. The use of PP-*g*-MA as surface active polymer contributes only
to a minor extent to the needle-like clay dispersion (Figure 12.9b); few isolated sepiolite
clays are alternated to bigger clusters. The use of PP-PEO and PP-acid, instead, leads to
much finer filler dispersions in the polymeric matrix, where aggregates of sepiolite are
no longer evident (Figure 12.9c and d).

Next, concerning the improvement of matrix/filler compatibility, there has been a number of interesting variations to the simple melt-processing method, with the aim of having
more efficient nanoparticle dispersion. Zhao et al. [36] incorporated an ultrasonic genera-
tor (maximum power output and frequency of 300 W and 20 kHz, respectively), connected
to a piezoelectric transducer, into the die of an extruder (Figure 12.10). It was shown that
the ultrasonic oscillations, in the direction parallel to the flow of the polymer melt, can
help attapulgite clay dispersion in PP.

Ma et al. [37] prepared a PP/sepiolite nanocomposite in an autoclave at 15 MPa and 200°C,
with a filler content between 1 and 10 wt%. CO_2 was pumped into the autoclave while a
pitched-bladed turbine was stirring the mixture for 30 min (Figure 12.11). It is known that
supercritical (sc) CO_2 can act as a lubricant, by reducing the polymer–polymer interactions
and increasing the polymer diffusion. Therefore, this method can be considered to be a
hybrid of solution processing and melt compounding.

Morphological investigations showed that the use of scCO_2 in the process improved
the sepiolite dispersion in PP. Moreover, since the mixing conditions were more gentle
than a twin screws extruder, the needle-like clay breakage due to processing was mini-
mized [37].

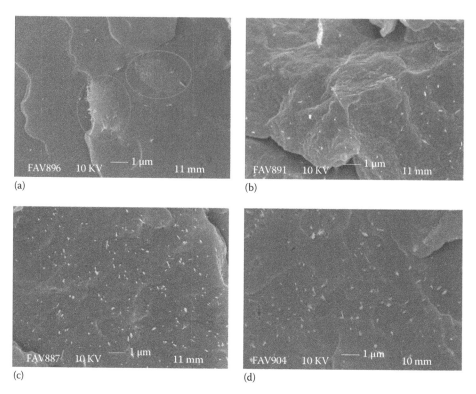

(a)   (b)

(c)   (d)

FIGURE 12.9
SEM micrographs of (a) PP + 2.5% Sep; (b) PP + PP-*g*-MA + 2.5% Sep; (c) PP + PP-acid + 2.5% Sep; and (d) PP + PP-PEO + 2.5% Sep. Red circles underline sepiolite clusters. A significant improvement in the dispersion of sepiolite in PP matrix is evident with the use of PP-PEO and PP-acid, where no agglomerates of sepiolite are found in nanocomposites at 2.5 wt% filler load. (Reproduced from Bilotti, E. et al., *J. Appl. Polym. Sci.*, 107, 1116, 2008. With permission from Wiley.)

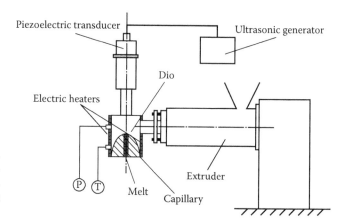

FIGURE 12.10
Schematic diagram of ultrasound-assisted melt compounding. (Reproduced from Zhao, L. et al., *J. Polym. Sci. B*, 45, 2300, 2007. With permission from Wiley.)

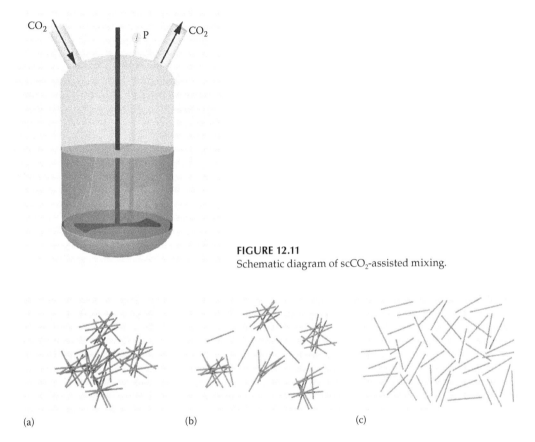

FIGURE 12.11
Schematic diagram of scCO₂-assisted mixing.

(a)   (b)   (c)

FIGURE 12.12
Schematic of needle-like clays dispersion states: (a) agglomerated (characteristic dimension of agglomerates: 10–100 μm), (b) partially agglomerated (characteristic dimension of agglomerates: 1–5 μm), and (c) well dispersed. Well-dispersed needle-like clays have also shown nematic ordering for volume fraction above 5 vol% [38].

12.4.2 Properties

The previous session has shown that the preparation method has a fundamental role in the dispersion of the nanofiller and the morphology of the polymer/clay nanocomposite. Different degrees of dispersion are obtained for different systems and, to a lesser extent, even within the same sample (Figure 12.12).

A good nanoclay dispersion is a prerequisite to obtain a nanocomposite—rather than a microcomposite—structure and therefore to achieve interesting physical properties. During the discussion of various nanocomposite properties, in the remaining part of this chapter, other factors of paramount importance will also need to be taken in account such as the content and orientation of nanoclays, the polymer–filler interaction, the degree of crystallinity, and the crystal morphology of the polymeric phase.

12.4.2.1 Crystallization and Crystallinity

Semicrystalline polymers like polyolefins can be significantly affected by the presence of nanofillers in their crystalline structure as well as the total amount of crystallinity [39–42]. In order to understand the relation between structure and physical–mechanical properties of nanocomposite materials, it is extremely important to evaluate the polymer crystallinity and crystallization behavior. Needle-like clays have been reported by several authors to act as nucleating agents for different polymers. Acosta et al. [32] first studied the crystallization kinetic of PP composites containing sepiolite surface treated with isobutyric acid. The authors conducted both isothermal and non-isothermal kinetics studies. Isothermal crystallization tests were carried out at three different temperatures: 126°C, 128°C, and 130°C. The results were interpreted with Avrami's equation [43,44], which describes the crystallization kinetics and can be written as

$$\ln(-\ln(1 - X_t)) = \mathrm{Ln}(K) + n\,\mathrm{Ln}(t) \tag{12.1}$$

where
   X_t is the progressive volume fraction transformed into crystal until the time t
   K is a scaling factor
   n is the Avrami constant

The constant n gives insight into the nucleation process (homogeneous/heterogeneous) and the shape of the growing crystal (rod/disc/sphere/sheaf). Non-isothermal crystallizations were carried out at three different cooling rates (10, 5, and 2.5 K/min) and studied in terms of the Harnish and Muschik's method. The above kinetic studies concluded that sepiolite treated with isobutyric acid acted as nucleating agent for PP, increasing both amount of crystal phase and crystallization rate (evaluated by Avrami's values for K). Two well-defined slopes of Avrami's plot were shown, which were claimed to derive from a second-ordered PP structure at the polymer/filler interface.

More recently, Wang and Sheng [45,46] studied the isothermal and non-isothermal crystallization of PP/org-attapulgite nanocomposites. Four temperatures (121°C, 122°C, 123°C, and 124°C) were chosen for the isothermal crystallization tests and the results explained in terms of Avrami's equation and Hoffman's theory (Figure 12.13). Org-attapulgite particles incorporated in PP matrix acted as heterogeneous nuclei, dramatically increasing the crystallization rate, with values of Avrami's exponent n increasing from 2.4–2.8 to 4.2–4.9.

Only minor effects were found on the equilibrium melting temperature and degree of crystallinity. Polarized optical microscopy observations revealed that the spherulites size decrease and the number increase with attapulgite content (Figure 12.14).

Polymer crystallization is also influenced by the nanoclay dispersion and/or by the use of any surface modification or compatibilizer. Pozsgay et al. [47] demonstrated that the ability of MMT to nucleate PP depends on the organo-treatment of clay and, consequently, on the alteration of the interlayer MMT distance rather than on the modification of the clay surface tension. The study concluded that the nucleation occurs not on the surfaces but rather in the interlayers of clay particles and attributed the nucleating effect to the collapsed MMT galleries of 1 nm distance. Svoboda et al. [48] found an increase in crystallization temperatures in PP-g-MA/MMT systems containing clay tactoids, but not in systems with well-dispersed MMT clays. An analogous behavior was found for PP/sepiolite nanocomposites [16]. Figure 12.15 shows the DSC cooling traces relative to two compatibilized systems, characterized by a better (PP/PP-g-MA/Sep) or a worse (PP/PP-acid/Sep) dispersion of sepiolite (refer to Section 4.1.2 and Figure 12.9). The agglomerated system (Figure 12.15a) shows a continuous and large increase in the crystallization temperature due to heterogeneous nucleation. On the other hand,

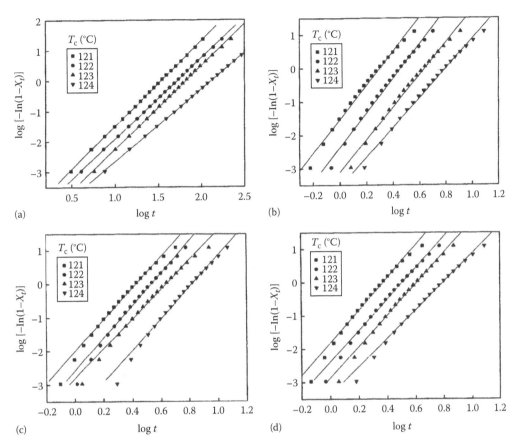

FIGURE 12.13
Avrami's plot for (a) neat PP and PP/ATP nanocomposites; (b) PP + 1% ATP; (c) PP + 3% ATP; and (d) PP + 5% ATP, at three different crystallization temperatures. (Reproduced from Wang, L.H. et al., *J. Macromol. Sci., Phys. B*, 43, 935, 2004. With permission from Taylor & Francis.)

the system with well-dispersed sepiolite (Figure 12.15b) shows only a small effect and a limiting concentration of crystallization nuclei is reached at 1 wt% of sepiolite, after which no significant changes in T_c are observed (Figure 12.16). This suggests that nanoclay aggregates are the predominant nucleating sites for PP, rather than the surfaces of individual sepiolite. However, the same experimental observation can also be explained by alteration in sepiolite surface energy. Inorganic fillers like clays have high-energy surfaces and adsorb the polymer preferentially along their crystal structure. When the filler surface is covered with an organic phase, the surface free energy is decreased, thus also the nucleating efficiency. A thick coating layer of polymer may also conceal topological features of the filler, which are effective nucleating sites. The different crystallization behavior can then be justified by a different tendency of the three compatibilizers used (PP-*g*-MA, PP-acid, and PP-PEO) to segregate the nanoclay surface, which is reasonable since PP-*g*-MA has a low degree of functionalization and a much higher molecular weight than PP-acid and PP-PEO.

Besides changing the overall crystallinity, the presence of an inorganic filler can also modify the crystal structure of a semicrystalline polymer. It is well known that PP is able to crystallize in different polymorphic forms: α, β, γ, and a mesomorphic crystal

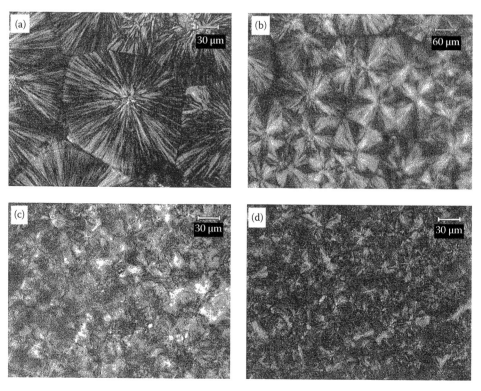

FIGURE 12.14
Polarized optical microscopy pictures of (a) PP, (b) PP + 1% ATP, (c) PP + 3% ATP, and (d) PP + 5% ATP.
(Reproduced from Wang, L.H. et al., *J. Macromol. Sci., Phys. B*, 43, 935, 2004. With permission from Taylor &
Francis.)

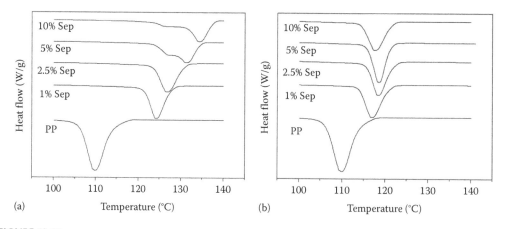

FIGURE 12.15
DSC cooling traces relative to two compatibilized systems: (a) PP/PP-*g*-MA/Sep and (b) PP/PP-acid/Sep.
The first is characterized by a bad nanoclay dispersion while the second by a good dispersion. (Reproduced
from Bilotti, E. et al., *J. Appl. Polym. Sci.*, 107, 1116, 2008. With permission from Wiley.)

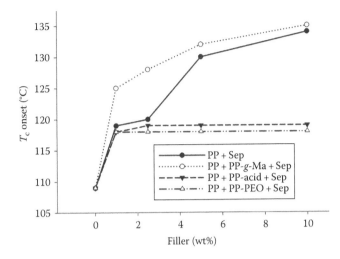

FIGURE 12.16
Onset temperatures of incipient crystallization for PP/Sep, PP/PP-*g*-MA/Sep, PP/PP-acid/Sep, and PP/PP-PEO/Sep nanocomposites, as a function of the filler content. (Reproduced from Bilotti, E. et al., *J. Appl. Polym. Sci.*, 107, 1116, 2008. With permission from Wiley.)

structure [49,50]. The presence of a different crystal phase can usually be detected by wide angle x-ray scattering (WAXS) techniques. For instance, the β phase of PP shows a characteristic diffraction peak at $2\theta = 16°$ while the γ phase at $2\theta = 20.3°$. To the best of the authors' knowledge, no effect of crystal phase change induced by the presence of needle-like clays in polyolefins has been reported so far in scientific literature.

12.4.2.2 Rheology

The rheology of particulate suspensions is governed by factors such as structure, size, and shape of the dispersed phase as well as its orientation and distribution and the strength of interaction between the dispersed and dispersant phase. Understanding the rheological properties of polymer/clay nanocomposites is crucial in gaining insights into the processability and the structure–property relation of these materials. The rheological properties of PP/sepiolite nanocomposites, with various loadings of fibrous clay, are presented in Figure 12.17 [8].

At a temperature of 200°C and a frequency range of 10^{-2} to 10^2 Hz, PP exhibits (Figure 12.17a) the usual behavior of homopolymer melts with narrow Mw distribution (i.e., $G' \propto \omega^2$ and $G'' \propto \omega$). The crossover of the curves G' and G'' separates a liquid-like behavior at low frequencies from a solid-like behavior at high frequencies. The complex viscosity (Figure 12.17b) and the storage modulus (Figure 12.17c) increase, as expected, with the addition of sepiolite clay. The increase relative to the virgin polymer, though, is much higher at low frequencies, while it completely converges at high frequencies, for concentrations as high as 2.5 wt%. The zero-shear viscosity plateau (Figure 12.17b) shifts toward lower frequency regions until it disappears, in the frequency range scanned, for filler loadings of 10 wt%. All the nanocomposites display significantly reduced frequency dependence of the storage modulus, G'. The terminal slope of G' rapidly decreases as the clay loading increases and the nanocomposite filled with 10 wt% of sepiolite shows almost a plateau at low frequencies (Figure 12.17c), which is characteristic of materials with solid-like behavior ($G', G'' \propto \omega^0$). This can be explained if it is assumed that the clay content has reached a threshold value to form a percolating physical clay network (Figure 12.18) [51]. In this case, the mobility and relaxation of polymer chains are seriously retarded in the conned space between nanoclays. Such a percolation threshold is between 5 and 10 wt% of filler (Figure 12.17b and c). Although the calculation of an accurate percolation threshold would require more experimental data

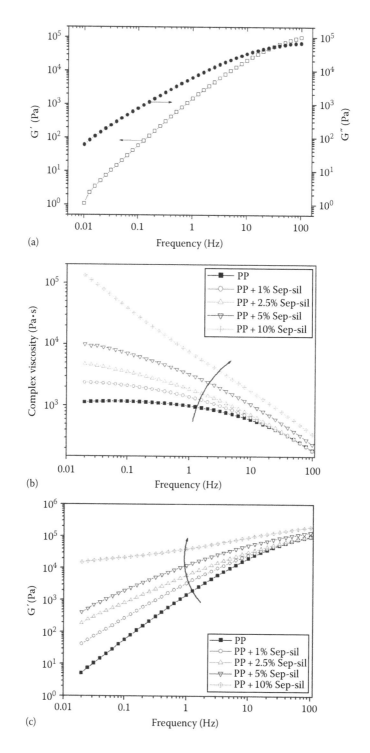

FIGURE 12.17
(a) Frequency sweep test on polypropylene melts at 200°C, (b) complex viscosity of PP/sepiolite (silane treated) nanocomposites as a function of the clay loading, and (c) storage modulus of PP/sepiolite (silane treated) nanocomposites as a function of the clay loading. (Reproduced from Bilotti, E. et al., *Macromol. Mater. Eng.*, 295, 15, 2010. With permission from Wiley.)

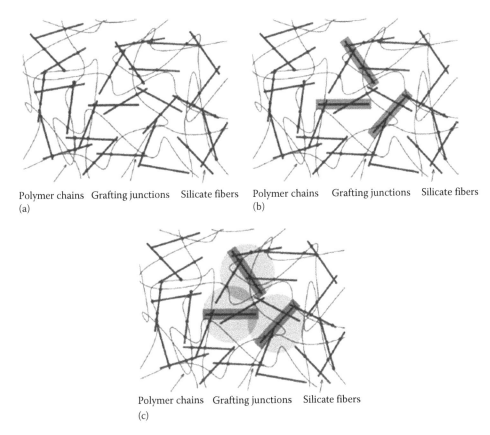

Polymer chains Grafting junctions Silicate fibers
(a)

Polymer chains Grafting junctions Silicate fibers
(b)

Polymer chains Grafting junctions Silicate fibers
(c)

FIGURE 12.18
Schematic description of the polymer/needle-like clay percolating structure. (Reproduced from Krishnamoorti, R. and Giannelis, E.P., *Macromolecules*, 30, 4097, 1997. With permission from American Chemical Society.)

(in particular above the percolation) to be fitted with a suitable percolation theory, the range of values given (5–10 wt%) is compatible with theoretical predictions [52].

Simplifying the lattice model in Isichenco's review [7], Shen et al. [52] formulated a "grafting-percolating" model able to describe the percolation thresholds of attapulgite sticks (aspect ratio $l/d = 30$–40) in a polymer matrix.

Applying a Monte Carlo simulation to the 3D lattice model (Figure 12.19), the authors were able to predict a percolation occurring at a volume fraction of attapulgite Φ_c of 0.02, or, alternatively, a weight fraction of 3–4 wt%. The slight divergence of this value from the one found for sepiolite nanoclays in PP can be explained by the lower aspect ratio of the latter. In fact, the percolation threshold is inversely proportional to the aspect ratio.

12.4.2.3 Thermal Behavior/Fire Retardancy

Among the nanoclay composites properties that have mostly attracted industrial interest are the enhanced barrier properties and thermal properties or fire retardancy. The ban of using traditional halogenous compounds as fire retardant additives for polymer (e.g., the in EU directives: EN 135501; CDP 89/106/EEC; EN 45545), due to their demonstrated carcinogenic effects, has driven the search for safer alternatives. Nanoclays have been widely investigated for this purpose and encouraging results have already been found for some specific systems, although a full understanding of the mechanisms is still lacking [53–55]. A popular physical

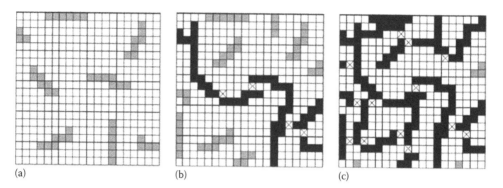

FIGURE 12.19
2D sketch of the percolation lattice model at volume fractions: (a) $\Phi < \Phi_c$, (b) $\Phi = \Phi_c$, and (c) $\Phi > \Phi_c$. Black occupied lattices represent the sticks percolated, while the gray occupied lattices represent the non-percolated sticks. The symbol "X" indicates the lattices occupied by grafted polymer chains. (Reproduced from Shen, L. et al., *Polymer*, 46, 5758, 2005. With permission from Elsevier.)

explanation is the formation of a barrier at the surface of the nanocomposite, which can prevent the diffusion of oxygen into the material and, at the same time, the evaporation of volatile reaction products originating during thermal degradation [56,57]. Important parameters are the *nanoclay dispersion* and its effects on the *rheological properties* [58] of the polymeric matrix but even the shape of the sample [59]. Marcilla et al. [60] studied the effect of high concentrations of sepiolite clays (25–30 wt%) on the degradation of different polymers in powder form. A strong catalytic effect was demonstrated in particular for the degradation of polyolefins. This result should not surprise if it is considered that some natural clays (zeolites in particular) are used as catalysts in the cracking of polymers, favoring their transformation into a fuel product and preventing the coke formation [61].

The TGA curves in Figure 12.20 show both a premature beginning of the degradation and a widening of the temperature range. This was explained by considering that some active (acid) sites on the surface of the sepiolite catalyze the degradation of the polyolefins but get deactivated as they participate to this process, decelerating the degradation.

In inert atmosphere (N_2), sepiolite caused a decrease of 70 K in the temperature of the single PP degradation step (the temperature corresponding to the maximum of mass loss rate) and 23 K in the temperature of PE decomposition (Figure 12.20a). The stronger effect in the case of PP, rather than PE, was justified by the larger amount of tertiary carbon atoms present. The same tests performed in oxidative atmosphere (Figure 12.20b) showed a more complicated degradation process with an increase in the number of reaction steps. Interestingly, however, the presence of sepiolite did not have any major effects on the temperature of the main decomposition step of PP.

Tartaglione et al. [62] have recently studied the thermal behavior of melt-compounded PP/sepiolite composites (3 wt% loading) and the effect of the clay surface modification. In inert atmosphere, the catalytic activity of sepiolite on PP pyrolysis was confirmed, except for the beginning of the degradation process (less than 20% of mass loss) where the presence of the clay improved the polymer stability (Figure 12.21a). This was explained by the adsorption of PP volatile products by the sepiolite external surface and internal zeolitic channels.

It was also noticed that the catalytic effect was reduced by the sepiolite surface functionalization (in particular the silanization), which could provide a shielding effect to the catalytically active hydroxyl groups [60]. In oxidative atmosphere, instead, the behavior

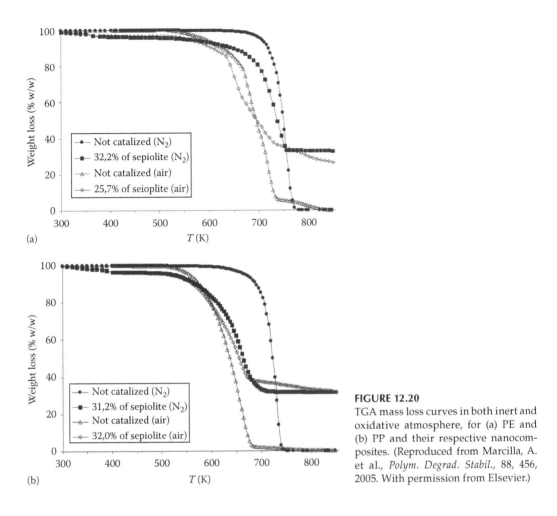

(a)

(b)

FIGURE 12.20
TGA mass loss curves in both inert and oxidative atmosphere, for (a) PE and (b) PP and their respective nanocomposites. (Reproduced from Marcilla, A. et al., *Polym. Degrad. Stabil.*, 88, 456, 2005. With permission from Elsevier.)

was different. The presence of sepiolite, and more significantly silane-grafted sepiolite, shifted the degradation temperature of PP of 20–40 K higher (Figure 12.21a). In conclusion, despite the catalytic effect of sepiolite on PP thermal degradation, a higher stability toward thermoxidation was achieved probably because of an efficient barrier to oxygen.

One may ask what the effects of nanoclay particle shape and sample thickness are. If barrier to oxygen and volatiles is the key factor, platelet-like clays, oriented perpendicularly to the gas direction, should be more efficient than needle-like clays. Similarly, thicker film samples should be more effective than thinner ones. Garcia et al. [59] recently studied the effect of three different nanoparticle geometries (spherical, fibrous, laminar) on the thermal degradation of low-density polyethylene (LDPE). The particles chosen were silica nanoparticles, silane-modified sepiolite, and ammonium salt–modified MMT. Sepiolite and MMT showed similar thermoxidative protection behavior, with a shift of 100 K of the degradation temperature of LDPE in air, while silica showed an inferior performance. No effect of sample thickness was observed, which suggested the formation of a protective layer on the sample surface. By studying the samples surfaces, a high concentration of inorganic phase was found. Sepiolite formed a thicker layer than MMT, which counterbalanced the effect of an a priori less favorable shape, and explained the overall similar reduction in thermal degradation.

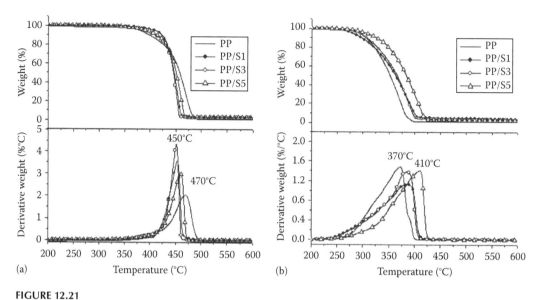

FIGURE 12.21

TGA mass loss curves and their derivative of PP/sepiolite nanocomposites in (a) inert and (b) oxidative atmosphere. Pristine sepiolite (S1) was modified with a quaternary ammonium salt (S3) or mercaptopropyl trimethoxysilane (S5). (Reproduced from Tartaglione, G. et al., *Compos. Sci. Technol.*, 68, 451, 2008. With permission from Elsevier.)

Two nanoclay shapes (needle and platelet) were also compared in the work of Marosfoi et al. [63] to study any synergism between them or in combination with magnesium hydroxide (MH), a more conventional micron-size fire retardant. The different composites were studied via cone calorimeter, which is the most recognized technique to evaluate the combustion characteristics of a material, and the heat release rate (HRR) values were reported (Figure 12.22). The addition of any of the inorganic filler investigated decreased the maximal heat release rate and the time to ignition of PP. The metal hydroxide has the effect of absorbing heat by water release upon degradation [64]. The clays were supposed to contribute, in agreement with current literature, to the formation of a protective surface layer. The use of micro- and nanofillers (either sepiolite or MMT) together considerably decreased the peak of HRR and the total HRR (area under the curve) compared to the binary formulations (PP + MH, PP + Sep, PP + MMT).

Figure 12.22b suggested a synergistic effect of clay combination. Addition of 5 wt% MMT/sepiolite mixture to the PP + 10% MH significantly decreased the maximum HRR (from 471 to 246 kW/m²) and increased the time of ignition in comparison with formulations with 5% MMT only.

Another synergistic behavior for the fire retardancy of PP was recently found by Hapuarachchi, who studied the combined use of sepiolite nanoclays and carbon nanotubes (CNTs) [65]. The ternary system PP/CNT/sepiolite showed an enhanced thermoxidation resistance in air (Figure 12.23a) and the highest residual char, when compared with the binary systems PP/CNT or PP/sepiolite. It was proposed that the CNT helped in bridging clay-rich area and therefore creating a tighter char network on the sample surface [65].

Hapuarachchi [65] also measured the heat release rates (HRR) of PP and its different nanocomposites by cone calorimeter (Figure 12.24). The time to ignition was slightly anticipated probably due to an increased thermal conductivity induced by the presence of CNTs. Nevertheless, the peak of HRR experienced a reduction of 80%, from 1933 to 355 kW/m². The similar values of total smoke release (CO and CO_2) suggested that the difference between

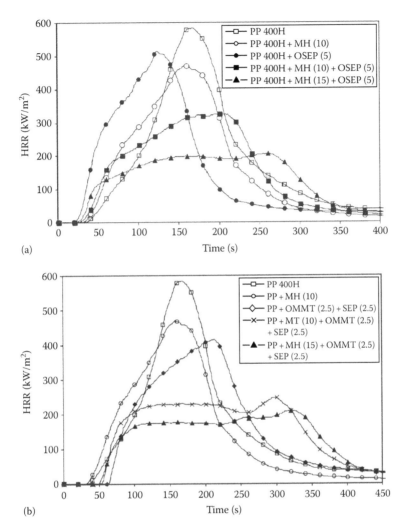

FIGURE 12.22
Effect of (a) sepiolite or MMT clays and (b) clay mixture (sepiolite + MMT) on the heat release rate (HRR) of PP, with and without magnesium hydroxide (MH). (Reproduced from Marosfoi, B.B. et al., *Polym. Adv. Technol.*, 19, 693, 2008. With permission from Wiley.)

the behavior of PP and PP nanocomposites had to be attributed to the condensed phase decomposition. In fact, the mass loss rate was reduced from 0.2 to 0.03 g/s by the addition of the two nanofillers. It was suggested that this effect was caused by a reinforcement of the carbonaceous char layer, which acted as an efficient thermal, mass, and gas transport barrier.

12.4.2.4 Mechanical Properties

The mechanical performances of polymer nanocomposites are influenced not only by several factors such as properties and amounts of the constituent phases (matrix and nanofiller), nanoparticle dispersion, morphology, and orientation, the matrix–filler interactions but also by the degree of crystallinity and crystalline phases of the polymer, as described

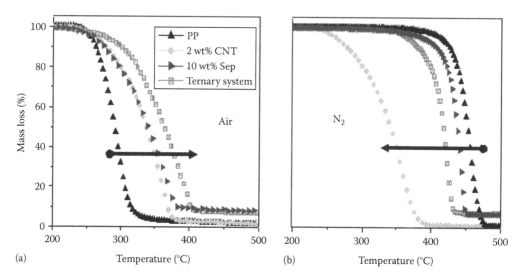

(a)   Temperature (°C)   (b)   Temperature (°C)

FIGURE 12.23
TGA mass loss curves and their derivative of PP nanocomposites, based on sepiolite clay, multi-wall carbon nanotube (CNT) or their combination, in oxidative (left) and inert (right) atmosphere [65].

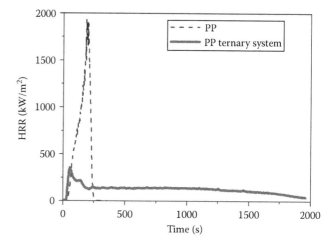

FIGURE 12.24
Heat release rate vs. time for unfilled PP and ternary PP nanocomposite (10 wt% sepiolite + 2 wt% CNT) [65].

earlier. The addition of a nanoclay to a polymeric matrix results almost univocally to an increase in Young's modulus and a corresponding decrease in elongation at break. The tensile strength and the impact strength of nanocomposite are instead less predictable and more sensitive to the filler/matrix interphase and the strength of interaction.

A good needle-like nanoclay dispersion in polyethylene was achieved by Du et al. [66] via an in situ polymerization (Section 12.4.1). The properties of the resulting nanocomposites are reported in Table 12.5.

Interestingly, the tensile strength was significantly enhanced with just 1 wt% of palygorskite without compromising the elongation at break. Even more remarkable was the increase in impact strength, in correspondence of the same nanoclay concentration, which suggests a positive matrix–filler interaction. It is reminded that a part of the PE macromolecules was directly grown from the palygorskite surface and is expected to compatibilize the palygorskite with the remaining PE matrix.

TABLE 12.5

Mechanical Properties of In Situ Polymerized PE/Palygorskite Nanocomposites

Sample	Filler wt [%][a]	Tensile Strength [MPa]	Elongation at Break [%]	Impact Strength [kJ/m²]
PE	0	31.4	180.2	50.7
PE/Palyg-1	1	38.1	172.1	82.9
PE/Palyg-2	1.7	32.3	170.2	70.9
PE/Palyg-3	2.2	26.1	166.7	60.2
PE/Palyg-4	3.3	22.4	160.8	42.8

Source: Reproduced from Du, Z.J., *J. Mater. Sci.*, 38, 4863, 2003. With permission.
[a] The final concentration of the PE/palygorskite nanocomposites has been obtained by diluting an in situ masterbatch with commercial PE, in a ratio of 2/8.

TABLE 12.6

Mechanical Properties of PP/Attapulgite Nanocomposites in Function of Filler Content and the Functionalization of the Needle-Like Clay [35]

Sample	Filler wt [%]	Young's Modulus [MPa]	Yield Stress [MPa]	Impact Strength [kJ/m²]
PP + ATP	0	788	32.1	3.6
	1	800	31.2	3.7
	2	836	29.8	3.9
	3	880	28.7	3.5
	5	904	26.5	3.4
	7	908	26.0	3.1
PP + Org-ATP	0	788	32.1	3.6
	1	860	33.4	5.3
	2	954	35.3	6.3
	3	980	36.6	6.8
	5	1038	33.6	5.7
	7	1122	32.7	5.4

Wang and Sheng [35] studied both pristine and organo-modified attapulgite clays as reinforcement for PP matrix. The mechanical properties of PP/attapulgite nanocomposites are reported in Table 12.6. All samples showed an increased Young's modulus, though the contribution of the organo-attapulgite is much larger than that of pristine attapulgite. The presence of non-modified attapulgite, independently from its weight fraction, led to a decrease in the yield stress value suggesting that the interfacial interaction between matrix and filler was weak. The yield stress of PP/org-attapulgite nanocomposites, instead, increased with the clay content, up to 3 wt%. The same concentration of attapulgite also showed the maximum increase (90%) in the impact strength. Higher org-attapulgite contents probably resulted in partial nanofiller agglomeration since all mechanical properties showed a negative trend.

The effect of the degree of nanoclay dispersion and polymer–filler interaction on the mechanical properties was also studied by Bilotti et al. [11–16]. Figure 12.25 shows the Young's modulus, yield stress, and strain at break of PP/sepiolite nanocomposites compatibilized by different functionalized polymers (Section 12.4.1). The results are also compared with a PP nanocomposite reinforced by an alkyl silane–coated sepiolite (Sep-sil), without any compatibilizers.

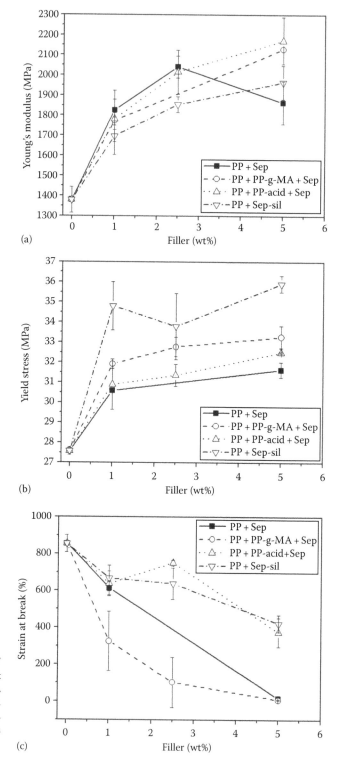

FIGURE 12.25
Mechanical tensile properties of PP/sepiolite nanocomposites at different filler loadings: (a) Young's modulus, (b) yield stress, and (c) strain at break. (Reproduced from Bilotti, E. et al., *J. Appl. Polym. Sci.*, 107, 1116, 2008. With permission from Wiley.)

As expected, an increase in Young's modulus was observed for all nanocomposite systems investigated (Figure 12.25a). An enhancement in yield stress could also be noticed (Figure 12.25b) for sepiolite content as high as 5 wt%, with the best results obtained for PP/Sep-sil nanocomposites. More differences among the nanocomposites were shown for the strain at break (Figure 12.25c). While nanocomposites with pristine clay or with PP-*g*-MA were dramatically embrittled at sepiolite concentrations as low as 2.5 wt%, the use of PP-acid or Sep-sil preserved the ductile nature of the polymer matrix that underwent yielding with necking stabilization and cold drawing even at filler concentrations above 5 wt%.

From the tensile tests presented, it appeared that the nanocomposites made by alkyl silane–functionalized sepiolite give the best mechanical performances, in particular for what concerns the yield stress. In fact, the sepiolite surface functionalization by silane is a reactive treatment, which decreases the interparticle aggregation (improved dispersion) and, at the same time, increases the matrix–filler interactions. The addition of functionalized polymers is, instead, a nonreactive surface treatment. It leads to a decrease of the particle–particle interaction but can also reduce the matrix–particle interaction, which leads to lower yield stress and ultimate tensile stress.

12.4.2.5 Oriented Nanocomposites

As discussed earlier in the chapter, the advantage of using a fiber-like nanoclay over a platelet-like nanoclay is particularly beneficial for the mechanical properties of aligned systems. In this case, needle-like fillers are expected to have a higher reinforcing efficiency than platelet-like fillers (Figure 12.3). Stretching a polymeric film containing nanoclays in the solid state is one of the simplest ways to induce nanofiller orientation. This process also promotes, at the same time, an orientation of the macromolecules and hence enhances the mechanical properties [67,68]. Highly drawn PP/sepiolite nanocomposite tapes have recently been prepared [8]. Figure 12.26 shows the stress–strain curves of pure PP tapes at different draw ratios. As expected, the solid-state drawing process dramatically changed the mechanical performance of PP, with

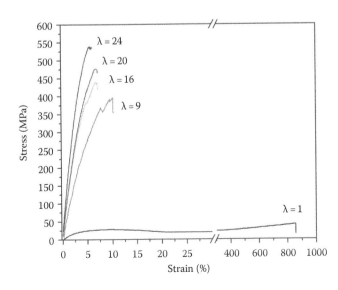

FIGURE 12.26

Stress–strain curves of PP tapes of different draw ratios λ. (Reproduced from Bilotti, E. et al., *Macromol. Mater. Eng.*, 295, 15, 2010. With permission from Wiley.)

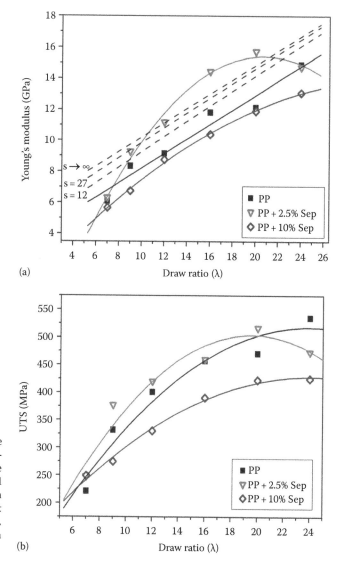

FIGURE 12.27

(a) Young's modulus and (b) ultimate tensile stress of PP/sepiolite tapes at different draw ratios. The dotted lines are Halpin-Tsai predictions of PP tapes filled with 2.5 wt% sepiolite, fully aligned in the direction of the tape, at three aspect ratios. (Reproduced from Bilotti, E. et al., *Macromol. Mater. Eng.*, 295, 15, 2010. With permission from Wiley.)

Young's moduli increasing from about 1.3 to 14 GPa and the ultimate tensile stress from about 50 to 550 MPa (for tapes drawn 24 times: $\lambda = 24$). In contrast, the strain at break is reduced to 5%.

The addition of sepiolite clay significantly modified the mechanical performance of these PP tapes (Figure 12.27a and b). Small amounts of sepiolite (up to 2.5 wt%) enhanced the stiffness of the PP tapes (for $10 < \lambda < 20$), surprisingly even above the optimistic predictions of the Halpin-Tsai micromechanical model (Figure 12.27a). Loadings above 5 wt% of sepiolite, instead, had a detrimental effect on the Young's modulus of PP tapes. Similar observations were made for the ultimate tensile stress of the nanocomposite tapes (Figure 12.27b). The negative effect of high loadings was explained by a worsening of the sepiolite dispersion above 2.5 wt% and/or by the formation of a nanofiller physical network, which prevented drawability and polymer alignment. Too high draw ratios ($\lambda > 20$) were, instead, believed to cause debonding between PP and sepiolite, due to an imperfect interfacial adhesion.

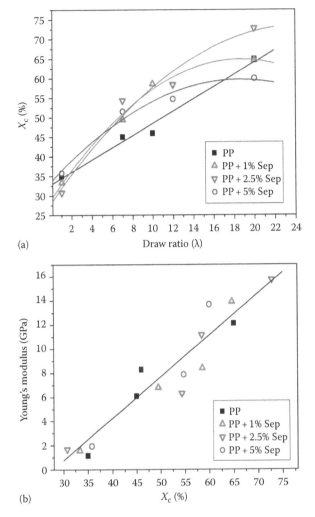

(a)

(b)

FIGURE 12.28

(a) Degree of crystallinity of different nano-composite tapes as a function of draw ratio λ and (b) Young's modulus of nanocomposite tapes as a function of the degree of polymer crystallinity. (Reproduced from Bilotti, E. et al., *Macromol. Mater. Eng.*, 295, 15, 2010. With permission from Wiley.)

An explanation for the exceptional reinforcement effect of small amounts of nanoclay (~2.5 wt%) at intermediate draw ratios ($10 < \lambda < 20$) could be found, if not in micromechanical composite theories, in the modified crystalline morphology.

The solid-state drawing causes an increase in the degree of crystallinity of PP, which directly relates with the increased Young's modulus of the tape (Figure 12.28a). It was also shown that the presence of sepiolite affects the crystallinity of the PP tapes (Figure 12.28a).

The nanocomposite tapes, with intermediate values of λ and low sepiolite content, showed a higher crystallinity than pure PP tapes (Figure 12.28a) with remarkable similarities between crystallinity and the mechanical properties shown in Figure 12.27a and b. A good correlation could be found between Young's modulus and crystallinity of the PP tapes (Figure 12.28b). This observation suggested that the deviations of the mechanical properties of the nanocomposite tapes from the pure PP tapes were to be attributed mainly to differences in polymer orientation in the composites tapes, which intimately relates to differences in degree of crystallinity and morphology developed upon solid-state drawing.

12.5 Conclusions

Needle-like clays have recently attracted increasing interest as an alternative nanofiller to the more commonly studied platelet-like nanoclays (e.g., MMT). Although the dispersion of needle-like clay is expected to be easier than for other forms of clays, it still remains a critical issue, in particular, for polyolefins. Nevertheless, different strategies to obtain homogeneous dispersions—at least for volume fractions as high as 3–4 wt%—have already demonstrated their validity. The in situ synthesis of polyolefins in presence of needle-like clays can produce good nanofiller dispersions while the simple melt-processing preparation method requires either the use of functional polymers, as compatibilizers between clay and matrix, or the surface functionalization of the clay. The second route is preferable if mechanical properties like yield stress and impact strength are of concern, since it can lead to an enhanced matrix–filler interaction, as well as a better nanoclay dispersion. Unidirectional nanocomposites based on needle-like clays show remarkable mechanical performances, even beyond the predictions of micromechanical models. The phenomenon could be explained by the effect of the nanoclay on the crystallinity and morphology of the oriented polymer matrix. Particularly interesting were the thermal and fire performances of the polyolefin/needle-like clay nanocomposites. Sepiolite was able to substantially retard the degradation of polyolefin in oxidative atmosphere. Synergisms are found between needle-like nanoclays and carbon nanotubes or a more conventional micron-size fire retardant and a combination of sepiolite and MMT.

References

1. Pinnavaia, T. J. and Beall, G. W. 2000. *Polymer–Clay Nanocomposites*. New York: Wiley.
2. Alexandre, M. and Dubois, P. 2000. Polymer-layered silicate nanocomposites: Preparation, properties and uses of a new class of materials. *Materials Science and Engineering R-Reports* 28:1–63.
3. Ray, S. S. and Okamoto, M. 2003. Polymer/layered silicate nanocomposites: A review from preparation to processing. *Progress in Polymer Science* 28:1539–1641.
4. Tjong, S. C. 2006. Structural and mechanical properties of polymer nanocomposites. *Materials Science and Engineering R-Reports* 53:73–197.
5. Van Es, M. 2001. Polymer-clay nanocomposites—The importance of particle dimensions. PhD thesis, Technical University of Delft, Delft, the Netherlands.
6. Fischer, H. 2003. Polymer nanocomposites: From fundamental research to specific applications. *Materials Science and Engineering C* 23:763–772.
7. Isichenko, M. B. 1992. Percolation, statistical topography, and transport in random-media. *Reviews of Modern Physics* 64:961–1043.
8. Bilotti, E., Deng, H., Zhang, R., Lu, D., Bras, W., Fischer, H. R., and Peijs, T. 2010. Synergistic reinforcement of highly oriented polypropylene tapes by sepiolite nanoclay. *Macromolecular Materials and Engineering* 295:15–16.
9. Halpin, J. C. and Kardos, J. D. 1976. The Halpin-Tsai equations: A review. *Polymer Engineering and Science* 16:344–352.
10. Hull, D. and Clyne, T. W. 1996. *An Introduction to Composite Materials*. 2nd edn. Cambridge, U.K.: Cambridge University Press.
11. Bilotti, E. 2009. Polymer/sepiolite clay nanocomposites. PhD thesis, Queen Mary University of London, London, U.K.

12. Grim, R. E. 1968. *Clay Mineralogy*. New York: McGraw–Hill.
13. Grim, R. E. 1962. *Applied Clay Mineralogy*. New York: McGraw–Hill.
14. https://www.webmineral.com
15. https://www.nrc-cnrc.gc.ca/obj/imi/images/english/nanocomposites1.gif
16. Bilotti, E., Fischer, H. R., and Peijs, T. 2008. Polymer nanocomposites based on needle-like sepiolite clays: Effect of functionalized polymers on the dispersion of nanofiller, crystallinity, and mechanical properties. *Journal of Applied Polymer Science* 107:1116–1123.
17. Galan, E. 1996. Properties and applications of palygorskite-sepiolite clays. *Clay Minerals* 31:443–453.
18. Aznar, A. J., Gutierrez, E., Diaz, P., Alvarez, A., and Poncelet, G. 1996. Silica from sepiolite: Preparation, textural properties, and use as support to catalysts. *Microporous Materials* 6:105–114.
19. Kuang, W. X., Facey, G. A., and Detellier, C. 2004. Dehydration and rehydration of palygorskite and the influence of water on the nanopores. *Clays and Clay Minerals* 52:635–642.
20. Wilson, I. *Special Clays*. Clays, March 2008.
21. http://www.nanoroadmap.it
22. Private communication with Tolsa (Spain).
23. http://www.hse.gov.uk/lau/lacs/37-2.htm
24. Suáreza, P., Quintana, M. C., and Hernández, L. 2008. Determination of bioavailable fluoride from sepiolite by "in vivo" digestibility assays. *Food and Chemical Toxicology* 46:490–493.
25. http://www.tolsa.com
26. Usuki, A., Kawasumi, M., Kojima, Y., Okada, A., Kurauchi, T., and Kamigaito, O. 1993. Swelling behavior of montmorillonite cation exchanged for omega-amino acids by epsilon-caprolactam. *Journal of Materials Research* 8:1174–1178.
27. Gul, R. J., Wan, P. S., Hyungsu, K., and Wook, L. 2004. Power ultrasound effects for in situ compatibilization of polymer–clay nanocomposites. *Journal of Materials Science and Engineering C* 24:285.
28. Rong, J., Li, H., Jing, Z., Hong, X., and Sheng, M. 2001. Novel organic/inorganic nanocomposite of polyethylene. I. Preparation via in situ polymerization approach. *Journal of Applied Polymer Science* 82:1829–1837.
29. Du, Z., Zhang, W., Zhang, C., Jing, Z., and Li, H. 2002. A novel polyethylene/palygorskite nanocomposite prepared via in-situ coordinated polymerization. *Polymer Bulletin* 49:151–158.
30. Acosta, J. L., Ojeda, M. C., Morales, E., and Linares, A. 1986. Morphological, structural, and interfacial changes produced in composites on the basis of polypropylene and surface-treated sepiolite with organic-acids. 1. Surface-treatment and characterization of the sepiolites. *Journal of Applied Polymer Science* 31:2351–2359.
31. Acosta, J. L., Ojeda, M. C., Morales, E., and Linares, A. 1986. Morphological, structural, and interfacial changes produced in composites on the basis of polypropylene and surface-treated sepiolite with organic-acids. 2. Thermal properties. *Journal of Applied Polymer Science* 31:1869–1878.
32. Acosta, J. L., Ojeda, M. C., Morales, E., and Linares, A. 1986. Morphological, structural, and interfacial changes produced in composites on the basis of polypropylene and surface-treated sepiolite with organic-acids. 3. Isothermal and non-isothermal crystallization. *Journal of Applied Polymer Science* 32:4119–4126.
33. Acosta, J. L., Morales, E., Ojeda, M. C., and Linares, A. 1986. Effect of addition of sepiolite on the mechanical properties of glass-fiber reinforced polypropylene. *Angewandte Makromolekulare Chemie* 138:103–110.
34. Wang, L. H. and Sheng, J. 2003. Graft polymerization and characterization of butyl acrylate onto silane-modified attapulgite. *Journal of Macromolecular Science—Pure and Applied Chemistry A* 40:1135–1146.
35. Wang, L. H. and Sheng, J. 2005. Preparation and properties of polypropylene/org-attapulgite nanocomposites. *Polymer* 46:6243–6249.
36. Zhao, L., Du, Q., Jiang, G., and Guo, S. 2007. Attapulgite and ultrasonic oscillation induced crystallization behavior of polypropylene. *Journal of Polymer Science B* 45:2300–2308.
37. Ma, J., Bilotti, E., Peijs, T., and Darr, J. A. 2007. Preparation of polypropylene/sepiolite nanocomposites using supercritical CO2 assisted mixing. *European Polymer Journal* 43:4931–4939.

38. Zhang, Z. X. and van Duijneveldt, J. S. 2006. Isotropic-nematic phase transition of nonaqueous suspensions of natural clay rods. *Journal of Chemical Physics* 124:4931–4939.
39. Pukanszky, B. 1995. In *Polypropylene: Structure, Blends and Composites*, ed. J. Karger-Kocsis. London, U.K.: Chapman & Hall.
40. Pukanszky, B., Belina, K., Rockenbauer, A., and Maurer, F. H. J. 1994. Effect of nucleation, filler anisotropy and orientation on the properties of PP composites. *Composites* 25:205.
41. Pukanszky, B., Mudra, I., and Staniek, P. 1997. Relation of crystalline structure and mechanical properties of nucleated polypropylene. *Journal of Vinyl Additive Technology* 3:53.
42. Van der Meer, D. W., Pukanszky, B., and Vancso, G. J. 2002. On the dependence of impact behavior on the crystalline morphology in polypropylenes. *Journal of Macromolecular Science—Physics* B41:1105.
43. Avrami, M. 1939. Kinetics of phase change. I: General theory. *Journal of Chemical Physics* 7:103.
44. Avrami, M. 1940. Kinetics of phase change. II: Transformation-time relations for random distribution of nuclei. *Journal of Chemical Physics* 8:212.
45. Wang, L. H., Sheng, J., and Wu, S. Z. 2004. Isothermal crystallization kinetics of polypropylene/attapulgite nanocomposites. *Journal of Macromolecular Science-Physics B* 43:935–946.
46. Wang, L. H. and Sheng, J. 2005. Non-isothermal crystallization kinetics of polypropylene/attapulgite nanocomposites. *Journal of Macromolecular Science-Physics B* 44:31–42.
47. Pozsgay, A., Frater, T., Papp, L., Sajo, I., and Pukanszky, B. 2002. Nucleating effect of montmorillonite nanoparticles in polypropylene. *Journal of Macromolecular Science-Physics B* 41:1249–1265.
48. Svoboda, P., Zeng, C. C., Wang, H., Lee, L. J., and Tomasko, D. L. 2002. Morphology and mechanical properties of polypropylene/organoclay nanocomposites. *Journal of Applied Polymer Science* 85:1562–1570.
49. Natta, G. and Corradini, P. 1960. Structure and properties of isotactic polypropylene. *Nuovo Cimento Suppl* 15:40.
50. Karger-Kocsis, J. 1995. *Polypropylene Structure, Blends and Composites: Structure and Morphology*. London, U.K.: Chapman & Hall.
51. Krishnamoorti, R. and Giannelis, E. P. 1997. Rheology of end-tethered polymer layered silicate nanocomposites. *Macromolecules* 30:4097–4102.
52. Shen, L., Lin, Y. J., Du, Q. G., Zhong, W., and Yang, Y. L. 2005. Preparation and rheology of polyamide-6/attapulgite nanocomposites and studies on their percolated structure. *Polymer* 46:5758–5766.
53. Gilman, J. W. 1999. Flammability and thermal stability studies of polymer layered-silicate (clay) nanocomposites. *Applied Clay Science* 15:31–49.
54. Gilman, J. W., Jackson, C. L., Morgan, A. B., Harris, R. H. Jr., Manias, E., Giannelis, E. P., Wuthenow, M., Hilton, D., and Phillips, S. H. 2000. Flammability properties of polymer-layered-silicate nanocomposites, polypropylene and polystyrene nanocomposites. *Chemistry of Materials* 12:1866–1873.
55. Morgan, A. B. 2006. Flame retarded polymer layered silicate nanocomposites: A review of commercial and open literature systems. *Polymers for Advanced Technologies* 17:206–217.
56. Bartholmai, M. and Schartel, B. 2004. Layered silicate polymer nanocomposites: New approach or illusion for fire retardancy? Investigations of the potentials and the tasks using a model system. *Polymers for Advanced Technologies* 15:355–364.
57. Gilman, J. W., Harris, R. H., Shields, J. R., Kashiwagi, T., and Morgan, A. B. 2006. A study of the flammability reduction mechanism of polystyrene-layered silicate nanocomposite: Layered silicate reinforced carbonaceous char. *Polymers for Advanced Technologies* 17:263–271.
58. Kashiwagi, T., Du, F., Douglas, J. F., Winey, K. I., Harris, R. H., and Shields, J. R. 2005. Nanoparticle networks reduce the flammability of polymer nanocomposites. *Nature Materials* 4:928–933.
59. Garcia, N., Hoyos, M., Guzman, J., and Tiembloet, P. 2009. Comparing the effect of nanofillers as thermal stabilizers in low density polyethylene. *Polymer Degradation and Stability* 94:39–48.
60. Marcilla, A., Gomez, A., Menargues, S., and Ruiz, R. 2005. Pyrolysis of polymers in the presence of a commercial clay. *Polymer Degradation and Stability* 88:456–460.

61. Manos, G., Yusuf, J., Papayannakos, N., and Gangas, N. H. 2001. Catalytic cracking of polyethylene over clay catalysts. Comparison with an ultrastable Y zeolite. *Industrial Engineering and Chemistry Research* 40:2220–2225.
62. Tartaglione, G., Tabuani, D., Camino, G., and Moisio, M. 2008. PP and PBT composites filled with sepiolite: Morphology and thermal behavior. *Composites Science and Technology* 68:451–460.
63. Marosfoi, B. B., Garas, S., Bodzay, B., Zubonyai, F., and Marosi, G. 2008. Flame retardancy study on magnesium hydroxide associated with clays of different morphology in polypropylene matrix. *Polymers for Advanced Technologies* 19:693–700.
64. Rothon, R. N. and Hornsby, P. R. 1996. Flame retardant effects of magnesium hydroxide. *Polymer Degradation and Stability* 54:383–385.
65. Hapuarachchi, T. D. 2010. Development and characterization of flame retardant nano-particulate bio-based polymer composites. PhD thesis, Queen Mary University of London, London, U.K.
66. Du, Z. J., Rong, J. F., Zhang, W., Jing, Z. H., and Li, H. Q. 2003. Polyethylene/palygorskite nanocomposites with macromolecular comb structure via in situ polymerization. *Journal of Materials Science* 38:4863–4868.
67. Peijs, T., Jacobs, M. J. N., and Lemstra, P. J. 2000. High performance polyethylene fibres. In *Comprehensive Composites, vol.1*, eds. T.-W. Cou, A. Kelly, and C. Zweben, pp. 263–302. Oxford, U.K.: Elsevier Science Publishers Ltd.
68. Ward, I. M. 1997. *Structure and Properties of Oriented Polymers*. London, U.K.: Chapman & Hall.

13

Polyolefin Nanocomposites with Functional Compatibilizers

Kiriaki Chrissopoulou and Spiros H. Anastasiadis

CONTENTS

13.1 Introduction

Polyolefins constitute the most widely used group of thermoplastics, often referred to as commodity thermoplastics. They are prepared by polymerization of simple olefins such as ethylene, propylene, butenes, isoprenes, and pentenes, as well as their copolymers, whereas they are the only class of macromolecules that can be produced catalytically with precise control of stereochemistry and, to a large extent, of (co)monomer sequence distribution. The term polyolefins derives from "oil-like" and refers to the oily or waxy feel that these materials have. They consist only of carbon and hydrogen atoms and are nonaromatic. An inherent characteristic common to all polyolefins is a nonpolar, nonporous, low-energy surface that is not receptive to inks and lacquers without special oxidative pretreatment. Polyolefin-based materials can be tailor-made for a wide range of applications: from rigid thermoplastics to high-performance elastomers. These vastly different properties are achieved by a variety of molecular structures, whose common features are low cost, excellent performance, long life cycle, and ease of recycling. Although polyolefins represent one of the oldest (if not the oldest) families of thermoplastic polymers, they are still characterized by innovations that provide new applications via stepwise and continuous technology renewal that reduce the eco-footprint during manufacture and use. The two most important and common polyolefins are polyethylene and polypropylene, and they are very popular due to their low cost and wide range of applications.

Since the first mass production of polyolefins with the development of Ziegler-type catalysts, commercial exploitation has been very rapid because of their attractive characteristics. The consumption growth rates have been high with the material becoming widely used in various industrial areas for fibers, films, and injection-molding articles. However, polyolefins are notch sensitive and brittle on exposure to severe conditions, such as low temperature or high rate of impact. In order to improve the competitiveness of polyolefines in engineering applications, it is important to simultaneously increase stability, heat distortion temperature, stiffness, strength, and impact resistance without sacrificing the easy processability. Modification of the polymers by the addition of fillers, reinforcements, or blends of special monomers or elastomers can render them more flexible with a variety of other properties, and their competitiveness in engineering resin applications can be greatly improved [1,2].

Polymer composites have been widely used in numerous applications since they offer an appealing improvement of the properties of the individual components [3]. Nevertheless, important compromises are often required in material design since, for example, an increase in strength is often accompanied by a loss in toughness and an increase in brittleness or a loss of optical clarity; it is anticipated that these problems can be overcome if the inorganic additive exists in the form of a fine dispersion within the polymeric matrix producing a *nano*composite [4]. In these cases, the final properties of the hybrid are determined mainly by the existence of many interfaces [5]. Moreover, the addition of nanoscopic fillers of high anisotropy in polymer matrices is even more interesting in the polymer industry with the enhanced reinforcement being a result of the much greater surface-to-volume ratios of these high-aspect-ratio additives.

A special case of nanocomposites is obtained [6–12] by mixing polymers with layered silicates (nanoclays) where three different types of structure can be identified depending on the interactions between the polymer and the inorganic nanoparticles: phase separated, where the polymer and the inorganic are mutually immiscible; intercalated, where the polymer chains intercalate between the layers of the inorganic material; and exfoliated, where the periodicity of the inorganic material is destroyed and the inorganic platelets are dispersed within the polymeric matrix. Achieving the desired structure is a scientific problem with numerous technological implications because it is the structure that controls the final properties of the micro- or nanocomposites [13,14]. Such nanocomposites exhibit remarkable improvement in a variety of properties like strength and heat resistance [15], gas permeability [16–19], flammability [20,21], and biodegradability [22]. In order to demonstrate all the advantages of polymer/layered silicate nanocomposites, it was anticipated that, large surface area, high aspect ratio and good interfacial interactions are essential; thus, the silicate should be exfoliated and the platelets should strongly adhere to the polymer [6,16,23]. In this chapter, we refer only to polyolefins/layered silicate nanocomposites and to the attempts that has been performed to achieve the desired nanohybrid structure.

The layered silicate nanoparticles are usually hydrophilic and their interactions with nonpolar polymers are not favorable. Thus, whereas hydrophilic polymers are likely to intercalate within Na-activated montmorillonite clays [24–29], hydrophobic polymers can lead to intercalated [23,30–32] or exfoliated [33] structures only with organophilized clays, i.e., with materials where the hydrated Na^+ within the galleries has been replaced by proper cationic surfactants (e.g., alkylammonium) by a cation exchange reaction. The thermodynamics of intercalation or exfoliation have been discussed [34–37] in terms of both enthalpic and entropic contributions to the free energy. It has been recognized that the entropy loss because of chain confinement is compensated by the entropy gain associated with the increased conformational freedom of the surfactant tails as the interlayer distance increases with polymer intercalation [34,38], whereas the favorable enthalpic interactions are extremely critical in determining the nanocomposite structure [39].

In the case of polyolefins, attempts to develop either intercalated or exfoliated nanocomposite structures with polypropylene or polyethylene have not been particularly successful due to the strong hydrophobic character of the polymers and the lack of favorable interactions with the silicate surfaces. Figure 13.1 shows x-ray diffractograms of

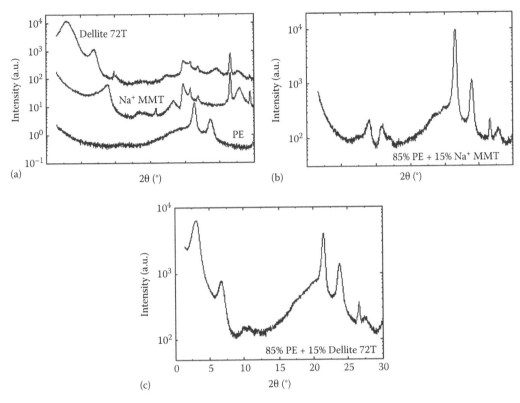

FIGURE 13.1
(a) X-ray diffractograms of polyethylene (PE), hydrophilic clay (Na⁺ MMT), and organophilic clay (Dellite 72T). The curves are shifted for clarity. (b) X-ray diffractogram of a micro-composite with 85 wt% PE + 15 wt% Na⁺ MMT. (c) X-ray diffractogram of a micro-composite with 85 wt% PE + 15 wt% Dellite 72T. (Reprinted from Chrissopoulou, K., et al., *Polymer*, 46, 12440, 2005. With permission from Elsevier.)

mixtures of a linear low-density polyethylene (LLDPE) with melt flow index (MFI) of 37 with a natural hydrophilic montmorillonite, Na⁺-MMT (Figure 13.1b) and with an organophilized analog, Dellite 72T, (Figure 13.1c), together with those of the individual components (Figure 13.1a) [40]. Two main diffraction peaks are evident in the diffraction curve of polyethylene at $2\theta_1 = 21.3°$ and at $2\theta_2 = 23.7°$ corresponding to the d_{110} and the d_{200} peaks of the orthorhombic crystal structure of polyethylene with the d_{110} possessing higher intensity [41]. The data for Na⁺-MMT show a main peak at $2\theta = 8.9°$ that corresponds to an interlayer distance of $d_{001} = 9.9\,\text{Å}$ whereas the ones for Dellite 72T shows its main peak at $2\theta = 2.9°$ which corresponds to $d_{001} = 30.4\,\text{Å}$. Comparison between the diffractograms of the hydrophilic and the organophilic clays shows that the incorporation of the surfactants leads to an increase of the interlayer spacing by more than $10\,\text{Å}$. In addition, the peaks corresponding to the nanoparticles appear in a different angular range than those corresponding to the polymer, which makes the identification of any intercalation obvious.

Figure 13.1b and c shows x-ray diffractograms for micro-composites prepared by melt mixing PE with the hydrophilic Na⁺-MMT and the organophilic Dellite 72T clay, respectively. The mixing was performed in a 0.5 cm³ DSM micro-extruder at 150°C [40]. It is evident that there is not any indication for either intercalation of the polymer chains or exfoliation of the clay layers with neither the natural clay nor the organophilized one.

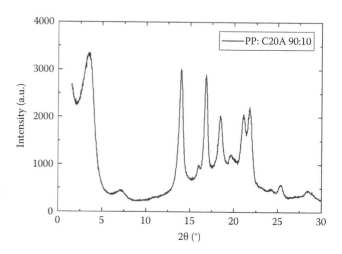

FIGURE 13.2
X-ray diffractograms of a PP:C20A 90:10 composite. (Reprinted from Chrissopoulou, K. et al., *J. Polym. Sci., Part B: Polym. Phys.*, 46, 2683, 2008. With permission from Wiley.)

The main diffraction peaks appear at the same angles with those of the pure inorganic materials. It is noted that variation of the concentration of the silicate from 15 to 5 wt% does not alter the results. The immiscibility of polyethylene with the hydrophilic clay is of course anticipated and it is shown here as a reference. Similar, with polyethylene, the diffractogram of a 90 wt% polypropylene + 10 wt% organophilic clay is shown in Figure 13.2 [42]. In this case, polypropylene was an impact copolymer that had MFI: 25 and the organoclay was Cloisite 20A. Polypropylene shows a number of diffraction peaks whose position and relative intensities indicate that the corresponding structure is that of the isotactic alpha form of polypropylene [41]. The data for the composite do not provide any indication for either intercalation of the polymer chains or exfoliation of the silicate platelets since the main diffraction peak appears at the same angle with that of the pure inorganic material. Thus, it is clear that the organophilization is not sufficient for polypropylene to intercalate and that the composites are phase-separated systems as expected from the nonpolar and very hydrophobic character of PP. Additionally, the peaks corresponding to PP appear very similar with those of the pure polymer, except from two small peaks that appear at 16.0° and 19.6° that could be related to the (300) reflection of the β-phase and the (130) plane of the γ-phase of polypropylene [41,43].

Despite the hope that polymer melt intercalation would be a promising way to fabricate polyolefin nanocomposites using conventional polymer processing techniques, these results verify that PE or PP even when mixed with hydrophobic alkylammonium surfactants modified layered inorganic material lead to phase-separated systems because of their nonpolar character and neither intercalation of the polymer chains nor exfoliation of the silicate platelets can be achieved. It is thus clear that for the synthesis of polyolefin/layered silicate nanocomposites, one has to modify the interactions between the polymer and the inorganic surfaces. As a solution, a polyolefin oligomer that has polar groups (like –OH, COOH, maleic anhydride) could be used as compatibilizer. In the approach to prepare a polyolefin nanocomposite using a functionalized compatibilizer, there seemed to be two important factors in terms of the structure of the oligomers. First, they should include a certain amount of polar groups to intercalate between silicate layers via hydrogen bonding to the oxygen groups of the silicate layers and, second, they should be miscible with the polymer. Since the content of polar functional groups in the oligomers will affect the miscibility, there must be an optimum content of polar functional groups in the compatibilizer. A plethora of investigations have been performed, employing various compatibilizers with different functionalities and the effect of parameters like their molecular weight, the type and the content of the functional groups, the compatibilizer-to-organoclay ratio, the processing method, etc., have been studied to optimize the structure and achieve the desired dispersion of the inorganic material. Among the different compatibilizers, maleic anhydride–functionalized polyolefins are most commonly used to improve the interfacial bonding between the fillers and the polymers. The rest of the chapter is divided in two main parts: in the first, the most important results obtained utilizing maleic anhydride–functionalized polyolefins are discussed, whereas in the second, a summary of the studies performed utilizing other functional oligomers/polymers is presented.

13.2 Maleic Anhydride–Functionalized Polyolefins as Compatibilizers

It is generally believed that the polar character of the anhydride group results in an affinity for the silicate surfaces, so that the maleated polyolefin can serve as a compatibilizer between the matrix and the filler. In one of the first works performed in this aspect, the capability of functional polyolefin oligomers to act as compatibilizers was examined. As mentioned before, there are two important factors that compete in order to achieve the complete homogeneous dispersion of the silicate layers [44,45]. Firstly, the oligomers should include a certain amount of polar groups to be intercalated between silicate layers via hydrogen bonding. Secondly, the oligomers should be well miscible with the polymer itself. Since the content of polar functional groups in the oligomers affects the miscibility with the polymer, there should be an optimum content to achieve exfoliation. Indeed, two maleic anhydride–modified polypropylene oligomers, PP-*g*-MA, with different number of polar groups were utilized and mixed with octadecyl amine–modified montmorillonite in binary mixtures [44]. The one which possessed one carboxyl group per approximately 25 units of propylene (PP-*g*-MA-1010) when mixed with the C18-MMT in 1:1 ratio was found to shift the angle of the 001 peak from $2\theta = 4°$ to $2\theta = 2.3°$ and, thus, to increase the interlayer distance from 21.7 to 38.2 Å. On the contrary, the second one that had only one carboxyl group per 190 units of propylene (PP-*g*-MA110TS) under the same conditions resulted in a phase-separated micro-composite. It is noted that the molecular weight of the oligomer was found not to affect the intercalation process. Moreover, in the case of PP-*g*-MA-1010, where an intercalated structure was observed, the interlayer distance was found to increase when the mixing weight ratio of PP-*g*-MA-1010/C18-MMT is changed from 1:3 to 3:1; in the case of the 3:1 ratio the peak appears at $2\theta = 1.2°$ that corresponds to an interlayer distance of $d_{001} = 72.2$ Å [45]. This behavior was attributed to a small amount of PP-*g*-MA-1010 that was intercalated into the silicate layers at the early step of the mixing process so that the silicate layers were coated by PP-*g*-MA-1010 and the interactions between the layers were weakened. Therefore, the remaining PP-*g*-MA-1010 could be easily intercalated and the interlayer distance could be further expanded as the weight of the compatibilizer increased.

At the same time, the effect of the miscibility of the polymer with the compatibilizer was also examined by utilizing similar PP-*g*-MA's with different number of polar groups. PP-*g*-MA-1001 (acid value: 26 mg KOH/g) and PP-*g*-MA-1010 (acid value: 52 mg KOH/g) were mixed with PP in a 77:23 ratio of PP:PP-*g*-MA [45]. Figure 13.3 shows optical polarized micrographs of these blends at their melt state (200°C) from which one can see that the

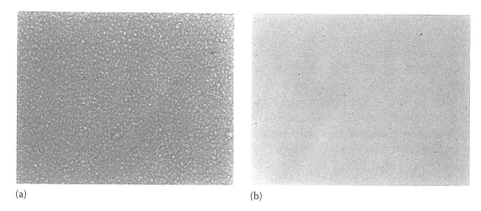

(a)   (b)

FIGURE 13.3
Optical polarized micrographs of the blends of PP with PP-*g*-MA at 200°C. (a) PP/PP-*g*-MA-1010 and (b) PP/PP-*g*-MA-1001. (Reprinted from Kawasumi, M. et al., *Macromolecules*, 30, 6333, 1997. With permission from American Chemical Society.)

compatibilizer with fewer polar groups is more miscible than the one with more. When ternary hybrids are synthesized composed of PP, PP-*g*-MA, and C18-MMT in a 3:1 ratio of compatibilizer: organoclay, an intercalated structure is observed for PP-*g*-MA-1010 but with an interlayer distance of d_{001} = 59 Å. Note that when the corresponding binary hybrid was prepared, a d_{001} = 72.2 Å was observed. This indicates that only PP-*g*-MA-1010 may have intercalated between the layers of C18-MMT. On the other hand, the XRD patterns of PP-*g*-MA1001, PP, C18-MMT with the same composition exhibit rather small peak with a gradual increase of the diffraction strength toward lower angles. Therefore, the latter hybrids should be more exfoliated compared to those with PP-*g*-MA-1010. A schematic explanation of what is happening is shown in Figure 13.4 [45]. There is a strong driving force for the intercalation of PP-*g*-MA via strong hydrogen bonding between the maleic anhydride group (or the COOH groups generated from the hydrolysis of the maleic anhydride group) and the oxygen atoms of the silicates. Thus, the interlayer spacing of the clay increases, and the interactions of the layers are weakened. The intercalated clays contact PP under a strong shear field. If the miscibility of PP-*g*-MA with PP is good enough to mix at the molecular level, the exfoliation of the intercalated clay takes place; otherwise, there is phase separation of the two polymers shown as an intercalated structure.

The effect of the ratio of the compatibilizer to organoclay on the final structure of the hybrid has been widely investigated. In one case, a maleic anhydride–grafted polyethylene, PE-*g*-MA, with 0.85% functional maleic anhydride groups and an MFI of 1.5 was mixed with polyethylene, PE, and a dimethyl dihydrogenated tallow quaternary ammonium chloride–modified montmorillonite (Dellite 72T) [40]. Ternary mixtures were prepared in a microextruder; the concentration of the layered silicate was kept constant at 10 wt% whereas the concentration of PE-*g*-MA (with respect to the total polymer content) was varied from 2.5 wt%, where PE-*g*-MA can be considered as an additive, up to 100 wt% where PE-*g*-MA is the polymer matrix. In this way, the importance of the ratio α of compatibilizer to organoclay became evident. Figure 13.5 shows the x-ray diffractograms of all the composites where the curves have been shifted vertically for clarity. For low concentrations of PE-*g*-MA, i.e., for low α's (below α = 0.45), the data exhibit the characteristic peak of the parent organoclay signifying a phase-separated system. As the relative ratio α increases above 0.72, the diffraction peak appears to shift in the range 2θ = 2.70°–2.45°, which indicates the existence of intercalated structures with interlayer distance of d_{001} = 32.7–36.0 Å. At low angles,

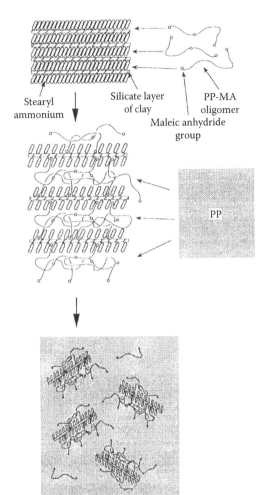

Stearyl ammonium

Silicate layer of clay

PP-MA oligomer

Maleic anhydride group

PP

FIGURE 13.4
Schematic representation of the dispersion process of the organophilized clay in a PP matrix with the aid of PP-*g*-MAs. (Reprinted from Kawasumi, M. et al., *Macromolecules*, 30, 6333, 1997. With permission from American Chemical Society.)

the intensity increases with decreasing scattering angle, which signifies the coexistence of exfoliated layers (together with intercalated ones). At $\alpha = 2.7$, there is only an indication of a shoulder in the data near $2\theta = 2.45°$. For even higher values of α, there is no indication for the existence of a scattering peak of either the montmorillonite or of an intercalated system. The diffracted intensity shows a continuous decrease with increasing scattering angle, which indicates that the structure has been destroyed due to the interactions of the polymer with the inorganic nanoparticles. This picture is valid for PE-*g*-MA concentrations corresponding to α values higher than 4.5 whereas the XRD results were further supported by TEM images. Moreover, an attempt was made to correlate the observed structure with the macroscopic rheological behavior of the blends at a temperature much higher than the PE crystallization temperature to ensure that the observed results derive only from the presence of the inorganic material and from the change of the structure [40]. A progressive change of the dependence of the elastic and loss moduli, G' and G'', respectively, as well as of the complex viscosity on frequency was observed with increasing the PE-*g*-MA content, i.e., by progressively changing the structure from a phase-separated to a completely exfoliated one. For the immiscible micro-composites, the G' and G'' moduli exhibit the expected ω^{-2} and ω^{-1} dependence on frequency in the flow regime, whereas the viscosity exhibits a

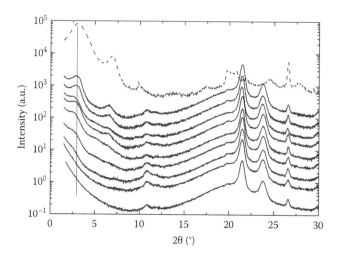

FIGURE 13.5
X-ray diffractograms of three-component PE hybrids containing PE-*g*-MA and 10 wt% organoclay Dellite 72T with different relative compositions of PE to PE-*g*-MA. The solid lines from top to bottom correspond to PE/PE-*g*-MA = 97.5/2.5, 95/5, 92/8, 85/15, 82/18, 80/20, 70/30, 51/49, 0/100. The data for the respective organoclay are shown for comparison with the dashed line. The curves are shifted for clarity. (Reprinted from Chrissopoulou, K. et al., *Polymer*, 46, 12440, 2005. With permission from Elsevier.)

low-frequency plateau; this behavior resembles the one of the pure polymer. In contrast, the behavior for the exfoliated system is entirely different corresponding to a solid-like response of both G′ and G″ with very weak frequency dependence.

The change in the behavior may be quantified by evaluating the shear-thinning exponent n by analyzing the low-frequency data in terms of a $\eta^* = A\omega^n$ dependence [46]. For a liquid-like behavior $n \sim 0$, whereas for a solid-like response, $n \sim -1$. Wagener and Reisinger [46] proposed to use the value of n as a measure of the degree of exfoliation since it was explicitly assumed that exfoliation leads to a percolated structure, which results in a solid-like behavior. Figure 13.6 shows the frequency dependence of the complex viscosity measured at 140°C for three different pairs of systems. In each pair, one specimen contains the organoclay Dellite 72T (filled symbols) while the other belongs to the neat polymer/s. The three different pairs correspond to an immiscible system, an exfoliated system (containing compatibilizer to organoclay ratio $\alpha = 4.4$), and a system where only partial exfoliation and slight intercalation can be inferred from the x-ray diffraction data (containing compatibilizer to organoclay ratio $\alpha = 1.4$). For comparison, the data for the respective polymer (or mixtures) in the absence of the organoclay is also included. Indeed, an increase of the exponent n is observed with the increase of the degree of exfoliation, when the rheological data are correlated to the x-ray diffraction results. n is zero for the homopolymer PE and $n = -0.1$ for the immiscible 85% PE/15% Dellite 72T hybrid whereas $n = -0.4$ for the slightly exfoliated 13.5% PE-*g*-MA + 76.5% PE + 10% Dellite 72T ($\alpha = 1.4$) and $n = -0.65$ for the completely exfoliated 44% PE-*g*-MA + 46% PE + 10% Dellite 72T ($\alpha = 4.4$). Care should be used in analyzing the data since the shear-thinning exponent does vary with the polymer matrix. The shear-thinning exponent is very low ($n = -0.1$) for 85 wt% PE/15 wt% PE-*g*-MA and it becomes $n = -0.39$ for 49 wt% PE/51 wt% PE-*g*-MA. However, in both cases, the frequency dependence is much weaker for the blend of polymers than that for the three-component hybrids. Therefore, the "shear-thinning exponent" can indeed be correlated with the structure of the system since it essentially describes the transition from

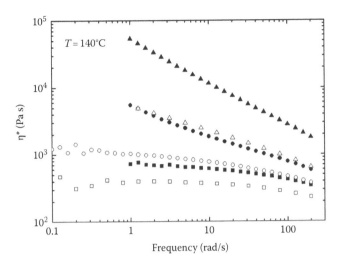

FIGURE 13.6

Frequency dependence of the complex viscosity of various systems at 140°C. (□) Polyethylene PE, (○) a binary mixture containing 85 wt% PE and 15 wt% PE-*g*-MA, (△) a binary mixture containing 51 wt% PE and 49 wt% PE-*g*-MA, (■) a binary hybrid containing 85 wt% PE + 15 wt% Dellite 72T, (●) a three-component hybrid containing PE-*g*-MA and 10 wt% Dellite 72T with PE/PE-*g*-MA = 85/15, and (▲) a three-component hybrid containing PE-*g*-MA and 10 wt% Dellite 72T with PE/PE-*g*-MA = 51/49. (Reprinted from Chrissopoulou, K. et al., *Polymer*, 46, 12440, 2005. With permission from Elsevier.)

a liquid-like behavior of the immiscible micro-composites to the solid- or gel-like behavior of the exfoliated nanocomposites due to the percolated structure of the nanohybrids.

In a continuation of that work, an attempt was made to quantify the role and the amount of the compatibilizer and to optimize the specimen preparation procedure in order to modify the structure of the nanohybrids from a phase-separated one to an exfoliated one in a controlled way. Thus, a polypropylene, PP, with MFI of 25 was used together with a maleic anhydride–grafted polypropylene, PP-*g*-MA, with an MFI of 115 and a maleic anhydride content $w_{MA} \sim 0.6\%$ and the widely used organoclay Cloisite 20A, C20A [42]. In order to examine the effectiveness of the specific PP-*g*-MA as a compatibilizer, binary hybrids of just the "compatibilizer" and the silicate were first prepared. Figure 13.7a shows the x-ray diffraction data of these composite materials containing PP-*g*-MA and C20A in order to examine the compatibilizer's ability either to intercalate between the layers of the inorganic material or to exfoliate the structure. The diffractograms of the pure materials are shown as well whereas the curves have been shifted vertically for clarity. The data of PP-*g*-MA shows a series of peaks that resemble the crystalline pattern of PP. When the ratio of PP-*g*-MA to organoclay, α, is lower than 1.5, the diffraction peak corresponding to the periodic structure of C20A is observed at exactly the same angle as in the pure clay signifying a phase-separated structure; it is only the increase of the scattered intensity at low angles that indicates some percentage of platelet exfoliation. As the ratio of PP-*g*-MA to organoclay increases, the characteristic peak decreases in intensity but remains at the same position and it is only for PP-*g*-MA to C20A ratios 9:1 by weight and higher that it vanishes. Figure 13.8 shows representative TEM images of two hybrids with ratios of PP-*g*-MA to C20A, 60:40 (Figure 13.8a) and 90:10 (Figure 13.8b) by weight. The dark lines represent the edges of the silicate layers and the white region the polymeric matrix. Clear differences are observed between the two systems. It is evident that a coexistence is observed in Figure 13.8a with clay particles retaining the layered structure and clay

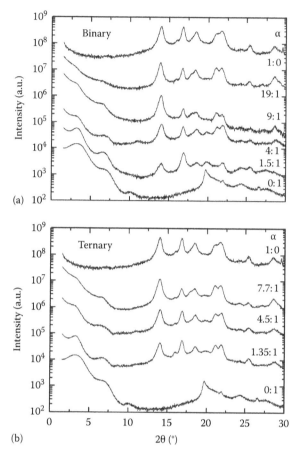

(a)

(b)

FIGURE 13.7
X-ray diffractograms of (a) binary composites of
PP-*g*-MA and C20A in different concentrations
and (b) ternary composites of PP, PP-*g*-MA,
and C20A at similar ratio α as in (a). (Reprinted
from Chrissopoulou, K. et al., *J. Polym. Sci., Part
B: Polym. Phys.*, 46, 2683, 2008. With permission
from Wiley.)

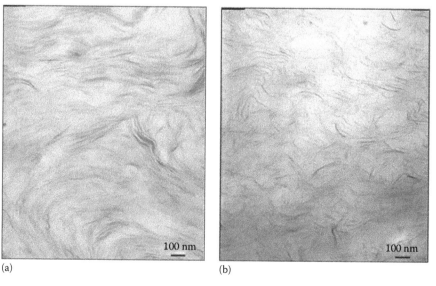

(a)   (b)

FIGURE 13.8
TEM images of (a) PP-*g*-MA:C20A 60:40 and (b) PP-*g*-MA:C20A 90:10. (Reprinted from Chrissopoulou, K. et al.,
J. Polym. Sci., Part B: Polym. Phys., 46, 2683, 2008. With permission from Wiley.)

platelets dispersed within the polymer matrix. For the higher concentration of PP-*g*-MA in Figure 13.8b, a uniform dispersion of exfoliated platelets is observed. These results are in excellent agreement with the x-ray diffraction data.

For the hybrids of Figure 13.7a, α ranges from 1.5 (for the PP-*g*-MA:C20A 60:40) to 19 (for the PP-*g*-MA:C20A 95:5). In order to examine whether it is this value of α that totally determines the structure, three-component hybrids were synthesized, utilizing PP, PP-*g*-MA, and C20A in such a way that the ratio α is similar to the one in the binary systems of Figure 13.7a. These hybrids were synthesized by mixing the three materials together in the micro-extruder under the same conditions that were used for the binary systems. The x-ray diffractograms of these hybrids with α = 1.35, 4.5, and 7.7 are shown in Figure 13.7b. Comparison between Figure 13.7a and b shows that each pair of specimens with similar α exhibits exactly the same behavior, with the diffraction curves having exactly the same shape. In particular, for the composite with α = 1.35, the main diffraction peak is at 2θ = 3.3°, at the same position with that of the corresponding binary mixture and that of the organoclay, leading to the conclusion that the lamellar structure is preserved and that the system is phase separated. An increase of the ratio α leads to an increase of the degree of exfoliation of the system, manifested by the decrease of the intensity of the main peak and its concurrent increase at low angles. In conclusion, it seems that indeed the structure is determined by the value of the ratio of PP-*g*-MA to C20A, i.e., by α, irrespective of the presence or absence of PP. It is noted that in this case no indications of intercalation are observed and the system goes progressively from the phase-separated structure to the exfoliated structure [42].

Since the development of the proper processing route in view of potential applications is sometimes almost as important as the understanding of the interactions in the blends, another preparation method was subsequently followed. Hybrids that comprised PP-*g*-MA and C20A organoclay with a specific α were utilized as a masterbatch and were further mixed with polypropylene in different compositions. Figure 13.9 shows the x-ray diffractograms of such composites with α = 4 (Figure 13.9a) and α = 9 (Figure 13.9b). It is obvious that in both cases mixing the masterbatch with the polymer results in even higher degrees of exfoliation evidenced by the disappearance of even small peaks present in the diffractogram of the masterbatch; this is reproducible and holds for every composition of the masterbatch. The given explanation was that achieving a certain degree of exfoliation utilizing the compatibilizer leads on one hand to the weakening of the interactions between the layers of the inorganic material and on the other hand to the creation of a friendly environment for the PP polymer. Thus, a "hairy particle" is formed with PP-*g*-MA being the "hair" chains, which can be homogeneously mixed with PP dispersing the silicate layers even more [42]. This explanation is in accordance with similar descriptions of the observed behavior in such systems [45].

It should be noted that results showing an increase of the interlayer distance that can even reach to full exfoliation with increasing PP-*g*-MA content (for a certain amount of organoclay) were observed in other cases as well [47]. However, those authors did not indicate the ratio α as the critical or one of the critical parameters that define the final structure.

The structure, rheological, and mechanical properties of PP/PP-*g*-MA/organoclay hybrids were investigated by Reichert et al. [48]. The PP had an MFI of 3.2 g/10 min, the maleic anhydride oligomer was Hostaprime, H (with 4.2 wt% MA content, M_n = 4000), and the inorganic material was SOMASIF ME100 organophilized with surfactants with different length. The synthesized hybrids consisted of 10 wt% organoclay and 20 wt% compatibilizer. The authors showed that the values of both the Young's modulus and the

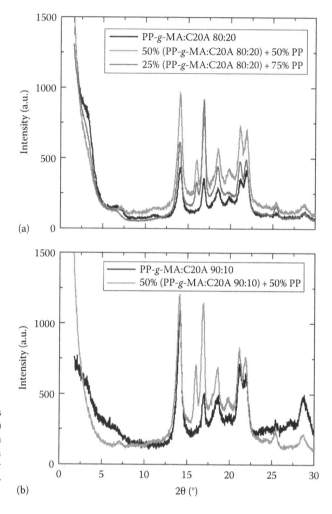

FIGURE 13.9
X-ray diffractograms of binary hybrids
with (a) PP-*g*-MA:C20A 80:20 and (b)
PP-*g*-MA:C20A 90:10 together with
their mixtures with PP. (Reprinted from
Chrissopoulou, K. et al., *J. Polym. Sci., Part
B: Polym. Phys.*, 46, 2683, 2008. With per-
mission from Wiley.)

yield stress made a jump when organoclays with surfactant chains carrying more
than 12 carbon atoms were utilized whereas they possessed low values for shorter surfactants. This increase was attributed to the increase in the degree of exfoliation due to the
enhanced interfacial coupling. In the absence of compatibilizer, both the Young's modulus
and the yield stress were very similar for all hybrids and independent from the surfactants
used for the organophilization. Thus they concluded that both alkyl chains with more
than 12 carbon atoms and the presence of a maleic anhydride–functionalized polypropylene compatibilizer were necessary in order to obtain some degree of exfoliation via
hydrogen bonding as well as covalent and ionic bond formation at the interfaces between
the filler and the polymer [48]. Nevertheless, the high amount of PP-*g*-MA resulted in a
rather poor impact resistance. At the same time, it was demonstrated that annealing of
the samples (200 min at a temperature of 220°C) is necessary in order for the hybrids to
reach their equilibrium structure, which is described by a higher degree of exfoliation and
a recovery of their thermorheological simplicity manifested via time–temperature super-
position (TTS) of the rheological data [49]. Indeed, $\delta = \tan^{-1}(G''/G')$ measured as a function
of $|G^*| = (G'^2 + G''^2)^{1/2}$ of a hybrid at different temperatures revealed that TTS does not

hold and that by decreasing temperature the hybrid becomes more and more elastic. It was only after thermal treatment that all isotherms merged at the same curve with even smaller δ values. X-ray diffraction measurements were utilized to capture the morphological changes that derive from the thermal annealing and showed that whereas a broad and less intense peak than the one of the organoclay was still detectable in the hybrid before annealing, no d_{001} peak of the silicate was observed after thermal treatment indicating that the layers became fully exfoliated.

The effect of PP-*g*-MA-to-organoclay ratio on the structure and properties of PP nanocomposites was investigated in other studies as well. In another case, a PP with melt index of 37 g 10/min, a PP-*g*-MA with 1 wt% MA content and a dimethyl di(hydrogenated tallow) modified MMT were melt blended with MMT content of 1, 3, 5, and 7 wt% and PP-*g*-MA-to-organoclay ratio of 0.5, 1, and 2 [50]. Complete exfoliation was not observed in any of these cases; however, quantitative analysis of TEM images reveals that the lengths of the particles decrease with addition of PP-*g*-MA but their thickness decreases more strongly, thus resulting in an increase of the aspect ratio, which is one of the main parameters for the polymer reinforcement. Figure 13.10 shows that the estimated aspect ratio increases with the ratio of PP-*g*-MA/organoclay for a constant 5 wt% MMT content whereas at the same time it decreases with clay content for a fixed PP-*g*-MA-to-organoclay ratio. However, the mechanical properties as well as the coefficient of thermal expansion do not seem to be seriously affected by the compatibilizer-to-organoclay ratio. On the other hand, rheological measurements demonstrated the formation of a percolated network with increasing PP-*g*-MA/organoclay ratio, indicated by the plateau in the storage modulus values and the increase of the complex viscosity at low frequencies. Similar results were obtained when maleated polypropylene was utilized as a compatibilizer in a thermoplastic olefin (TPO)–based nanocomposite containing approximately 25 wt% of elastomer [51].

Hotta and Paul [52] utilized LLDPE (MFI = 2 g/10 min), maleic anhydride–grafted polyethylene (LLDPE-*g*-MA, MFI = 1.5 g/10 min), and dimethyl-di(hydrogenated-tallow)-modified montmorillonite to synthesize nanocomposites with either varying the inorganic content or varying the compatibilizer-to-organoclay ratio, α. Figure 13.11 shows x-ray diffractograms of nanocomposites with constant amount of clay and varying α, which demonstrate that by increasing α from all LLDPE (α = 0) to complete LLDPE-*g*-MA (α = 11) composites the peak at 2θ = 3.6° gradually disappears indicating a change of the

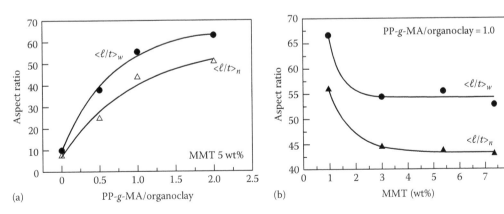

FIGURE 13.10
Effect of (a) MMT content and (b) PP-*g*-MA/organoclay ratio on number and weight average aspect ratio of PP/PP-*g*-MA/MMT nanocomposites. (Reprinted from Kim, D.H. et al., *Polymer*, 48, 5308, 2007. With permission from Elsevier.)

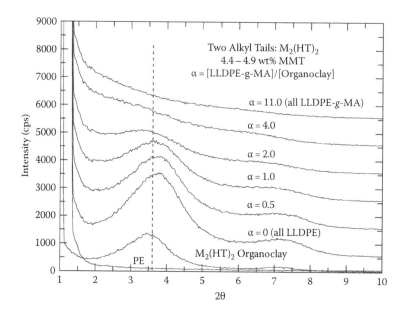

FIGURE 13.11

X-ray diffractograms of MMT organoclay and its LLDPE nanocomposites for different α values. (Reprinted from Hotta, S. and Paul, D.R., *Polymer*, 45, 7639, 2004. With permission from Elsevier.)

structure from phase separated to completely exfoliated. These results are further verified by TEM measurements, as shown in Figure 13.12. Moreover, analysis of these TEM images showed that both the average particle length and particle thickness decrease in such a way that result in an increasing aspect ratio from a value of 5 ($\alpha = 0$) to the value of 42 ($\alpha = 11$).

In addition to the number density of the functional groups of the compatibilizer used and the compatibilizer-to-organoclay ratio, the effect of the molecular weight of PP-*g*-MA

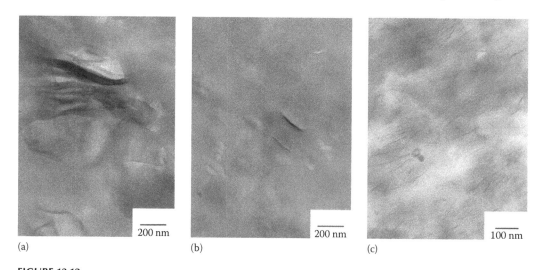

FIGURE 13.12

TEM micrographs of LLDPE nanocomposites for ratios (a) $\alpha = 0$, (b) $\alpha = 1$, and (c) $\alpha = 11$. The MMT content is always 4.5–4.9 wt%. (Reprinted from Hotta, S. and Paul, D.R., *Polymer*, 45, 7639, 2004. With permission from Elsevier.)

on the resulting structure was investigated [53]. PP with molecular weight of $M_w = 250\,kg/mol$ was mixed with Cloisite 15A and with two PP-*g*-MAs with different molecular weights and different grafting content. The results showed that the use of PP-*g*-MA with low molecular weight and high grafting density ($M_w = 9.1\,kg/mol$, MA content 3.8%) leads to relatively good and uniform intercalation evidenced by a main diffraction peak at XRD pattern and a regularity in the stacking in the TEM images but not to any profound signs of exfoliation. This was attributed to the interactions of PP-*g*-MA with the clay particles and to the lack of miscibility with the matrix polymer. On the other hand, when PP-*g*-MA with high molecular weight and low grafting content ($M_w = 330\,kg/mol$, MA content 0.5%) was utilized, no intercalation was observed but there were signs of exfoliation indicated by TEM images showing more disordered and distanced layered structure. It was proposed that this compatibilizer could interact less with the clay but had enhanced miscibility with PP resulting in a higher degree of exfoliation. It is noted that in this case the ratio of compatibilizer to organoclay was always kept equal to 2.

Along the same framework, a series of four PP-*g*-MA with different molecular weights and different MA content (two of which are the same with the ones studied above [53]) was utilized as compatibilizers [54]. Moreover, in this case the ratio of compatibilizer to organoclay was varied from 1 to 10. The authors found that of all the compatibilizers, but the one with the lower M_w and higher MA content were equally efficient in compatibilizing PP and organoclay especially in ratios $\alpha > 3$. Nevertheless, they did not report any indication for intercalation. Similarly, two different PP-*g*-MAs were utilized with different molecular weights and different maleic anhydride groups grafted on the chains [55]. The compatibilizer with the higher molecular weight was ineffective in improving the interactions and in changing the interlayer distance of the organoclay. On the contrary, the one with the lower molecular weight and the high MA content showed an increase of d_{001}. It is noted that the interlayer spacing did not increase linearly as a function of the weight fraction of PP-*g*-MA for constant amount of organoclay but it seemed that there is a threshold amount of compatibilizer above which the weakly held platelets are intercalated significantly.

In addition to the formation of hydrogen bonds between the oxygen groups of the silicate and the functionalized polymer, an alternative explanation concerning the type of interaction between the maleated polyolefins and the organoclays was proposed [56]. In this work, organophilized montmorillonite was utilized; the surfactants used were in one case 1-hexadecylamine (HDA) and in another cetylpyridinium chloride monohydrate (CPCl). The authors utilized FTIR, DSC, and MALDI-TOF to show that a reaction between hexadecylamine and PP-*g*-MA takes place, anhydride groups are consumed, and amide groups are mainly formed. Figure 13.13 shows the FTIR spectra of PP-*g*-MA, HAD-modified-MMT, and their mixture with 20 wt% filler content. Although the intensity of the vibrations' characteristic for the reaction product are small due to the presence of the silicate, the spectrum resembles the one observed in the case where the PP-*g*-MA and HDA reacted in the absence of the clay and shows characteristic vibrations of the anhydride (asymmetric stretching of the C=O group at 1856 cm^{-1}, symmetric C=O stretching at 1775 cm^{-1}, carboxylic acid at 1712 cm^{-1}), whereas the vibration of the amine groups of HDA appear at 1550 and 1650 cm^{-1}. On the contrary, CPCl that contains no active hydrogen atoms does not react with the specific compatibilizer. XRD spectra showed that in the case of the HDA surfactants, the organoclay characteristic diffraction peak was not observed in the spectra (at least in the case that the hybrid comprised of 20 wt% PP-*g*-MA and 2 wt% silicate) while in the case of CPCl, a decrease in the thickness of the interlayer galleries was observed. The authors explained the results with the assumption that chemical reactions remove the surfactant from the surface of MMT and hydrogenated silicate sites are left behind. The high

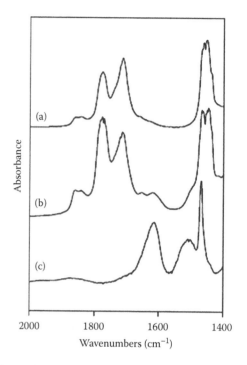

FIGURE 13.13
FTIR spectra of (a) PP-*g*-MA, (b) its reaction product obtained at 20wt% filler content and (c) org-MMT modified with hexadecyl amine. (Reprinted from Szàzdi, L. et al., *Polymer*, 46, 8001, 2005. With permission from Elsevier.)

energy surface interacts either with the anhydride or the amide groups by dipole–dipole interactions, whereas even the unmodified polypropylene may adhere stronger to the surface by London dispersion forces than to the silicate covered by aliphatic chains.

PP-*g*-MA was utilized as compatibilizer even for the synthesis of PA6/PP/organoclay composites [57]. Whereas an intercalated/exfoliated structure in the PA6/PP nanocomposites was observed, the enhancement in the mechanical properties was rather limited. This was attributed to the lack of interactions between the organoclay and PP and missing compatibility between the PA6 and PP phases [58]. Adding the PP-*g*-MA compatibilizer to the blend decreased the MFI and increased the strength and ductility parameters. This was attributed to the formation of a grafted copolymer (PA6-*g*-PP) creating an interphase between PA6 and PP. Incorporation of the organoclay improved the stiffness and reduced the ductility, as expected, whereas it did not affect the crystallinity of the PA6 phase. The dispersion of PP became finer due to the compatibilizer. The organoclay was partially exfoliated, partially intercalated, and partially aggregated, and preferentially located in the PA6 and PA6-*g*-PP phases. It was speculated that hydrogen bonding between the amine groups of the octadecylamine intercalant of the clay and carbonyl groups of the PA6 and PA6-*g*-PP favors the exfoliation of the organoclay.

PP-*g*-MA has been utilized as a compatibilizer even in the case of thermoplastic polyolefin/organoclay hybrids. The polymer has a ratio of PP to rubber in the TPO of 25:100 by weight, whereas the compatibilizer was utilized in a 3:1 ratio to the organoclay. The XRD measurements indicated intercalated structure whereas an increase in the tensile and storage moduli was observed [59].

It should be noted that the above review does not cover all the works that have been published in this area. A great number of studies have appeared, which attempted to synthesize polyolefin/organoclay nanocomposites utilizing maleated additives: depending on the individual components, phase-separated, intercalated, or exfoliated structures have been reported [60–63].

The usefulness of utilizing PP-*g*-MA or PE-*g*-MA has been doubted in the past; it was claimed that when such an oligomer is used as a compatibilizer, the polymer and the MA-pretreated organoclay are effectively at theta conditions, and the extrusion is the only parameter that promotes the mixing because of the imposed mechanical shear [64]. As a result, the structure and the properties of the resulting hybrid materials would depend strongly on the processing conditions and range from very moderate dispersions and property improvements to good dispersions and better performing hybrids. This explanation together with the many parameters (e.g., the molecular weight of the components, the amount of polar groups, the miscibility of the two polymers, the kind of surfactants modifying the inorganic surface, the processing conditions, etc.) that play a role when synthesizing such three-component hybrids may partly explain the sometimes contradictory results. At the same time, however, it can be safely concluded that a certain ratio of compatibilizer to organoclay and adequate miscibility of the maleated polyolefin and the polymer are the necessary parameters that should be obeyed in order to accomplish a certain degree of exfoliation.

13.3 Other Functional Compatibilizers

In addition to maleic anhydride–functionalized polyolefin oligomers or polymers, a major category of compatibilizers utilized are specifically synthesized homopolymers and/or copolymers that could modify the interactions and control the structure in polyolefin/organoclay nanocomposites.

In one of the first attempts to synthesize polypropylene/clay nanocomposites, a distearyldimethylammonium-modified montmorillonite was first mixed in toluene solutions with a polyolefinic oligomer (polyolefin diol, carbon number = 150–200) possessing two –OH end groups in different ratios of oligomer to organoclay [65]. X-ray diffraction measurements of these composites demonstrated that a 1:1 ratio of additive to organoclay simply increased the interlayer distance, whereas values of this ratio higher than three resulted in difractograms exhibiting no obvious diffraction peaks. This was attributed to the interactions of the OH groups with the silicate layer through hydrogen bonding. Moreover, TEM images show that further blending of binary hybrids even with ratio 1:1 with PP resulted in the exfoliation of the silicate and in the dispersion of the platelets in the polymer matrix.

A comparison between the compatibilizing efficiency of two different compatibilizers was reported concerning LLDPE/organoclay nanocomposites. A low-molecular-weight oxidized polyethylene (M_n = 2950, acid number = 30 mg KOH/g) was utilized and compared with the more frequently used PE-*g*-MA (MFI = 4 g/10 min, 1.6 wt% maleic anhydride, acid number = 18.3 KOH/g) [66]. Oxidized polyethylene is a modified PE that contains various functionalities such as mainly carboxylic acids, esters, and ketones, and is being used as a processing agent in certain polymer and coating formulations. Comparison of the obtained morphology of the compatibilized hybrids (compatibilizer-to-organoclay ratio kept constant at α = 3) showed that the use of oxidized polyethylene results in intercalated nanocomposites with a 1 nm increase of the interlayer distance of the organoclay whereas PE-*g*-MA led to a higher degree of exfoliation despite its lower functionality. All nanocomposites showed a solid-like behavior with increasing clay content whereas the estimated percolation threshold was higher in the hybrids with oxidized polyethylene. It is noted, however, that the corresponding aspect ratio values of clay tactoids was smaller. The use of oxidized polyethylenes with varying polarity did not have any further influence on the observed intercalated structure whereas it affected differently the measured properties [67].

An alternative approach to the use of maleic anhydride polyolefin oligomers has been introduced by Manias et al. [64]. It was proposed to synthesize random copolymers of PP with typically 1 mol% of functionalized monomers; the monomers and the molecular characteristics of the copolymers are given in Table 13.1. All the functionalized PPs were derived from the same random PP copolymer synthesized by metallocene catalysis, which contained 1 mol% *p*-methylstyrene (p-MS) comonomers. Subsequently, the p-MS's were interconverted to functional groups containing hydroxyl (OH) and maleic anhydride by lithiation and free-radical reactions, respectively [68–70]. These functionalized polypropylenes were melt intercalated under static conditions with dimethyldioctadecylammonium-modified MMT. Figure 13.14 shows x-ray diffractograms of the resulted hybrids, which indicate an intercalated structure with ~10 Å increase of the interlayer distance for all the differently functionalized polymers. Moreover, bright field TEM images show the coexistence of both intercalated tactoids and disordered/exfoliated stacks of layers. Besides the random functionalization of PP, the authors utilized polypropylene-block-poly(methyl methacrylate) diblock copolymers containing 1 and 5 mol% of PMMA. This synthesis involved PP preparation by metallocene catalysis, hydroboranation of the olefinic chain end, and free-radical polymerization of the PMMA block [71]. The characteristics of the diblocks are included in Table 13.1 as well. Mixing this polymer with octadecylammonium-modified montmorillonite resulted in composites containing approximately 20% exfoliated/disordered layers and the rest as intercalated tactoids, as can be seen in Figure 13.15.

Altering the type of surfactants used to make the clay organophilic was proposed to modify the polymer–clay interactions [64]. Instead of modifying the polymer, the

TABLE 13.1

Chemical Formulae and Characteristics of Functionalized PP

Functionalized PP: Random Copolymers		x (mol %)	M_w	T_m (°C)
A:	PP-r-(PP-MS)$_x$	1.0	200 000	154
B:	PP-r-(PP-MA)$_x$	0.5	200 000	155
C:	PP-r-(PP-OH)$_x$	0.5	200 000	155

Functionalized PP: Dibtock Copolymer

| D: | PP-b-(PPMA)$_x$ | 1.0 | 200 000 | 155 |
| | PP-b-(PPMA)$_x$ | 5.0 | 220 000 | 154 |

Source: Reprinted from Manias, E. et al., *Chem. Mater.*, 13, 3516, 2001. With permission from American Chemical Society.

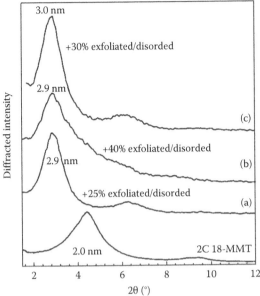

FIGURE 13.14

X-ray diffraction patterns of the dimethyldiocta-decylammonium-modified montmorillonite (2C18-MMT) and all of the functionalized-PP/2C18-MMT nanocomposites. The functional groups used are (a) methylstyrene 1 mol%, (b) maleic anhydride 0.5 mol%, and (c) hydroxyl-containing styrene 0.5 mol%. (Reprinted from Manias, E. et al., *Chem. Mater.*, 13, 3516, 2001. With permission from American Chemical Society.)

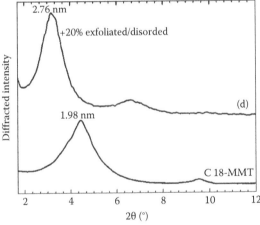

FIGURE 13.15

XRD patterns of C18-MMT and the PP-*b*-PMMA/C18-MMT nanocomposite. (Reprinted from Manias, E. et al., *Chem. Mater.*, 13, 3516, 2001. With permission from American Chemical Society.)

authors attempted to utilize semifluorinated surfactants and further mix it with neat PP. Specifically, they first exchanged all the native MMT cations by octadecylammoniums, and, subsequently, they introduced a second semifluorinated alkyltrichlorosilane surfactant (CF_3–$(CF_2)_5$–$(CF_2)_2$–Si–Cl_3). This second surfactant was tethered to the surface through a reaction of the trichlorosilane groups with hydroxyl groups in the cleavage plane of the MMT. The resulted organoclay contained octadecyl ammoniums at full CEC and ~60% additional semifluorinated surfactants. Following the organophilization, hybrids were synthesized by melt intercalation that possessed an intercalated structure with a 12 Å increase of the interlayer distance of the fluorinated montmorillonite. Moreover, use of mechanical shear promoted further the dispersion. All the hybrids showed enhanced Young's modulus with only a small decrease in the maximum strain at break, increased heat deflection temperature, and reduced flammability while preserving the polymer optical clarity [64].

Amphiphilic block and random copolymers of different chemical structures were utilized as dispersing agents in nanocomposites of high-density polyethylene and (dimethyldioctadecyl ammonium) modified montmorillonite, as well [72]. Polyethylene-block-poly(ethylene glycol), PE-*b*-PEG, had 33 methylene groups and 2.6 ethylene oxide units per molecule on average; hence it was an end-functionalized PE with a small polar head group rather than a block copolymer. Poly(ethylene-*co*-vinyl alcohol), PE-*r*-VOH, and poly(ethylene-*co*-methacrylic acid), PE-*r*-MAA were random copolymers, possessing several polar groups randomly distributed along the chain, chosen because of their ability to form hydrogen bonds with aluminosilicates. It is noted, that the hydroxyl groups of PE-*r*-VOH are of basic nature while those of PE-*r*-PMAA have an acidic character. A poly(ethylene-graft maleic anhydride), PE-*g*-MA, with low anhydride content so that there is one anhydride moiety per two molecules, which can be considered as end-functionalized PE, was utilized for comparison. Figure 13.16a shows that the morphology varies between phase separated and intercalated depending on the dispersing agent. It is noted, however, that the copolymer concentration was kept very low (copolymer/

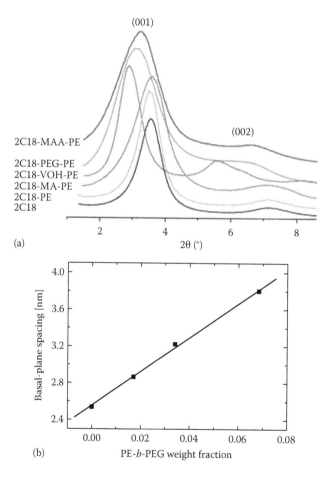

(a)

(b)

FIGURE 13.16
(a) XRD patterns of the 2C18-MMT and its HDPE composites with and without the different compatibilizers. (b) Influence of the PE-*b*-PEG compatibilizer concentration on the *d*-spacing of the nanocomposites. (Reprinted from Osman, M.A. et al., *Polymer*, 46, 8202, 2005. With permission from Elsevier.)

organoclay weight ratio of 0.17). The best copolymer in terms of increasing interlayer distance was proved to be PE-*r*-VOH; nevertheless, in all hybrids but this, a decrease in oxygen permeability was measured. The best hybrid in terms of reduced permeability and optimization of mechanical properties was the hybrid containing PE-*b*-PEG. Moreover, Figure 13.16b shows that as far as PE-*b*-PEG is concerned, increase of the copolymer concentration leads to a linear dependence of the interlayer distance on the PE-*b*-PEG weight fraction.

In another work, polyethylene-based model macromolecules were especially designed and synthesized to act as surfactants and/or compatibilizers to alter the interactions and control the structure in PE/organoclay nanocomposites [40]. The synthesis of three types of additives was performed utilizing anionic polymerization under high vacuum [73] followed by the appropriate post-polymerization reactions in order to introduce or reveal the desired functional moieties. Polyethylene chains functionalized by dimethyl ammonium chloride either as a single end-group, NPE, or as multiple functional groups, PE-*g*-NPE, along the chain were synthesized by anionic polymerization of butadiene and subsequent hydrogenation to produce polyethylene. The third type of additive is a diblock copolymer of polyethylene-block-poly(methacrylic acid), PE-*b*-PMAA, also synthesized anionically followed by hydrogenation and deprotection of the methacrylic acid. A widely used maleic anhydride–grafted polyethylene with a low degree of functionalization was utilized for comparison, as well. The molecular characteristics of all the synthesized compatibilizers are shown in Table 13.2. In the case of the polyethylene chains with quaternized amine end-groups, all the attempts to synthesize a composite either utilizing hydrophilic or organophilic montmorillonite led to phase-separated structures, which can be possibly attributed to the high molecular weight of the chains in conjunction with the single functional group. Nevertheless, similar were the results utilizing the polymer with the multiple functional groups grafted along the chain. It should be noted, however, that in all cases the ratio of the compatibilizer/organoclay was kept low due to the specialized synthesis that produced a limited amount of the polymers. As far as the effect of the diblock copolymer on the hybrid structure is concerned, it was anticipated that a diblock copolymer of PE-*b*-PMAA will intercalate into the galleries of the montmorillonite due to the polarity of the carboxyl groups of poly(methacrylic acid). This would bring the polyethylene block into the galleries, making the environment more friendly for the polyethylene homopolymer. Figure 13.17 shows x-ray diffraction results for hybrids where the organic/inorganic

TABLE 13.2

Molecular Characteristics of the Polymeric Surfactants/Compatibilizers

Type of Additive	Code	M_w	$I = M_w/M_n$
N-PBd	NPBd-1	9,700	1.04
	NPBd-2	22,900	1.04
PBd-*b*-PtBuMA	PBd-block	4,800	1.05
	PBd-*b*-PtBuMA	13,800	1.06
PBd-*g*-NPBd	PBd backbone	110,600	1.03
	NPBd branch	0.500	
	PBd-*g*-NPBd	118,100	1.03

Source: Reprinted from Chrissopoulou, K. et al., *Polymer*, 46, 12440, 2005. With permission from Elsevier.

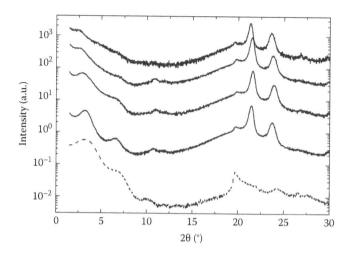

FIGURE 13.17

X-ray diffractograms of three-component PE hybrids containing PE-*b*-PMAA and 13 wt% Cloisite 20A. The PE-*b*-PMAA composition in the polymer is (from top to bottom) 15, 10, 6, and 2 wt%. The data for the respective organoclay are shown with the dashed line. The curves are shifted for clarity. (Reprinted from Chrissopoulou, K. et al., *Polymer*, 46, 12440, 2005. With permission from Elsevier.)

composition is kept constant but the amount of the copolymer additive is varied between 2 and 15 wt%. This way the ratio of copolymer to organoclay was varied from 0.13:1 to 1:1. It can be seen that for the lower copolymer concentration there is not any change of the interlayer distance of the organoclay (Cloisite 20A). The main peak is shown at $2\theta = 3.15°$ leading to $d_{001} = 28.0$ Å. As the concentration of the additive increases, there is a gradual shift of the main diffraction peak to lower angles. At the higher concentration of the additive, i.e., 15 wt% (ratio 1:1), a very weak peak observed at $2\theta = 2.45°$ corresponds to $d_{001} = 36.0$ Å, which means an increase of the interlayer distance by ~8 Å. This increase is accompanied by a significant decrease of the intensity of the peak and by an increase of the scattering at lower angles. Based on these results, it was concluded that, most probably, as the concentration of the PE-*b*-PMAA increases, there exist both intercalated and exfoliated regions. The degree of exfoliation is even higher than the corresponding result utilizing PP-*g*-MA at the same α, although such a conclusion is totally qualitative.

Similar results were found when a polypropylene-block-poly(propylene glycol), PP-*b*-PPG, diblock copolymer was utilized as compatibilizer; its ability to increase the interlayer distance of PP/dimethyl dioctadecyl ammonium–modified montmorilonite hybrids was studied and compared with the corresponding behavior of PP-*g*-MA [55]. The ratio of compatibilizer to organoclay used was low but nevertheless 2 wt% of PP-*b*-PPG resulted in a ~4 Å increase of interlayer distance, which was better than what was observed utilizing maleated PP under the same conditions.

Wang et al. [74] utilized end-functionalized (ammonium-terminated) polyolefines containing a terminal functional group as compatibilizers for PP nanocomposites; they were synthesized via a facile route using reactive chain transfer agents while maintaining a well-controlled molecular weight and narrow molecular weight distribution [75]. The PP-*t*-NH$_3^+$ polymers were prepared by the combination of *rac*-Me$_2$Si[2-Me-4-Ph(Ind)]$_2$ZrCl$_2$/MAO catalyst and p-NSi$_2$-St/H$_2$ (p-NSi$_2$-St: 4-{2-[N,N-bis(trimethylsilyl)amino]ethyl}styrene) chain transfer agent. Both pristine Na$^+$ MMT clay and dioctadecylammonium-modified organophilic clay were utilized. Mixtures with compositions 90:10 were prepared by melt intercalation and in both cases featureless XRD patterns were obtained indicating the formation of exfoliated structures through the alkali or dioctadecylammonium cations exchange by the ammonium-terminated PP. The exfoliated structure, as indicated by TEM, was maintained after further mixing with isotactic PP, i-PP, which is compatible with the backbone of the largely isotactic PP-*t*-NH$_3^+$ polymer. Apparently, the i-PP chains serve as diluents in the ternary PP-*t*-NH$_3^+$/MMT/i-PP system, with the thermodynamically stable PP-*t*-NH$_3^+$/MMT exfoliated structure dispersed within the i-PP matrix.

Metallocene catalysts are being used to polymerize propylene with one of several different hydroxyl, amino, or carboxylic acid functional co-monomers. When metallocene catalysts are applied, co-monomer distribution is uniform along the PP backbone, and stereospecificity is excellent. In such a case, a compatibilizer with hydroxyl functionality was prepared, and its ability to penetrate the narrow galleries of the organoclays was evaluated [43]. The polymerization catalyst for the preparation of hydroxyl-functionalized PP was dimethylsilanylbis(2-methyl-4-phenyl-1-indenyl)zirconium dichloride. Maleic anhydride–grafted polypropylene (PP-*g*-MA) with a maleic anhydride content of 0.5 wt% was used as a reference compatibilizer. Composites with compatibilizer-to-organoclay ratio 1:1 and 2:1 were prepared. It was only the latter that showed exfoliated structure for both compatibilizers according to XRD measurements but rheological measurements showed that the composites with hydroxyl-functionalized PP exhibited smaller viscosity compared to the pure polymer indicating probably improved processability despite the addition of the filler; thus it was concluded that hydroxyl-functionalized PP was an effective compatibilizer. Nevertheless, composites with PP-*g*-MA showed higher viscosity compared with the pure polymer and a deviation from the low-frequency plateau indicating a higher degree of exfoliation.

In an alternative route to increase polypropylene polarity in order to make it more compatible with clay, Cl and SO$_2$Cl groups were introduced by reaction with sulfuryl chloride under UV irradiation in the presence of small amounts of pyridine [76]. Two different compounding procedures were used in synthesizing the nanocomposites: direct melt mixing and a masterbatch mixing process; the masterbatch was afterward blended with commercial isotactic polypropylene. An organophilized silicate (Cloisite 15A) and three chlorosulfonated polypropylenes with different degrees of functionalization were utilized in this case whereas the morphology of the obtained nanocomposites was examined using TEM and x-ray diffraction. A mixture of intercalated and exfoliated structure was found independently of the preparation method. The highest degree of intercalation was observed for systems with the compatibilizer possessing a medium amount of SO$_2$Cl, but a distinctly higher amount of Cl indicating that chlorine is more efficient in organoclay delamination. Nevertheless, full exfoliation of clay platelets has not been achieved. As far as the mechanical properties are concerned, the presence of the dispersed clay improves the stiffness and tensile strength of the nanocomposites but markedly reduces their strain at break.

In summary, the need for effective compatibilizers for the preparation of polyolefin/layered silicate nanocomposites has led to the synthesis of various model functional polymers. Molecules with functional ammonium or hydroxyl groups as well as block or random copolymers among others have been tested for their ability to compatibilize the components. Phase-separated, intercalated, exfoliated, or even mixed structures have been observed depending on the kind, the molecular characteristics, and the degree of functionalization, as well as the concentration of the additive. In some cases, enhanced dispersion in comparison to the widely used maleated polyolefins has been reported. Nevertheless, despite the number and the quality of the reported studies, it is not clear yet which is the best compatibilizer for such systems.

13.4 Conclusions

Nanocomposites based on polyolefins and layered silicates are synthesized in order to improve the properties of these widely used polymers. Their synthesis necessitates the use of functional oligomers/polymers to modify the interactions between the two constituents especially because of their strong incompatibility. To achieve the dispersion of the inorganic material in the polymer matrix and obtain the desired structure, compatibilizers such as maleic anhydride–functionalized polyolefins, polymers with functional hydroxyl or ammonium groups, as well as block and graft copolymers have been utilized. X-ray diffraction and transmission electron microscopy have been mainly used to identify the structure, whereas, in certain cases, complementary techniques like rheology can provide complementary information. Many of the compatibilizers used were found effective in exfoliating the clay whereas the most important parameters that control the structure are the ratio of the compatibilizer to organoclay as well as the miscibility of the compatibilizer to the polyolefin; organophilization of the inorganic material by utilizing surfactants with long enough alkyl chains is certainly a prerequisite whereas attention should be paid to whether equilibrium is established. It is understood that the strong interaction between the functional groups of the compatibilizer and the organoclay particles can lead to the formation of "hairy particles," which can in turn favorably mix with the polyolefin matrix polymer. Nevertheless, a direct correlation between the nanohybrid structure and optimized mechanical, thermal, and/or barrier properties of the nanocomposite has not been established.

Acknowledgments

We would like to acknowledge I. Altintzi, I. Andrianaki, R. Shemesh, H. Retsos, M. Pitsikalis, N. Hadjichristidis, E. P. Giannelis, and N. Theophilou for their involvement in the experimental work of our group reviewed herein as well as the European Union in the frame work of GROWTH programme (NANOPROP Project No. GSRD-CT-2002-00834), the Greek General Secretariat for Research and Technology (PENED Programme 03ED581) and NATO Scientific Affairs Division (Science for Stability Programme) for financial support.

References

1. Milewski, J. V. and Katz, H. S., eds. 1987. *Handbook of Reinforcements for Plastics*. New York: Van Nostrand Reinhold.
2. Karger-Kocsis, J., ed. 1995. *Polypropylene: Structure, Blends and Composites*, vol. 3. London, U.K.: Chapman & Hall.
3. Mitchell, C. A., Bahr, J. L., Arepalli, S., Tour, J. M., and Krishnamoorti, R. 2002. Dispersion of functionalized carbon nanotubes in polystyrene. *Macromolecules* 35:8825–8830.
4. Sharp, K. G. 1998. Inorganic/organic hybrid materials. *Advanced Materials* 10:1243–1248; Bockstaller, M. R., Mickiewicz, R. A., and Thomas, E. L. 2005. Block copolymer nanocomposites: Perspectives for tailored functional materials. *Advanced Materials* 17:1331–1349; Fischer, H. 2003. Polymer nanocomposites: From fundamental research to specific applications. *Materials Science and Engineering C* 23:763–772.

5. Granick, S., Kumar, S. K., Amis, E. J. et al. 2003. Macromolecules at surfaces: Research challenges and opportunities from tribology to biology. *Journal of Polymer Science, Part B: Polymer Physics* 41:2755–2793.
6. Pinnavaia, T. J. and Beall, G. W. 2000. *Polymer-Clay Nanocomposites*. West Sussex, U.K.: John Wiley & Sons.
7. Giannelis, E. P., Krishnamoorti, R., and Manias, E. 1999. Polymer-silicate nanocomposites: Model systems for confined polymers and polymer brushes. *Advances in Polymer Science* 138:107–147.
8. Giannelis, E. P. 1996. Polymer layered silicate nanocomposites. *Advanced Materials* 8:29–35.
9. Alexandre, M. and Dupois, P. 2000. Polymer-layered silicate nanocomposites: Preparation, properties and uses of a new class of materials. *Materials Science and Engineering R* 28:1–63.
10. Sinha Ray, S. and Okamoto, M. 2003. Polymer/layered silicate nanocomposites: A review from preparation to processing. *Progress in Polymer Science* 28:1539–1641.
11. Okada, A. and Usuki, A. 2006. Twenty years of polymer-clay nanocomposites. *Macromolecular Materials and Engineering* 291:1449–1476.
12. Usuki, A., Hasegawa, N., and Kato, M. 2005. Polymer-clay nanocomposites. *Advances in Polymer Science* 179:135–195.
13. Vaia, R. A., Price, G., Ruth, P. N., Nguyen, H. T., and Lichtenhan, J. 1999. Polymer/layered silicate nanocomposites as high performance ablative materials. *Applied Clay Science* 15:67–92.
14. Biswas, M. and Sinha Ray, S. 2001. Recent progress in synthesis and evaluation of polymer-montmorillonite nanocomposites. *Advances in Polymer Science* 155:167–221.
15. Schmidt, D., Shah, D., and Giannelis, E. P. 2002. New advances in polymer/layered silicate nanocomposites. *Current Opinion in Solid State and Materials Science* 6:205–212.
16. Giannelis, E. P. 1998. Polymer-layered silicate nanocomposites: Synthesis, properties and applications. *Applied Organometallic Chemistry* 12:675–680.
17. Bharadwaj, R. K. 2001. Modelling the barrier properties of polymer-layered silicate nanocomposites. *Macromolecules* 34:9189–9192.
18. Messersmith, P. B. and Giannelis, E. P. 1995. Synthesis and barrier properties of poly(ε-caprolactone)-layered silicate nanocomposites. *Journal of Polymer Science, Part A: Polymer Chemistry* 33:1047–1057.
19. Yano, K., Usuki, A., Okada, A., Kurauchi, T., and Kamigaito, O. 1993. Synthesis and properties of polyimide-clay hybrid. *Journal of Polymer Science, Part A: Polymer Chemistry* 31:2493–2498.
20. Gilman, J. W. 1999. Flammability and thermal stability studies of polymer/layered silicate (clay) nanocomposites. *Applied Clay Science* 15:31–49.
21. Gilman, J. W., Jackson, C. L., Morgan, A. B. et al. 2000. Flammability properties of polymer-layered-silicate nanocomposites. Polypropylene and polystyrene nanocomposites. *Chemistry of Materials* 12:1866–1873.
22. Sinha Ray, S., Yamada, K., Okamoto, M., and Ueda, K. 2002. Polylactide-layered silicate nanocomposite: A novel biodegradable material. *Nano Letters* 2:1093–1096.
23. Vaia, R. A., Jandt, K. D., Kramer, E. J., and Giannelis, E. P. 1996. Microstructural evolution of melt intercalated polymer-organically modified layered silicates nanocomposites. *Chemistry of Materials* 8:2628–2635.
24. Vaia, R. A., Sauer, B. B., Tse, O. K., and Giannelis, E. P. 1997. Relaxation of confined chains in polymer nanocomposites: Glass transition properties of poly(ethylene oxide) intercalated in montmorillonite. *Journal of Polymer Science, Part B: Polymer Physics* 35:59–67.
25. Hackett, E., Manias, E., and Giannelis, E. P. 2000. Computer simulation studies of PEO/layered silicate nanocomposites. *Chemistry of Materials* 12:2161–2167.
26. Shen, Z., Simon, G. P., and Cheng, Y. 2003. Saturation ratio of poly(ethylene oxide) to silicate in melt intercalated nanocomposites. *European Polymer Journal* 39:1917–1924.
27. Elmahdy, M. M., Chrissopoulou, K., Afratis, A., Floudas, G., and Anastasiadis, S. H. 2006. Effect of confinement on polymer segmental motion and ion mobility in PEO/layered silicate nanocomposites. *Macromolecules* 39:5170–5173.

28. Chrissopoulou, K., Afratis, A., Anastasiadis, S. H., Elmahdy, M. M., Floudas, G., and Frick, B. 2007. Structure and dynamics in PEO nanocomposites. *European Physical Journal Special Topics* 141:267–271.

29. Fotiadou, S., Chrissopoulou, K., Frick, B., and Anastasiadis, S. H. 2010. Structure and dynamics of polymer chains in hydrophilic nanocomposites. *Journal of Polymer Science, Part B: Polymer Physics* 48:1658–1667.

30. Vaia, R. A., Jandt, K. D., Kramer, E. J., and Giannelis, E. P. 1995. Kinetics of polymer melt intercalation. *Macromolecules* 28:8080–8085.

31. Chrissopoulou, K., Anastasiadis, S. H., Giannelis, E. P., and Frick, B. 2007. Quasielastic neutron scattering of poly(methyl phenyl siloxane) in the bulk and under severe confinement. *Journal of Chemical Physics* 127:144910-1–144910-13.

32. Anastasiadis, S. H., Chrissopoulou, K., and Frick, B. 2008. Structure and dynamics in polymer/layered silicate nanocomposites. *Materials Science and Engineering B* 152:33–39.

33. Usuki, A., Kojima, Y., Kawasumi, M. et al. 1993. Synthesis of nylon-6 clay hybrid. *Journal of Materials Research* 8:1179–1184.

34. Vaia, R. A. and Giannelis, E. P. 1997. Lattice model of polymer intercalation in organically modified layered silicates. *Macromolecules* 30:7990–7999.

35. Balazs, A. C., Singh, C., and Zhulina, E. 1998. Modelling the interactions between polymers and clay surfaces through self-consistent field theory. *Macromolecules* 31:8370–8381.

36. Zhulina, E., Singh, C., and Balazs, A. C. 1999. Attraction between surfaces in a polymer melt containing telechelic chains: Guidelines for controlling the surface separation in intercalated polymer-clay composites. *Langmuir* 15:3935–3943.

37. Bains, A. S., Boek, E. S., Coveney, P. V., Williams, S. J., and Akbar, M. V. 2001. Molecular modeling of the mechanism of action of organic-clay swelling inhibitors. *Molecular Simulation* 26:101–145.

38. Vaia, R. A. and Giannelis, E. P. 1997. Polymer melt intercalation in organically modified layered silicates: Model predictions and experiment. *Macromolecules* 30:8000–8009.

39. Vohra, V. R., Schmidt, D. F., Ober, C. K., and Giannelis, E. P. 2003. Deintercalation of a chemically switchable polymer from a layered silicate nanocomposite. *Journal of Polymer Science, Part B: Polymer Physics* 41:3151–3159.

40. Chrissopoulou, K., Altintzi, I., Anastasiadis, S. H., Giannelis, E. P., Pitsikalis, M., Hadjichristidis, N., and Theophilou, N. 2005. Controlling the miscibility of polyethylene/layered silicate nanocomposites by altering the polymer/surface interactions. *Polymer* 46:12440–12451.

41. Clark, E. S. 1996. Unit cell information on some important polymers. In *Physical Properties of Polymers Handbook*, ed. J. E. Mark. Woodbury, NY: AIP Press, pp. 409–415.

42. Chrissopoulou, K., Altintzi, I., Andrianaki, I., Shemesh, R., Retsos, H., Giannelis, E. P., and Anastasiadis, S. H. 2008. Understanding and controlling the structure of polypropylene/layered silicate nanocomposites. *Journal of Polymer Science, Part B: Polymer Physics* 46:2683–2695.

43. Ristolainen, N., Vainio, U., Paavola, S., Torkkeli, M., Serimaa, R., and Seppälä, J. 2005. Polypropylene/organoclay nanocomposites compatibilized with hydroxyl-functional polypropylenes. *Journal of Polymer Science, Part B: Polymer Physics* 43:1892–1903.

44. Kato, M., Usuki, A., and Okada, A. 1997. Synthesis of polypropylene oligomer-clay intercalation compounds. *Journal of Applied Polymer Science* 66:1781–1785.

45. Kawasumi, M., Hasegawa, N., Kato, M., Usuki, A., and Okada, A. 1997. Preparation and mechanical properties of polypropylene-clay hybrids. *Macromolecules* 30:6333–6338.

46. Wagener, R. and Reisinger, T. J. G. 2003. A rheological method to compare the degree of exfoliation of nanocomposites. *Polymer* 44:7513–7518.

47. Százdi, L., Ábrányi, Á., Pukánszky, B. Jr., Vancso, J. G., and Pukánszky, B. 2006. Morphology characterization of PP/clay nanocomposites across the length scales of the structural architecture. *Macromolecular Materials and Engineering* 291:858–868.

48. Reichert, P., Nitz, H., Klinke, S., Brandsch, R., Thomann, R., and Mülhaupt, R. 2000. Poly(propylene)/organoclay nanocomposite formation: Influence of compatibilizer functionality and organoclay modification. *Macromolecular Materials and Engineering* 275:8–17.

49. Reichert, P., Hoffmann, B., Bock, T., Thomann, R., Mülhaupt, R., and Friedrich, C. 2001. Morphological stability of poly(propylene) nanocomposites. *Macromolecular Rapid Communications* 22:519–523.

50. Kim, D. H., Fasulo, P. D., Rodgers, W. R., and Paul, D. R. 2007. Structure and properties of polypropylene-based nanocomposites: Effect of PP-*g*-MA to organoclay ratio. *Polymer* 48:5308–5323.

51. Kim, D. H., Fasulo, P. D., Rodgers, W. R., and Paul, D. R. 2007. Effect of the ratio of maleated polypropylene to organoclay on the structure and properties of TPO-based nanocomposites. Part I: Morphology and mechanical properties. *Polymer* 48:5960–5978.

52. Hotta, S. and Paul, D. R. 2004. Nanocomposites formed from linear low density polyethylene and organoclays. *Polymer* 45:7639–7654.

53. Perrin-Sarazin, F., Ton-That, M.-T., Bureau, M. N., and Denault, J. 2005. Micro- and nano-structure in polypropylene/clay nanocomposites. *Polymer* 46:11624–11634.

54. Wang, Y., Chen, F.-B., and Wu, K.-C. 2005. Effect of molecular weight of maleated polypropylenes on the melt compounding of polypropylene/organoclay nanocomposites. *Journal of Applied Polymer Science* 97:1667–1680.

55. Mittal, V. 2008. Mechanical and gas permeation properties of compatibilized polypropylene-layered silicate nanocomposites. *Journal of Applied Polymer Science* 107:1350–1361.

56. Százdi, L., Pukánszky, B. Jr., Földes, E., and Pukánszky, B. 2005. Possible mechanism of interaction among the components in MAPP modified layered silicate PP nanocomposites. *Polymer* 46:8001–8010.

57. Chow, W. S., Mohd Ishak, Z. A., Karger-Kocsis, J., Apostolov, A. A., and Ishiaku, U. S. 2003. Compatibilizing effect of maleated polypropylene on the mechanical properties and morphology of injection molded polyamide 6/polypropylene/organoclay nanocomposites. *Polymer* 44:7427–7440.

58. Chow, W. S., Ishiaku, U. S., Mohd Ishak, Z. A., Karger-Kocsis, J., and Apostolov, A. A. 2004. The effect of organoclay on the mechanical properties and morphology of injection-molded polyamide 6/polypropylene nanocomposites. *Journal of Applied Polymer Science* 91:175–189.

59. Mishra, J. K., Hwang, K.-J., and Ha, C.-S. 2005. Preparation, mechanical and rheological properties of a thermoplastic polyolefin (TPO)/organoclay nanocomposite with reference to the effect of maleic anhydride modified polypropylene as a compatibilizer. *Polymer* 46:1995–2002.

60. Galgali, G., Ramesh, C., and Lele, A. 2001. A rheological study on the kinetics of hybrid formation in polypropylene nanocomposites. *Macromolecules* 34:852–858.

61. Xu, W., Liang, G., Zhai, H., Tang, S., Hang, G., and Pan, W.-P. 2003. Preparation and crystallization behaviour of PP/PP-g-MAH/Org-MMT nanocomposite. *European Polymer Journal* 39:1467–1474.

62. Lertwimolnun, W. and Vergnes, B. 2005. Influence of compatibilizer and processing conditions on the dispersion of nanoclay in a polypropylene matrix. *Polymer* 46:3462–3471.

63. Tjong, S. C., Bao, S. P., and Liang, G. D. 2005. Polypropylene/montmorillonite nanocomposites toughened with SEBS-g-MA: Structure-property relationship. *Journal of Polymer Science, Part B: Polymer Physics* 43:3112–3126.

64. Manias, E., Touny, A., Wu, L., Strawhecker, K., Lu, B., and Chung, T. C. 2001. Polypropylene/montmorillonite nanocomposites. Review of the synthetic routes and materials properties. *Chemistry of Materials* 13:3516–3523.

65. Usuki, A., Kato, M., Okada, A., and Kurauchi, T. 1997. Synthesis of polypropylene-clay hybrid. *Journal of Applied Polymer Science* 63:137–139.

66. Durmus, A., Kasgoz, A., and Macosko, C. W. 2007. Linear low density polyethylene (LLDPE)/clay nanocomposites. Part I: Structural characterization and quantifying clay dispersion by melt rheology. *Polymer* 48:4492–4502.

67. Durmuş, A., Woo, M., Kaşgöz, A., Macosko, C. W., and Tsapatsis, M. 2007. Intercalated linear low density polyethylene (LLDPE)/clay nanocomposites prepared with oxidized polyethylene as a new type compatibilizer: Structural, mechanical and barrier properties. *European Polymer Journal* 43:3737–3749.

68. Lu, B. and Chung, T. C. 2000. Synthesis of maleic anhydride grafted polyethylene and polypropylene, with controlled molecular structures. *Journal of Polymer Science, Part A: Polymer Chemistry* 38:1337–1343.

69. Lu, H. L., Hong, S., and Chung, T. C. 1999. Synthesis of polypropylene-co-p-methylstyrene copolymers by metallocene and Ziegler Natta catalysts. *Journal of Polymer Science, Part A: Polymer Chemistry* 37:2795–2802.

70. Lu, H. L., Hong, S., and Chung, T. C. 1998. Synthesis of new polyolefin elastomers, poly(ethylene-*ter*-propylene-*ter*-p-methylestyrene) and poly(ethylene-*ter*-1-octene-*ter*-p-methylstyrene), using metallocene catalysts with constrained ligand geometry. *Macromolecules* 31:2028–2034.

71. Chung, T. C., Lu, H. L., and Jankivul, W. 1997. A novel synthesis of PP-b-PMMA copolymers via metallocene catalysis and borane chemistry. *Polymer* 38:1495–1502.

72. Osman, M. A., Rupp, J. E. P., and Suter, U. W. 2005. Effect of non-ionic surfactants on the exfoliation and properties of polyethylene-layered silicate nanocomposites. *Polymer* 46:8202–8209.

73. Hadjichristidis, N., Iatrou, H., Pispas, S., and Pitsikalis, M. 2000. Anionic polymerization: High vacuum techniques. *Journal of Polymer Science, Part A: Polymer Chemistry* 38:3211–3234.

74. Wang, Z. M., Nakajima, H., Manias, E., and Chung, T. C. 2003. Exfoliated PP/clay nanocomposites using ammonium terminated PP as the organic modification for montmorillonite. *Macromolecules* 36:8919–8922.

75. Chung, T. C. 2002. Synthesis of functional polyolefin copolymers with graft and block structures. *Progress in Polymer Science* 27:39–85.

76. Kotek, J., Kelnar, I., Studenovský, M., and Baldrian, J. 2005. Chlorosulfonated polypropylene: Preparation and its applications as a coupling agent in polypropylene-clay nanocomposites. *Polymer* 46:4876–4881.

Index

Milton Keynes UK
Ingram Content Group UK Ltd.
UKHW051943071024
449327UK00026B/2151